The Palgrave Handbook of ESG and Corporate Governance

Paulo Câmara · Filipe Morais
Editors

The Palgrave Handbook of ESG and Corporate Governance

Editors
Paulo Câmara
Faculty of Law/Governance Lab
Portuguese Catholic University (UCP)
Lisbon, Portugal

Filipe Morais
Henley Business School
University of Reading
Henley-on-Thames, UK

ISBN 978-3-030-99467-9   ISBN 978-3-030-99468-6 (eBook)
https://doi.org/10.1007/978-3-030-99468-6

© The Editor(s) (if applicable) and The Author(s), under exclusive license to Springer Nature Switzerland AG 2022
This work is subject to copyright. All rights are solely and exclusively licensed by the Publisher, whether the whole or part of the material is concerned, specifically the rights of translation, reprinting, reuse of illustrations, recitation, broadcasting, reproduction on microfilms or in any other physical way, and transmission or information storage and retrieval, electronic adaptation, computer software, or by similar or dissimilar methodology now known or hereafter developed. The use of general descriptive names, registered names, trademarks, service marks, etc. in this publication does not imply, even in the absence of a specific statement, that such names are exempt from the relevant protective laws and regulations and therefore free for general use.
The publisher, the authors and the editors are safe to assume that the advice and information in this book are believed to be true and accurate at the date of publication. Neither the publisher nor the authors or the editors give a warranty, expressed or implied, with respect to the material contained herein or for any errors or omissions that may have been made. The publisher remains neutral with regard to jurisdictional claims in published maps and institutional affiliations.

Cover credit: Xinzheng

This Palgrave Macmillan imprint is published by the registered company Springer Nature Switzerland AG
The registered company address is: Gewerbestrasse 11, 6330 Cham, Switzerland

To António Borges
Founding Chairman of the European Corporate Governance Institute
(2002–2010)
In Memoriam

Foreword

The term *corporate government* can be found in English language books that are searchable online since 1800. It refers to the delegation of power inside of organizations once they grow beyond a certain size. It was used to discuss the government of university colleges, banks, and other early business corporations, like canals and railway companies. The word *governance* started to be used in the 1960s, initially in political science. It is defined as "the action or manner of governing a state or organization". The term corporate governance was first used in the 1970s and its use rose sharply until 2003. Its increasing prominence was closely associated with scandals and crisis, like the Lockheed bribery scandal, Enron, the Russia–Asia–Brazil financial crisis and the collapse of Lehman Brothers. It reshaped the investment industry, corporations, regulation, and enforcement. It became a discipline in its own right, but always kept a negative connotation.

The term *global warming* started its ascent in 1980, plateaued in the 90s, and sharply rose again from 2003 onward. It coincided with the approximate first use of the term *environmental, social, governance* (ESG). It marks the latest development and an inflection point in the corporate government debate. It encompasses previously disparate topics like corporate social responsibility (CSR) and socially responsible investment (SRI). Corporate governance has become synonymous with a debate about the future of the capitalist system and the role of finance and business in shaping the future of society. It has turned from a negative to a positive, from a concept associated with scandals, crisis, and failure to improving the economic system for the benefit of all.

The Handbook of Corporate Governance and ESG perfectly captures this trend. It brings together a group of international authors to shed light on ESG from a multi-disciplinary perspective. It spans the full array of subjects that are now discussed under the ESG heading, while linking them back to fundamental corporate governance concepts and insights. The Handbook also shows that there remain many open questions.

What is the role of corporations and finance in reshaping capitalism? For example, there is a tension between fostering market competition and the type of coordination needed to address environmental and social challenges. How can companies adopt costly environmental or social standards when their competitors do not? The principle of "doing well by doing good" does not apply in all situations. Does the market for corporate control promote or undermine the implementation of ESG objectives? Is divestment a substitute or a complement for engagement?

More fundamentally, the profit motive itself has been called into question, even when defining profits long-term. What is the role of investors? Should investors engage with companies to bring about change or invest selectively? How should institutional investors aggregate the social preferences of their clients, or is this an exclusive remit for politics and government? What should be the role of government? "Enlightened" shareholder capitalism relies on government and regulation to define the rules, but who appoints the government? Are government officials motivated or competent in implementing policies without help from investors and/or business? Carbon emissions and climate are forcing us to reconsider paradigms, including corporate governance. The publication of a Handbook on corporate governance and ESG could not be more timely.

Brussels, Belgium   Marco Becht

Introduction

Sustainability is quickly becoming the Holy Grail for governments, businesses, and society. It represents a fundamental shift in human development arguably more significant than the industrial revolution. But unlike many other revolutions, this time, no one can be left behind. Therefore, this time we require collaboration and cooperation (not competition) among and between developed, developing, and under-developed countries if we are to succeed. Unlike previous revolutions where the impetus came from mankind natural desire to enhance their quality of life through innovation, the challenge is now qualitatively different. It is about the very survival of mankind—in other words, we are facing a collective threat and we must therefore change together. It is a rather unique circumstance that self-interest and common good are aligned in such a global scale. While all types of organizations have an important role to play, governments and companies have special relevance in enabling a successful response to this unprecedented challenge.

Governments are required to create an effective legal and regulatory framework that encourages—and even mandates—investment and innovation aligned with the sustainability imperative, and the phasing out of investments that are harmful to the pursuit of the sustainability agenda. Moreover, states are to set the example and to adopt Environment, Social and Governance (ESG) principles as cornerstones of public policy guidelines.

But the quest for sustainability has thus far been grossly incomplete. The recent COP26 was labeled by the UN Secretary-General, Antonio Guterres, as disappointing and a reflection of "the interests, the contradictions, and the state of political will in the world today". The last-minute change to the Glasgow pact by China and India to soften language on coal power, was met with deep disappointment and it is a sad icon of how much is still to be done at political global level.

States must also deploy consistent efforts to achieve uniform international solutions, namely in urgent issues such as climate change. In the EU, some progress has been made but not without intense controversy. As an example, the preparation of the EU Taxonomy Regulation and its respective delegated regulations—a fundamental instrument for European transition to a green economy—has been subject to intense discussion and lobbying by EU-member states and the end result of this legislative package will likely not meet unanimity from climate experts and industries. Similarly, the 2020 Consultation document presented by the European Commission in respect to sustainable corporate governance faced fierce criticism in terms of the methodology underlying the initiative, which anticipates further division in subsequent steps of the legislative process.

These tensions and pressures at the political and regulatory levels contrast with the narrative that is today present around the responsible investment movement, especially ESG (Environmental, Social and Governance). While at the highest level of international policy development is patchy, the ESG movement is quickly expanding and claims significantly more progress. This is despite the focus being placed very much on ESG disclosure than on significant corporate reforms, and there remaining significant issues on materiality assessment and comparability across ESG reports and rating providers. On the other hand, environmental issues remain at the front of corporate attention and, in spite of some advances on a proposed Directive on corporate due diligence and corporate accountability, relevant aspects such as human rights or a social taxonomy clearly lag behind the ESG methodologies and debate.

While the quest for global sustainability requires significant shifts in global governance in pursuit of a common agenda and multipolar collaboration, the slow pace of international harmonization and the intricacies and deficiencies of transnational decision-making turn the focus to the private sector. It therefore remains a fact that companies are the main vehicle for wealth creation. Companies are our best hope of producing the kind of technological innovation that will enable transition to a new economic model that is socially and environmentally sustainable. They are also our biggest concern if they fail to truly transform themselves in a timely fashion.

This explains why corporate governance is placed at the forefront in the path to a more sustainable economy, at environment and social levels. The ESG movement shows the relevance both of investor-led changes and of board-led reforms, as a multipolar trend, and it indicates that companies must live up to these new challenges. The economic transition must be followed by changes of business models and that implies a fresh view at performance metrics, risk management, board duties, stakeholder dialogue and disclosure practices. The role of stewardship codes, governance codes and governance policies therefore become paramount. The whole corporate governance system will be decisive to prepare, adopt, execute, monitor, and enforce decisions in ESG matters.

Companies are also at various stages of development. Many companies—especially small and medium sized, but also a significant number of large ones—continue to have piecemeal approaches to integrating ESG with disparate and unarticulated initiatives. Others are making progress in integrating ESG into their strategy. But very few companies are actually changing their business models. Boards remain unclear as to how to discharge new responsibilities, how to create accountability metrics, and which tools to use to govern and manage for sustainability. Overall, the risk of defective disclosure (greenwashing) is significant. Furthermore, many directors remain stuck in the past and captives of a narrow and outdated interpretation of shareholder value pursuit. This means that the ESG ecosystem is in a state of flux and requires clarification and decisive action to create clarity and certainty for both investors and companies.

The focus of attention of this Handbook is therefore on the adoption of ESG by companies and the legal and regulatory developments in the European Union in relation to ESG and Corporate Governance.

This Handbook is a unique collection of empirical articles and essays on Corporate Governance and ESG developments, prepared by experts from seven jurisdictions—Germany, Italy, Luxembourg, Netherlands, Norway, Portugal, and the United Kingdom. It is divided into three fundamental parts.

Part I is about general aspects of ESG. In this part the reader can find rich analysis and discussions on a range of topics: the systematic interaction between corporate governance and ESG (*Paulo Câmara*); sustainable governance and corporate due diligence (*Guido Ferrarini*); a discussion on reforming company law in a way that European Businesses can thrive (*Beate Sjåfjell*); a discussion on the development of smart regulation of sustainable finance (*Dirk A. Zetzsche and Linn Anker-Sørensen*); board duties around societal responsibility and learning anxiety (*Jaap Winter*); a reconciliation of ESG and shareholder primacy (*Luca Enriques*); an empirical paper on climate finance (*Miguel A. Ferreira*); an overview and discussion of the ESG bond markets (*Manuel Requicha Ferreira*) and; the role of companies in promoting human rights (*Ana Rita Campos*).

Part II of this Handbook takes the reader through critical analysis of some of the key regulatory and legal developments with a focus in European initiatives, including a discussion of ESG and EU law and a move to harder forms of regulation (*António Garcia Rolo*); a regulator's perspective on sustainability and sustainable finance (*Gabriela Figueiredo Dias*); a rich overview and discussion of current ESG reporting instruments, limitations, and opportunities (*Julien Froumouth and Joana Frade*); a discussion of the business judgment rule in the context of ESG (*Bruno Ferreira and Manuel Sequeira*); a discussion of ESG and executive remuneration (*Inês Serrano de Matos*); and how compliance can be an instrument at the service of ESG performance (*Lara Reis*).

Part III provides the reader with perspectives and empirical evidence of the progress on ESG progress in different types of companies: ESG in listed companies (*Abel Sequeira Ferreira*); the ESG adoption and challenges in UK growth companies (*Filipe Morais et al.*); ESG and banks (*Mafalda de Sá*);

ESG and asset managers (*Tiago dos Santos Matias*); and ESG, state-owned companies, and smart cities (*José Miguel Lucas*).

Finally, Part IV includes a concluding chapter, co-authored by Paulo Câmara and Filipe Morais, that aims at pointing out the ESG challenges ahead for future action.

The editors believe that the breadth of this book contributions means that every reader—students, practitioners, and academics—will certainly be able to extract value and new knowledge from its different chapters and reflect further on this evolving challenge of sustainability, of which corporate governance and ESG are key components.

This edited Palgrave Handbook of ESG and Corporate Governance is a Governance Lab initiative. Governance Lab is a non-profit, independent group of academics and practitioners whose members share a common interest in the research and practice of governance across private, public, and governmental organizations. Its mission is to contribute to the improvement of governance practice, law, and regulation in different jurisdictions, including Portuguese-speaking countries such as Portugal, Brazil, Angola, Mozambique, Cape Verde, Sao Tome, and Principe, where most of its members are located.

The Editors are very grateful to all colleagues from Governance Lab and beyond who have contributed to this Handbook. Special thanks are owed to Matilde Azevedo Perez, Erik Oioli, Sofia Santos, and Luís Amado. We would like to also thank Palgrave Macmillan for all the support and patience during the preparation of the manuscript.

Paulo Câmara
Filipe Morais

Contents

Part I General Aspects of ESG

1 The Systemic Interaction Between Corporate Governance and ESG 3
Paulo Câmara

2 Sustainable Governance and Corporate Due Diligence: The Shifting Balance Between Soft Law and Hard Law 41
Guido Ferrarini

3 Reforming EU Company Law to Secure the Future of European Business 59
Beate Sjåfjell

4 Towards a Smart Regulation of Sustainable Finance 87
Dirk A. Zetzsche and Linn Anker-Sørensen

5 The Duty of Societal Responsibility and Learning Anxiety 115
Jaap Winter

6 ESG and Shareholder Primacy: Why They Can Go Together 131
Luca Enriques

7 Climate Finance 137
Miguel A. Ferreira

8 The New ESG Bond Markets 149
Manuel Requicha Ferreira

9 The Role of Companies in Promoting Human Rights 167
Ana Rita Campos

Part II ESG Regulatory Developments

10	ESG and EU Law: From the Cradle of Mandatory Disclosure to More Forceful Steps António Garcia Rolo	191
11	Sustainability and Sustainable Finance: A Regulator's Perspective and Beyond Gabriela Figueiredo Dias	217
12	ESG Reporting Joana Frade and Julien Froumouth	231
13	The EU Taxonomy Regulation and Its Implications for Companies Rui de Oliveira Neves	249
14	Business Judgement Rule as a Safeguard for ESG Minded Directors and a Warning for Others Bruno Ferreira and Manuel Sequeira	267
15	ESG and Executive Remuneration Inês Serrano de Matos	287
16	How Can Compliance Steer Companies to Deliver on ESG Goals? Lara Reis	305

Part III ESG in Particular Types of Companies

17	ESG and Listed Companies Abel Sequeira Ferreira	329
18	ESG in Growth Listed Companies: Closing the Gaps Filipe Morais, Jenny Simnett, Andrew Kakabadse, Nada Kakabadse, Andrew Myers, and Tim Ward	359
19	ESG and Banks: Towards Sustainable Banking in the European Union Mafalda de Sá	375
20	The EU Asset Managers' Run for Green Tiago dos Santos Matias	399
21	ESG, State-Owned Enterprises and Smart Cities José Miguel Lucas	415

Part IV Final Conclusions

22 Conclusion: ESG and the Challenges Ahead 441
Paulo Câmara and Filipe Morais

Index 451

Notes on Contributors

Linn Anker-Sørensen is Senior Manager and the Head of Decentralized Finance at EY Tax and Law, Norway. She also regularly lectures at the University of Oslo's Faculty of Law. Dr. Anker-Sørensen is the author of numerous publications in the field of financial, corporate, and securities law, with a special view on corporate groups, accounting law, financial technologies, regulatory technologies, and sustainable finance. Her latest book on *Corporate Groups and Shadow Business Practices* (Cambridge University Press) is scheduled for publication in Spring 2022. Dr. Anker-Sørensen is regularly invited to speak at regulators, institutions, and universities. In February 2021, Dr. Anker-Sørensen was appointed as Member of the European Securities and Markets Authority's Consultative Working Group on Financial Innovations.

Paulo Câmara is Professor at the Faculty of Law of the Catholic University of Portugal (Lisbon), Managing Partner of Sérvulo & Associados, and Head of Governance Lab, an independent research group dedicated to corporate governance. He is also the Chairman of the General Meeting of the Portuguese Association of Banks (since 2016), and of other financial institutions, and also Chairman of the General Meeting of the Portuguese Compliance and Regulatory Observatory (since 2017). Member of Drafting Committee of the *Financial Law and Capital Markets Journal*, member of the Scientific Committee of the *Competition & Regulation Journal*, and member of the Monitoring Committee of the IPCG Corporate Governance Code. He was Vice-President of the Public Company Practice and Regulation Subcommittee of the IBA—International Bar Association, from 2010 to 2012. He was a member of the European Securities Committee, from 2006 to 2008. He was also Head of the Regulatory Policy and International Department and of the Issuers Department of the Securities Market Commission (CMVM - Comissão do Mercado de Valores Mobiliários), from 1998 to 2008.

Ana Rita Campos is Legal expert at the European Banking Association. Ana holds a Law Degree (FDUL), LL.M. from the Portuguese Catholic University (2007/2008). She worked previously as a lawyer at VdA—Vieira de Almeida.

Inês Serrano de Matos is Assistant Professor at the Faculty of Law of Universidade de Coimbra and a Ph.D. Candidate at the same university.

Rui de Oliveira Neves is Partner at Morais Leitão, Galvão Teles, Soares da Silva & Associates. Rui has been practicing law for 22 years with particular emphasis on the areas of energy and capital markets, being acknowledged by the market as a leading lawyer in such areas of practice. Before spending the last 8 years as General Counsel at Galp, he had practiced law for 14 years also at Morais Leitão, Galvão Teles, Soares da Silva & Associates, where he attained a partner position. In the academic side he is a visiting lecturer at the Faculties of Law of the Universidade Católica Portuguesa in energy law and of the Universidade de Lisboa on the governance and capitals markets legal fields. He is also a member of the Governance Lab since its inception and the author of various specialized publications in the areas of corporate law, corporate governance, and capital markets.

Mafalda de Sá holds a Law degree from the Faculty of Law of the University of Coimbra, LL.M. in European Law from the College of Europe, Bruges, and Ph.D. candidate at the Faculty of Law of the University of Coimbra.

Gabriela Figueiredo Dias is Chairperson at International Ethics Standards Board for Accountants and former Chairperson at the CMVM—Portuguese Securities Commission (2016–2021). She joined the CMVM in 2007, having held management positions since 2008 and advising the Management Board since 2013. She was Head of the international affairs and relations, regulatory policy and issuers, markets and financial information between 2013 and 2015, and Vice-Chair of the Management Board between 2015 and 2016. From 2016 to 2021 she was Vice-Chair of the OECD Corporate Governance Committee and has represented the CMVM at the Board of Supervisors of ESMA.

She is also the author of several publications on Civil Law, Commercial Law, Company Law, Corporate Governance, and Securities Law. Speaker at various symposia and seminars in academic institutions, professional organizations, and companies.

Tiago dos Santos Matias is Director of the International Affairs and Regulatory Policy Department at the Securities Market Commission (CMVM). Formerly, Tiago was Director of the Ongoing Supervision Department, at the CMVM; PMO and Legal Director of Banco CTT; Senior Associate at PLMJ—Sociedade de Advogados RL; Deputy Legal Director of the Banif Group; Legal Director and Deputy General Counsel of Banif—Banco de Investimento, S.A., Banif Gestão de Ativos—SGFIM, S.A., Banif Capital—SCR, S.A., Banif Açor Pensões - SGFP, S.A. and Gamma—STC, S.A.; Senior consultant at KPMG and consultant at Ernst & Young. He holds a Law degree (Lisbon Faculty of Law, Portuguese Catholic University); Post-Graduate courses in Taxation (ISCTE), Administrative Law (Lisbon Faculty of Law, Portuguese Catholic University), and Accounting (Lisbon Faculty of Management, Portuguese Catholic University), and was a Ph.D. student (Nova School of Law).

Luca Enriques is Professor of Corporate Law in Oxford, in association with Jesus College. He studied law at the University of Bologna before completing his LL.M. at Harvard Law School and working at the Bank of Italy while at the same time earning a Doctorate degree in Business Law at Bocconi University. He then became a member of the University of Bologna Faculty of Law (1999–2012). During that period, he was a consultant to Cleary Gottlieb Steen & Hamilton and an advisor to the Italian Ministry of the Economy and Finance on matters relating to corporate, banking, and securities law with a special focus on European Union policy initiatives. He was a Commissioner at Consob, the Italian Securities and Exchange Commission between 2007 and 2012 and Professor of Business Law at LUISS University, Department of Law, in Rome in 2013–2014.

He has held visiting posts at various academic institutions including Harvard Law School, where he was Nomura Visiting Professor of International Financial Systems (2012–2013), Cornell Law School (where he was a John Oling Fellow for short stays in 1999 and 2000), the Instituto de Impresa in Madrid (2005), the Radzyner School of Law at the Interdisciplinary Center Herzliya (2013–2014), the University of Cambridge Faculty of Law (2015), Columbia Law School (2016–2019), and the University of Sydney Law School (2019).

He has published widely in the fields of company law, corporate governance, and financial regulation. He is a European Corporate Governance Institute (ECGI) Research Fellow, a member of ECGI's board, the chair of its Research Committee, a Fellow Academic Member of the European Banking Institute, where he co-chairs its FinTech Task Force, and one of the founding academic editors of the Oxford Business Law Blog.

Guido Ferrarini is Professor of Business Law and Capital Markets Law at the University of Genoa, Department of Law and Director of Centre for Law and Finance.

He graduated from Genoa Law School in 1972; LL.M., Yale Law School, 1978. He was awarded a Dr. jur. h.c. from Ghent University in 2009. He held Visiting Professorships at several universities in Europe (Bonn, Frankfurt, Ghent, Hamburg, LSE, UCL, Tilburg, and Duisenberg) and the United States (Columbia, NYU, and Stanford), teaching courses on comparative corporate governance and financial regulation.

Miguel A. Ferreira holds the BPI | Fundação "la Caixa" Chair in Responsible Finance at Nova School of Business and Economics. He is the Dean of Faculty and Research at Nova School of Business and Economics. He is also a research associate of the European Corporate Governance Institute (ECGI) and a research fellow of the Centre for Economic Policy Research (CEPR). He has a Ph.D. in Finance from the University of Wisconsin-Madison, a Master in Economics from Nova School of Business and Economics, and a Degree in Business from ISCTE. His research interests include corporate finance and governance. His research has been published in academic journals including the *Journal of Finance*, *Journal of Financial Economics*, *Review of Financial Studies* and *Management Science*. He is associate editor for the *Journal of Banking and Finance*. He has been a recipient of several research grants and awards including a European Research Council (ERC) grant. He teaches corporate finance in the undergraduate program. He also has an extensive experience of teaching in executive education programs, consulting, and expert witness.

Bruno Ferreira is PLMJ's Managing Partner and Partner in the Banking and Finance and Capital Markets practices. Bruno joined PLMJ in 2016. He holds a Law degree (Faculty of Law, University of Coimbra, 2001) and a Master's in Law (Faculty of Law, University of Lisbon, 2010).

Abel Sequeira Ferreira is member of the Board and Executive Director of AEM—the Portuguese Issuers Association, and member of the Board of Directors of the European Issuers federation.

Abel is a member of the High-Level Think Tank for Sustainable Finance, coordinated by the Portuguese Ministry of Environment and Energy Transition, and a member of the Task Force for the Development of the Portuguese Capital Market, coordinated by the Ministry of Finance.

Among other positions, he is the only Portuguese member of the European IPO Task Force, a member of the Executive Monitoring Committee of the Portuguese Corporate Governance Code, and was a member of the Advisory Board to the Mission Structure for the Capitalization of Portuguese Companies.

His extensive experience in the areas of capital markets, corporate governance, and sustainable finance, includes board level management positions at BVLP—Lisbon and Porto Stock Exchange, at Euronext Lisbon, at Interbolsa, at FESE—Federation of European Stock Exchanges, and at ECSDA—European Central Securities Depositories Association.

Abel holds a Law Degree and a Master's Degree in Law from the Faculty of Law of the University of Lisbon and is developing Ph.D. research in the area of corporate governance at the Catholic University of Portugal.

He is a Lawyer, the author of several articles on corporate governance and capital markets, and for the past two decades has regularly taught graduate courses, as Visiting Professor, at the Faculty of Law of the University of Lisbon, at the Faculty of Law of the Nova University of Lisbon, at ISG—Business & Economics School and at INDEG-ISCTE Business School.

Joana Frade is Law graduate of the Faculty of Law of the University of Lisbon. Joana Frade practiced law in a private capacity between 1999 and 2017. She was a legal advisor at Caixa Geral de Depósitos and an in-house lawyer at the CMVM—Portuguese Securities Market Commission, where she also served as a legal advisor in the Investment Fund Management Supervision Department. She was head of compliance at the Credit Suisse (Luxembourg), S.A. branch in Portugal between 2017 and 2020, and is currently a legal advisor to the Management Board of the Fundação Oriente.

Julien Froumouth graduated in Business administration and economics from the Toulouse Business School. Julien is currently Sustainable Finance Advisor at the Luxembourg Bankers' Association ('ABBL'), where he is coordinating activities around Sustainable Finance, Corporate Social Responsibility as well as Financial Education and Training. He was previously the Head of Regulatory, Tax and Compliance projects at Credit Suisse (Luxembourg) S.A., after having headed the Business Project Manager at the Luxembourg Stock Exchange where he especially played a key role in the launch and the development of the Luxembourg Green Exchange, the first platform dedicated to Sustainable securities, including but not limited to Green, Social, and Sustainability Bonds. Finally, Julien has been working as a consultant at Deloitte Luxembourg managing large transformation, regulatory, and operational efficiency projects within the Banking sector (Private, Retail, and Corporate and Investment Banking) as well as in the asset servicing industry.

António Garcia Rolo holds a Law degree from the Faculty of Law of the University of Lisbon in 2013 and LL.M. in European Legal Studies from the College of Europe (Bruges Campus) in 2014. He has been a Ph.D. candidate in Private Law at the Faculty of Law of the University of Lisbon since 2018.

Guest Assistant at the Faculty of Law of the University of Lisbon since 2016, where he lectures securities law, corporate finance, listed company law, and civil procedural law, and is also a Researcher at the Private Law Research Center of the aforementioned Faculty. He is a legal advisor in the Resolution Department of the Portuguese Central Bank, having previously completed a professional internship at Linklaters LLP Lisbon Office.

He has published research on company, corporate, and securities' law. Member of the Governance Lab since July 2019

Professor Andrew Kakabadse is B.Sc., M.A., Ph.D., AAPSW, FBPS, FIAM, FBAM, Henley Business School and Emeritus Professor, Cranfield School of Management. He is Advisor to numerous governments and corporations. He has written 45 books, 90 book chapters, and well over 230 articles. He is conducting global studies on the qualities of effective boards, the role of Ministers of State, and the nature of collaborative leadership. He is life member of Thinkers 50 and elected member of European Academy of Sciences and Arts. He is listed in Who's Who for over 15 years. He is Advisor to the UK Parliament and numerous corporations, NGOs, and other governments.

Professor Nada Kakabadse of Henley Business School has extensive experience in researching governance, leadership, top teams, boards, and directorship. She has published 22 books, over 210 articles, and 100 book chapters. She is a consultant to numerous international corporations and has advised governments and global third-sector entities, and Member of the Governing Council of the Empress Theophano Foundation. She is both an elected member to and a UK representative of the European Academy of Sciences and Arts. She is listed in Who's Who and lectures internationally.

José Miguel Lucas is Legal expert at the Prudential Supervision Department of Banco de Portugal. Master in Law from the University of Coimbra, in Legal and Political Sciences, with a dissertation in the area of Administrative Law. Post-Graduate in Administrative Litigation at the Faculty of Law of the University of Lisbon. Formerly, José Miguel was a lawyer at VdA—Vieira de Almeida, in Lisbon, between 2007 and 2013, in the areas of Public Law and Intellectual Property.

Filipe Morais is Lecturer in Governance/Programme Director of the M.Sc. Management for Future Leaders, Henley Business School, University of Reading, UK. Filipe published research papers, reports, chapters, and books in the areas of corporate governance, strategic management, and sustainability. He sits on the editorial board of the *Journal of Business Governance & Ethics* and is a member of the ECGI.

Since 2021 he serves as Independent Member of the Ratings Ratification Committee (RRC) of Johannesburg-based company Risk Insights (Pty), Ltd, a World Economic Forum (WEF) awarded company that works in the areas of sustainability and risk.

Andrew Myers has over 30 years' experience as a management researcher, gaining his skills as an academic, and then working as a research consultant on projects for over 70 organizations in both private and public sectors. His main areas of interest include leadership and board performance, the role of NEDs in society, stakeholder management, and ESG.

Lara Reis is Managing Director at Haitong Bank since 2017, currently responsible for the Compliance Department and also Data Protection Officer, having previously been the Bank's legal director. Before starting her in-house career, she practiced law for 13 years, having dedicated herself mainly to the areas of banking and financial law and capital markets. Between 2010 and 2017 she was a senior associate at Clifford Chance, in London, where she was part of the Structured Finance and Real Estate Finance departments. Previously, she was an associate in the banking and financial area at VdA and a trainee lawyer at PLMJ.

Lara holds has a Law degree from the Faculty of Law of the University of Coimbra and completed a postgraduate degree in Governance, Risk & Compliance, from the Alliance Manchester Business School (University of Manchester) in partnership with the International Compliance Association

(ICA). She is also a lawyer qualified in common law and registered as a solicitor in the United Kingdom. In 2021 she was awarded a scholarship for the Santander Women Emerging Leaders program at London School of Economics.

Manuel Requicha Ferreira holds a degree in Law (Faculty of Law, University of Coimbra, 2004) and is a Master's in Law (Faculty of Law, University of Lisbon, 2012). He is also partner (Cuatrecasas, Gonçalves Pereira) and author of several publications in the area of Securities Law and Insolvency Law.

Manuel Sequeira is Lawyer at PLMJ Advogados, SP, RL, integrating the M&A, Private Equity and Capital Markets Practice Areas. Manuel Sequeira is an author in publications in the field of Commercial Law, namely, in Financial Instruments Law and Banking Law. Manuel holds a Law degree from the Faculty of Law of Universidade Nova de Lisboa (2011) and a Master's degree in Commercial Law from the Faculty of Law of the University of Lisbon (2015). He has been a member of the Governance Lab since March 2016.

Jenny Simnett is an experienced executive and non-executive director with over 35 years' business experience, including 15 years as an interim professional in transformation and change across diverse B2B and B2C organizations. Early career in sales and marketing management with Reuters financial information services, in the United Kingdom, Sweden, Denmark, Mexico, and Austria. A past trustee of charities: Samaritans, Relate, and Carer Support. Currently a board director and chair of the nomination and remuneration committee for Tower Hamlets Community Housing in London and a doctoral researcher in corporate governance at Henley Business School.

Beate Sjåfjell is Professor at the Faculty of Law, University of Oslo in Norway, coordinates the Research Group Companies, Markets and Sustainability, and several international research projects and networks. These include the EU-funded project Sustainable Market Actors for Responsible Trade (SMART, 2016–2020) and Daughters of Themis: International Network of Female Business Scholars.

Tim Ward is CEO of the Quoted Companies Alliance, the independent membership organization championing the interests of small- to mid-size quoted companies. His past roles have included Head of Issuer Services at the London Stock Exchange, Finance Director at FTSE International, the index company, and various management roles at a smaller quoted company. Tim is a Chartered Accountant, has a MBA from Henley Business School, and is a qualified executive coach and mentor.

Jaap Winter is President of the Vrije Universiteit Amsterdam and Chairman of the Supervisory Board of Van Gogh Museum. He is a professor of international company law at the University of Amsterdam. Jaap Winter was the chairman of the High Level Group of Company Law Experts that advised the EU Commission and Finance Ministers on the developments of corporate governance in the EU in 2001–2002. Since then, he was a member of the European Corporate Governance Forum set up by the EU Commission to advise it on a regular basis on governance developments. He was a member of the Tabaksblat Committee that drafted the Dutch Corporate Governance Code in 2003. He received the prestigious ICGN International Corporate Governance Award in 2004.

Jaap Winter has worked with many boards (executive and non-executive) in both advisory and educational capacities. Together with Erik van de Loo he conducts board reviews, focusing on the interplay between the system and the roles that need to be played on the one hand, and group dynamics and member's individual contribution on the other. They have also initiated academic research into this interplay.

Jaap Winter has published widely on matters of corporate law and corporate governance and often speaks at conferences and other public occasions. He was a member of the Supervisory Board of the Dutch securities regulator AFM and is a member of the Supervisory Board of Het Koninklijke Concertgebouw N.V. and the Supervisory Board of Randstad Holding N.V.

Dirk A. Zetzsche holds the ADA Chair for Financial Law since March 2016 and heads the House of Sustainable Governance & Markets at the University of Luxembourg. Zetzsche has served as advisor to and participated in expert working groups initiated by public institutions, such as the European Commission, the European Parliament, the European Securities & Markets Authority, the European Banking Authority, as well as various governments and supervisory authorities throughout Europe and Asia. His more than 300 academic publications deal with corporate and financial law, in particular alternative fintech, collective investment schemes, asset management, family offices, shareholder rights, corporate governance, institutional banking law, secured transactions, international financial regulation as well as sustainable and inclusive finance. Prof. Zetzsche is listed by SSRN consistently since 2019 among the top ten legal scholars worldwide measured by downloads in the last twelve months. Prof. Zetzsche has also worked in the private sector, as a lawyer licensed under German law and board member of regulated entities.

ABBREVIATIONS

BHR	UN Business and Human Rights Treaty
BJR	Business Judgment Rule
CEO	Chief Executive Officer
CLDL	The Dutch Child Labour Due Diligence Law (*Wet Zorgplicht Kinderarbeid*)
CRD V	Capital Requirements Directive V—Directive (EU) 2019/878 of the European Parliament and of the Council of 20 May 2019
CRR	Capital Requirements Regulation—Regulation (EU) No 575/2013 of the European Parliament and of the Council of 26 June 2013
CSR	Corporate Social Responsibility
EBA	European Banking Authority
EC	European Commission
ECB	European Central Bank
EFRAG	European Financial Reporting Advisory Group
EP	European Parliament
EPI	Enacting Purpose Initiative
ESG	Environment, Social and Governance
ESMA	European Securities Markets Authority
EU	European Union
GAR	Green Asset Ratio
GFANZ	Glasgow Financial Alliance for Net Zero
GIIN	Global Impact Investing Network
GRI	Global Reporting Initiative
GSCA	German Supply Chain Act (*Lieferkettengesetz* 2021)
IIRC	International Integrated Reporting Council
IORP II	Directive (EU) 2016/2341 of the European Parliament and of the Council of 14 December 2016 on the activities and supervision of institutions for occupational retirement provision (IORPs)
IRBC agreements	Agreements on International Responsible Business Conduct

KPI	Key Performance Indicators
MSA	UK Modern Slavery Act
NFRD	Directive 2014/95/EU of the European Parliament and of the Council of 22 October 2014 amending Directive 2013/34/EU as regards disclosure of non-financial and diversity information by certain large undertakings and groups
NGFS	Network of Central Banks and Supervisors for Greening the Financial System
OECD Guidelines	The OECD Guidelines for Multinational Enterprises 2011
RTS	Regulatory Technical Standards
SASB	Sustainability Accounting Standards Board
SDG	Sustainable Development Goals
SFAP	Sustainable Finance Action Plan
SFDR	Regulation (EU) 2019/2088 of the European Parliament and of the Council of 27 November 2019 on sustainability-related disclosures in the financial services sector
SME	Small and Medium-Sized Enterprises
SMID	Small and Mid-Cap Companies
SOE	State-Owned Enterprise
SREP	Supervisory Review and Evaluation Process
SRI	Socially Responsible Investment
SSRN	Social Science Research Network
Taxonomy Regulation	Regulation (EU) 2020/852 of the European Parliament and of the Council of 18 June 2020 on the establishment of a framework to facilitate sustainable investment
TCFD	Task Force on Climate-related Financial Disclosures
UN	United Nations
UNEP	United Nations Environment Programme
UN GP Interpretative Guide	UN Corporate responsibility to respect human rights—an interpretative and practical guide
UN Guiding Principles	UN Guiding Principles on Business and Human Rights
UNPRI	United Nations Principles of Responsible Investment
ZLT	Portuguese Innovation Zones

LIST OF TABLES

Chapter 18

Table 1	ESG gaps in UK small and mid-caps by sector	366
Table 2	Closing the ESG gaps: questions for boards	371

PART I

General Aspects of ESG

CHAPTER 1

The Systemic Interaction Between Corporate Governance and ESG

Paulo Câmara

1 Introduction

In recent years a growing number of companies have adopted ESG (Environment, Social and Governance) objectives in their investment activities. Asset managers, banks,[1] insurers and other financial institutions have taken the lead in this respect, showing concern about the social and environmental impact of their investments and promoting their alignment with United Nations Sustainable Development Goals (UN-SDGs) and the Paris Agreement on climate change.

One important indicator relates to the signatories of the United Nations Principles of Responsible Investment (UNPRI), the pioneer ESG standard. The number of signatories to UNPRI has reached more than 3.800 organizations, with total of over 100 Trillion USD assets represented. Large

---

[1] Mazars, *Responsible Banking Practices Benchmark Study* (2021) shows that in 2020 74% of the banks adopted ESG measures, while in 2019 the percentage was of 49% (based on a sample of 37 banks based in Africa, the Americas, Asia–Pacific and Europe); Mafalda de Sá, *ESG and Banks*, Chapter 19 in this book.

[2] BlackRock, Vanguard, and State Street Global Advisors. See Lucian Bebchuk/Scott Hirst, *The Spectre of the Giant Three*, Boston University Law Review, Vol. 99 (2019), 721–741.

P. Câmara (✉)
Faculty of Law/Governance Lab, Portuguese Catholic University (UCP), Lisbon, Portugal
e-mail: pc@servulo.com

© The Author(s), under exclusive license to Springer Nature Switzerland AG 2022
P. Câmara and F. Morais (eds.), *The Palgrave Handbook of ESG and Corporate Governance*, https://doi.org/10.1007/978-3-030-99468-6_1

international fund managers (including the 'Big Three'[2]) have identically voiced their support to ESG.[3] Consequently, the number of ESG financial products (namely ESG investment funds and ESG pension funds) and their inflows rose considerably.[4] Moreover, there is a growing number of funds rebranding to ESG.[5]

In a stricter sense, ESG is the broad term that refers to the inclusion of environmental (E), social (S) and governance (G) criteria into investment decisions taken by companies as a manifestation of responsible or sustainable investment practices. While at its core ESG relates to investors' portfolio decisions, some extensions are to be considered to other financial decisions, namely investment advice decisions, lending decisions and underwriting decisions.[6] In its turn, financial institutions' decisions are aimed at generating successive and lasting impact in invested companies and in other organizations (*'cascade effect'*[7]). Therefore, ESG can also be viewed in a broader sense, where it relates to the influence of environmental, social and governance criteria in organizational decision-making at any level.[8]

ESG clearly marked a turning point of the evolution of the financial system, as it rapidly became an international movement of investors.[9] Hart/Zingales call it a *'new mantra'*,[10] while Mark Carney admits that *'there is real*

---

[3] Larry Fink, *Letter to CEOs* (2018–2021). José Azar/Miguel Duro/Igor Kadach/Gaizka Ormazabal, *The Big Three and Corporate Carbon Emissions Around the World*, ECGI—Finance Working Paper 715/2020 have shown that 'firms under the influence of the Big Three are more likely to reduce corporate carbon emissions'; Jill Fisch/Asaf Hamdani/Steven Davidoff Solomon, The New Titans of Wall Street, A Theoretical Framework for Passive Investors, 168 Penn. L. Rev. 17 (2019). Critically, arguing for the existence of a 'rational hypocrisy' from the Big Three, see Anna Christie, *The Agency Costs of Sustainable Capitalism*, University of Cambridge Faculty of Law Research Paper No. 7/2021; Lucian Bebchuk/Scott Hirst, Index Funds and the Future of Corporate Governance: Theory, Evidence, and Policy, 119 Colum. L. Rev. 2029 (2019).

[4] Morningstar, *Sustainable Fund Flows Reach New Heights in 2021's First Quarter* (30 April 2021).

[5] Dana Brakman Reiser/Anne M. Tucker, *Buyer Beware: Variation and Opacity in ESG and ESG Index Funds*, Cardozo Law Review, Vol. 41 (2020), 1923; Attracta Mooney, *Greenwashing in Finance: Europe's Push to Police ESG Investing*, FT (10-Mar.-2021), using Morningstar data.

[6] See namely Hao Liang/Luc Renneboog, *Corporate Social Responsibility: A Review of the Literature* (2020), ECGI WP 701/2020, 4 ("*incorporation of Environmental, Social and Governance considerations into corporate management and investor's portfolio decisions*").

[7] Regarding ESG cascade effect, see further below, 4.4. and 9.

[8] Alan Palmiter, *Capitalism, Heal Thyself* (2021), available at SSRN 3950395. Regarding the influence of ESG in SOE and smart cities, see José Miguel Lucas, Chapter 21 in this book.

[9] John Hale, *A Broken Record: Flows for U.S. Sustainable Funds Again Reach New Heights*, Morningstar (21-Jan.-2021) referring \$51 billion in 2020 of sustainable funds in the US in contrast to \$5.4 billion in 2018.

[10] The *"new mantra", especially in Europe, is ESG*: Oliver Hart/Luigi Zingales, *Serving Shareholders Doesn't Mean Putting Profit Above All Else*, Promarket.org (Oct.-2017).

momentum behind sustainable investing'[11] and Rebecca Henderson[12] and Guido Ferrarini[13] refer to ESG as *'a game changer'*.

Historically, the term ESG was coined at a United Nations 2004 joint initiative of financial institutions[14] which were invited by United Nations Secretary-General Kofi Annan to develop guidelines and recommendations on how to better integrate environmental, social and corporate governance issues in asset management, securities brokerage services and associated research functions. The topic was later developed in other United Nations initiatives[15]—and most notably the Principles of Responsible Investment, that promotes ESG criteria in asset management. The main Principles address the need to incorporate ESG issues into investment analysis and decision-making processes, the promotion of active ownership and incorporation of ESG issues into ownership policies and practices and the push for appropriate disclosure on ESG issues by the invested entities[16].

Therefore, at the essence of ESG lies the recognition of an inextricable link between environmental and social sustainability and corporate governance.[17] In other words, ESG expresses the connection between corporate governance and social and environmental sustainability. Badly governed companies cannot be sustainable.

This movement erupted as a market-led initiative encouraged by the UN, but recently, this trend has also been amplified by regulatory pressure. The European Union began the route[18] mainly[19] through the European Commission Action Plan on Financing Sustainable Growth (2018), followed by the

---

[11] Mark Carney, *Value(s). Building a Better World for All*, London (2021), 420.

[12] Rebecca Henderson, *Reimagining Capitalism. How Business Can Save the World* (2020), 141.

[13] Guido Ferrarini, *Corporate Purpose and Sustainability*, ECGI—Law Working Paper #559/2020 ('sustainability as a game changer'), 58–61 and Chapter 2 in this book.

[14] Global Compact, *Who Cares Wins Connecting Financial Markets to a Changing World Recommendations by the Financial Industry to Better Integrate Environmental, Social and Governance Issues in Analysis, Asset Management and Securities Brokerage* (2004).

[15] See Freshfields, *A Legal Framework for the Integration of Environmental, Social and Corporate Governance Issues Into Institutional Investment*, UNEP Finance Initiative (2005).

[16] For further information, see unpri.org.

[17] Dorothy Lund/Elisabeth Pollman, *The Corporate Governance Machine*, ECGI Working Paper n. 564 (2021), 35; Beate Sjåfjell/Benjamin J. Richardson, *Company Law and Sustainability. Legal Barriers and Opportunities* (eds.), Cambridge (2015), 313.

[18] Signalling a 2018 shift in respect to previous legislative interventions, see Luca Enriques/Paulo Câmara, *The Portuguese Securities Code at Twenty: Some Comments on the Expansion, Goals and Limits of EU Financial Market Law*, Caderno do Mercado de Valores Mobiliários (2021). For further analysis, see António Garcia Rolo, *ESG and EU Law*, in Chapter 10 in this book.

[19] The 2008 amendment to Article 11 of the Treaty on the functioning of the European Union also paved the way to a subsequent legislative policy re-direction (see Beate Sjåfjell/Benjamin J. Richardson, *Company Law and Sustainability. Legal Barriers and Opportunities* (eds.), cit., 313).

European Green Deal (2019), and the European Green Deal Investment Plan (2020). Then came a succession of important legislative interventions facilitating ESG activism, such as the Shareholder Rights Directive II[20] (that namely fosters shareholders' engagement), the Pension Funds Directive II/IORP II[21] (allowing and encouraging pension funds to take into account the potential long-term impact of investment decisions on environmental, social, and governance factors), the Benchmarking Regulation (concerning Paris-aligned Benchmarks)[22], the Sustainable Finance Disclosure Regulation (SFDR) (imposing disclosure duties in respect to financial market participants),[23] the Taxonomy Regulation (establishing a taxonomy of sustainable objectives)[24] and the Proposed Corporate Sustainability Reporting Directive (CSRD) (that expands sustainability-related information duties). Other legislative initiatives are expected to be approved soon.[25] Some relevant measures have also been announced in the US, both by the Biden Presidency and by the SEC.[26]

---

[20] Directive (EU) 2017/828 of the European Parliament and of the Council of 17 May 2017 amending Directive 2007/36/EC as regards the encouragement of long-term shareholder engagement.

[21] Directive (EU) 2016/2341 of the European Parliament and of the Council of 14 December 2016 on the activities and supervision of institutions for occupational retirement provision (IORPs).

[22] Regulation (EU) 2019/2088 of the European Parliament and of the Council of 27 November 2019 on sustainability-related disclosures in the financial services sector. For further analysis, see Joana Frade/Julien Froumouth, *ESG Reporting*, in Chapter 12 in this book.

[23] Regulation (EU) 2019/2089 of the European Parliament and of the Council of 27 November 2019 amending Regulation (EU) 2016/1011 as regards EU Climate Transition Benchmarks, EU Paris-aligned Benchmarks and sustainability-related disclosures for benchmarks.

[24] Regulation (EU) 2020/852 of the European Parliament and of the Council of 18 June 2020 on the establishment of a framework to facilitate sustainable investment. For further analysis, see Rui de Oliveira Neves, *The EU Taxonomy Regulation*, Chapter 13 in this book.

[25] Reference is made namely to the EC Proposal for a Directive to strengthen the application of the principle of equal pay for equal work or work of equal value between men and women through pay transparency and enforcement mechanisms, to the Proposal on due diligence requirements to protect human rights and the environment in the supply chain and to the level 2 measures related to SFDR and Taxonomy Regulation (regarding the latter, see Guido Ferrarini, Chapter 2 in this book).

[26] Securities and Exchange Commission, *Release N.os. 33-11042; 34-94478, The Enhancement and Standardization of Climate-Related Disclosures for Investors* (2022); White House, A Roadmap to Build a Climate-Resilient Economy (October 14, 2021); Suzanne Smetana, *ESG and the Biden Presidency*, Harvard Law School Forum on Corporate Governance (19-Feb.-2021); Paul Mahoney/Julia Mahoney, *The New Separation of Ownership and Control: Institutional Investors and ESG*, Virginia Law and Economics Research Paper n. 2021–09.

This chapter intends to be setting the scene and to analyse, in different angles, how corporate governance interacts with ESG (environment, sustainability and governance) matters. A corporate governance system involves connectivity, complementarity and interaction of all its elements.[27] The inclusion of ESG objectives affects the whole governance system, as it impacts decision-making processes, ownership policies, product governance strategies, internal control procedures and disclosures. Therefore, the influence of ESG in corporate governance, and vice-versa, is not peripherical, but it is systemic. We therefore propose to briefly present the topics under discussion by addressing the systemic nature of the interference between corporate governance and ESG.

2 CORPORATE GOVERNANCE AND ESG: LEVELS OF IMPACT

The sharp rise of ESG activity occurs at a time when some major trends have transformed the global corporate governance landscape.

On the one hand, climate change concerns have escalated at global level. Transition to a net zero economy appears as inevitable but the progress on meeting the 2015 Paris Agreement targets has been unsatisfactory so far. The COP 26 Glasgow meeting (2021) also confirmed some key States' difficulties in translating words into concrete action. This context reinforced the role of the private sector in addressing climate change and the need to clarify and to strengthen the role of companies in the mitigation and adaptation to the environmental crisis.

On the other hand, institutional investors have been more active and vocal in ESG matters. In the past decade, globally shareholder ownership of listed companies suffered a major shift towards the formation of shareholder blocks owned by large institutional investors ("reconcentration of equity ownership").[28] This trend[29] propelled a growing role to be played by institutional investors. The large scale of portfolios of these institutional investors (by some coined the 'universal owners'[30]), with diversification and investment in several

---

[27] Cristina Mele/Jaqueline Pels/Francesco Polese, *A Brief Review of Systems Theories and Their Managerial Applications*, Service Science, Vol. 2, No. ½ (2010), 130–131.

[28] Jay Cullen/Jukka Mähönen, *Taming Unsustainable Finance. The Perils of Modern Risk Management*, in Beate Sjåfjell/Christopher M. Bruner (eds.), *The Cambridge Handbook of Corporate Law, Corporate Governance and Sustainability*, Cambridge (2019), 103.

[29] Ronald J. Gilson/Jeffrey Gordon, *The Agency Costs of Agency Capitalism: Activist Investors and the Revaluation of Governance Rights* (2013), ECGI—Law Working Paper n. 197; Ronald J. Gilson/Jeffrey Gordon, *Agency Capitalism: Further Implications of Equity Intermediation* (2014), ECGI Working Paper n. 239/2014; Ronald J. Gilson, *Leo Strine's Third Way: Responding to Agency Capitalism*; Journal of Corporation Law, Vol. 33 (2007), 47–56.

[30] Frederick Alexander, *An Honorable Harvest: Universal Owners Must Take Responsibility for Their Portfolios*, Journal of Applied Corporate Finance, Vol. 32, 2 (Spring 2020), 24–30.

countries, implied a priority in addressing systemic risks, such as climate change risks. Against this backdrop, the largest asset managers and banks have been widely supportive in ESG matters.[31]

In several jurisdictions such a trend was amplified by a proliferation of stewardship codes, which encouraged a clarification of institutional investors' stewardship duties and paved the way for their further engagement with invested companies.[32] The original driver for this trend was the influential UK Stewardship Code, whose 2020 version expanded the scope of stewardship by embracing ESG concerns.[33] Such Code states that *signatories systematically integrate stewardship and investment, including material environmental, social and governance issues, and climate change, to fulfil their responsibilities*.[34] The same pattern was followed by the 2020 version of the ICGN Global Stewardship Principles. This approach irradiated globally. In a recent account it is estimated that 84% of the stewardship codes refer to ESG topics.[35]

Furthermore, following a longstanding debate on corporate social responsibility, an increased attention is given to the inclusion of environmental and social concerns in the purpose of the companies,[36] taking into account not only shareholder interests, but also stakeholder interests, such as workers, clients, creditors, business partners and the community.[37] This broader concept of corporate purpose is not contrary to profit-making. The gist of purposeful business implies that companies must be profitable, but profit is not in itself the purpose of companies.[38] This vision of purpose beyond profit was namely embodied in a widely publicized US Business Roundtable 2019

---

[31] Wolf-Georg Ringe, *Investor-led Sustainability in Corporate Governance*, ECGI WP n. 615/2021, 13–16; Jeffrey Gordon, *Systematic Stewardship*, ECGI WP n. 566/2021.

[32] Dionysia Katelouzou/Alice Klettner, *Sustainable Finance and Stewardship: Unlocking Stewardship's Sustainability Potential*, ECGI—Law Working Paper No. 521/2020.

[33] Paul Davies, *The UK Stewardship Code 2010–2020 from Saving the Company to Saving the Planet?*, ECGI WP n. 506/2020.

[34] FRC, *The UK Stewardship Code* (2020), Principle 7.

[35] Dionysia Katelouzou/Dan W. Puchniak, *Global Shareholder Stewardship: Complexities, Challenges, and Possibilities*, ECGI WP 595/2021 9, 47.

[36] See Beate Sjåfjell, Chapter 3 in this book. See also Colin Mayer, *Firm Commitment*, Oxford (2013); Id., *Prosperity. Better Business Makes the Greater Good*, Oxford (2018); Colin Mayer/Bruno Roche (ed.), *Putting Purpose into Practice. The Economics of Mutuality*, Oxford (2021); Colin Mayer/Ronald Gilson/Martin Lipton, *Corporate Purpose and Governance*, Journal of Applied Corporate Finance, Vol 31, n. 3 (Summer 2019); Jill Fisch/Steven Davidoff Solomon, *Should Corporations Have a Purpose?* ECGI—Law Working Paper n.º 510/2020 (2020); Martin Lipton/Steven A. Rosenblum/Karessa L. Cain/Sebastian V. Niles/Amanda S. Blackett/Kathleen C. Iannone, *It's Time to Adopt The New Paradigm*, Harvard Law School Forum on Corporate Governance (2019); Nuno Moreira da Cruz/Filipa Pires de Almeida/Manon Blom-El Nayal, *Responsible Business Leadership and the Path Towards Purpose*, Catolica Lisbon Center For Responsible Business & Leadership (abr.-2020).

[37] See Guido Ferrarini, *Sustainability as a Game Changer*, Chapter 2 in this book.

[38] Colin Mayer/Bruno Roche (ed.), *Putting Purpose into Practice. The Economics of Mutuality*, Oxford (2021), 11–12.

Statement[39] subscribed by 183 US CEO's, and in Davos Manifesto (2020) as well as in several important interventions by the British Academy.[40]

This purposeful business movement implies a focus placed on the foundational reason why each company exists. As Alex Edmans states: '*a purpose defines who the enterprise is and why it exists*'. Professor Colin Mayer and the British Academy further sustain that '*the purpose of the corporation is to do things that address the problems confronting us as customers and communities, suppliers and shareholders, employees and retirees*',[41] while the Davos Manifesto stated that the purpose of a company is '*to engage all its stakeholders in shared and sustained value creation. In creating such value, a company serves not only its shareholders, but all its stakeholders—employees, customers, suppliers, local communities and society at large*'.[42] Moreover, the COVID-19 pandemic crisis also forced a rethink from boards and investors in terms of the core corporate values and increased the attention to social and environmental priorities.[43]

The evolution regarding corporate purpose is relevant under ESG because it places a wider range of interests (including climate, social and governance targets) at the forefront of the underlying objective of a company. Furthermore, it has a great potential to be explored both from investment companies and from invested companies. On the one hand, as a matter of internal and external coherence it is expected that the ESG strategy of financial institutions is aligned with their self-determined purpose.[44] On the other hand, due to the pressure of investors, ESG decisions taken at invested companies level are expected to match their respective corporate purpose.

---

[39] Business Roundtable, *Statement on the Purpose of the Corporation* (2019). Regarding the relevance of this document, with contrasting views, see: (in a negative sense) Lucian Bebchuk/Roberto Tallarita, *The Illusory Promise of Stakeholder Governance* (2020); Id., Lucian Bebchuk/Kobi Kastiel/Roberto Tallarita, *For Whom Corporate Leaders Bargain*, Southern California Law Review, Vol 93 (2021); Luca Enriques, *The Business Roundtable CEOs' Statement: Same Old, Same Old*, Promarket (9-set.-2019); and (in a positive sense) Margaret Blair, *Two Years After the Business Roundtable Statement: Pointing in the Right Direction*, ProMarket (13-sept.-2021); Colin Mayer, *Shareholderism Versus Stakeholderism—A Misconceived Contradiction. A Comment on 'The Illusory Promise of Stakeholder Governance' by Lucian Bebchuk and Roberto Tallarita*, ECGI Law Working Paper No. 522/2020.

[40] British Academy, *Reforming Business for the 21st Century. A Framework for the Future of the Corporation* (2018); Id., *Principles for Purposeful Business. How to Deliver the Framework for the Future of the Corporation* (2020); Id., *Policy & Practice for Purposeful Business. The Final Report of the Future of the Corporation Programme* (2021).

[41] Colin Mayer, *Prosperity. Better Business Makes the Greater Good*, cit., 40; Id., *Corporate Purpose and Governance*, Journal of Applied Corporate Finance, Vol 31 n. 3 (Summer 2019), 14; British Academy, *Principles for Purposeful Business. How to Deliver the Framework for the Future of the Corporation* (2020), 8.

[42] World Economic Forum, Davos Manifesto 2020.

[43] Mark Carney, *Value(s). Building a Better World for All* (2021), 211–260.

[44] See however below the analysis regarding product-specific strategies.

The interaction between corporate governance and ESG is therefore structural, reciprocal and multifaceted. Two main levels are to be considered: the investors' level and the invested companies' level.

On the one hand, at investors' level, corporate governance represents the pillar 'G' under the acronym ESG (environment, social and governance) and therefore corporate governance stands as one of the main criteria for responsible investment. A plethora of corporate governance indicators is therefore used by asset managers and other financial institutions in order to guide their investment strategy. The recent Proposal for a corporate sustainability reporting Directive (CSRD) lists the following governance factors: *(i) the role of the undertaking's administrative, management and supervisory bodies, including with regard to sustainability matters, and their composition; (ii) business ethics and corporate culture, including anti-corruption and anti-bribery; (iii) political engagements of the undertaking, including its lobbying activities; (iv) the management and quality of relationships with business partners, including payment practices; (v) the undertaking's internal control and risk management systems, including in relation to the undertaking's reporting process*.[45] Such catalogue should not be deemed as exhaustive. Other corporate governance indicators—such as remuneration and disclosure practices—are also to be considered.

The role and significance of governance in this context, however, goes beyond being the third ESG pillar. In effect, at the financial institutions' level, ESG also implies involvement of the whole system of governance of the institutional investors namely from the board, the investment function, compliance, HR and the risk management functions. This is relevant in order to effectively channel ESG guidelines into investment decisions, dialogue with stakeholders, due diligence and risk management exercises and finally to provide accurate internal and external information regarding their execution.

On the other hand, financial firms usually manage and distribute financial products with different degrees of ESG involvement. In fact, each financial product may incorporate a distinct ESG strategy and relevant differences may be shown between each fund or portfolio managed. Non-ESG financial products differ substantially from ESG financial products, and the latter can also imply different levels of sustainability commitment, as we will see further below.[46] Particularities may derive from the different nature of financial instruments, from diverse type of underlying management (e.g. active vs. passive management) or from diverse investment policies. Therefore, the product-specific features also impact the relationship between corporate governance and ESG.

---

[45] Proposal for a Directive of the European Parliament and of the Council amending Directive 2013/34/EU, Directive 2004/109/EC, Directive 2006/43/EC and Regulation (EU) No 537/2014, as regards corporate sustainability reporting.

[46] The EU Regulation SFDR provides for a classification of the ESG degree of involvement of financial products in terms of environmental objectives, as we will analyse below. See infra, 4.2.

Finally, at invested firm level, the growing investor pressure in ESG matters will determine further scrutiny in respect to the ESG options taken by each invested company. Therefore, in its turn, this will impact the companies' policies, its culture and its actions—in what is further below described as a part of a 'cascade effect' of ESG.[47] This also leads to a need of accurate information regarding ESG, both for internal and external purposes.

In sum, corporate governance operates not only as a criterion of sound investment (under the 'G' pillar), but also as an enabler of ESG-related decisions. At both investor-level and invested-level, the corporate governance system serves to prepare, adopt, execute, monitor and enforce decisions in ESG matters.

3 ESG Scope and Main Components

As its scope tends to be increasingly wider, ESG encompasses different financial institutions and therefore investment decisions are impacted by distinct degrees of regulation. The central element of ESG deals with asset managers—including investment fund managers, pension fund managers, private equity managers, portfolio managers and insurance companies who manage insurance-based investment products. The common denominator is that these entities take investment decisions on behalf of their clients: they are empowered to buy and sell financial instruments and to exercise rights (including voting rights) attached to the financial instruments under management. Asset managers have therefore fiduciary duties in respect to the beneficiaries of the funds and portfolios managed. Asset management is precisely a financial service where fiduciary duties are more intense.[48]

The EU Sustainable Finance Disclosure Regulation (SFDR) also includes in its scope investment advisory firms,[49] due to their influence in investment decisions. Furthermore, ESG has also the potential of being a reference for banks and other financial institutions in terms of other financial products and services—namely green deposits, green loans and sustainability-linked loans.[50]

The expansion of ESG scope of application does not follow a 'one size fits all' logic. In this respect ESG is a more pressing matter to asset managers than, e.g. to investment consultants, because the former take investment decisions

---

[47] See below, at 4.4 and 9.

[48] Under MiFID II, suitability duties are more intense regarding asset management. See Max Mathew Schanzenbach/Robert Sitkoff, *Reconciling Fiduciary Duty and Social Conscience: The Law and Economics of ESG Investing by a Trustee*, Stanford Law Review, Vol. 72 (2020), 381-ff; Tiago Santos Matias, *EU Asset Managers' Run for Green*, Chapter 20 in this book.

[49] Articles 2(11), 3, n. 2, 4 n. 5 and 6 n.2 of Regulation (EU) 2019/2088.

[50] See Mafalda de Sá, *ESG and Banks*, Chapter 19 in this book; Sofia Santos/Tânia Duarte, *O Setor financeiro e o crescimento sustentável. A nova finança do século XXI* (2019), 115–132.

on behalf of their clients (decision-making role), while the consultants' activity is to render investment advice (decision-influencing role).

In respect to asset managers, the impact of ESG is central and decisive, because it shapes investment strategies, ownership guidelines and stewardship policies. In the words of Finance Professor Rebecca Henderson *'the widespread use of material, replicable, comparable ESG metrics is a game changer, potentially enabling investors to develop a much richer understanding of the relationship between a firm's investment in social and environmental performance and returns to the individual firm (...) and returns to the portfolio as a whole'*.[51]

For asset managers, ESG presents itself as an integrated tool for investment assessment. According to the options and strategies of financial companies, investments are analysed not only in the financial dimension, but also in the dimensions of social and environmental sustainability and governance it presents, in a long-term perspective.

ESG also proposes a redefinition of the measure of value creation. In other words, the metrics of value and economic growth are seen at a wider matrix.[52]

Finally, while it improves resource allocation/portfolio choice criteria, ESG leads also to the promotion of good practices. Financial institutions are sought to be promoters of environmental and social sustainability and to engage in improving the governance and sustainability standards of invested companies.

ESG impacts on different decisions, such as investment decisions, investment advice, risk management decisions and stewardship decisions (such as the exercise of voting rights). The areas of impact are wide and each of them deserves separate analysis.

Investment in shares accounts for the higher potential governance influence in ESG matters, because it combines voice and exit strategies altogether. The analysis on other financial assets is much scarcer.[53] On the other extreme, the impact of ESG in sovereign bonds[54] or real estate assets are some examples of areas in which the 'G' pillar suffers deeper adjustments. Even in these cases, however, the UNPRI has considered that governance-related topics are to be identically under scrutiny—namely in matters related to anti-bribery, money laundering, cybersecurity and general compliance with the law.[55]

---

[51] Rebecca Henderson, *Reimagining Capitalism. How Business Can Save the World* (2020), 141; Mark Carney, *Value(s). Building a Better World for All*, London (2021), 419.

[52] Mark Carney, *Value(s). Building a Better World for All*, 418–453.

[53] Pedro Matos, *ESG and Responsible Institutional Investing Around the World. A Critical Review*, CFA Institute Research Foundation (2020), 53–54.

[54] Raphaël Semet/Thierry Roncalli/Lauren Stagnol, *ESG and Sovereign Risk: What is Priced in by the Bond Market and Credit Rating Agencies?*, available at SSRN 3940945.

[55] Specific factors related to sovereign bonds are country's political stability, government and regulatory effectiveness, institutional strength. Specific elements related to real estate funds are ESG clauses in leases. See UNPRI, *A Practical Guide To ESG Integration In Sovereign Debt* (2019); Id., *ESG Engagement for Sovereign Debt Investors* (2020); UNEPFI,

4   INVESTOR-LEVEL RELATIONSHIP OF ESG AND CORPORATE GOVERNANCE

The motives behind institutional investors' decisions to follow an ESG approach are multiple. Some firms adopt an ethics perspective, being committed to solving a civilizational problem (*the 'doing well by doing good' approach*) and thriving to contribute to the transformation process into a more sustainable economy. Other institutions use a financial foundation for ESG policies, seeking to take advantage of sustainable investments. As an example, in this context, Larry Fink stated that 'Environmental, Social and Governance (ESG) factors can provide essential insights into management effectiveness and thus a company's long-term prospects'.[56] Reputational concerns also come into play when adopting ESG investment criteria—namely because younger generation of investors take ESG more seriously.[57]

Regardless of its motivation, a growing number of institutional investors find sufficient incentives to structure efficient governance instruments to execute its ESG policy.

Below we present and describe three main features of ESG at investor-level: (i) ESG as a set of investment criteria; (ii) ESG as a commitment; and (iii) ESG as a method.

ESG as a Set of Investment Criteria

ESG refers to the set of responsible investment criteria, used by investors, according to environment, social and governance features of the invested companies.

The range of ESG factors to be considered is not harmonized, but at its core it includes climate change mitigation; climate change adaptation; sustainable use and protection of water and marine resources; transition to a circular economy; pollution prevention and control; the protection and restoration

---

Sustainable Real Estate Investment Implementing The Paris Climate Agreement: An Action Framework, (February 2016); UNPRI, *An Introduction to Responsible Investment: Real Estate,* at unprii.org; ISS, *Winning the Net Zero Arms Race—Commitment vs Action as Investors Seek Answers on Sovereign Climate Performance* (29 April 2021).

[56] Larry Fink, Letter to CEOs (2017).

[57] Afdhel Aziz, Playing for the Planet: How Playmob Helped the UN Conduct the Largest Climate Chance Survey Ever Using the Power of Gaming, Forbes (Jan. 28, 2021); Alex Edmans, *Grow the Pie* (2020), 35–36; Sergio Gramitto Ricci/Christina M. Sautter, *Corporate Governance Gaming,* Nevada Law Journal, Vol. 22 (2021), 23–25; Michal Barzuza/Quinn Curtis/David Webber, *Shareholder Value(s): Index Fund ESG Activism and the New Millennial Corporate Governance,* 93 Southern California Law Review (2020), ECGI Law Working Paper 545/2020 (2020) 1283–1312; Id., *The Millennial Corporation* (2021), SSRN 3918443.

of biodiversity and ecosystems (as E factors),[58] human rights; labour standards in the supply chain; child and slave labour; workplace health and safety; human capital management and employee relations; diversity; relations with local communities; health and access to medicine; consumer protection (as S factors) and board structure, size, diversity, skills and independence; executive remuneration; shareholder rights; disclosure of information; business ethics; bribery and corruption; internal controls and risk management (as G factors).[59]

Regarding the G factors, although they represent sound indications of corporate governance, they nevertheless may oversimplify the complexity of governance assessment and run the risk of involving purely mechanic box ticking exercises in their respective scrutiny. Important qualitative elements, albeit more difficult to measure and to compare, such as corporate culture or risk culture, should also integrate these G factors lists.[60] Furthermore, the list of relevant governance factors has expanded over time: cybersecurity is an example of a new governance indicator, namely due to the increase of remote working following the pandemic.

The concept of ESG is very broadly designed, as an 'umbrella term',[61] so that it encompasses different choices of sustainability and governance indicators from companies. Some institutions may opt to give more granular description to th4.2 e pillar factors, the S pillar or to the G pillar.

The underlying objective of ESG is therefore to give the chosen degree of attention or prevalence to investment in sustainable and responsible companies, in order to maximize returns from sustainable economy and to avoid risks underlying non-sustainable and poorly governed companies. The aim is to redirect capital flows to sustainable investments.

ESG as a Commitment; Optionality Arrangements

Under client or peer pressure or regulatory influence, each financial company will ultimately have to take a stand in terms of the ESG approach to be taken.

---

[58] Article 9 of Regulation (EU) 2020/852 of the European Parliament and of the Council of 18 June 2020 (Taxonomy Regulation).

[59] EIOPA, *Opinion on the Supervision of the Management of Environmental, Social and Governance Risks Faced by IORPs* (2019), 3; Mark Carney, *Value(s). Building a Better World for All*, cit., 419.

[60] Alex Edmans, *Response to the European Commission Study on Sustainable Corporate Governance* (2020), 3.

[61] John Hill, *Environmental, Social, and Governance (ESG) Investing: A Balanced Analysis of the Theory and Practice of a Sustainable Portfolio* (2020), 13; Peter Mülbert/Alexander Sajnovits, *The Inside Information Regime of the MAR and The rise of the ESG Era*, ECGI WP n. 548/2020, 7; Pedro Matos, *ESG and Responsible Institutional Investing Around the World. A Critical Review*, CFA Institute Research Foundation (2020), 1.

The ESG commitment may have a voluntary or a mandatory nature.[62] The promotion of ESG guidelines has been pushed from multiple sources in the last years—namely regulatory interventions (e.g. EU Regulations), stewardship codes,[63] international corporate standards[64] and voluntary initiatives adopted by financial firms. Other initiatives also deserve to be mentioned, such as the Net Zero Asset Managers Commitment, that commits asset managers to help deliver the goals of the Paris Agreement of net zero greenhouse gas emissions by 2050.[65] ESG entails therefore a multi-actor and multi-level commitment.[66]

The approach that each company takes in respect to ESG implies consistency and commitment in terms of the investment options.

The mentioned commitment does not exclude autonomy in choosing the ESG priorities that relate more to the purpose and sector of each company. Some investors may focus on climate-related criteria in their investments, while others may follow mainly social sustainability objectives. For instance, a pharmaceutical company may opt to give priority to ESG objectives related to health system, while an energy company may tend to fight climate change at the forefront of its ESG objectives. On the other hand, the scale of organizations may also be relevant. Smaller financial institutions may face difficulties in achieving very ambitious ESG goals.

This explains why optionality arrangements are important. In the UK, the Stewardship Code, originally approved in 2010 makes use of a set of 'apply and explain' Principles for asset managers and asset owners.[67] As its preamble clarifies, '*the Code does not prescribe a single approach to effective stewardship. Instead, it allows organisations to meet the expectations in a manner that is aligned with their own business model and strategy*'.[68] Similarly,

---

[62] As an illustration of mandatory ESG commitment, in Portugal pension fund managers are forced to incorporate ESG guidelines into their investment policies (article 53 n. 4 Law 27/2020).

[63] Dionysia Katelouzou, *Shareholder Stewardship. A Case of (Re)embedding the Institutional Investors and the Corporation?*, Beate Sjåfjell/Christopher M. Bruner (eds.), *The Cambridge Handbook of Corporate Law, Corporate Governance and Sustainability*, Cambridge (2019), 585, 595; Dionysia Katelouzou/Alice Klettner, *Unlocking Stewardship Sustainability Potential*, ECGI Law Working Paper No. 521/2020.

[64] In this regard, see Guido Ferrarini, Chapter 2 in this book.

[65] See www.netzeroassetmanagers.org.

[66] In respect to the multi-level, multi-actor and multi-instrumental of CSR narrative, see Birgit Spiesshofer, *Responsible Entreprise. The Emergence of a Global Economic Order*, München (2018), 370, 498.

[67] Bobby Reddy, *The Emperor's New Code? Time to Re-Evaluate the Nature of Stewardship Engagement Under the UK's Stewardship Code*, University of Cambridge Faculty of Law Research Paper No. 10/2021.

[68] Financial Reporting Council, *The UK Stewardship Code* (2020), 4.

most stewardship codes in other jurisdictions also involve comply or explain approaches.[69]

The EU Sustainable Finance Disclosure Regulation (SFDR) identically entails two types of comply or explain options. On the one hand, financial institutions may opt to consider principal adverse impacts of investment decisions on sustainability factors and disclose a statement on due diligence policies with respect to those impacts. Alternatively, such companies may opt to not consider adverse impacts of investment decisions on sustainability factors, if they disclose clear reasons for why they do not do so, including, where relevant, information as to whether and when they intend to consider such adverse impacts.[70]

From a pre-contractual disclosure point of view, the SFDR also allows for a comply or explain alternative. Financial institutions may opt for (i) disclosing a description of the way sustainability risks are integrated into their investment decisions and the results of the assessment of the likely impacts of sustainability risks on the returns of the financial products they make available; or (ii) disclosing a statement according to which they deem sustainability risks not to be relevant, with a clear and concise explanation of the reasons therefor.[71] However, this option is not available for larger institutions, herein defined as exceeding the average number of 500 employees.

Moreover, different financial products may entail different levels of ESG commitment. This is also clear under the EU classification of financial products according to its ESG involvement: such classification distinguishes between products with no particular ESG focus (the so-called 'article 6 products'); financial products that promote, among other characteristics, environmental and/or social characteristics, provided that the companies in which the investments are made follow good governance practices ('article 8 products'); and financial product with sustainable investment as its objective and an index has been designated as a reference benchmark ('article 9 products').[72]

Optionality arrangements and product-specific options are therefore central components of the current ESG landscape.[73]

The binding facet of ESG has two other implications, to be addressed below: the importance of remuneration and the rules in force to prevent misleading information related to the ESG commitment.[74]

---

[69] Dionysia Katelouzou/Alice Klettner, *Unlocking Stewardship Sustainability Potential*, cit., 25.

[70] Article 4 (1) and (2) of Regulation (EU) 2019/2088. See Sebastiaan Niels Hooghiemstra, *The ESG Disclosure Regulation—New Duties for Financial Market Participants & Financial Advisers* (March 2020), 4–5, available at SSRN.

[71] Article 6 (1) and (2) of Regulation (EU) 2019/2088.

[72] See articles 6, 8 and 9 of Regulation (EU) 2019/2088.

[73] Considering that it '*at least potentially impacts negatively the degree of harmonisation*', see Danny Busch, *Sustainability Disclosure in the EU Financial Sector*, European Banking Institute Working Paper Series 2021, n. 70 (2021).

[74] See *infra*, 6a and 8.

ESG as a Governance Method

ESG implies a governance method with multiple tools, both general and specific. Firstly, general tools of investor engagement are common in ESG approaches—namely voting guidelines, exercise of voting rights, annual letters to CEO's and direct (both formal and informal) communication with boards.[75]

Moreover, specific ESG tools have been developed, mainly under the influence of the UNPRI, and those include removal from portfolio of companies that do not meet ESG criteria (negative screening); choosing companies that meet ESG factors (positive screening), either following a norm-based screening or opting for companies that are benchmark examples and can in turn serve as an inspiring example for others to follow (*'best in class'*)[76]; or adopting a strategy with a single objective—e.g. emissions reduction or gender diversity (*'single-theme'* funds).

As the UNPRI states, 'screening uses a set of filters to determine which companies, sectors or activities are eligible or ineligible to be included in a specific portfolio'.[77]

Negative screening, however, deserves further analysis. This is an exclusionary approach as it involves leaving aside from the investment radar companies that are dedicated to activities or based on specific ESG criteria that are legal but unethical or unsustainable.[78] Common examples of these usually excluded issuers' activities (originally coined in jargon as 'sin stocks') are alcohol, tobacco, gambling, pornography or military weapons. This exclusionary exercise is at times subject to criticism, because in some cases it runs the risk of leaving aside ESG-compliant companies (i.e. companies that in spite of operating in these sectors in other metrics score high in ESG terms) or conglomerates that predominantly operate in mainstream sectors and only residually in sensible areas.[79] On the other hand, selling 'blacklisted stocks' is only viable because there is a market for such stocks. As Edmans reminded, 'an investor can only sell shares if someone else buys, so divestment doesn't deprive a polluting company of capital'.[80] Finally, and most importantly, in

---

[75] Benjamin J. Richardson, *Aligning Social Investing with Nature's Timescales*, in Beate Sjåfjell/Christopher M. Bruner (eds.), *The Cambridge Handbook of Corporate Law, Corporate Governance and Sustainability*, Cambridge (2019), 573.

[76] Hao Liang/Luc Renneboog, *Corporate Social Responsibility and Sustainable Finance: A Review of the Literature*, ECGI—Finance Working Paper No. 701/2020, 13.

[77] Toby Belsom/Catie Wearmouth, *Screening*, UNPRI, at unpri.org.

[78] John Hill, *Environmental, Social, and Governance (ESG) Investing: A Balanced Analysis of the Theory and Practice of a Sustainable Portfolio*, London (2020), 14.

[79] Birgit Spiesshofer, *Responsible Entreprise. The Emergence of a Global Economic Order*, München (2018), 288–289; critically: John Hill, *Environmental, Social, and Governance (ESG) Investing: A Balanced Analysis of the Theory and Practice of a Sustainable Portfolio*, cit., 16, 181.

[80] Alex Edmans, *Is Sustainable Investing Really a Dangerous Placebo?*, Medium (30-set.-2021).

governance terms, negative screening is a pure 'exit' approach while investment followed by engagement in blacklisted companies has the potential of shareholder activism being a driver for change towards greener companies. This is why some investors (namely activist investor pressure group Carbon Action100+ [81]) prefer to focus on blacklisted companies and try to engage with those companies in view of re-directing their activities to a more ESG-friendly pace. Other relevant illustration focused on emission-reduction rather than blacklisting is the Glasgow Financial Alliance for Net Zero (GFANZ), that assembles over 160 firms aimed at accelerating the transition to net zero emissions by 2050 at the latest.

The 'Cascade Effect'

ESG governance affects several types of entities and persons in successive waves of influence—in a manner that we label the 'cascade effect'. We define the ESG cascade effect as the potential aptitude for companies to engage in ESG-based decisions and to systemically influence others to do so, including investors, investee companies and their respective supply chain and community.

The ESG binding effect starts at the level of the asset manager, which represents ESG's first layer of impact. At this level, corporate governance is an important tool to enforce decisions at investors' sphere and its product governance policy, its investment policy and its risk management approach. ESG does not serve merely to refine screening methods, it mainly implies the involvement of governance methods and structures that lead to analysis, procedures, decisions and initiatives taken in ESG matters.

Decisions taken by asset managers will in turn pressure invested companies to act in a more sustainable manner. The decision-making structure of invested companies will inevitably be affected. ESG factors will be relevant to assess risks, impacts and the corporate purpose. Such is the second layer of impact.

Companies will also influence their supply chain (i.e. outsourced companies and other business partners) and other stakeholders affected by ESG decisions, in a third layer of impact.

Moreover, ESG decisions and reports have a wider audience and will be relevant not only to large investors and companies, but also to non-professional investors, to consumers and to the public.[82] The workforce will also be paying attention to ESG factors, namely in recruitment processes. This is the fourth layer of impact derived from ESG.

On the one side, this 'cascade effect' reflects the potential effectiveness of ESG guidelines, decisions and initiatives. It represents a very relevant dimension of responsible investment in terms of its transformative potential.

---

[81] Carbon Action 100+, *Progress Report* (2020).

[82] Iris H-Y Chiu, *Corporate Reporting and the Accountability of Banks and Financial Institutions*, in Iris H-Y Chiu (ed.), *The Law on Corporate Governance in Banks*, Cheltenham (2015), 198–199.

On the other hand, this cascade effect also mirrors, mainly at third and fourth level, the influence based on social mechanisms and the importance of reputational incentives in this context. For this dynamic to operate properly it is very important to have clear, truthful and objective information along each of the cascade levels.

Finally, taking into account the current funding gap for the achievement of SDGs,[83] the ESG framework also presents a huge potential to be used in public funds' management.[84] International finance institutions, central banks,[85] development banks, public infrastructure funds, sovereign wealth funds[86] are starting to also use the ESG approach and method, as a direct way of pursuing their public purpose. That may be taken as another layer of impact of the 'cascade effect' described above.

This presentation of the cascade effect does not mean we can take for granted that its consequences are always fully accomplished in each case. Intrinsic and extrinsic variables must be considered in this respect. On the one hand, as we have seen, the degree of ESG commitment varies from investor to investor and may vary within the same investor considering the type of financial product in question[87]. Furthermore, each investor will choose and give prevalence to the stakeholder issues that are most material to its business model. The sector of activity of invested companies may also be very relevant in shaping ESG priorities. On the other hand, there are general external variables that also bear relevance. It has been studied that a stronger level of ESG investment incorporation is positively related to stronger environmental and social norms prevailing in the investor home country.[88] ESG influence also manifests more clearly in more competitive markets.[89] The quality of ESG data also plays a major influence in this regard: in case of defective disclosure from the financial firm the cascade effect might not even operate at first level. The cascade effect therefore points at the potential far-reaching ESG interactions but naturally does not preclude scrutiny on the extent to which its consequences are effectively achieved. Still, one can predict that as the flow

---

[83] Emilios Avgouleas, *Resolving the Sustainable Finance Conundrum: Activist Policies and Financial Technology*, 84 Law and Contemporary Problems 55–73 (2021).

[84] Regarding SOE's, see José Miguel Lucas, Chapter 21 in this book.

[85] Regarding the potential expansion of the Taxonomy Regulation: Luna Romo, *Una taxonomía de actividades sostenibles para Europa* (January 2021), Banco de España Occasional Paper No. 2101, 21–22.

[86] John Hill, *Environmental, Social, and Governance (ESG) Investing: A Balanced Analysis of the Theory and Practice of a Sustainable Portfolio* (2020), 227–246.

[87] See above 4.2.

[88] Rajna Gibson/Simon Glossner/Phillip Krueger/Pedro Matos/Tom Steffen, *Do Responsible Investors Invest Responsibly?* (May 25, 2021). Swiss Finance Institute Research Paper No. 20–13, European Corporate Governance Institute – Finance Working Paper 712/2020.

[89] Thomas Chemmanur/Dimitrios Gounopoulos/Panagiotis Koutroumpis/Yu Zhang, *CSR and Firm Survival: Evidence from the Climate and Pandemic Crises*, SSRN 3928806.

of ESG investment continues to increase, the massive extension of the cascade effect will become more and more visible.

In the next sections we will analyse in more detail in what terms ESG impacts corporate governance and in which manners it contributes to redefine board duties and skills, disclosure, risk management and remuneration.

5 REDEFINING BOARD DUTIES AND SKILLS; THE 'KNOW YOUR STAKEHOLDER RULE'

Recent years have intensified the debate on whether and to what extent ESG factors determine an extension to board member duties. Focusing on the E pillar, the G7 namely referred to a *duty to safeguard the planet for future generations*.[90] Similarly, Beate Sjäfjell refers to a duty of environmental care[91] while in this book Jaap Winter proposes to introduce a duty of societal responsibility of the board.[92]

The debate gravitates around two structural questions: on the one hand, on the extent to which existing board member duties (namely, duties of care) already include or imply ESG-related duties; on the other hand, there is a vivid and much divided discussion on the merit of having future legislative action to expressly expand the current set of board duties.

Regarding the first question, the OECD Corporate Governance principles clearly state that: '*The board is not only accountable to the company and its shareholders but also has a duty to act in their best interests. Boards are expected to take due regard of, and deal fairly with, other stakeholder interests including those of employees, creditors, customers, suppliers and local communities. Observance of environmental and social standards is relevant in this context*'. But this is not a harmonized field of law and it is ultimately dependent upon each jurisdiction position in relation to expectations of board members' conduct and to the balance between shareholder and stakeholders' interests in that respect.

Nevertheless, there are additional factors that push for change at board level and that influence the current interpretation of board duties in climate, social and governance issues.

Firstly, ESG-related disclosure duties increase the pressure for recognizing that the spectre of board member duties is expanded in respect to ESG

---

[90] Carbis Bay G7 Summit Communiqué, *Our Shared Agenda for Global Action to Build Back Better* (June 2021).

[91] Beate Sjäfjell/Benjamin J. Richardson, *Company Law and Sustainability. Legal Barriers and Opportunities* (eds.), cit., 329.

[92] Jaap Winter, *The Duty of Societal Responsibility and Learning Anxiety*, Chapter 5 in this book.

matters.[93] As below addressed, the area of disclosure duties has rapidly developed in the EU and in other countries, as well as the flow of voluntary ESG disclosure has expanded considerably. Both these trends bear implications in terms of the range of duties of the directors that are owed to the company in terms of scrutiny and assessment of the process regarding the preparation of ESG-related information. This impacts the board, that must ensure oversight of ESG risks and opportunities, ESG disclosures and of general compliance of ESG commitments.

Secondly, a consensus is emerging in respect to the board duties to identify, assess and manage ESG-related risks, and most notably climate risks.[94] The World Economic Forum has namely recommended that '*the board should be accountable for the company's long-term resilience in respect to potential shifts in the business landscape that may result from climate change*'.[95] We will examine this topic further below.[96]

Thirdly, some recent cases of ESG-related litigation increased the pressure in terms of board effective commitment in ESG matters. Climate litigation is on the rise and a UN report found that in 2020 the number of climate-related cases reached at least 1550 cases filed in 38 countries.[97] Interestingly, two of the leading cases refer to Dutch court rulings: in 2019, the Supreme Court of the Netherlands required the state to take measures against climate change[98]; in 2021 the District Court in The Hague issued a ruling that Royal Dutch Shell must reduce its global net carbon emissions by 45% by 2030 compared to 2019 levels (2021).[99] Other notable court cases have been presented globally in gender pay gap matters.[100]

Finally, the flow of shareholder proposals related to ESG also increased the importance of board members to live up to their ESG-related duties.[101]

---

[93] Ellie Mullholland, *UK Directors' Duties in a Changing Climate and the Net Zero Transition*, in Andreas Engert/Luca Enriques/Wolf-Georg Ringe/Umakanth Varottil/Thom Wetzer, *Business Law and the Transition to a Net Zero Economy*, München (2022), 56–57.

[94] Brett McDonnell/Hari Osofsky/Jacqueline Peel/Anita Foerster, "*Green Boardrooms?*", Connecticut Law Review (2021), 517–518.

[95] World Economic Forum (with PWC), *How to Set Up Effective Climate Governance on Corporate Boards. Guiding Principles and Questions* (2019), Principle 1.

[96] See *infra*, 7.

[97] UNEP, *Global Climate Litigation Report. 2020 Status Review* (2020).

[98] Urgenda Foundation v. The State of the Netherlands C/09/456689 (2015).

[99] Milieudefensie et al. v Royal Dutch Shell PLC C/09/571932 (2021). See Benoit Mayer, Milieudefensie et al. *v Royal Dutch Shell: Do Oil Corporations Hold a Duty to Mitigate Climate Change?*, in Andreas Engert/Luca Enriques/Wolf-Georg Ringe/Umakanth Varottil/Thom Wetzer, *Business Law and the Transition to a Net Zero Economy*, 79–83.

[100] Alexia Fernández Campbell, *They Did Everything Right—And Still Hit the Glass Ceiling. Now, These Women are Suing America's Top Companies for Equal Pay*, Vox (10-Dec.-2019).

[101] Brett McDonnell/Hari Osofsky/Jacqueline Peel/Anita Foerster, *Green Boardrooms?*, cit.

One preeminent example is Engine n. 1 success case in appointing three ESG-minded directors on the board of Exxon Mobil (2021).

In order to take stock of ESG-related duties, both at investor-level and invested-level, boards should know who their relevant stakeholders are and how they are impacted by the company, in order to be able to understand their needs and approaches. This is what we call the 'know your stakeholder rule' and it stands as a prerequisite for any ESG strategy.

The literature distinguishes between primary stakeholders and secondary stakeholders.[102] The first group comprises customers, employees, supply chain partners and the communities. Secondary stakeholders include regulators, special-interest groups, consumer-advocate groups, NGO's, the media and the competitors. For ESG purpose, the core lies on primary stakeholders, because they are the ones connected to the value-creation process of the firm.

The selection of relevant stakeholders is closely related to the purpose of each company. In fact, evidence shows that companies with good ratings on material sustainability issues significantly outperform firms with poor ratings on these issues, while companies with good ratings on immaterial sustainability issues do not significantly outperform competitors with poor ratings on the same issues.[103] This is one of the reasons why corporate purpose statements should be clearly articulated, disclosed and monitored.

ESG criteria imply a long-term view of the investments. It therefore becomes part of what Mariana Mazzucato describes as a 'mission-oriented approach'.[104] Although most ESG funds reach excellent short-term performance,[105] the full benefits they bring should also be viewed in a long-term perspective, as a component of an inter-generational sustainability strategy. The assessment of ESG board duties is more complex precisely by taking this long-term metric into consideration.

One of the topics that is still under development is the integration of ESG skills in selecting the board composition.[106] The relevance of ESG-related risks, commitments, initiatives and disclosures clearly indicates that the board should have proper knowledge of these subject matters. Moreover, ESG is becoming increasingly technical and involves a granular analysis of data, which

---

[102] R. Edward Freeman/Jeffrey S. Harrison/Stelios Zyglidopoulos, *Stakeholder Theory: Concepts and Strategies*, Cambridge (2018), 1.

[103] Khan Mozaffar/George Serafeim/Aaron Yoon, *Corporate Sustainability: First Evidence on Materiality*, The Accounting Review (2016) 91 (6): 1697–1724.

[104] Missions require long-term thinking and patient finance: Mariana Mazzucato, *Mission Economy: A Moonshot Guide to Changing Capitalism*, Dublin (2021), 181.

[105] Quinn Curtis, Jill Fisch/Adriana Robertson, *Do ESG Mutual Funds Deliver on Their Promises?*, Michigan Law Review, ECGI WP n. 586/202.

[106] CERES, *Lead from the Top: How Corporate Boards Engage on Sustainability Performance* (2015) (only 17% of Fortune 200 board members with ESG credentials); Tensie Whelan, *U.S. Corporate Boards Suffer from Inadequate Expertise in Financially Material ESG Matters*, NYU Stern Center for Sustainable Business (January 2021) (29% of Fortune 100 with ESG credentials).

bears a necessary reflection in the board's capabilities. Therefore, there is an increasing need for climate literacy and ESG literacy at board level. That should namely (but not exclusively) be reflected in the profile of non-executive directors.

In a recent public address, the fund manager Vanguard underlined this point, by sustaining that *disclosures should provide enough information so that an investor can assess the climate competency of a company's board.*[107] This statement should be read in a broader sense, as referring to the relevance of ESG competencies of the board, including social and governance matters.

Accordingly, the European authorities ESMA and EBA, in their 2021 redraft of suitability guidelines for management body have included within risk management skills the following: identifying, assessing, monitoring, controlling and mitigating the main types of risk of an institution including environmental, governance and social risks and risk factors. Moreover, environmental, governance and social risks are now included in the catalogue of the matters regarding which the management body collectively must have an appropriate understanding of and for which the members are collectively accountable.[108]

ESG skills bear relevance both in the recruitment process and in subsequent training programmes for board members. Regarding the first component, it is relevant to note that the ICGN Corporate Governance Principles recommends that *there should be a formal induction for all new board directors to ensure they have a comprehensive understanding of the company's purpose.*[109]

In larger companies, this evolution may lead to some structural changes at board organization level such as the appointment of ESG committees (internal or external[110]), ESG working groups or a chief ESG officer.[111] Each of these options has merits in terms of facilitating an integrated analysis and a fluid flow of information regarding ESG matters and its embedment in the governance structure and the company culture. Among these possibilities, the choice of the correct governance solution depends upon the specific features of each company and should be solidly anchored in the proportionality principle.

---

[107] Vanguard, *Letter to SEC. Public Input Welcomed on Climate Change Disclosures* (11-june-2021).

[108] EBA/ESMA, *Joint Guidelines on the Assessment of the Suitability of Members of the Management Body and Key Function Holders Under Directive 2013/36/EU and Directive 2014/65/EU*, ESMA 35-36-2319/ EBA/GL/2021/06, 63 (d), 70 (c).

[109] ICGN, *Global Governance Principles* (2021) 1.5.

[110] Regarding the merits of external ESG advisory committees: Alex Edmans, *Grow the Pie*, 233–234. Stating a 17% increase in the number of sustainability board committees across the 100 largest of the Forbes Global 2000 companies, see *The Sustainability Board Report* 2020, 2–3.

[111] Mervyn King/Jill Atkins, *Chief Value Officer: Accountants Can Save the Planet*, 72–114, New York (2016) (proposing a chief value officer); António Gomes da Mota, *Corporate Governance in the New Multi-Stakeholder World: Realities and Challenges*, Prémio, 26 March 2021 (proposing a Chief Stakeholder Officer).

Finally, regarding board composition it has been noticed that the growth of ESG movement also impacted the push for further gender balance at the boards, both at financial firms' level and at invested companies' level. In the UK, the FCA presented a proposal seeking to ensure that disclosure is provided, on a comply or explain basis, on whether at least 40% of board directors of each listed company are women.[112] In the EU, a Directive proposal on the subject has been under discussion since 2012[113] and recently the European Commission has pledged to make a new push for that Directive to be finally approved, but there is uncertainty as to its outcome.

6 Redefining Disclosure

One of the foundational documents of ESG, the United Nations Principles of Responsible Investment states, in its Principle 3, that signatories 'will seek appropriate disclosure on ESG issues by the entities in which we invest'. Furthermore, one of the objectives of the European Commission's Action Plan on Sustainable Finance (2018) is to *'foster transparency and long-termism in financial economic activity'*. Finally, the European Green Deal explicitly indicated that *'companies and financial institutions will need to increase their disclosure on climate and environmental data so that investors are fully informed about the sustainability of their investments'*.[114]

Disclosure is therefore at the heart of the relationship of corporate governance and ESG issues. In order to have an efficient ESG orientation and investment selection, the flow of information to the financial institutions is of critical relevance.[115] Moreover, the lack of a firm-level disclosure may lead to the potential mispricing of assets and financial instruments.[116] Finally, disclosure is also critical for the scrutiny of investor and invested companies.

The disclosure ecosystem has therefore been gradually changing to meet the expectations of institutional and non-institutional investors. In 2019, 90%

---

[112] Financial Conduct Authority, *Diversity and Inclusion on Company Boards and Executive Committees*, CP 21/24.

[113] European Commission, *Proposal for a Directive of the European Parliament and of the Council on Improving the Gender Balance Among Non-executive Directors of Companies Listed on Stock Exchanges and Related Measures*, COM (2012) 614 final.

[114] European Commission, *Communication to the European Parliament, The European Council, The Council, The European Economic and Social Committee and The Committee of the Regions, The European Green Deal*, COM/2019/640 final (2019), 2.2.1.

[115] Some authors argue that the disclosure approach is insufficient: see namely Jay Cullen/Jukka Mähönen, *Taming Unsustainable Finance. The Perils of Modern Risk Management*, in Beate Sjåfjell/Christopher M. Bruner (eds.), *The Cambridge Handbook of Corporate Law, Corporate Governance and Sustainability*, Cambridge (2019), 101, 105–107.

[116] Zacharias Sautner/Laurence van Lent/Gregory Vilkov/Ruishen Zhang, *Firm-Level Climate Change Exposure*, ECGI Finance Working Paper n. 686/2020, 2. For the opposite viewpoint see Paul Mahoney/Julia Mahoney, *The New Separation of Ownership and Control: Institutional Investors and ESG*, cit., 5–8, 22–29.

of the S&P 500 Index have published a sustainability report—while in 2020 the rate was of 20%.[117]

The impetus for ESG disclosure has also been boosted by regulatory interventions.[118] In 2014, the European Union adopted a Non-Financial Reporting Directive (NFRD),[119] requiring all large public-interest companies to provide information about relating to environmental, social and employee matters, respect for human rights, anti-corruption and bribery matter.[120] The NFRD is currently under revision and it will be amended by the Corporate Sustainability Reporting Directive (CSRD). Moreover, the EU Prospectus Regulation dictates that *ESG circumstances can also constitute specific and material risks for the issuer and its securities and, in that case, should be disclosed* in the prospectus.[121] Finally, the European Commission published a Capital Markets Union New Action Plan namely comprising the establishment of a European Single Access Point for financial and non-financial information publicly disclosed by companies.[122]

One of the key criteria for ESG disclosure under this EU regime is the 'double materiality perspective'. This means that the concerned companies will have to report not only about how ESG topics affect their business (outside-in perspective) but also about their own external impact on people, the society and the environment (inside-out perspective).[123] This approach has been further developed by the Proposal for a Corporate Sustainability Reporting Directive (CSRD), that requires the management report to include information necessary to understand the undertaking's impacts on sustainability matters, and information necessary to understand how sustainability

---

[117] See Governance & Accountability Institute, *Flash Report S&P 500 2020. Trends on the Sustainability Reporting Practices of S&P 500 Index Companies* (2020).

[118] Regarding the SEC projects in the US, see Dana Brakman Reiser, *Progress Is Possible. Sustainability in US Corporate Law and Corporate Governance*, in Beate Sjåfjell/Christopher M. Bruner (eds.), *The Cambridge Handbook of Corporate Law, Corporate Governance and Sustainability*, Cambridge (2019), 143–145.

[119] Directive 2014/95/EU of the European Parliament and of the Council of 22 October 2014 amending Directive 2013/34/EU as regards disclosure of non-financial and diversity information by certain large undertakings and groups.

[120] David Monciardini, *Conflicts and Coalitions. The Drivers for European Corporate Sustainability Reforms*, in Beate Sjåfjell/Christopher M. Bruner (eds.), *The Cambridge Handbook of Corporate Law, Corporate Governance and Sustainability*, Cambridge (2019), 617; Birgit Spiesshofer, *Responsible Entreprise. The Emergence of a Global Economic Order*, München (2018), 468–469.

[121] Recital 54 of Regulation (EU) 2017/1129 of the European Parliament and of the Council of 14 June 2017.

[122] European Commission, *A Capital Markets Union for People and Businesses. New Action Plan*, COM/2020/590 final.

[123] European Commission, *Guidelines on Non-Financial Reporting: Supplement on Reporting Climate-Related Information* (2019/C 209/01), 4–5; EBA/ESMA/EIOPA, *Response to IFRS Foundation's Consultation on Sustainability Reporting* (16-dec.-2020).

matters affect the undertaking's development, performance, and position.[124] The EU took the lead in this respect, and therefore it remains to be confirmed that this solution will be followed by other international standard setters.

In relation to pension funds, according to EU Law the statement on investment policy principles, to be publicly available, must include how the investment policy takes environmental, social and governance factors into account.[125] Such information must also be disclosed to prospective members.[126]

At present ESG disclosure also poses important challenges, which are namely: (i) Defective disclosure; (ii) Fragmentation of the information; (iii) Excessive reliance on ESG ratings or other third-party service providers. These points are addressed below in further detail.

Defective Disclosure ('Greenwashing')

ESG disclosure presents a risk of the information presented being defective—exaggerated, selective, deceptive or false.

There are two different basic forms of defective ESG disclosure: manipulative disclosure and selective disclosure. In both cases, we may distinguish entity-level and product-level defective disclosure.[127]

There are important causes for defective information. On the one hand, defective disclosure is mainly rooted in the fact that ESG data is usually unaudited.[128] At EU level, the new CSRD intends to change this, by imposing mandatory audit to non-financial information, although only for large companies. Moreover, the fragmentation of ESG metrics and disclosure frameworks—to be analysed below—also increases the risk of disclosing misleading ESG information. Finally, due to PR pressure, companies sometimes overestimate their respective accomplishments in ESG matters and/or embark in rhetoric exercises with no full adherence to effective action.[129]

---

[124] Proposed new article 19b of Directive 2013/34/EU.

[125] Article 30 Directive 2016/2341 (IORP II).

[126] Article 41 (1) c) and (3) c) Directive 2016/2341 (IORP II).

[127] This classification adapts the three-types matrix presented in Ellen Pei-yi Yu/Bac Van Luu/Catherine Huirong Chen, *Greenwashing In Environmental, Social and Governance Disclosures*, cit., 3.

[128] Ellen Pei-yi Yu/Bac Van Luu/Catherine Huirong Chen, *Greenwashing in Environmental, Social and Governance Disclosures*, Research in International Business and Finance, Vol. 52(C) (2020), 3.

[129] Anna Christie, *The Agency Costs of Sustainable Capitalism*, University of Cambridge Faculty of Law Research Paper No. 7/2021, 5 (arguing '*a significant volume of rhetoric emanating from the Big Three in relation to climate change*'); Tariq Fancy, *Financial World Greenwashing the Public with Deadly Distraction in Sustainable Investing Practices*, USA Today (march 2021): the former Blackrock CIO claims that '*sustainable investing boils down to little more than marketing hype, PR spin and disingenuous promises from the investment community*'.

Defective ESG disclosure is generally labelled as *greenwashing*. While the use of this term is very popular among market institutions and regulators, it is clearly inaccurate as it solely points to environmental (the green—E-pillar) and not also to social and governance defective disclosure. The need for truthful, trustworthy and objective disclosure covers all the three ESG pillars, and not just one of them.

The significance of the risk of misleading information may be confirmed by the screening exercise made in 2021 by the European Commission and national consumer authorities. This sweep exercise in online markets concluded that '*in 42% of cases the ESG claims were exaggerated, false or deceptive and could potentially qualify as unfair commercial practices under EU rules*'.[130]

Additional concerns in terms of disclosure are brought by ESG index funds, that in many cases have an opaque structure.[131]

In the EU, the articulation between the Sustainability Finance Disclosure Regulation (SFDR) and the Taxonomy Regulation is precisely directed at ensuring reliable information in respect to ESG practices from financial institutions.[132] The latter prescribes a much-needed taxonomy of environmentally sustainable economic activities.[133] Under the SFDR, financial market participants will namely be requested to classify their financial products in one of three categories: general products, with no particular ESG focus (the so-called 'article 6 products'); financial products that promote, among other characteristics, environmental and/or social characteristics, provided that the companies in which the investments are made to follow good governance practices ('article 8 products'); and financial product with sustainable investment

---

[130] European Commission, *Screening of Websites for 'Greenwashing': Half of Green Claims Lack Evidence* (28-jan.-2021).

[131] Dana Brakman Reiser/Anne M. Tucker, *Buyer Beware: Variation and Opacity in ESG and ESG Index Funds*, Cardozo Law Review, Vol. 41 (2020), 2003; Hester M. Peirce, *Statement on the Staff ESG Risk Alert* (April 2021); Gary Gensler, *Remarks at the Asset Management Advisory Committee Meeting*, CLS Blue Sky Blog (8 July 2021).

[132] Danny Busch, *Sustainability Disclosure in the EU Financial Sector*, European Banking Institute Working Paper Series 2021, n. 70 (2021); Sebastiaan Niels Hooghiemstra, *The ESG Disclosure Regulation—New Duties for Financial Market Participants & Financial Advisers* (March 2020), 10–12.

[133] Regarding the importance of taxonomy, in general, see OECD, *Developing Sustainable Finance Definitions and Taxonomies*, Paris (2020).

as its objective[134] and an index has been designated as a reference benchmark ('article 9 products').

Article 8 and article 9 products are frequently labelled respectively as 'light green' and 'dark green' financial products. This terminology is incorrect as environmental factor is but one among three ESG factors and social and governance matters are also relevant for the SFDR.

Each of these financial products have distinct disclosure obligations. In respect to article 9 financial products, this namely implies the duty to disclose: (i) information on how the designated index is aligned with that product objective; (ii) an explanation as to why and how the designated index aligned with that objective differs from a broad market index; and (iii) information regarding the methodology used for the calculation of the indices and the benchmarks used.

The responsibility to prevent and deter defective ESG disclosure rests mainly within each company board, as a central component of its directors' fiduciary duties. As the UK Competition Market Authority stated: '*Businesses should be able to back up their claims with robust, credible and up to date evidence*'.[135] Fact-based or data-based ESG information will have to inevitably prevail. Companies that disclose incorrect ESG statements will face significant litigation and reputational risks. Therefore, greenwashing will growingly become more costly.[136] Furthermore, the role of supervisory authorities will be decisive in terms of effective green-washing prevention.

---

[134] The SFDR defines 'sustainable investment' as 'an investment in an economic activity that contributes to an environmental objective, as measured, for example, by key resource efficiency indicators on the use of energy, renewable energy, raw materials, water and land, on the production of waste, and greenhouse gas emissions, or on its impact on biodiversity and the circular economy, or an investment in an economic activity that contributes to a social objective, in particular an investment that contributes to tackling inequality or that fosters social cohesion, social integration and labour relations, or an investment in human capital or economically or socially disadvantaged communities, provided that such investments do not significantly harm any of those objectives and that the invested companies follow good governance practices, in particular with respect to sound management structures, employee relations, remuneration of staff and tax compliance'. See article 2 (17) of Regulation (EU) 2019/2088.

[135] Competition Market Authority, '*Green' Claims: CMA Sets Out the Dos and Don'ts for Businesses* (21-may-2021).

[136] Alessio Pacces, *Will the EU Taxonomy Regulation Foster a Sustainable Corporate Governance?*, ECGI WP 611/2021, 7–8.

Fragmentation of Information

Sustainability is a global problem that requires global harmonization of legal responses.[137] Nevertheless, the information regarding ESG factors is still fragmented and asymmetric, which makes it difficult for asset managers, investors and the public at large.

There in terms of ESG disclosure, there remains a big gap between EU and non-EU companies. On the one hand, there is a proliferation of disclosure templates—namely the Global Reporting Initiative, the SASB, the TFCD, the IFRS and IIRC.[138] As the British Academy states: '*There is considerable confusion, inconsistency and cost associated with the variety of information being produced*'.[139] In the same vein, the OECD alerts that '*current market practices, from ratings to disclosures and individual metrics, present a fragmented and inconsistent view of ESG risks and performance*'.[140]

It is however noteworthy that some convergence initiatives are already in place. On the one hand, some of the leading standard setters on ESG reporting (including the Global Reporting Initiative ["GRI"], Climate Disclosure Standards Board ["CDSB"], Sustainability Accounting Standards Board ["SASB"], International Integrated Reporting Council ["IIRC"] and CDP ["Carbon Disclosure Project"]) have announced a commitment to the creation of a single reporting system. On the other hand, the World Economic Forum, in collaboration with the international audit firms (Big Four) released its recommended system of universal metrics to measure ESG performance ('Stakeholder Capitalism Metrics'). Finally, the International Financial Reporting Standards (IFRS) Foundation announced the intention to create a new Sustainability Standards Board, a world standard-setter in the field.

Currently, at EU level, we are at a transition phase where the most relevant indications on key non-financial performance indicators relevant to the specific activity are still based on a document of a recommendatory nature (European Commission, Guidelines on reporting climate-related information under the Directive 2014/95/EU). Moreover, the European Non-financial Reporting Directive does not mandate integrated reporting of financial and non-financial information. However, the European regime is under review, with a view to greater standardization, better comparability of information and lower costs for issuers. In 2021, the European Commission presented a proposal for a

---

[137] Dana Brakman Reiser, *Progress Is Possible. Sustainability in US Corporate Law and Corporate Governance*, in Beate Sjåfjell/Christopher M. Bruner (eds.), *The Cambridge Handbook of Corporate Law, Corporate Governance and Sustainability*, Cambridge (2019), 131.

[138] Iris H-Y Chiu, *Corporate Reporting and the Accountability of Banks and Financial Institutions*, in Iris H-Y Chiu (ed.), *The Law on Corporate Governance in Banks*, Cheltenham (2015), 228–231.

[139] British Academy, *Principles for Purposeful Business. How to Deliver the Framework for the Future of the Corporation* (2020), 25.

[140] OECD, *Business and Finance Outlook 2020: Sustainable and Resilient Finance* (2020).

Corporate Sustainability Reporting Directive, aiming at a larger universe of companies and with a general EU-wide audit requirement for reported sustainability information. Moreover, the European Financial Reporting Advisory Group published technical recommendations and a roadmap for the development of EU sustainability reporting standards.[141] We have therefore to wait for a refinement of the European regime and for a global unified approach in terms of ESG reporting.

Excessive Reliance on ESG External Advice

As noted, any ESG assessment involves the gathering and analysis of a large amount of data. For most middle and small sized asset managers, it is very hard or not possible to prepare proprietary models of ESG assessment. Financial institutions will have to rely on information provided by third parties—such as ESG ratings,[142] ESG benchmarks and ESG indexes. The importance of proxy advisors also grows exponentially, in relation to ESG activism in voting matters. Moreover, in the EU, the Taxonomy Regulation will arguably dictate the increasing need for external labelling or certification providers.[143] The importance of these service providers becomes therefore crucial.

However, there remain causes for concern regarding the disparate range of ESG ratings—namely different scope of categories, different measurement of categories, and different weights of categories[144]—, which in part is due to the lack of uniformity of ESG reporting.

This context determines a concern on the potential lack of accuracy and on the overreliance in these providers.[145] The discussion lies on the one hand, on the importance that each institutional investor is a true owner of its ESG strategy and monitors its execution (without blank cheques to third parties)

---

[141] EFRAG, *Final Report Proposals For a Relevant and Dynamic EU Sustainability Reporting Standardsetting* (February 2021).

[142] Ingo Walter, *Sense and Nonsense in ESG Ratings*, Journal of Law, Finance, and Accounting, Vol. 5, No. 2 (2020), pp. 307–336.

[143] Article 19 of the EU Taxonomy Regulation indicates the need for future technical screening criteria regarding EU labelling and certification schemes, methodologies for assessing environmental footprint, and statistical classification systems (see Christos Gortsos, *The Taxonomy Regulation: More Important Than Just as an Element of the Capital Markets Union*, European Banking Institute Working Paper Series n. 80 (2020), 21–22). Discussing assurance in the green bond context: Stephen Kim Park, *Green Bonds and Beyond*, in Beate Sjåfjell/Christopher M. Bruner (eds.), *The Cambridge Handbook of Corporate Law, Corporate Governance and Sustainability*, Cambridge (2019), 605.

[144] Florian Berg/Julian Kölbel/Roberto Rigobon, *Aggregate Confusion: The Divergence of ESG Ratings* (May 17, 2020), available at SSRN.

[145] Pedro Matos, *ESG and Responsible Institutional Investing Around the World. A Critical Review*, CFA Institute Research Foundation (2020), 53. The same concern is reflected in the public consultation that preceded the US Fiduciary Duties Regarding Proxy Voting and Shareholder Rights (2021) whose final version was later suspended (U.S. Department Of Labor, *Statement Regarding Enforcement of its Final Rules on ESG Investments and Proxy Voting by Employee Benefit Plans* [10 March 2021]).

and, on the other hand, on whether these service providers have the knowledge and the large resources required for ESG analysis of thousands to million companies worldwide. Proxy advisors are regulated both in Europe and in the US but other ESG service providers are not.

This also serves as a reminder that ESG involvement through these providers comes at a cost. The issue of costs has nevertheless to be weighed against the cost of non-disclosure.[146] As mentioned above, currently the ESG global investment landscape suffers from lack of accessible, accurate and comparable information, and not from excessive information.

7 Redefining Risk

It is scientifically well documented that climate change and other environmental failures determine a wide myriad of risks.[147] One of the conclusions of the Glasgow Climate Pact lies precisely on the recognition that *'climate change has already caused and will increasingly cause loss and damage and that, as temperatures rise, impacts from climate and weather extremes, as well as slow onset events, will pose an ever-greater social, economic and environmental threat'*.[148]

The recommendations of the Financial Stability Board's Task Force on Climate-Related Financial Disclosure also made very clear the potential financial impacts of climate-related events and a wide collection of scientific research is available to confirm it.[149] Moreover, climate change is a 'multiplier of threats', as it increases exponentially, and over the long term, other sources of risk, such as the risk of conflicts, the risk of massive, disorganized immigration and the risk of national security.[150] These risks affect companies on a global scale.

Social crises and governance flaws are equally causal determinants to important risks. As namely the scandals at Enron (2001), Worldcom (2001), VW (2015), Deepwater Horizon (2010) and Shell (2021) show, ESG risks can

---

[146] John Coates, *ESG Disclosure—Keeping Pace with Developments Affecting Investors, Public Companies and the Capital Markets*, Harvard Law School Forum on Corporate Governance (13-mar.-2021).

[147] Stefano Giglio/Bryan Kelly/Joannes Stroebel, Annual Review of Financial Economics, Vol. 13, 15–36 (2021); Jonathan Jona/Naomi Soderstrom, *Evolution of Climate-Related Disclosure Guidance and Application of Climate Risk Measurement in Research*, in Carol Adams (ed.), Handbook of Accounting and Sustainability (2022).

[148] Glasgow Climate Pact (2021), VI.

[149] Beate Sjåfjell, *The Financial Risks of Unsustainability: A Research Agenda*, University of Oslo Faculty of Law Research Paper No. 2020–18, Nordic & European Company Law Working Paper No. 21-05.

[150] Michael E. Mann, *The New Climate War. The Fight to Take Back Our Planet*, London/Victoria (2021), 180–181.

be financially material and may lead to very significant losses.[151] Moreover, the pandemic resulting from COVID-19 also demonstrated the importance of adequately managing social risks.[152]

In this context, following the pandemic period, the OECD recognized the need to companies to improve the management of environmental, social and governance (ESG) risk.[153] Furthermore, it is being prepared an amendment to the EU banking prudential regime (CRD IV and CRR) to require banks to systematically identify, disclose and manage ESG short-, medium- and long-term risks as part of their risk management. Such risks are to be included in credit institutions' strategies and processes for evaluating internal capital needs as well as adequate internal governance.[154]

Other EU legislative measures have also been approved mandating UCITS and AIF fund managers to integrate sustainability risks in the management activity, taking into account the nature, scale and complexity of the business of the investment companies.[155] Moreover, investment firms are required to review the investment products they offer or recommend and the services they provide on a regular basis, taking into account any event that could materially affect the potential risk to the identified target market.

The risk management matrix should therefore integrate risks related to environmental, social sustainability and governance. This bears implications in terms of companies' duties, as they are forced to systematically identify, assess, manage, in the short, medium and long term, and to communicate ESG risks.

In order to be effective, ESG risk management also implies a method in gathering, assessing and reviewing information. It implies a flux of information to ensure access to complete, objective, accurate and timely non-financial information from invested companies. It also implies good stakeholder governance—and namely establishing a sound dialogue with stakeholders—as a way of mitigating social risks.

---

[151] UNEP, *The Materiality of Social, Environmental and Corporate Governance Issues to Equity Pricing* (2004).

[152] Paulo Câmara, *COVID-19, Administração e Governação de Sociedades*, in Paulo Câmara (coord.), *Administração e Governação de Sociedades* (2020); Id., *Coronavirus e Corporate Governance*, Ver (20-mar.-2020).

[153] OECD, *The Future of Corporate Governance in Capital Markets Following the COVID-19 Crisis* (2021), 1.5.

[154] See European Commission, *Proposal for a Directive of the European Parliament and of the Council Amending Directive 2013/36/EU as Regards Supervisory Powers, Sanctions, Third-Country Branches, and Environmental, Social and Governance Risks*, COM (2021) 663 final (27.10.2021); EBA, *Report on Management and Supervision of ESG Risks for Credit Institutions and Investment Firms* (2021). See also Mafalda de Sá, *ESG and Banks*, Chapter 19 in this book.

[155] See Delegated Regulation (EU) 2021/1255 (AIFMD) and Delegated Regulation (EU) 2021/1270 (UCITS).

The types of risks are different in each of the ESG pillars. In a recent EU proposal,[156] 'environmental risk' is defined as the risk of losses arising from any negative financial impact on the institution stemming from the current or prospective impacts of environmental factors on the institution's counterparties or invested assets, including factors related to the transition towards the following environmental objectives: (a) climate change mitigation; (b) climate change adaptation; (c) the sustainable use and protection of water and marine resources; (d) the transition to a circular economy; (e) pollution prevention and control; (f) the protection and restoration of biodiversity and ecosystems. This presentation follows the Taxonomy Regulation structure.

Environmental risk includes physical risks, liability risks, transition risks, reputational risks, regulatory risks[157] and systemic risks.[158] These risks can have long-term effects.[159] For governance matters, risks may arise from any part of the governance system (e.g. ineffective financial controls, tunnelling, defective remuneration structures) and may represent the source of liability risks, regulatory risks and reputational risks.

A greater difficulty arises when mapping social risks. Social vulnerabilities are extremely dependent upon the context of each company, its dimension, its activity and the community it affects. Therefore, the preparation of social vulnerability indexes must deal with firm-specific variations and spatial variations.[160] It is therefore disappointing that the EU has approved a Taxonomy Regulation that solely covers environmental issues and social minimum standards.[161] In other words, the EU still lacks a Social Taxonomy Regulation. This is a cause for concern as it implies a relevant asymmetry in identifying and managing social risks, and it is a matter to be inevitably addressed in the near future.

---

[156] Article 4 (1), 52 e of Regulation (EU) No 575/2013, as amended by European Commission Proposal for a Regulation of the European Parliament and of the Council amending Regulation (EU) No 575/2013 as regards requirements for credit risk, credit valuation adjustment risk, operational risk, market risk and the output floor, 27.10.2021 COM (2021) 664 final.

[157] Zacharias Sautner/Laurence van Lent/Gregory Vilkov/Ruishen Zhang, *Firm-Level Climate Change Exposure*, ECGI Finance Working Paper No. 686/2020, 2.

[158] Edith Ginglinger/Quentin Moreau, *Climate Risk and Capital Structure*, ECGI Finance Working Paper n. 737/2021 (2021); Barnali Choudhury, *Climate Change as Systemic Risk* (October 2020), SSRN; Eva Micheler/Coraline Jenny, *Sustainability and Systemic Risk—A Conference Report*, LSE Law—Policy Briefing Paper No. 44 (2020).

[159] As Larry Fink stated: 'Climate change is different. Even if only a fraction of the projected impacts is realized, this is a much more structural, long-term crisis' (*Annual Letter to CEO's* [2020]).

[160] For an example of strong spatial variation of social vulnerability, see Ivan Frigerio/Mattia De Amicis, *Mapping Social Vulnerability to Natural Hazards in Italy: A Suitable Tool for Risk Mitigation Strategies*, Environmental Science & Policy, Vol. 63 (September 2016), 187–196.

[161] Marleen Och, *Sustainable Finance and the EU Taxonomy Regulation—Hype or Hope?*, Jan Ronse Institute for Company & Financial Law Working Paper No. 2020/05 (November 2020), 6.

In general, ESG risks can be either short-term or long-term. In particular, climate change is considered as a problem of extreme risk with both short-term and long-term impact, in the sense that it may have physical as well as systemic and irreversible effects.[162]

ESG must also be embedded in the risk culture, both at investors' level and at invested company's level. Some institutional investors are faced with specific regulatory frameworks in this respect. The EU pension fund Directive forces the system of governance of such funds to include consideration of environmental, social and governance factors related to investment assets in investment decisions, and to be subject to regular internal review.[163] Its risk management function also must assess environmental, social and governance risks relating to the investment portfolio and the management thereof.[164]

Finally, in respect to banking, the Basle Committee has been active in publishing several documents regarding climate-related risk[165] and is preparing a set of Principles for the effective management and supervision of climate-related financial risks.[166] Furthermore, banks are beginning to be faced with stress-testing exercises against sustainability risks, to assess their resilience against a catalogue of plausible climate-related events and to determine the impact of climate-related risk drivers on their risk profile.[167]

In conclusion, the pressure to adequately identify, manage and report ESG risks is here to stay, both at the level of institutional investors and of invested companies.

8 Redefining Remuneration Policies

Remuneration practices have also been affected by ESG objectives.[168] A WTW report documented that 51% of the S&P 500 companies already incorporate

---

[162] The Economist Intelligence Unit, *The Cost of Inaction: Recognising the VALUE at risk from Climate Change* (2015); Filipe Duarte Santos, *Alterações Climáticas*, Lisbon (2021), 49–50.

[163] Article 21 (1) Directive 2016/2341 (IORP II).

[164] Articles 25 (2) g) and 28 (2) g) Directive 2016/2341 (IORP II).

[165] Basel Committee on Banking Supervision, *Climate-Related Financial Risks: A Survey on Current Initiatives* (30 April 2020); Id., *Climate-Related Risk Drivers and their Transmission Channels* (14 April 2021); Id., *Climate-Related Financial Risks—Measurement Methodologies* (14 April 2021).

[166] Basel Committee on Banking Supervision, *Consultative Document Principles for the Effective Management and Supervision of Climate-Related Financial Risks* (2021).

[167] Mark Carney, *Foreword*, in Herman Bril/Georg Kell/Andreas Rasche (ed.), *Sustainable Investing. A Path to a New Horizon* (2021), xxxii; Patrizia Baudino/Jean-Philippe Svoronos, *Stress-Testing Banks for Climate Change—A Comparison of Practices*, FSI (2021).

[168] See Inês Serrano de Matos, Chapter 15 in this book. In general: Paulo Câmara, *Remunerações e Governo das Sociedades: Uma nova agenda, em Instituto Português de Corporate Governance, Volume Comemorativo do XV Aniversário*, (2018), 267–284.

ESG metrics in their incentive plans.[169] Deloitte also reported that almost 40% of the Fortune 100 companies incorporated ESG measures in their remuneration plans[170] and signalled a prospect of increase in the next 1–2 years.[171] The most popular ESG metrics are GHG emissions, diversity and inclusion metrics, customer satisfaction and worker safety.[172]

The European Commission Action Plan on *Financing Sustainable Growth* (2018) directly addressed this issue, by stating that '*The governance of public and private institutions, including (…) executive remuneration, plays a fundamental role in ensuring the inclusion of social and environmental considerations in the decision-making process*'.[173] Furthermore, the revised version of the EU Shareholders' Rights Directive imposes the inclusion of financial and non-financial performance criteria in the remuneration policy, including, where appropriate, criteria relating to corporate social responsibility.[174] The European Commission Proposal on Due Diligence also prescribes that climate action plans take into account, when setting variable remuneration, if variable remuneration is linked to the contribution of a director to the company's business strategy and long-term interests and sustainability. The role of incentives is also recognized under Principle 6 of the Climate Governance Principles, that namely states that it could be considered to extend variable incentives to non-executive directors.[175] Some institutional investors[176] have also supported the inclusion of ESG measures in remuneration policies.

The remuneration policy is a central component of the corporate strategy and as such it is instrumental to the ESG strategy of each firm. Such policy is also a key pillar of the governance structure of a company and therefore must be consistent with the options taken in ESG policies, risk management policies

---

[169] Willis Towers Watson, *ESG Incentive Metrics S&P 500 Highlights* (March 2020), noting that only 4% include ESG metrics in long-term incentive programmes. According to Bloomberg, 9% of the 2,684 companies in the FTSE All World Index tracked by researcher Sustainalytics in a 2020 study had tied executive pay to ESG (Kevin Orland, Canadian banks tie CEO pay to ESG, setting them apart from the crowd (18 March 2021).

[170] Kristen Sullivan/Maureen Bujno, *Incorporating ESG Measures into Executive Compensation Plans* (April 2021).

[171] Deloitte, *Road to Net Zero… Incentivising Leadership* (September 2021), at 4.

[172] WEF, *Measuring Stakeholder Capitalism Towards Common Metrics and Consistent Reporting of Sustainable Value Creation* (2020), 9; Semler Brossy, *ESG + Incentives 2021 Report* (2021), 4.

[173] European Commission, *Action Plan on Financing Sustainable Growth*, COM (2018) 97 final (8-mar.-2018), at 1.

[174] Article 9a introduced by Directive (EU) 2017/828 of the European Parliament and of the Council of 17 May 2017.

[175] World Economic Forum, *How to Set Up Effective Climate Governance on Corporate Boards. Guiding Principles and Questions* (January 2019).

[176] See namely BlackRock, *Incentives Aligned with Value Creation* (2021); Cevian Capital, *Cevian Capital Requires ESG Targets in Management Compensation Plans* (3-march-2021).

and engagement/stewardship policies. Full and coherent articulation between these policies becomes therefore of critical importance.

Both at institutional investors' level and at invested company level, the main concern is to ensure alignment between ESG objectives and the incentives that are in place. And, in fact, it is widely recognized that remuneration can be a very powerful tool to enforce ESG strategies.[177] Furthermore, recent research has shown that the integration of corporate social responsibility (CSR) criteria into executive compensation is associated with greater firm innovation.[178]

Remuneration policies are also important tools to promote sound and effective risk management of financial institutions.[179] Therefore, KPI must be articulated with the risk management matrix, as discussed above. The preamble text of SFDR states that the '*structure of remuneration [must] not encourage excessive risk-taking with respect to sustainability risks and is linked to risk-adjusted performance*'. It is now clear that it must include risks related to environmental sustainability, social sustainability and governance.

Three main aspects of remuneration policy deserve particular attention: (i) the structure of remuneration policy; (ii) disclosure; and (iii) the decision-making process. These will be dealt with below.

The main ESG implications relate to the structure of variable component of the remuneration of the financial institutions (investor companies) and listed companies (invested companies).[180]

The topic involves some degree of complexity.[181] Firstly, there is a debate on whether ESG-linked pay KPI might lead to short-term focus from the board.[182] In response to this question, it is important to note that ESG-related metrics can affect short-term, medium-term and long-term incentives. The European Directive on alternative fund managers, for instance, forces an assessment of remuneration indicators in a longer period '*appropriate to the*

---

[177] Lucian A. Bebchuk/Holger Spamann, *Regulating Bankers' Pay*, GEO. L.J., Vol. 98 (2010), 247; Mark J. Roe/Holger Spamann/Jesse Fried/Charles Wang, *The European Commission's Sustainable Corporate Governance Report: A Critique* (14-out.-2020), 14.

[178] Albert Tsang/Kun Tracy Wang/Simeng Liu/Li Yu, *Integrating Corporate Social Responsibility Criteria into Executive Compensation and Firm Innovation: International Evidence*, Journal of Corporate Finance, Vol. 70 (2021) (covering a sample of firms from 30 countries).

[179] Iris Chiu, *Corporate Governance of Financial Institutions*, in Helmut K. Anheier/Theodor Baums (eds.), *Advances in Corporate Governance. Comparative Perspectives*, Oxford (2020), 71.

[180] Please bear in mind that in Europe financial institutions (investor-level) remuneration is subject to tighter regulation whereas listed companies may adapt a comply or explain approach.

[181] PWC/LBS/ccg, *Paying Well by Paying for Good* (2021).

[182] See Alex Edmans/Luca Enriques/Steen Thomsen, *Call for Reflection on Sustainable Corporate Governance* (2020), available at ecgi.org: 'tying pay to stakeholder targets may lead to short-term behaviour to hit the targets'. In the same sense, see PWC/LBS/CCG, *Paying Well by Paying for Good*, cit., 30.

fund life-cycle.[183] In Germany, there is an explicit rule that mandates listed companies to have their remuneration policy aligned with their respective long-term development.[184]

This serves as a caution in terms of the way ESG-linked KPI are drafted. On the one hand, ESG-linked KPI should be involved with long-term assessment in order to avoid a short-term focus from management.[185] On the other hand, KPI should be not only quantitative but also qualitative. Moreover, there are tail events that require adaptation and that may not be captured in standard KPI (e.g. safety risk).[186] These indicators should also be drafted in a precise way and avoid vague and undetermined formulations, namely in terms that are too easy to achieve.[187] Finally, ESG-linked KPI are part of a mix of performance indicators and should not be isolated (in order to avoid what Alex Edmans call the '*hit the target, miss the point*' effect).[188] These observations, however, should not deter companies from using ESG metrics in their remuneration policies. In Europe and in the US, as an additional argument, the say on pay regime serves as a tool for shareholder scrutiny in respect to the inclusion of ESG elements in remuneration policies.[189]

The core underlying objective is to align key performance indicators with ESG targets. Companies are to adopt a clear strategy to identify ESG metrics that are relevant to its business and are compatible with its long-term business interest and vision, as well as with sustainable investment.

In respect to the 'E' pillar, the most frequent metrics relate to carbon emissions. A distinction is drawn here between Scope 1, Scope 2 and Scope 3 metrics. Scope 1 reports to direct emissions from owned and controlled companies, while Scope 2 concerns indirect emissions from sources of purchased electricity and Scope 3 includes all indirect emissions along the

---

[183] AIFMD, Annex II. See Dirk Zetzsche, *The Alternative Investment Fund Managers Directive* (ed.), 3rd edition (2020), 149.

[184] § 87a Aktiengesetz. See Christian Arnold/Julia Herzberg/Ricarda Zeh, *Das Vergütungssystem börsennotierter Gesellschaften nach§87a AktG*, AG 9/2020, 313.

[185] ICGN, *Integrating ESG into Executive Compensation Plans* (2020) ('a move towards longer-term incentives is now needed').

[186] PWC/LBS/CCG, *Paying Well by Paying for Good* (2021).

[187] The case of Honeywell inevitably comes to mind, whose KPI was merely to 'drive a robust ESG programme' (Andrew Hill, *Executive Pay and Climate: Can Bonuses be Used to Reduce Emissions?* FT [14-nov.2021]).

[188] Alex Edmans, *Response to the European Commission Study on Sustainable Corporate Governance* (2020), 3; PWC/LBS/CCG, *Paying Well by Paying for Good*, cit., 30; PWC/Phillippa O'Connor/Lawrence Harris/Tom Gosling, *Linking Executive Pay to ESG Goals* (2021).

[189] Illaria Capelli, *La sostenibilità ambientale e sociale nelle politiche di remunerazione degli amministratori delle società quotate: la rilevanza degli interessi degli stakeholder dopo la SHRD II, Orizzonti del Diritto Commerciale*, 2|2020 (2020), 575–588.

company's value chain, including suppliers, customers and partners.[190] The latter is clearly more demanding and harder to implement and monitor.

On the other hand, these environment-related indicators will inevitably push for longer term indicators. Recent reports even give evidence for the existence of hyper-long-term incentive plans, that have effects long after board mandate termination.[191]

KPI in this context need to be meaningful, measurable and its structure should be subject to disclosure.[192] In 2009, the European Commission already recommended to listed companies that *performance criteria should promote the long-term sustainability of the company and include non-financial criteria that are relevant to the company's long-term value creation*.[193] In order to adapt their performance indicators, each financial institution must: (i) identify ESG objectives; (ii) set relevant measurement indicators; (iii) measure and validate.[194] In this exercise, when setting up objectives, the remuneration policy must be articulated with the company's purpose, both at investors' level and invested-level. As previously said, ESG approach must be adapted to each firm. The challenge therefore is to transform Key Performance Indicators into Key Purpose Indicators.

On the other hand, the introduction of claw back and malus clauses related to ESG may be considered to enforce ESG objectives.[195] Claw back clauses are apt to respond to longer term objectives[196] but in many cases may be poor substitutes for long-term deferral clauses and restricted stock, which are easier to enforce.[197] Any of these remuneration techniques, however, avoid that ESG-linked KPI lead to short-term focus from the board.

---

[190] Shai Ganu, *Climate Issues 'Heat Up' in Boardrooms*, Willis Towers Watson (2021), that states that '*Analysis by WTW shows that while around 11 per cent of top 350 European companies had tCO2e emission reduction targets in management goals and incentives, only 2% of the US S&P 500 companies did*'.

[191] Shai Ganu/Philipp Geiler, *Combating Climate Change Through Executive Compensation*, Willis Towers Watson (2020).

[192] Beate Sjåfjell, *The Role of Business Law in the Jigsaw Puzzle of Sustainability*, University of Oslo Faculty of Law Research Paper No. 2015-20, Nordic & European Company Law Working Paper No. 15-07 (2015); Mark J. Roe/Holger Spamann/Jesse Fried/Charles Wang, *The European Commission's Sustainable Corporate Governance Report: A Critique* (14-out.-2020), 14; Andrew Johnston/Jeroen Veldman/Robert G. Eccles/Simon F. Deakin et al., *Corporate Governance for Sustainability Statement* (11-dec.-2019).

[193] European Commission, Recommendation of 30 April 2009, 2009/385/EC, at 3.2.

[194] Ruth Simsa/Olivia Rauscher et al., *Methodological Guideline for Impact Assessment*, TSI Working Paper Series No. 1, Brussels (2014), 12.

[195] In this sense, PRI, *Integrating ESG issues into executive pay. An investor initiative in partnership with UNEP Finance Initiative and UN Global Compact A Review Of Global Utility And Extractive Companies* (2016), 6, 19–10.

[196] In general: Iris Chiu, *Corporate Governance of Financial Institutions*, cit., 72–73.

[197] Alex Edmans, *Grow the Pie. How Great Companies Deliver Both Purpose and Profit*, Cambridge (2020), 126.

In terms of disclosure, the EU Sustainability Finance Disclosure Regulation (SFDR) imposes financial institutions to include in their remuneration policies information on how those policies are consistent with the integration of sustainability risks.[198] SFDR has a principles-based approach with a focus on disclosure. No details are imposed as to which elements of the remuneration policy must be adapted. The SFDR also mandated disclosure of such information on their websites.[199]

On the other hand, the revised EU Shareholders Rights Directive requires a remuneration report that namely includes information on how the remuneration policy contributes to the long-term performance of the company, and information on how the performance criteria were applied.[200] Some sustainability metrics may be complex and therefore in some cases its disclosure should be supplemented with qualitative information.[201]

Finally, remuneration committees will also have to adapt to the evolving ESG remuneration implications. One of the main topics relates to ESG qualifications of remuneration committee members, that are not mandatory by law, but will increasingly be important in practice.

9 Conclusion

Corporate governance has ever been considered as an organizational tool for a better future. In the ESG context, this can be manifested in a very tangible sense. Indeed, ESG is ultimately a vehicle for boosting climate, social and governance-based decisions.

As we have seen, the intersection of corporate governance and ESG is apt to produce a 'cascade effect'. We have defined ESG cascade effect as the potential aptitude for companies to engage in ESG-based decisions and to systemically influence others to do so. Such is a metric, with effects and consequences that can be assessed at the investors' level, at invested companies' level, at supply chain level and at the community at large.[202] At any of these levels, ESG potential impact is systemic, and its degree of influence is variable and depends upon the ESG policies and upon product-specific arrangements in place.

The cascade effect bears cross-border implications, and that is particularly important in terms of climate change-related policies and behaviours. As

---

[198] Article 5 I of Regulation (EU) 2019/2088 of the European Parliament and of the Council of 27 November 2019 on sustainability-related disclosures in the financial services sector.

[199] Article 5 II of Regulation (EU) 2019/2088 of the European Parliament and of the Council of 27 November 2019. Further disclosures will be required by the level 2 EC Regulation.

[200] Article 6b introduced by Directive (EU) 2017/828 of the European Parliament and of the Council of 17 May 2017.

[201] Alex Edmans, *The Dangers of Sustainability Metrics*, VOX (11 February 2021).

[202] See *supra*, 4.3.

climate change problems are global by nature, they require global responses. Therefore, the cross-border 'cascade effect' is also an important and necessary effect that stems from ESG.

The reciprocal influence of corporate governance and ESG determines a double and reciprocal empowerment. On one side, the governance reach is extended to ESG issues; and, on the other side, ESG decisions are adopted, implemented and enforced due to the governance structure. This double perspective also shows that the relevance of corporate governance for ESG goes beyond the financial sector and decisively impacts the whole economic landscape.

The analysis presented therefore confirms the need for a systemic analysis of corporate governance that places sustainability goals of financial institutions at its centre.[203] ESG namely shows that there can be an alignment between investor value and stakeholder value—what Mark Carney coins as the 'divine coincidence'.[204] The core priority of the forthcoming ESG agenda lies precisely on boosting the chances for such alignment, namely in critical areas such as board duties and skills, disclosure, risk and remuneration.

---

[203] Similarly, Beate Sjåfjell/Christopher M. Bruner, *Corporations and Sustainability*, in *The Cambridge Handbook of Corporate Law, Corporate Governance and Sustainability*, Cambridge (2019), 4.

[204] Mark Carney, *Value(s). Building a Better World for All*, cit., 426, 432, 453. See also Luca Enriques, Chapter 6 in this book. Regarding shareholder alignment paved through green financing, see also Julian Nyarko/Eric Talley, *Corporate Climate: A Machine Learning Assessment of Climate Risk Disclosures*, in Andreas Engert/Luca Enriques/Wolf-Georg Ringe/Umakanth Varottil/Thom Wetzer (eds.), *Business Law and the Transition to a Net Zero Economy*, (2022), 3–5.

CHAPTER 2

Sustainable Governance and Corporate Due Diligence: The Shifting Balance Between Soft Law and Hard Law

Guido Ferrarini

1 INTRODUCTION

I recently argued that sustainability can be seen as a game changer in corporate governance,[1] to the extent that not only regulation but also conduct guidelines and ethical standards operate as sustainability constraints on the behaviour of enterprises and their pursuit of profits. In the present paper, I

---

[1] G. Ferrarini, 'Redefining Corporate Purpose: Sustainability as a Game Changer', in D. Busch, G. Ferrarini and S. Grünewald (eds.), *Sustainable Finance in Europe*. Corporate Governance, Financial Stability and Financial Markets, Palgrave MacMillan, 2021, Chapter 4. An earlier version of that chapter was published as 'Corporate Purpose and Sustainability' (December 7, 2020), European Corporate Governance Institute - Law Working Paper #559/2020, available at SSRN: https://ssrn.com/abstract=3753594 or http://dx.doi.org/10.2139/ssrn.3753594.

My paper does not consider the proposal for a Directive on corporate sustainability due diligence which was adopted by the Commission on 23 February 2022.

G. Ferrarini (✉)
Business Law, University of Genoa, Genoa, Italy
e-mail: guido.ferrarini@unige.it

EUSFiL - Jean Monnet Center of Excellence on Sustainable Finance and Law, Genoa, Italy

ECGI, Brussels, Belgium

© The Author(s), under exclusive license to Springer Nature Switzerland AG 2022
P. Câmara and F. Morais (eds.), *The Palgrave Handbook of ESG and Corporate Governance*, https://doi.org/10.1007/978-3-030-99468-6_2

further analyse the regulatory and ethical constraints to value maximization motivated by sustainability concerns. In addition, I show that the borders between soft law and hard law in this area are shifting, as a result of EU regulatory initiatives on corporate due diligence which are directed to significantly reduce the impact of business activities on the environment and society. In "Sustainability as a Game Changer" section, I explain the role of regulation and international standards in making firms internalize their negative externalities as to the environment and society. I also highlight the role of non-financial disclosure in promoting compliance with international standards. In "The International Principles on Corporate Responsibility" section, I consider the main standards followed by international firms as to environmental and social sustainability, with particular regard to those on corporate due diligence. In "The European Parliament's Draft Directive on Corporate Due Diligence and Accountability" section, I examine recent EU proposals to transplant some of these standards into hard law through a directive like the one recently suggested by the European Parliament. In "Problems and Limits of the Draft Directive" section, I emphasize the problems and limits of the due diligence obligations envisaged by the proposed directive. In "Concluding Remarks" section I conclude.

2   Sustainability as a Game Changer

In the present section, I summarize the main outcomes of my previous paper by focusing on two topics: the role played by sustainability in the definition of corporate purpose; the regulatory and ethical constraints to shareholder wealth maximization which are motivated by sustainability. In addition, I underline the role of non-financial disclosure in creating incentives to corporate sustainability.

Corporate Purpose and Sustainability

An increasing number of firms make reference to the pursuit of environmental and social goals in the definition of their purpose. This raises important issues with respect to the way in which the trade-offs between profit maximization and social value are solved. As shown in my previous paper, corporate purpose has been analysed from different perspectives with different aims in mind.[2] Lawyers look at corporate purpose mainly to establish for whom the corporation is run and what are the duties of directors. The legal systems diverge on definitions, but not very much on substance, given the limited relevance of

---

[2] Ibidem. See also E. Rock, For Whom is the Corporation Managed in 2020?: The Debate over Corporate Purpose (May 1, 2020), European Corporate Governance Institute—Law Working Paper No. 515/2020, NYU School of Law, Public Law Research Paper No. 20–16, NYU Law and Economics Research Paper, available at SSRN: https://ssrn.com/abstract=3589951 or http://dx.doi.org/10.2139/ssrn.3589951.

corporate purpose in the practice of law.[3] Moreover, the discussion on corporate purpose generally extends to the definition of the company's interest, which grounds the duty of loyalty of directors and the rules on conflicts of interest.

Economists focus on corporate purpose to define the role of firms in a market economy and the incentives—including the pursuit of profit—through which business corporations efficiently serve their productive function. Finance scholars are especially interested in valuation issues and mainly think of corporate purpose in terms of either shareholder value or firm value maximization.[4] Management studies show how corporate purpose and its derivatives (like corporate mission, vision and values) can be resorted to in orienting the corporate organization towards the goals that the directors and managers choose to follow in the strategy and activities of firms. Clearly these goals are not identified exclusively with the pursuit of profit but extend to social responsibility issues. Moreover, the definition of purpose in detail depends on management style, corporate culture and the specificities of the industry concerned. Recent works by finance and management scholars argue, however, that the value to maximize is not only shareholder value (or firm value), but also (and for some predominantly) social value.[5] Similar works implicitly vindicate the importance of CSR and stakeholder management, which have been largely neglected by economists and finance scholars until the beginning of this century.[6]

Amongst existing theories, presumably the dominant one today is enlightened shareholder value (ESV), which requires stakeholder interests to be satisfied subject to shareholder value maximization.[7] After being suggested by economics and finance scholars, ESV has been widely adopted in policy discussions and in corporate practice, possibly with variations such as those suggested by the theory of 'shared value'.[8] However, ESV needs refinement today to take account of some of the criticisms and insights found in recent

---

[3] See H. Fleischer, Corporate Purpose: A Management Concept and its Implications for Company Law (January 21, 2021), European Corporate Governance Institute—Law Working Paper No. 561/2021, available at SSRN: https://ssrn.com/abstract=3770656 or http://dx.doi.org/10.2139/ssrn.3770656.

[4] See M. Jensen, 'Value Maximization, Stakeholder Theory, and the Corporate Objective Function' (2010) 22 Journal of Applied Corporate Finance 32, and (2002) 12 Business Ethics Quarterly 235.

[5] See C. Mayer, *Prosperity*. Better Business Makes the Greater Good, Oxford University Press, 2018; A. Edmans, *Grow the Pie*. How Great Companies Deliver Both Purpose and Profit, Cambridge University Press, 2020; R. Henderson, *Reimagining Capitalism*. How Business Can Save the World, Penguin Business, 2020.

[6] See O. Hart and L. Zingales, 'Companies Should Maximize Shareholder Welfare not Market Value' (2017) Journal of Law, Finance, and Accounting, 247.

[7] See Jensen, note 194.

[8] M. Porter and M. Kramer, 'Creating Shared Value: How to Reinvent Capitalism—And Unleash a Wave of Innovation and Growth' (2011) Harvard Business Review 3.

scholarly works stressing the social values that should be pursued by corporations.[9] Stakeholder protection should not be seen exclusively as instrumental to long-term value maximization—as narrowly suggested by ESV—but also as an outcome of the compliance with legal rules and ethical standards, which apply to different types of firms and aim at controlling externalities that either directly or indirectly derive from their activities. In a rising number of situations firms internalize externalities not only because it is profitable in the long run or at least suitable to reduce their risk exposures, but also to comply with the regulatory and ethical standards that protect relevant stakeholders.

Interestingly, these regulatory and ethical constraints on firm behaviour do not necessarily determine a reduction in firm value. Some empirical studies on the relationship between CSR and economic performance rather prove the opposite. A. Ferrell, H. Liang and L. Renneboog in particular find that well-governed firms that suffer less from agency concerns engage more in CSR and have higher CSR ratings.[10] They also find that a positive relation exists between CSR and value, suggesting at least that CSR is not inconsistent with shareholder value maximization.[11] Their general argument is interesting for present purposes: 'Corporate social responsibility need not to be inevitably induced by agency problems but can be consistent with a core value of capitalism, generating more returns to investors, through enhancing firm value and shareholder wealth'.[12]

Regulatory and Ethical Constraints to Value Maximization

The role of regulation in constraining shareholder wealth maximization is easily understood. Environmental protection, to make an obvious example, largely depends on government regulation, which is binding on firms and influences their actions. No doubt, firms comply with this type of regulation not only for ethical reasons, but also to avoid the administrative and criminal sanctions which would derive from violations of the relevant rules and would negatively affect their economic value. Stakeholder protection in similar cases cannot be seen as directly instrumental to firm value maximization, for it is primarily required by regulation. No matter what corporate managers think about the merits of regulation and its effectiveness in protecting the relevant stakeholders, they have to comply with the prescriptions in question.

---

[9] For a recent account of the centrality of value, see M. Carney, Value(s). Building a Better World for All, William Collins, 2021, 379 ff.

[10] A. Ferrell, H. Liang and L. Renneboog, 'Socially Responsible Firms' (2016) 122 Journal of Financial Economics 585. These authors consider well governed firms as represented by lower cash hoarding and capital spending, higher pay-out and leverage ratio and stronger pay-for-performance.

[11] Ibidem, 602.

[12] Ibidem, 605.

In many cases, however, the need to comply generates either organizational or technological innovation, reducing operational costs and enhancing corporate profitability. Moreover, many actions are performed by firms, particularly the largest ones, in compliance with ethical standards that are globally recognized in statements and guidelines issued by international organizations and subscribed by firms for the protection of given stakeholders. These documents are not binding per se, but their principles are often reflected in the applicable national laws and for the rest may be followed voluntarily by the corporations concerned, especially when their managers are officially committed to respect the relevant standards.

Notwithstanding the non-binding nature of similar standards and their limited enforcement, companies' policies and practices increasingly comply with them and respond to investors' growing attention towards the ESG performance of investee companies, including the formal adoption of due diligence, environmental and human rights policies in line with international standards. In the sustainable investment strategies usually followed by institutional investors, the 'norm-based screening'—which screens issuers against minimum standards of business practice based on international frameworks, such as the UN treaties, the UN Global Compact, the OECD Guidelines for Multinational Enterprises and the International Labour Organization standards—is one of the most commonly used for portfolio selection.[13] Moreover, common voluntary standards have been developed targeting investor stewardship obligations (such as the ICGN Global Stewardship principles and the EFAMA Stewardship Code)[14] or sustainable investment (such as the Principles for Responsible Investing),[15] which put further pressure on investors with regard to the sustainability-related initiatives and policies of investee companies.

The voluntary application of international standards might be motivated by reputational concerns or by the personal conviction of the managers about the morality of the actions undertaken. Therefore, like in the case of regulation, the calculus of instrumentalism may be 'indirect' in similar cases and the protection of stakeholders may simply derive from the compliance with

---

[13] See https://www.unpri.org/an-introduction-to-responsible-investment/an-introduction-to-responsible-investment-screening/5834.article. See also Eurosif, '2018 SRI Study for an overview of trends related to SRI strategies in Europe' (2018). See also ISS ESG, 'Norm-based Research Evaluation of ESG Controversies. Research Methodology' (2020), for an overview on the methodological process adopted by ISS ESG to evaluate corporate compliance/failure to comply with international principles (in particular, the Principles of the UN Global Compact and the OECD Guidelines for Multinational Enterprises).

[14] S. Alvaro, M. Maugeri, and G. Strampelli, 'Institutional Investors, Corporate Governance and Stewardship Codes: Problems and Perspectives' (2019), CONSOB Legal Research Papers (Quaderni Giuridici), 19.

[15] S. Kim and A. Yoon, 'Analyzing Active Managers' Commitment to ESG: Evidence from United Nations Principles for Responsible Investment' (March 17, 2020), available at SSRN: https://ssrn.com/abstract=3555984 or http://dx.doi.org/10.2139/ssrn.3555984.

the relevant standards. As a result, the managers do not compare the shareholders' interests with those of given stakeholders, nor ask to what extent protecting the latter will enhance the long-term value of the firm—as theoretically required under the ESV approach—given that their action is required per se under the international standards. Of course, to the extent that discretion is left to the managers under the individual standard—particularly if the latter is broadly formulated and there are no implementing provisions—the managers will also refer to the impact of their actions on the long-term value of the firm. But they may also decide on similar actions on purely moral grounds, filling their discretion in a way that they deem consistent with the content and spirit of the standard to apply.

Once more, reputational concerns will also be at play, in addition to the ethical beliefs of the managers, to the extent that either the consumers or the investors monitor the firm's compliance with the relevant standards. The increasing importance of sustainability multiplies this type of situations, given that not all aspects of sustainable growth are specifically dealt with by regulation, while the urgency of the problems involved requires the active cooperation of corporations, which increasingly follow (or simply declare to follow) the international guidelines and standards both in environmental and social matters. Sustainability can therefore be seen as a game changer, to the extent that not only regulation, but also conduct guidelines and ethical standards operate as constraints on the behaviour of enterprises and their pursuit of profits.

Non-Financial Disclosure and Incentives

Non-financial disclosure enhances the reputational incentives for firms to follow sustainability standards. Article 2 of the Non-financial Reporting Directive (NFRD) provides that 'the Commission shall prepare non-binding guidelines on methodology for reporting non-financial information, including non-financial KPIs, general and sectoral, with a view to facilitating relevant, useful and comparable disclosure of non-financial information by undertakings'. In addition, Recital 17 of the Directive states that, when preparing the non-binding guidelines, 'the Commission should take into account current best practices, international developments and the results of related Union initiatives'.

To this effect, the Commission issued Communication (2017/C 215/01) including 'Guidelines on non-financial reporting (methodology for reporting non-financial information)'. Under Article 1 a. of the NFRD, the non-financial statement contains information including 'a brief description of the undertaking's business model'. As specified in the Guidelines, 'a company's business model describes how it generates and preserves value through its products or services over the longer term'. Moreover, 'companies may consider including appropriate disclosures relating to their business environment; their organization and structure; the markets where they operate; their objectives and strategies; and the main trends and factors that may affect their future development'.

Furthermore, under Article 1 b. of the NFRD, the non-financial statement contains information including 'a description of the policies pursued by the undertaking in relation to those matters, including due diligence processes implemented'. According to the Guidelines, 'due diligence processes relate to policies, to risk management and to outcomes… They help identify, prevent and mitigate existing and potential adverse impacts'. Companies should provide material disclosures on due diligence processes implemented, including on its suppliers and subcontracting chains. Companies may also consider providing relevant information on setting targets and measuring progress. The Commission specifies that OECD Guidance documents for several sectors, UN Guiding Principles on Business and Human Rights, the Tripartite Declaration of Principles concerning Multinational Enterprises and Social Policy, or ISO 26000 provide useful guidance on this.

3 THE INTERNATIONAL PRINCIPLES ON CORPORATE RESPONSIBILITY

The growing importance and diffusion of the principles and guidelines issued by international organizations and standard setters (including the IMF, the OECD, the World Bank and the United Nations) have led an author to identify a new field of the law significantly dubbed as 'international corporate law' (ICL).[16] The emergence of ICL has partially responded to the 'interjurisdictional externalities and nationalist bias of domestic regimes'. With specific reference to corporate responsibility towards the environment and society, it has the potential to fill the gaps in national legislations, by establishing new standards for corporate behaviour that take into account the negative effects of company activities on third parties.

The UN Guiding Principles on Business and Human Rights

The main guidelines addressing corporate responsibility are the UN Guiding Principles on Business and Human Rights [UN Guiding Principles] which provide standards for both States and business enterprises to prevent, address and remedy human rights abuses committed in business operations. The UN Guiding Principles include 14 principles specifically addressing the responsibilities of business enterprises in relation to the respect of human rights, providing also a set of operational recommendations going from the issuance of a specific policy on human rights to the performance of a human rights due diligence and the provision of remedies to the adverse impacts the company

---

[16] M. Pargendler, 'The Rise of International Corporate Law' (2020), European Corporate Governance Institute—Law Working Paper, 555/2020, FGV Direito SP Research Paper Series n. Forthcoming.

has caused or has contributed to generate with its actions. The Human Rights Council formally endorsed the Principles in 2011 and to date at least 377 large companies adopted a formal statement explicitly referring to human rights in compliance with Principle 16 of the UN Guiding Principles on Business and Human Rights.[17] Unlike the UN Guiding principles, the UN Global Compact is an initiative that global corporations can commit to by respecting 10 key principles of business behaviour in human rights, labour, the environment and corruption.[18] Currently, the UN Global Compact counts more than 12,000 signatories in over 160 countries covering all business sectors.[19]

The UN Guiding Principles deal extensively with the corporate responsibility to respect human rights. Amongst the 'foundational principles', Principle 11 states that business enterprises should respect human rights, while Principle 12 specifies that their responsibility refers to internationally recognized human rights. Under Principle 13, business enterprises are required to '(a) Avoid causing or contributing to adverse human rights impacts through their own activities, and address such impacts when they occur; (b) Seek to prevent or mitigate adverse human rights impacts that are directly linked to their operations, products or services by their business relationships, even if they have not contributed to those impacts'. Principle 15 further specifies that 'business enterprises should have in place policies and processes appropriate to their size and circumstances, including: (a) A policy commitment to meet their responsibility to respect human rights; (b) A human rights due diligence process to identify, prevent, mitigate and account for how they address their impacts on human rights; (c) Processes to enable the remediation of any adverse human rights impacts they cause or to which they contribute'.

Amongst the 'operational principles', Principle 16 deals with the 'policy commitment' of business enterprises,[20] while Principle 17 provides for 'human rights due diligence' which is directed to 'identify, prevent, mitigate and account for how [business enterprises] address their adverse human rights impacts'. Human rights due diligence should cover, in particular, 'adverse human rights impacts that the business enterprise may cause or contribute to through its own activities, or which may be directly linked to its operations,

---

[17] See https://old.business-humanrights.org/en/company-policy-statements-on-human-rights.

[18] See https://www.unglobalcompact.org/what-is-gc/mission/principles.

[19] See https://www.unglobalcompact.org/what-is-gc/participants.

[20] 'As the basis for embedding their responsibility to respect human rights, business enterprises should express their commitment to meet this responsibility through a statement of policy that: (a) Is approved at the most senior level of the business enterprise; (b) Is informed by relevant internal and/or external expertise; (c) Stipulates the enterprise's human rights expectations of personnel, business partners and other parties directly linked to its operations, products or services; (d) Is publicly available and communicated internally and externally to all personnel, business partners and other relevant parties; (e) Is reflected in operational policies and procedures necessary to embed it throughout the business enterprise'.

products or services by its business relationships'. Interestingly, the commentary to this Principle states what follows: 'Human rights due diligence can be included within broader enterprise risk-management systems, provided that it goes beyond simply identifying and managing material risks to the company itself, to include risks to rights-holders'.

The OECD Guiding Principles and the ILO Tripartite Declaration

The OECD Guidelines for Multinational Enterprises, firstly adopted in 1976, are also important. They consist of a set of voluntary standards and principles for responsible business conduct addressed to multinational enterprises operating in or from the adhering countries. Specifically, the latest version of the OECD Guidelines was adopted in 2011 by the 42 OECD and non-OECD governments adhering to the OECD Declaration on International Investment and Multinational Enterprises, and today 49 governments have established a National Contact Point with the duty of ensuring the effectiveness of the OECD Guidelines by undertaking promotional activities, handling enquiries and providing a grievance mechanism to resolve cases with regard to the non-observance of the recommendations. The OECD Guidelines cover a diverse range of topics related to business behaviour, from company disclosure and reporting on financial, social and environmental material information to the respect of employees, human rights, the environment, consumers interest and the fight against bribery and other illicit conducts, as well as the promotion of science and technology development, fair competition and tax compliance. To complement the standards of behaviour established by the OECD Guidelines, in 2018, the OECD Due Diligence Guidance for Responsible Business Conduct was adopted,[21] with the aim of providing practical support to business enterprises on the implementation of the OECD Guidelines. Moreover, the OECD has developed sector-specific due diligence guidance and good practice documents for the minerals,[22] agriculture[23] and garment and footwear supply chains,[24] as well as for the extractive sector.[25]

The OECD Guidelines for Multinational Enterprises rely extensively on the UN Guiding Principles on Business and Human Rights, but have a broader scope also including employment and industrial relations, environment, combating bribery, bribe solicitation and extortion, consumer interests,

---

[21] OECD (2018), OECD Due Diligence Guidance for Responsible Business Conduct.

[22] OECD (2016), OECD Due Diligence Guidance for Responsible Supply Chains of Minerals from Conflict-Affected and High-Risk Areas: Third Edition, OECD Publishing, Paris. http://dx.doi.org/10.1787/9789264252479-en.

[23] OECD, Recommendation of the Council on the OECD-FAO Guidance for Responsible Agricultural Supply Chains, OECD/LEGAL/0428.

[24] OECD (2017), OECD Due Diligence Guidance for Responsible Supply Chains in the Garment and Footwear Sector.

[25] OECD (2016), Recommendation of the Council on the Due Diligence Guidance for Meaningful Stakeholder Engagement in the Extractive Sector.

science and technology, competition and taxation. In Chapter 2 on General Policies, they state that 'Enterprises should: 11. Avoid causing or contributing to adverse impacts on matters covered by the Guidelines, through their own activities, and address such impacts when they occur. 12. Seek to prevent or mitigate an adverse impact where they have not contributed to that impact, when the impact is nevertheless directly linked to their operations, products or services by a business relationship'. These two paragraphs reflect the 'protect, respect and remedy framework' of the UN Guiding Principles, extending it beyond human rights to areas such as the environment and employment relations. In a similar vein, para. 14 states that 'due diligence is understood as the process through which enterprises can identify, prevent, mitigate and account for how they address their actual and potential adverse impacts as an integral part of business decision-making and risk management systems. Due diligence can be included within broader enterprise risk management systems, provided that it goes beyond simply identifying and managing material risks to the enterprise itself, to include the risks of adverse impacts related to matters covered by the Guidelines. Potential impacts are to be addressed through prevention or mitigation, while actual impacts are to be addressed through remediation'.

The Tripartite Declaration of Principles concerning Multinational Enterprises and Social Policy (MNE Declaration), which was approved by the International Labour Office (ILO) in 1977 and later amended (the last time in 2017) similarly refers to the UN Guiding Principles on Business and Human Rights, extending however their reach to the fundamental rights set out in the ILO Declaration on Fundamental Principles and Rights at Work.

4 THE EUROPEAN PARLIAMENT'S DRAFT DIRECTIVE ON CORPORATE DUE DILIGENCE AND ACCOUNTABILITY

The European Commission recently suggested that legal requirements for corporate due diligence could strengthen a practice already widespread in the market.[26] Moreover, their introduction in EU legislation would be in line with the Regulation (EU) 2020/852 on the establishment of a framework to facilitate sustainable investment [Taxonomy Regulation].[27] Article 3 of this Regulation requires business activities to comply with the minimum safeguards set out in Article 18 in order to be considered as 'environmentally sustainable', i.e. to establish procedures 'to ensure the alignment with the OECD Guidelines for Multinational Enterprises and the UN Guiding Principles on Business and Human Rights, including the principles and rights set

---

[26] See Sect. 3 of the Commission's questionnaire on sustainable governance recently submitted to Consultation at https://ec.europa.eu/info/law/better-regulation/have-your-say/initiatives/12548-Sustainable-corporate-governance/public-consultation

[27] Regulation (EU) 2020/852 of the European Parliament and of the Council of 18 June 2020 on the establishment of a framework to facilitate sustainable investment, and amending Regulation (EU) 2019/2088.

out in the eight fundamental conventions identified in the Declaration of the International Labour Organisation on Fundamental Principles and Rights at Work and the International Bill of Human Rights'. All this means that companies should adopt a specific human rights policy, establish human rights due diligence processes and provide a system of remedies for adverse impacts.

The European Parliament's Resolution on Corporate Due Diligence

The European Parliament recently approved a resolution including a draft Directive seeking to transplant international guidelines such as the UN Guiding Principles and the OECD Guidelines at EU level.[28] As stated in the 10th recital of the draft Directive's Preamble, 'in order to ensure a level playing field, the responsibility for undertakings to respect human rights under international standards should be transformed into a legal duty at Union level. By coordinating safeguards for the protection of human rights, the environment and good governance, this Directive should ensure that all Union and non-Union large undertakings and high-risk or publicly listed small and medium-sized undertakings operating in the internal market are subject to harmonized due diligence obligations, which will prevent regulatory fragmentation and improve the functioning of the internal market'.

As a consequence, the draft Directive foresees *due diligence obligations*, which are grounded on the duty of undertakings to respect human rights, the environment and good governance (Art. 1 (1)). The draft Directive leaves the obligations to comply with under the due diligence procedures regulated by the Directive to different legal texts of either hard law or soft law. It is different therefore to the UN Guiding Principles, where the 'duty to respect' includes both the duty to avoid infringements of human rights and the duty to prevent them. Indeed, under Principle 13 the responsibility to respect human rights requires that business enterprises (a) avoid causing or contributing to adverse human rights impacts through their own activities, and (b) seek to prevent or mitigate adverse human rights impacts that are directly linked to their operations, products or services by their business relationships. Letter (b) essentially refers to the due diligence duty, while letter (a) includes the duty not to cause adverse human rights impacts. The proposed Directive is not directly concerned with (a). The reason for its more limited scope may depend on the fact that it aims to transform soft law of international origin into hard law of the Union and the Member States. This makes it more difficult to define the duties of enterprises—other than the due diligence ones—and the responsibility deriving from their infringement. Moreover, in the case of

---

[28] See European Parliament resolution of 10 March 2021 with recommendations to the Commission on corporate due diligence and corporate accountability (2020/2129(INL)), available at https://www.europarl.europa.eu/doceo/document/TA-9-2021-0073_EN.html. The resolution carries an Annex including recommendations for drawing up a Directive of the European Parliament and of the Council on Corporate due diligence and corporate accountability.

human rights it is relatively easy to define them with respect to international law, whereas it is more difficult to do something similar with respect to the environment and the harms which may be caused to it by business activities.

Therefore, the draft Directive is mainly concerned with the preventative measures required for companies to avoid adverse impacts and with the remedies applicable if such impacts materialize. Indeed, according to Art. 4 (1) the Member States 'shall lay down rules to ensure that undertakings carry out effective due diligence with respect to potential or actual adverse impacts on human rights, the environment and good governance in their operations and business relationships'. Under these rules, the undertakings concerned shall 'take all proportionate and commensurate measures and make efforts within their means to prevent adverse impacts on human rights, the environment and good governance from occurring in their value chains', and shall be required 'to identify, assess, prevent, cease, mitigate, monitor, communicate, account for, address and remediate the potential and/or actual adverse impacts on human rights, the environment and good governance that their own activities and those of their value chains and business relationships may pose' (Art. 1 (2)).

Due Diligence Strategy

In order to comply with their due diligence duty, undertakings shall adopt a 'due diligence strategy', which includes some of the key characteristics of a compliance and risk management programme. Under this strategy, undertakings shall 'in an ongoing manner make all efforts within their means to identify and assess, by means of a risk based monitoring methodology that takes into account the likelihood, severity and urgency of potential or actual impacts on human rights, the environment or good governance, the nature and context of their operations, including geographic, and whether their operations and business relationships cause or contribute to or are directly linked to any of those potential or actual adverse impact'(Art. 4 (1)). However, if a large undertaking, whose direct business relationships are all domiciled within the Union, or a small or medium-sized undertaking concludes that it does not cause, contribute to, or that it is not directly linked to any potential or actual adverse impact on human rights, the environment or good governance, it shall publish a statement to that effect and shall include its risk assessment containing the relevant data, information and methodology that led to this conclusion (Art. 4 (3)). Otherwise, it shall establish and effectively implement a due diligence strategy (Art. 4 (3)).

As part of their due diligence strategy, undertakings shall: (i) specify their potential or actual adverse impacts on human rights, the environment and good governance identified and assessed in conformity with Art. 4 (2); (ii) map their value chain and publicly disclose relevant information about it; (iii) adopt and indicate all proportionate and commensurate policies and measures with a view to ceasing, preventing or mitigating potential or actual adverse impacts on human rights, the environment or good governance; (iv) set up a prioritization strategy in the event that they are not in a position to deal with all the potential or actual adverse impacts at the same time. As to value chain due diligence, undertakings shall ensure that their business relationships put in place and carry out human rights, environmental and good governance policies that are in line with their due diligence strategy. Undertakings shall ensure that their purchase policies do not cause or contribute to potential or actual adverse impacts on human rights, the environment or good governance (Art. 4 (7) and (8)).

Adverse Impact, Business Relationships and Value Chain

One of the core concepts of the draft Directive is that of 'potential or actual adverse impact' of business activities. The relevant definitions are offered in Art. 3 of the draft with regard to the different types of harm which can be caused by companies to society and the environment. Firstly, '"potential or actual adverse impact on human rights" means any potential or actual adverse impact that may impair the full enjoyment of human rights by individuals or groups of individuals in relation to human rights, including social, worker and trade union rights, as set out in Annex xx to this Directive'. Secondly, '"potential or actual adverse impact on the environment" means any violation of internationally recognised and Union environmental standards, as set out in Annex xxx to this Directive'. Thirdly, '"potential or actual adverse impact on good governance" means any potential or actual adverse impact on the good governance of a country, region or territory, as set in Annex xxxx to this Directive'. The three Annexes shall be reviewed on a regular basis by the Commission and be consistent with the Union's objectives on human rights, on environmental protection and climate change mitigation, and on good governance.

Two other core concepts are those of business relationship and value chain, which are also defined in Art. 3 of the draft. The first concept 'means subsidiaries and commercial relationships of an undertaking throughout its value chain, including suppliers and sub-contractors, which are directly linked to the undertaking's business operations, products or services'. The second concept 'means all activities, operations, business relationships and investment chains of an undertaking and includes entities with which the undertaking has a direct or indirect business relationship, upstream and downstream, and which either: (a) supply products, parts of products or services that contribute to the undertaking's own products or services, or (b) receive products or services from the undertaking'. The value chain, therefore, includes the supply chain, but also the customers who buy the firm's products or services.

Enforcement

The draft Directive provides for both public and private enforcement. Art. 18 (1) requires Member States to 'provide for proportionate sanctions applicable to infringements of the national provisions adopted in accordance with this Directive and shall take all the measures necessary to ensure that those sanctions are enforced. The sanctions provided for shall be effective, proportionate and dissuasive and shall take into account the severity of the infringements committed and whether or not the infringement has taken place repeatedly'. Furthermore, Art. 19 (2) requires Member States to adopt a *civil liability regime* for any harm arising out of potential or actual adverse impacts on human rights, the environment or good governance that undertakings have caused or contributed to by acts or omissions. National law should therefore define the wrongs from which the civil liability will arise. It is not clear however if the duties in general to respect human rights, the environment and good governance should be covered, or only the due diligence duties specifically foreseen by the Directive. The text is unclear, but only the latter duties should be relevant for the civil liability regime at issue. Indeed, Art. 19 (1) specifies that 'the fact that an undertaking respects its due diligence obligations shall not absolve the undertaking of any liability which it may incur pursuant to national law'. Moreover, Art. 19 (3) provides that 'undertakings that prove that they took all due care to avoid the harm in question, or that the harm would have occurred even if all due care had been taken, are not held liable for that harm'.

These two provisions appear to contradict each other. In order to solve this potential conflict, one should assume that para. 3 refers to the liability for breach of the due diligence obligations foreseen by the national legislation implementing the Directive, while para. 1 refers to the liability for breach of the legal entitlements foreseen under the substantive law of the Member State. The Directive only specifies the due diligence obligations, so that the States would be free to identify the 'duties to respect' that ground similar obligations through substantive law provisions. Once more, the distinction between organizational law and substantive law provisions is relevant and helps solving the civil liability problems originated by adverse impacts in the areas covered by the draft Directive. However, the draft should be amended to clarify the grounds and scope of the liability provisions that Member States should adopt in implementing the Directive.

5 Problems and Limits of the Draft Directive

The proposed directive is to some extent imprecise and open to criticism from the perspective of legal certainty, as it refers to numerous texts of soft law in a hard law context. No doubt, the directive tries to be specific as to the types of standards with respect to which corporations should be accountable. For instance, Recital 23 clarifies the type of environmental standards that will be relevant under Art. 3: 'Annex xxx sets out a list of types of business-related adverse impacts on the environment, whether temporary or permanent, that are relevant for undertakings. Such impacts should include, but should not be limited to, production of waste, diffuse pollution and greenhouse emissions that lead to a global warming of more than 1.5 °C above pre-industrial levels, deforestation, and any other impact on the climate, air, soil and water quality, the sustainable use of natural resources, biodiversity and ecosystems. The Commission should ensure that those types of impacts listed are reasonable and achievable. To contribute to the internal coherence of Union legislation and to provide legal certainty, this list is drawn up in line with Regulation (EU) 2020/852 of the European Parliament and of the Council'.

Similarly, Recital 24 circumstantiates adverse impacts on governance such as corruption by reference to several sources of international law and standards: 'Annex xxxx sets out a list of types of business-related adverse impacts on good governance that are relevant for undertakings. They should include non-compliance with OECD Guidelines for Multinational Enterprises, Chapter 7 on Combatting Bribery, Bribe Solicitation and Extortion and the principles of the OECD Convention on Combating Bribery of Foreign Public Officials in International Business Transactions and situations of corruption and bribery where an undertaking exercises undue influence on, or channels undue pecuniary advantages to, public officials to obtain privileges or unfair favourable treatment in breach of the law, and including situations in which an undertaking becomes improperly involved in local political activities, makes illegal campaign contributions or fails to comply with the applicable tax legislation. The Commission should ensure that those types of impacts listed are reasonable and achievable'.

Clearly, if all the soft law principles and standards just mentioned were transformed into binding legal rules serious problems would arise at the level of legal certainty and compliance. It is enough to consider that the relevant principles and standards were originally formulated to be included in non-binding legal instruments, so that they often are rather generic and not always rigorous on a technical level. Compliance with them could therefore be difficult to firms and public authorities would encounter serious difficulties in supervising them. However, I think that the directive should not transform soft law standards into hard law obligations. Rather, it should introduce due diligence obligations through hard law and require companies to take preventative measures, mainly of an organizational character, in order to avoid or reduce their adverse impacts. These impacts may consist of breaches of hard law rules by the company, but also of deviations from soft law standards that are general in character. Rather than transforming soft law standards into hard law, the directive should foresee due diligence obligations which expose the company to sanctioning only to the extent that the necessary preventative/organizational measures have not been adopted.

If the applicable standards are not sufficiently defined, the managers should have discretion as to the measures to adopt and should not be sanctioned if their discretion is reasonably exercised. Otherwise, the rule of law would be violated. Art. 1 of the proposed directive consistently distinguishes between the compliance with the substantive rules (e.g. protecting human rights under either European or national law) and that with organizational rules such as the due diligence obligations. Para. 1 of this Article specifies that the 'Directive is aimed at ensuring that undertakings under its scope operating in the internal market fulfil their duty to respect human rights, the environment and good governance ...', while para. 2 provides: 'This Directive lays down the value chain due diligence obligations of undertakings under its scope, namely to take all proportionate and commensurate measures and make efforts within their means to prevent adverse impacts ...'. These two paragraphs should be read in the sense that the directive does not create new substantive rules, which only derive from existing texts of international, European or national law. Rather it gives rise to organizational rules, which mainly require risk management measures and activities.

Nonetheless, I believe that the above issues should be made more explicit in the final text of the Directive and that a clearer distinction should be made between what I have called as substantive rules and organizational ones. In other words, companies should be subject to sanctions for failing to abide by their own due diligence strategy, but not for failing to abide by the international standards themselves.[29]

6 Concluding Remarks

As shown in this paper, regulation and international standards constrain value maximization on sustainability grounds by requiring firms to internalize their negative externalities as to the environment and society. In addition, the borders between soft law and hard law in this area are shifting as a result of legislative initiatives of the EU Commission and the European Parliament, which aim to transplant the international standards on corporate due diligence into EU law so as to reduce the impact of business activities on the environment and society. A similar shift towards public regulation will improve firms' compliance with international sustainability standards but may cause uncertainty as to the firms' precise obligations. A clearer distinction should therefore be made in the proposed directive between general standards, substantive law rules that companies should comply with and organizational rules which serve the purposes of risk management and compliance. However, the Directive should be focused on corporate due diligence obligations and accountability, while the substantive law rules that firms must comply with should be left

---

[29] See, for a wider treatment, The ECLE Group (P. Davies, S. Emmenegger, G. Ferrarini, K. Hopt, A. Opalski, A. Pietrancosta, A. Recalde Castells, M. Roth, M. Schouten, R. Skog, M. Winner, E. Wymeersch), Commentary: The European Parliament's Draft Directive on Corporate Due Diligence and Corporate Accountability, available at https://ecgi.global/news/commentary-european-parliament's-draft-directive-corporate-due-diligence-and-corporate.

to other texts of either European or national law. Furthermore, the Directive should specify that the due diligence obligations do not per se transform the international soft law standards into binding prescriptions, except to the extent that such standards are referred to and possibly specified in the company's due diligence strategy.

CHAPTER 3

Reforming EU Company Law to Secure the Future of European Business

Beate Sjåfjell

1 INTRODUCTION

The world faces a complex convergence of social and ecological crises: climate change, biodiversity loss, resource scarcity, human rights violations, rising

---

This text was first published as an article in the European Company and Financial Law Review, volume 18 (2021) issue 2, and it is republished in this volume with the kind permission of the journal. The chapter draws on the work done in the H2020-funded project Sustainable Market Actors for Responsible Trade (SMART), 2016–2020, grant agreement 693642, and I am grateful to my colleagues in the SMART project, notably in the context of this chapter, Jukka Mähönen, Tonia Novitz, Clair Gammage, Hanna Ahlström, Mark B. Taylor, and Sarah Cornell.

B. Sjåfjell (✉)
Faculty of Law, University of Oslo, Oslo, Norway
e-mail: b.k.sjafjell@jus.uio.no

Faculty of Economics and Management, Norwegian University of Science and Technology, Trondheim, Norway

European Legal Studies Department, College of Europe, Bruges, Belgium

inequality and societal instability.[1] The United Nations adopted Sustainable Development Goals (SDGs) for 'the future of humanity and of our planet', calling on business to contribute to solving these pressing challenges. Yet business in aggregate is a driver of the current convergence of crises and the discussion of how to promote sustainable business is therefore high up also on the agenda of the European Union (EU).

The adoption of the United Nations Sustainable Development Goals (SDGS) in 2015,[2] together with the Paris Agreement on Climate Change in the same year,[3] has given a new impetus to the public discourse concerning what we need to do to achieve sustainability. The EU's commitment to implementing the SDGs is elaborated on in the European Commission's 2016 communication 'Next steps for a sustainable European future—European action for sustainability', and the EU's 2017 Consensus on Development.[4]

The EU increasingly shows recognition of the need for regulatory initiatives to promote the integration of sustainability into European business, resonating with the EU's high-level commitment to sustainability. The EU's recognition of the need to change the way business operates reflects an emerging understanding of the weaknesses of the siloed approach to law and policy, where environmental law and policy has perceived to be sufficient to ensure adequate environmental protection, where labour issues could be left to labour

---

[1] This is also recognised in the EU Green Deal, which in its first paragraph states: 'The atmosphere is warming and the climate is changing with each passing year. One million of the eight million species on the planet are at risk of being lost. Forests and oceans are being polluted and destroyed', with reference in footnote 1 to these sources: '(i) Intergovernmental Panel on Climate Change (IPCC): Special Report on the impacts of global warming of 1.5 °C; (ii) Intergovernmental Science-Policy Platform on Biodiversity and Ecosystem Services: 2019 Global assessment report on biodiversity and ecosystem services; (iii) The International Resource Panel: Global Resources Outlook 2019: Natural Resources for the Future We Want; (iv) European Environment Agency: The European Environment—State and Outlook 2020: Knowledge for Transition to a Sustainable Europe', The European Commission, Communication from the Commission to the European Parliament, the European Council, the Council, the European Economic and Social Committee and the Committee of the Regions: The European Green Deal, 11.12.2019, COM (2019) 640 final.

[2] See also *The UN General Assembly*, General Assembly Resolution 70/1. Transforming Our World: The 2030 Agenda for Sustainable Development, A/RES/70/1, (25 September 2015), www.undocs.org/A/RES/70/1.

[3] *The Paris Agreement*, Paris, 12 December 2015, in force 4 November 2016, U.N. Doc. FCCC/CP/2015/L.9/Rev/1.

[4] *The European Commission*, '2019 EU Report on Policy Coherence for Development', SWD (2019) 20 Final, 28.1.2019; Joint Statement by the Council and the representatives of the governments of the Member States meeting within the Council, the European Parliament and the Commission, 'The New European Consensus on Development, Our World, Our Dignity, Our Future', (2017/C 210/01), p. 3; *The European Commission*, 'Next Steps for a Sustainable European Future', COM (2016) 739 final, 22.11.2016.

law and human rights issues to human rights law, and so on.[5] There is ample research on the limitations of, for example, environmental law,[6] and of how business law currently reinforces these inherent limitations and is associated with negative environmental and social business impacts.[7]

This chapter is a contribution to the discourse on how to regulate European business so that it contributes to a sustainable future for all, including for European business itself. A sustainable future is defined, based on sustainability research, as one that secures social foundations for humanity now and for the future within planetary boundaries.[8]

Section 2 briefly outlines the basis in the EU treaties for reform of EU company law. Section 3 is the main part of the chapter, starting out with a brief discussion of the risks of continued unsustainability, moving on to the argument for including company law in the legislative toolbox, and outlining ideas for how such a reform could be shaped. This is based on the reform proposals presented by the H2020-funded project Sustainable Market Actors for Responsible Trade (SMART), which was concluded in the spring of 2020.[9] Section 4 offers some concluding reflections.

2 THE EU LEGAL BASIS FOR SUSTAINABILITY REFORMS

The EU's commitment to sustainability is anchored in the EU Treaties. Sustainability is an overarching objective of the European Union and meant to be the guiding principle for the EU's policies and activities within Europe and in its relations with the rest of the world, to promote 'peace, its values

---

[5] Beate Sjåfjell/Mark B. Taylor, 'Clash of Norms: Shareholder Primacy vs. Sustainable Corporate Purpose', International and Comparative Corporate Law Journal 13 (2019), 40.

[6] Stephan Wood/Georgia Tanner/Benjamin J. Richardson, 'What Ever Happened to Canadian Environmental Law?', Ecology Law Quarterly 2010, 981.

[7] Christopher M. Bruner/Beate Sjåfjell, 'Corporate Law, Corporate Governance and the Pursuit of Sustainability' in: Beate Sjåfjell/Christopher M. Bruner (eds.), The Cambridge Handbook of Corporate Law, Corporate Governance, and Sustainability, 2019, p. 713–720; Lynn A. Stout, The Shareholder Value Myth: How Putting Shareholders First Harms Investors, Corporations, and the Public, 2012.

[8] Melissa Leach/Kate Raworth/Johan Rockström, 'Between Social and Planetary Boundaries Navigating Pathways in the Safe and Just Space for Humanity', World Social Science Report 2013, 84; see further Beate Sjåfjell/Tiina Häyhä/Sarah Cornell, 'A Research-Based Approach to the UN Sustainable Development Goals. A Prerequisite to Sustainable Business' University of Oslo Faculty of Law Research Paper 2 (2020), available at SSRN: https://ssrn.com/abstract=3526744 or http://dx.doi.org/10.2139/ssrn.3526744 (last accessed 21.2.2021).

[9] See smart.uio.no. The author has subsequently become a member of the European Commission's Informal Group of Company Law Experts (ICLEG), for the period 2020–2024. The reform proposals presented in this article draw on the SMART project results, independently of the ongoing work in the ICLEG.

and the wellbeing of its peoples'.[10] The EU's treaty-based values and objectives further include respect for human dignity and human rights, social policy, minority peoples and rights of the child.[11] Together with the legal requirement for policy coherence for development (PCD), requiring that any area of EU law and policy must not work against developmental policies, this reinforces the sustainability aim of 'leaving no-one behind'.[12]

Any area of EU law and policy is as a matter of EU law meant to contribute to the overarching objectives of the EU as set out in the EU Treaties.[13] To reinforce this, the EU Treaties contain crosscutting rules such as the environmental integration duty in Article 11 TFEU[14]:

> Environmental protection requirements *must* be integrated into the definition and implementation of the Union policies and activities, in particular with a view to promoting sustainable development.[15]

This rule encapsulates a legal principle that constitutes one of the most important elements of EU environmental law.[16] Article 11 TFEU entails that

---

[10] Treaty on the European Union (TEU), Article 3(1), with the values set out in Article 2: 'respect for human dignity, freedom, democracy, equality, the rule of law and respect for human rights, including the rights of persons belonging to minorities'. See further Article 3(3) and 3(5) TEU and Article 21 TEU.

[11] Article 2 TEU. See also the Preamble of the Treaty, where the Member States confirm their 'attachment to fundamental social rights as defined in the European Social Charter signed at Turin on 18 October 1961 and in the 1989 Community Charter of the Fundamental Social Rights of Workers'.

[12] Article 208 in the Treaty on the Functioning of the European Union (TFEU). See Clair Gammage, 'The EU's Evolving Commitment to Promoting Sustainability in its External Actions: Policy (In)Coherence for Development?' SMART working paper on file with current author, University of Oslo (2020). See also General Assembly Resolution A/RES/70/1.

[13] The system of the Treaties as well as the case law of the Court of Justice shows that the general objectives function as a framework for EU law and thereby for the institutions of the EU, see Beate Sjåfjell, The Legal Significance of Article 11 TFEU for EU Institutions and Member States, in: Beate Sjåfjell/Anja Wiesbrock (ed.), The Greening of European Business Under EU Law: Taking Article 11 TFEU Seriously, 2015, p. 51.

[14] *Sjåfjell*, The Legal Significance (fn. NOTEREF _Ref64996330 \h * MERGEFORMAT 13) See also *David Grimeaud*, 'The Integration of Environmental Concerns into EC Policies: A Genuine Policy Development?' European Energy and Environmental Law Review 2000, 207; *Garcia M. Durán/Elisa Morgera*, Environmental Integration in the EU's External Relations: Beyond Multilateral Dimensions, 2012; *Javier Solana*, 'The Power of the Eurosystem to Promote Environmental Protection', European Business Law Review 2019, 547.

[15] Emphasis added. *Beate Sjåfjell*, Towards a Sustainable European Company Law: A Normative Analysis of the Objectives of EU Law, 2009, p. 204–214 and 217–228; inter alia *Ludwig Krämer*, The Genesis of EC Environmental Principles, in: Richard Macrory (ed.), Principles of European Environmental Law, 2004, p. 29–47.

[16] See *Christina Voigt*, Article 11 TFEU in the Light of the Principle of Sustainable Development in International Law, in: Beate Sjåfjell/Anja Wiesbrock (eds.), The Greening of European Business under EU Law, 2014, p. 31–50.

any legal basis in the Treaties is also a basis for environmental protection requirements, with its aim of contributing to sustainability. It constitutes a core tool to implement the concept of sustainability in EU policies and to facilitate the transition towards sustainability. With its explicit aim of 'sustainable development', complying with the duty contained in Article 11 TFEU entails integrating environmental protection requirements in such a way as to achieve sustainable development.[17]

In addition to the environmental integration principle in Article 11 TFEU and the principle of social policy integration in Article 9 TFEU,[18] we have a general principle of integration of policy objectives contained in Article 7 TFEU requiring that the EU must ensure consistency between its policies and activities.[19]

Economic development and social welfare, or as formulated as an objective of EU law: achieving 'a highly competitive social market economy, aiming at full employment and social progress',[20] is in the long run fully dependent on the stability of our ecosystems. Likewise, societal stability is dependent on ensuring fundamental social rights. Thereby the social dimension is also included in the aim of a sustainable development, while a number of other EU sources, including the values and aims expressed in Articles 2 and 3(5) TEU, the Charter of Fundamental Rights and the EU's own case law, provide further bases for the inclusion and promotion of fundamental human and social rights.[21]

The subsidiarity principle of the EU entails that in areas where the EU does not have exclusive competence, it may only act when the objectives of an action cannot be sufficiently achieved by the Member States, but can be better achieved at EU level, 'by reason of the scale and effects of the proposed action'.[22] Under the principle of proportionality, the 'content and form' of

---

[17] Arguably even more clearly expressed in the Charter of Fundamental Rights of the European Union, OJ 2000 C-364/1 Article 37: 'A high level of environmental protection and the improvement of the quality of the environment must be integrated into the policies of the Union and ensured in accordance with the principle of sustainable development'.

[18] Article 9 TFEU: 'In defining and implementing its policies and activities, the Union shall take into account requirements linked to the promotion of a high level of employment, the guarantee of adequate social protection, the fight against social exclusion, and a high level of education, training and protection of human health'.

[19] Article 7 TFEU: 'The Union shall ensure consistency between its policies and activities, taking all of its objectives into account'.

[20] Article 3(3) TEU.

[21] For a discussion of the legal status of human rights protection after the Lisbon Treaty, which, inter alia, gives binding, primary law status to the Charter of Fundamental Rights, see S. Douglas-Scott, 'The Court of Justice of the European Union and the European Court of Human Rights after the Treaty of Lisbon', in: Stephen Weatherill/Sybe de Vries/Ulf Bernitz (eds.), The Protection of Fundamental Rights in the EU After Lisbon, 2013, p. 153–180.

[22] Article 3(5) TEU.

EU action shall not 'exceed what is necessary to achieve the objectives of the Treaties'.[23]

The transnational nature of business and its unsustainability makes it clear that action on EU level is necessary. Individual initiatives by Member States can be inspiring examples and also stimulate EU action, and initiatives such as the French vigilance law are laudable.[24] However, they also bring with them challenges including questions of scope and of legal certainty for businesses with cross-border operations and activities. To ensure the contribution of business to the Treaty objectives of sustainability, action on EU level is necessary.

All this forms the framework also for the specific legal bases for company law. The EU regulation on company law, accounting law and auditing law is based on Article 50 of the TFEU, especially 50(1). The provision is complemented by Article 50(2)(g) stating that the European Parliament, the Council and the Commission shall carry out the duties devolving upon them under the preceding provisions, in particular:

> by coordinating to the necessary extent the safeguards which, for the protection of the interests of members and others, are required by Member States of companies or firms within the meaning of the second paragraph of Article 54 with a view to making such safeguards equivalent throughout the Union.

With this provision, the scope of Article 50(1) is enlargened to encompass all legal business forms, both public and private:

> Companies or firms formed in accordance with the law of a Member State and having their registered office, central administration or principal place of business within the Union shall, for the purposes of this Chapter, be treated in the same way as natural persons who are nationals of Member States.
>
> 'Companies or firms' means companies or firms constituted under civil or commercial law, including cooperative societies, and other legal persons governed by public or private law, save for those which are non-profit-making.[25]

Article 50 is the basis of major European company, accounting and auditing legislation, such as the Company Law Directive 2017, the Accounting Directive, the Transparency Directive, and the Shareholder Rights Directive I and II. Article 50 does not set other limitations to the type, purpose or size of the undertaking, except excluding entities with pure non-profit purpose.

---

[23] Article 3(4) TEU.

[24] See *Véronique Magnier*, Old-Fashioned Yet Innovative: Corporate Law, Corporate Governance and Sustainability in France, in: Beate Sjåfjell/Christopher M. Bruner (eds.), The Cambridge Handbook of Corporate Law, Corporate Governance and Sustainability, 2019, p. 276–289.

[25] Article 54 TFEU.

The EU Treaties show that there is a general duty for the EU institutions to act to achieve the overarching objectives of the European Union.[26] The unsustainability of business activities in Europe and across global value chains present threats to the achievement of the overarching goals, both as set out in the Treaties and those flowing from international policy commitments as set out in the SDGs and the Paris Agreement. Dealing with these cross-border threats require action on EU level, and the EU has the legal bases to do so.

3   Integrating Sustainability into EU Company Law

The Risks of Continued Unsustainability

The economic, corporate and financial risks of continued unsustainability[27] should bring the question of how to reform European business squarely onto the radar of all who are concerned with the resilience of European societies. The argument is accordingly not just one of business contributing to sustainability. It is as much about securing the future of European business. From a systemic perspective, this may come across as an unnecessary and formalistic point to make. Of course, securing a sustainable future for all is also relevant and important for business. Without a sustainable future for humanity on this planet, there is no sustainable future for business either. Business—and the economy in which it operates—does not exist as a bubble separate from society and the environment. Yet, the point is necessary to make in light of the false dichotomy between economic issues on the one hand, and 'ethical' issues, encompassing the interest of people and the environment on the other. The emphasis on the financial risks of climate change is one example of an impactful initiative seeking to bridge this dichotomy. I take this as my starting point, and develop this further in a summary of a broader and research-based approach to the corporate and financial risks of unsustainability.

Much of the EU's sustainability-oriented policy initiatives since the adoption of Paris Agreement on Climate Change in 2015 has concerned the mitigation of the risks of climate change. The EU's Sustainable Finance Initiative[28] has a strong emphasis on climate change and the EU's Green Deal is

---

[26] Articles 1(1) and 3(6) TEU.

[27] *World Economic Forum*, The Global Risks Report 2020; *Patrick W. Keys/Victor Galaz* et al., 'Anthropocene Risk', Nature Sustainability 2019, 667; *Beate Sjåfjell*, 'The Financial Risks of Unsustainability: A Research Agenda', University of Oslo Faculty of Law Legal Studies Research Paper 18 (2020), available at SSRN: https://ssrn.com/abstract=3637969 (last accessed 18.2.2021). This Section draws notably on the latter paper.

[28] *The European Commission*, Communication from the Commission. Action Plan: Financing Sustainable Growth, COM/2018/097 final, 8.3.2018.

marketed under the slogan of 'Striving to be the first climate-neutral continent'.[29] The recommendations of the Financial Stability Board's Task Force on Climate-Related Financial Disclosure[30] have done much to contribute to awareness of potential financial impacts of climate-related risks and opportunities. The 2019 additional guidelines to the EU's so-called Non-Financial Reporting Directive are based on these recommendations.[31] Yet, the Task Force's report has shortcomings when analysed in the context of a research-based sustainability perspective, as we outline in this subsection, where we discuss the financial and business risks of unsustainability.[32] First, concerning the Task Force report's scoping of climate change, it excludes important aspects of climate risks. The report does not speak about societal risks, including risk of societal breakdown. Although the report discusses both acute and chronic physical risks of climate change, it only explores the risks to human beings to a very limited extent. There is only a very brief mention of negative impacts on the 'workforce' from acute physical risks,[33] but business-relevant risks also include the financial risks associated with increased likelihoods of spread of diseases and of unmanageable heatwaves. The report mentions the danger of 'catastrophic environmental and social consequences' in its introduction,[34] but it stops short of explaining the potential severity of the financial risks of climate change. The report does not spell out what climate change, insufficiently mitigated, may entail in the form of global catastrophic risks.[35]

A broader approach is called for.[36] Climate change, albeit a crucial issue, is just one of the hitherto identified *planetary boundaries*, whose breaching

---

[29] See https://ec.europa.eu/info/strategy/priorities-2019-2024/european-green-deal_en (last accessed 21.2.2021).

[30] *Task Force on Climate-Related Financial Disclosures*, Final Report: Recommendations of the Task Force on Climate-Related Financial Disclosures, June 2017, available at www.fsb-tcfd.org/publications/final-recommendations-report (last accessed 15.2.2021).

[31] *The European Commission*, Communication from the Commission—Guidelines on Non-financial Reporting: Supplement on Reporting Climate-Related Information, C/2019/4490, OJ C 209, 20.6.2019, p. 1–30.

[32] The Task Force concentrates on financial risks, i.e. risks from an investor or financier perspective, while the risks of unsustainability that we discuss here are equally business risks—risks that affect the viability of the business itself. Business and financial risks are naturally closely linked albeit not always overlapping. I will not go further into this distinction here.

[33] *Task Force on Climate-Related Financial Disclosures* (fn. NOTEREF _Ref65006521 \h * MERGEFORMAT 32), 10.

[34] *Task Force on Climate-Related Financial Disclosures* (fn. NOTEREF _Ref65006521 \h * MERGEFORMAT 32), 1 (Background).

[35] See e.g. *Seth D. Baum/Itsuki C. Handoh*, 'Integrating the Planetary Boundaries and Global Catastrophic Risk Paradigm', Ecological Economics 107 (2014), 13; Global Challenges Foundation., 'Global Catastrophic Risks 2018' (2018), https://globalchallenges.org/initiatives/analysis-research/reports/ (last accessed 1.2.2021).

[36] *Keys/Galaz* et al. (fn. NOTEREF _Ref64997961 \h * MERGEFORMAT 27). This article also illustrates how risks span across scales as a consequence of social-ecological interconnectivity.

has potential to fundamentally change how the world functions.[37] Biodiversity loss, natural resource use and the release of novel entities are examples of other environmental categories that should be included in assessments of financial risks.[38] Further, risks concerning social aspects should be included. As examples of the undermining of social foundations, I have selected human rights violations, lack of decent work and tax evasion that undermines the welfare state. Tax evasion is intrinsically linked to an undermining of the economic bases for our societies, the increasing inequality between and within countries, and the rise of populism and the risk of societal instability that this entails. Some of the most disturbing trends in major industrialised countries reflect such a lack of social stability, and corporations and associated financial markets have a role in this.[39]

The Task Force report divides climate-related risks into two major categories: (1) *transition risks*, i.e. risks related to the transition to a lower-carbon economy and (2) *physical risks*, i.e. related to the physical impacts of climate change. In a broader approach, the risk of *business model change* may be added to the transition risks, and *global catastrophic risk* to the physical risks. Further, a third major risk category is proposed: *Societal risks*, including risk of unrest, risk of authoritarianism and societal breakdown risk.[40]

The category of *transition risks* are accordingly financial and business risks for businesses that are not taking part in the transition to sustainability—which we have begun to see the contours of—or not transitioning quickly enough. The risks are caused by action or expected action from other actors or institutions: policy-makers, victims of environmental harm or human rights violations, investors or other financiers, or competitors.

Policy risks, as the first category of transition risks, may be seen as increasing now in the EU with its Green Deal and willingness to legislate to promote sustainability. Some business actors may then see lobbying against such initiatives as a protection against risk. However, businesses aiming to mitigate the policy risk will need to see this in connection with other aspects of the financial and business risks of unsustainability, including other transition risks as well as the physical and societal risks. These risks will tend to increase if policy risks are

---

[37] *Johan Rockström/Will Steffen* et al., 'Planetary Boundaries: Exploring the Safe Operating Space for Humanity', Ecology and Society 14 (2009); *Will Steffen/Katherine Richardson* et al., 'Planetary Boundaries: Guiding Human Development on a Changing Planet', Science 347 (2015), 736.

[38] *PricewaterhouseCoopers (PwC)/WWF*, Nature is Too Big to Fail. Biodiversity: The Next Frontier in Financial Risk Management, 2020, www.pwc.ch/en/insights/regulation/nature-is-too-big-to-fail.html (last accessed 20.2.2021). We see this reflected also in the new initiative for a Task Force on Nature-Related Financial Disclosures, https://tnfd.info/ (last accessed 19.2.2021).

[39] See further *Beate Sjåfjell/Christopher M. Bruner*, Corporations and Sustainability, in: Beate Sjåfjell/Christopher M. Bruner (eds.), Cambridge Handbook of Corporate Law, Corporate Governance and Sustainability, 2019, p. 3–12.

[40] *Sjåfjell*, Financial Risks of Unsustainability (fn. NOTEREF _Ref64997961 \h * MERGEFORMAT 27).

low; i.e. if policy-makers do not regulate efficiently to support the transition to sustainability and to mitigate impacts of unsustainability.

An illustration of this is the *liability risk:* the increase in lawsuits against corporations, including parent corporations, for environmental or social harm allegedly caused by their subsidiaries, and against lead corporations for negative environmental or social impacts in their global value chains, shows that the liability risk of unsustainability is materialising.[41] While many cases are rejected for procedural reasons, and many lost, some are likely to be won, and the sheer multitude of cases makes them a risk to be reckoned with.

A further illustration is *reputation risk*, in the form of customers or clients choosing not to purchase products or services from the business, employees and job seekers looking for work elsewhere, and contractual parties, private and public, not wishing to renew or sign up with the business. With social norms and expectations gradually changing, there is an indication of a shift in consumer preferences,[42] as well as in the preferences of job seekers.[43] Media also plays an important role here, in revealing unsustainable business activities, as illustrated for example through the Panama Papers.[44]

Further related to the above, is the *technology risk*. The risk of 'stranded assets'—assets that no longer have value—to any corporation involved in exploiting fossil fuels is the obvious example here. However, this risk is also borne by corporations indirectly relying on these resources, such as manufacturers of fossil-fuelled cars. Further, with the emerging recognition of the impact of, e.g. increasing biodiversity loss, businesses continuing with products that either are based on exploitation of biodiversity resources that are

---

[41] Mark B. Taylor, 'Litigating Sustainability—Towards a Taxonomy of Counter-Corporate Litigation', University of Oslo Faculty of Law Research Paper 8, 2020, available at SSRN: https://ssrn.com/abstract=3530768 (last accessed 5.1.2021). See also the *Grantham Research Institute on Climate Change and the Environment*, Global Trends in Climate Change Legislation and Litigation: 2017 Update, 2017, www.lse.ac.uk/GranthamInstitute/wp-content/uploads/2017/ (last accessed 12.1.2021).

[42] Julia Wilson, 'Consumer Preferences Continue to Shift Toward Sustainability, Market Research Shows', TriplePundit, 2018, www.triplepundit.com/story/2018/consumer-preferences-continue-shift-toward-sustainability-market-research-shows/55496 (last accessed 12.12.2020). However, limitations of relying on 'consumer power' are well-established, *Eléonore Maitre-Ekern/Carl Dalhammar*, 'Towards a Hierarchy of Consumption Behaviour in the Circular Economy', Maastricht Journal of European and Comparative Law 26 (2019), 394; *Julian Kirchherr/Laura Piscicelli* et al., 'Barriers to the Circular Economy: Evidence From the European Union (EU)', Ecological Economics 150 (2018), 264.

[43] María Del Mar Alonso-Almeida/Josep Llach, 'Socially Responsible Companies: Are They the Best Workplace for Millennials? A Cross-National Analysis', Corporate Social Responsibility and Environmental Management 26 (2019), 238.

[44] *International Consortium of Investigative Journalists*, 'Explore the Panama Papers Key Figures', 31.1.2017, https://www.icij.org/investigations/panama-papers/explore-panama-papers-key-figures/ (last accessed 12.2.2021); *Amy Wilson-Chapman/Antonio Cucho/Will Fitzgibbon*, 'What Happened After the Panama Papers?', 3.4.2019, https://www.icij.org/investigations/panama-papers/what-happened-after-the-panama-papers/ (last accessed 7.12.2020).

becoming scarce, may find themselves outcompeted by products developed in new ways. The shift from unsustainable linear business models to sustainable circular models involves financial and business risks for businesses not anticipating and adapting to this shift.[45] *Business model change* is therefore proposed as a risk category to be taken into account.

The two next broad risk categories, *physical risks* and *societal risks*, distinguish themselves from the transition risks category above, in that they affect businesses also seeking to transition to sustainability, and quickly. The severity of the risks depends on the speed with which the global society transitions to sustainability, and will, of course, have local variations. This emphasises the importance for business to work together to mitigate the physical and societal risks as far as possible. This could take the form of lobbying for and not against necessary legislative and policy reforms. It could take the form of transitioning in their businesses and whole sectors towards sustainability, without waiting for policy reforms. Of course, this is not only relevant for business. It also places the onus on the government at national levels as well as on the EU to act to mitigate physical and social risks to citizens.

Ultimately, the *physical risks* may go beyond that which can be managed through anticipation and adaptation. This is reflected through changes in insurance premiums for certain areas, 'with some insurers simply withdrawing from the market'.[46] Unmitigated environmental degradation and continued overshoot of planetary boundaries brings with it global catastrophic risks, defined as the risk of a scenario which takes 'the lives of a significant portion of the human population, and may leave survivors at enhanced risk by undermining global resilience systems'.[47] The risks posed to most businesses in global catastrophe scenarios are existential.[48]

Societal risks include risk of social unrest caused by social inequality, human rights violations and the corporate undermining of the economic basis of our welfare systems. Tax evasion, or other forms of undermining the economic

---

[45] A linear business model is one that takes responsibility for, e.g. a product, only until it is sold, and typically does not encompass the supply chain of the product and all its components. Conversely, a circular business model takes responsibility for a product from cradle to cradle, encompassing each stage from design to recycling/upcycling or management of the waste.

[46] See also *World Economic Forum*, Global Risk Report 2020, 15 January 2020, www.weforum.org/reports/the-global-risks-report-2020/ (last accessed 20.2.2021), 32: 'More common extreme weather events could make insurance unaffordable or simply unavailable for individuals and businesses: globally, the 'catastrophe protection gap'—what should be insured but is not—reached US$280 billion in 2018'.

[47] *Shahar Avin/Bonnie C. Wintle* et al., 'Classifying Global Catastrophic Risks', Futures 102 (2018), 20.

[48] See also *World Economic Forum*, Global Risk Report 2020 (fn. NOTEREF _Ref64999257 \h * MERGEFORMAT 46).

basis for good welfare societies,[49] may negatively impact on these societies' ability to protect their population against physical risks due to environmental degradation, human rights violations and lack of decent work, thereby causing the physical risks to materialise or to be strengthened. Similarly, lack of economic resources in a country undermines its possibility to put into place relevant measures to adapt to environmental change.

Businesses involved in undermining the economic basis of societies, in human rights violations, exploitation of workers, or manipulating the public discourse and democratic processes, may find that these bring a spectrum of societal risks. These may range from societal unrest, via paving the way for increased authoritarianism, to societal collapse. Societal unrest may take the form of rioting and lack of safety for workers, customers and creditors, with negative impacts for the business. An increase in authoritarianism can materialise through sudden regime changes with increased risk of nationalisation or instability in the country that make it difficult to continue with business as planned.[50]

There are a number of scenarios that can lead to global catastrophic risks, including climate change and other environmental degradation.[51] Ultimately, there is a risk of societal collapse, which may be caused by a combination of the factors discussed here.

All these risks are not only relevant for business involved in wrongdoings, although they may be more directly at risk from social unrest which targets those perceived to have been involved in wrongdoings. The financial and business risks of unsustainability are also relevant for other businesses in the same country or region. In case of societal collapse, practically all businesses are affected. This strengthens the argument for business recognising these risks and working together, for example in a sector, to alleviate them. It underlines the importance of responsible business behaviour, and the significance of business supporting appropriate policy reforms.

The categories of global catastrophic risk, and the risk of societal collapse, underline that we cannot settle for a mainstream 'business case' approach. Recognising the risks of unsustainability does not mean that it is sufficient to only internalise environmental, social and broader governance issues to the extent that a clear cause-and-effect line can be drawn from ignoring an issue to the risk it entails for the corporation. Identifying the risk is not intended as a boundary of what issues are relevant to corporate sustainability. The point is to challenge the dichotomy of profits versus sustainability and show that

---

[49] *Jason Hickel*, The Divide: A Brief Guide to Global Inequality and its Solutions, 2017; *Jason Hickel*, 'The Sustainable Development Index: Measuring the Ecological Efficiency of Human Development in the Anthropocene', Ecological Economics 167 (2020).

[50] Involving different types of policy—or political—risks to that discussed above.

[51] See *World Economic Forum*, Global Risk Report 2020 (fn. NOTEREF _Ref64999257 \h * MERGEFORMAT 46).

however little a business may care about 'ethics' and 'corporate social responsibility', (un)sustainability will sooner or later, in one way or another, affect most businesses. This is the grand challenge facing our economies, including business and financial markets, and more broadly, our societies.

The Argument for Reform of EU Company Law

Company law is currently the missing piece in EU's regulatory reform to promote sustainable business. Starting from the paradigm shift of Corporate Social Responsibility (CSR) in the Commission Communication of 2011,[52] which was followed up notably in accounting law, with the adoption of the so-called Non-Financial Reporting Directive in 2014,[53] the next steps were taken in financial market law. Firstly, the revision of the Shareholder Rights Directive in 2017[54] and moving on to the broader Sustainable Finance Initiative, with its 2018 Action Plan,[55] already followed up, inter alia, with the Disclosure Regulation in 2019[56] and Taxonomy Regulation in 2020.[57] Further work is ongoing under the auspices of the Renewed Sustainable Finance Strategy.[58] All of these Sustainable Finance actions aim to shift the investments and the governance activities of investors to promoting more sustainable business activities. Currently, the so-called Non-Financial Reporting Directive is being revised,[59]

---

[52] *The European Commission*, A Renewed EU Strategy 2011–2014 for Corporate Social Responsibility. COM (2011) 681 final, Section 3.1.

[53] Directive 2014/95/EU of the European Parliament and of the Council of 22 October 2014 amending Directive 2013/34/EU as regards disclosure of non-financial and diversity information by certain large undertakings and groups, OJ 2014 No. L 330/1.

[54] Directive 2007/36/EC of the European Parliament and of the Council of 11 July 2007 on the exercise of certain rights of shareholders in listed companies, OJ L 184, 14.7.2007, p. 17–24, as amended by Directive (EU) 2017/828 of the European Parliament and of the Council of 17 May 2017 amending Directive 2007/36/EC as regards the encouragement of long-term shareholder, OJ L 132, 20.5.2017, p. 1–25.

[55] *The European Commission*, Communication from the Commission. Action Plan: Financing Sustainable Growth, COM/2018/097 final, 8.3.2018.

[56] Regulation (EU) 2019/2088 of the European Parliament and of the Council of 27 November 2019 on sustainability-related disclosures in the financial services sector (Text with EEA relevance), PE/87/2019/REV/1.
OJ L 317, 9.12.2019, p. 1–16.

[57] Regulation (EU) 2020/852 of the European Parliament and of the Council of 18 June 2020 on the establishment of a framework to facilitate sustainable investment, and amending Regulation (EU) 2019/2088, (Text with EEA relevance), PE/20/2020/INIT, OJ L 198, 22.6.2020, p. 13–43.

[58] *The European Commission*, 'Renewed Sustainable Finance Strategy and Implementation of the Action Plan on Financing Sustainable Growth', 5.8.2020, https://ec.europa.eu/info/publications/sustainable-finance-renewed-strategy_en (last accessed 21.2.2021).

[59] *The European Commission*, 'Non-Financial Reporting', https://ec.europa.eu/info/business-economy-euro/company-reporting-and-auditing/company-reporting/non-financial-reporting_en#review (last accessed 21.2.2021).

which is expected to see it closely aligned to the Taxonomy,[60] and shifting from its rather misleading language of 'non-financial' to the more accurate 'sustainability reporting'.

The EU Green Deal's ambition that 'sustainability should be further embedded into the corporate governance framework', is followed up most recently in its Sustainable Corporate Governance initiative.[61] In this initiative, the environmental and notably climate change focus of the Sustainable Finance Initiative is merged with the push for mandatory human rights due diligence, informed by the UN Guiding Principles, supported by the European Parliament and a range of national legislative initiatives.[62]

Company law is crucial to business and to the governance of business. Company law provides the dominant legal form of the company for organising business, and it sets out the organisation and the structure for the decision-making in companies. Many of the above very briefly outlined initiatives aim to influence the corporate board directly or indirectly. As also the European Commission has observed earlier, boards have a 'vital part to play in the development of responsible companies'.[63] The EU has in its Sustainable Finance Initiative, in Action 10 of its Action Plan, indicated that it sees a role for legislative intervention in the rules concerning corporate boards.[64]

As a matter of company law, the corporate board has a crucial role in determining the strategy and the direction of the undertaking, and supervising how this plays out.[65] The core duty of the board is to promote the interests of the company. The definition of the interests of the company, as a

---

[60] See also regarding plans for delegated act by June 2021 'specifying the information companies subject to the non-financial reporting directive will have to disclose on how, and to what extent, their activities align with those considered environmentally sustainable in the taxonomy', *European Commission*, 'Implementing and Delegated Acts', https://ec.europa.eu/info/law/sustainable-finance-taxonomy-regulation-eu-2020-852/amending-and-supplementary-acts/implementing-and-delegated-acts_en (last accessed 21.2.2021).

[61] The EU Green Deal Action Plan, section 2.2.1, and see the European Commission's recently concluded public consultation on the topic at https://ec.europa.eu/info/law/better-regulation/have-your-say/initiatives/12548-Sustainable-corporate-governance (last accessed 21.2.2021). The legislative proposal is expected in the second quarter of 2021.

[62] *Lise Smit/Claire Bright* et al. for the European Commission, Study on Due Diligence Requirements Through the Supply Chain, final report, 20.2.2020, https://op.europa.eu/sv/publication-detail/-/publication/8ba0a8fd-4c83-11ea-b8b7-01aa75ed71a1/language-en (last accessed 21.2.2021).

[63] *European Commission*, The EU Corporate Governance Framework (Green paper). COM(2011) 164 final, 5.4.2011, p. 5.

[64] *European Commission*, Communication from the Commission. Action Plan: Financing Sustainable Growth, COM/2018/097 final, 8.3.2018, p. 11.

[65] The 'board' is used in this article as a general term encompassing both levels of boards in countries where there is a supervisory board and an administrative or management board, such as the German *Aufsichtsrat* and *Vorstand*.

matter of company law, varies across European countries,[66] from the monistic, concentrating on the economic interest, with more or less emphasis on the shareholders—which we may denote 'shareholder value' jurisdictions[67]—to the pluralistic, including a variety of other involved or affected parties (often misleadingly denote 'stakeholder value' jurisdictions).[68] Company legislation rarely expressly stipulates what is included in the interests of the company. The interpretation is thereby left to the boards. In light, inter alia, of the business judgement rule,[69] the question of whether the board has interpreted its duty correctly rarely comes to a head in case law.[70]

No company system insists on boards focusing only on returns for shareholders,[71] and certainly not requiring that returns be maximised. In addition to the obvious point that jurisdictions expect boards to ensure legal compliance, company law provides—across this spectrum—a large latitude to the board and by extension the management to shape business in a sustainable manner.[72] However, as is also evident from the state of unsustainability we are in, boards in aggregate do not predominantly choose sustainability-enhancing options even within the realm of the business case, let alone challenge the outer boundaries of the scope to pursue profit in a sustainable manner by going beyond the business case.[73]

At the heart of the problem we find the social norm of *shareholder primacy*, a systemically entrenched barrier to the contribution of business to sustainability. Denoting *shareholder primacy* as a barrier of such significance is a short form for a complex mix of perceived market signals and economic incentives,

[66] Beate Sjåfjell/Andrew Johnston et al., Shareholder Primacy: The Main Barrier to Sustainable Companies, in: Beate Sjåfjell/Benjamin J. Richardson (eds.), Company Law and Sustainability: Legal Barriers and Opportunities, 2015, p. 79–147.

[67] Andrew Johnston, Market-led Sustainability through Information Disclosure: The UK Approach, in: Beate Sjåfjell/Christopher Bruner (eds.), Cambridge Handbook of Corporate Law, Corporate Governance and Sustainability, 2019, p. 204–217.

[68] Andreas Rühmkorf, Stakeholder Value versus Corporate Sustainability: Company Law and Corporate Governance in Germany, in: Beate Sjåfjell/Christopher Bruner (eds.), Cambridge Handbook of Corporate Law, Corporate Governance and Sustainability, 2019, p. 232–245.

[69] Which exists in some form in all European jurisdictions, albeit not necessarily employing the terminology of 'business judgment rule'.

[70] Sjåfjell/Johnston et al. (fn. NOTEREF _Ref65000616 \h * MERGEFORMAT 66).

[71] Even the shareholder value bastion of the UK sets out in the UK Companies Act Section 172 that the board in promoting the success of the company 'for the benefit of its members as a whole', shall have regard to, among other things, 'the impact of the company's operations on the community and the environment'.

[72] E.g. *Andrew Johnston/Jeroen Veldman* et al., 'Corporate Governance for Sustainability', 7.1.2020, Available at SSRN: https://ssrn.com/abstract=3502101 or http://dx.doi.org/10.2139/ssrn.3502101 (last accessed 27.1. 2021).

[73] The lack of cases challenging the boundaries for how far corporate boards can go in promoting long-term sustainability in their decision-making, is a striking feature in the multijurisdictional comparative analysis presented in *Sjåfjell/Johnston* et al. (fn. NOTEREF _Ref65000616 \h * MERGEFORMAT 66).

informed by path-dependent corporate governance assumptions and postulates from legal-economic theories.[74] Shareholder primacy should be distinguished from the legal norm denoted *shareholder value*, which we find notably in the UK.[75] That this distinction often is not made is symptomatic of the dominance of the shareholder primacy thinking, conflating what is seen as practice and what still dominant legal-economic theories describe as efficient, with what company law actually sets out.[76]

Shareholder primacy, combined with a lack of understanding of the scope the law gives the board, and by extension management, has given rise to legal myths inspired by law-and-economics postulates, dictating that the board and senior managers are the 'agents' of the shareholders and must maximise returns to shareholders as measured by the current share price.[77] This is contrary to a proper analysis of company law, which shows that any legislation allowing for companies to exist and become a dominant form of business, has done so based on the assumption that this is positive for society. The idea is that companies create value, for themselves, for their employees, their business partners, their local communities and the broader society. There is no company law in any jurisdiction that promotes companies as a business form based on the assumption that this will maximise returns to shareholders to the detriment of society. Yet, that is far too often the consequence.

The capital markets function to funnel and exacerbate the shareholder primacy drive, supported by securities regulation and stock exchange rules that have as their primary aim to protect investors, not the various other interest affected by corporate activity.[78] The normative impact of the shareholder primacy drive goes beyond the listed corporations, and is exacerbated by the chasm between corporate law's approach to corporate groups and the dominance and practice of such groups,[79] and the extensive use of global value

---

[74] *Sjåfjell/Johnston* et al. (fn. NOTEREF _Ref65000616 \h * MERGEFORMAT 66).

[75] In earlier work, David Millon uses 'radical' and 'traditional' shareholder primacy to distinguish between the social norm and the legal norm; *David Millon*, 'Radical Shareholder Primacy', University of St. Thomas Law Journal 10 (2013), 1013. On UK law, see *Johnston* (fn. NOTEREF _Ref65000826 \h * MERGEFORMAT 67).

[76] *Sjåfjell/Johnston* et al. (fn. NOTEREF _Ref65000616 \h * MERGEFORMAT 66).

[77] Along with that of shareholders owning corporations, which they as a matter of corporate law clearly do not; e.g. *Paddy Ireland*, 'Company Law and the Myth of Shareholder Ownership', Modern Law Review 62 (1999), 32; *Lorraine Talbot*, Critical Company Law, 2nd ed., 2015.

[78] *Jay Cullen/Jukka Mähönen*, Taming Unsustainable Finance: The Perils of Modern Risk Management, in: Beate Sjåfjell/Christopher M. Bruner (ed.), Cambridge Handbook of Corporate Law, Corporate Governance and Sustainability, 2019, p. 100–113; *Christopher M. Bruner*, 'Corporate Governance Reform in a Time of Crisis', Journal of Corporation Law, 36 (2011), 309.

[79] *Blanaid Clarke/Linn Anker-Sørensen*, The EU as a Potential Norm Creator for Sustainable Corporate Groups, in: Beate Sjåfjell/Christopher M. Bruner (eds.), Cambridge Handbook of Corporate Law, Corporate Governance and Sustainability, 2019, p. 190–203.

chains, and other non-equity modes of control,[80] allowing for an intensified externalisation of environmental, social and economic costs.

This is not to say that shareholder primacy is in the interest of shareholders or that it is supported by shareholders. Rather, as outlined in Sect. 3.1 above, continuing on the track of 'business as usual', brings with it systemic risks affecting all investors, as we see expressed also in the emerging recognition of the financial risks of climate change. However, having become such a deeply entrenched norm, it is intrinsic to the financial market system that shareholders operate within. Without comprehensive reforms of the whole system, it is not easy to break out from or properly understand the consequences of the system. The legislative initiatives very briefly outlined in the beginning of this Section are important steps towards such a comprehensive reform, based on some degree of recognition that the current system is detrimental for shareholders, for businesses and for society more broadly. Yet, without connecting them to and including the core corporate governance rules in company law, the outlined initiatives have not had their desired effect. The unrealised potential of the Non-Financial Reporting Directive is amply illustrated through the Alliance for Corporate Transparency's study of 1000 companies' reports, which shows very limited follow-up.[81] In addition to lack of EU-level rules on verification,[82] the gap between the dominant view of the duty of the boards being to maximise returns for shareholders and the broader sustainability issues that the board is being asked to report on, has left sustainability reporting susceptible to green-washing, blue-washing and—since 2015—SDG-washing.[83]

To push back against the shareholder primacy drive, company law needs to take back the power to define what the purpose of the company is and what the duties of the board are. This has to be done in a way that dismantles the legal myth that shareholder primacy has developed into, clarifying that the purpose of business and the duty of the board is actually not to maximise returns for shareholders. However, that is not enough. In light of the convergence of crises we face, a legislative reform should also facilitate and ensure that business partakes in the transition to sustainability.

The argument for including company law in the legislative toolbox, is not one against the relevance of other areas of law. On the contrary, it is one

---

[80] *Jaakko Salminen*, Sustainability and the Move from Corporate Governance to Governance through Contract, in: Beate Sjåfjell/Christopher M. Bruner (eds.), Cambridge Handbook of Corporate Law, Corporate Governance and Sustainability, 2019, p. 57–70.

[81] *The Alliance for Corporate Transparency*, 2019 Research Report. An analysis of the sustainability reports of 1000 companies pursuant to the EU Non-Financial Reporting Directive, 2019, www.allianceforcorporatetransparency.org/ (last accessed 12.2.2021).

[82] *Jukka Mähönen*, 'Comprehensive Approach to Relevant and Reliable Reporting in Europe: A Dream Impossible?', Sustainability (12) 2020.

[83] *SDG Knowledge Hub*, Responsible Business Report Finds High Risk of 'SDG Washing', 29.5.2018, http://sdg.iisd.org/news/responsible-business-report-finds-high-risk-of-sdg-washing/ (last accessed 12.1.2021).

that argues for coherence in policy-making and legislation, connecting the key governance role of the corporate board to business compliance with legislation on environmental protection, product safety and labour, and promoting business action beyond legal compliance. Competing social norms to the shareholder primacy drive, notably the OECD Guidelines for Multinational Enterprises and the UNGPs, have been crucial in opening up the public debate on what responsible business is and how it should act, and they are relevant sources in the further discussion of how business should be regulated and governed. Yet, these competing social norms are not enough in themselves to change the prevailing system, and the self-regulation or voluntary improvement by business of their activities has proven insufficient.[84] In such a fragmented regulatory system, it is difficult, time-consuming and even irrational for the individual sustainability-oriented company to attempt to report openly on its actions, with the risk of being outcompeted by businesses whose publicly available reporting is neither relevant nor reliable as concerns its sustainability impacts. This is the backdrop for the reforms proposed in the Section below.

Redefining Corporate Purpose and the Duties of the Board

Corporate purpose has recently re-emerged as a topical issue, disrupting the established truism of corporate purpose being to maximise returns for shareholders. Even the US 'Business Roundtable' has, in its statement of August 2019, allegedly moved away from shareholder primacy and towards a broader 'stakeholder' approach.[85] It seems an open question whether this was a strategic step to pre-empt support for US senator Elisabeth Warren's 2018 proposal for an Accountable Capitalism Act,[86] or a recognition of the risks of continuing with business as usual, but it certainly has gained a lot of attention, including from Lucian Bebchuck and Roberto Tallarita with their critique of stakeholder approaches.[87] However, as Colin Mayer points out, 'the stakeholder-shareholder debate is vacuous and misses the point. It

---

[84] *Charlotte Villiers*, Global Supply Chains and Sustainability: The Role of Disclosure and Due Diligence Regulation, in: Beate Sjåfjell/Christopher M. Bruner (eds.), Cambridge Handbook of Corporate Law, Corporate Governance and Sustainability, 2019, p. 551–565.

[85] *Business Roundtable*, 'Business Roundtable Redefines the Purpose of a Corporation to Promote "An Economy That Serves All Americans"', 19.8.2019, https://www.businessroundtable.org/business-roundtable-redefines-the-purpose-of-a-corporation-to-promote-an-economy-that-serves-all-americans (last accessed 23.2.2021).

[86] *Elizabeth Warren*, 'Warren Introduces Accountable Capitalism Act', https://www.warren.senate.gov/newsroom/press-releases/warren-introduces-accountable-capitalism-act, 15.8.2018 (last accessed 23.2.2021).

[87] *Lucian A. Bebchuk/Roberto Tallarita*, 'The Illusory Promise of Stakeholder Governance', Cornell Law Review 106 (2020), 91.

is the wrong framing of the question'.[88] There may be a danger that stakeholder approaches are used as a deflection device away from the discussion of what changes of the current system are required to achieve sustainable business, as Carol Liao has highlighted in the context of the B Lab-driven Canadian law reform.[89] Certainly taking some kind of stakeholder approach is not synonymous with business contributing to strong, or real, sustainability.[90]

Rather, what is the topic here, is corporate purpose understood as the overarching purpose set out for companies in company law. Company law does generally not explicitly set out such an overarching purpose of companies, which has, together with the historically understandable emphasis on the relationship between shareholders and companies in company legislation, given ample space for the development of the law-and-economics based conceptualisation of the purpose as maximisation of returns to shareholders. The first aim of proposing a legislative and explicit redefinition of corporate purpose is therefore to dismantle this legal myth that has dominated so much of also the corporate governance discussion over the last decades.[91] Secondly, and giving direction to the content of such a redefinition, it should be done in a way that supports the transition towards sustainability, securing the future of European businesses, by facilitating the shift in business models and stimulating creative innovation.

Together with Jukka Mähönen, and drawing on contributions of the research teams across two international research projects,[92] I have developed ideas on how to redefine corporate purpose and the duties of the board, to achieve these aims.[93] Corporate purpose is, as set out in the newly revised

---

[88] *Colin Mayer*, 'The Future of the Corporation and the Economics of Purpose', Journal of Management Studies 2020, Section 1, https://doi-org.ezproxy.uio.no/10.1111/joms.12660 (last accessed 23.2.2021).

[89] *Carol Liao*, 'A Critical Canadian Perspective on the Benefit Corporation: "The Benefit Corporation and the Firm Commitment Universe"', Seattle University Law Review 40 (2017), 2, 683.

[90] See e.g. *Nigel Roome*, Looking Back, Thinking Forward: Distinguishing Between Weak and Strong Sustainability, in: Pratima Bansal/Andrew J. Hoffman (eds.), The Oxford Handbook of Business and the Natural Environment, 2011. See also *Sjåfjell/Bruner* (fn. NOTEREF _Ref65001764 \h * MERGEFORMAT 39), and further *Beate Sjåfjell/Jukka Mähönen*, 'Corporate Purpose and the Misleading Shareholder vs Stakeholder Dichotomy' (2021) working paper on file with current author.

[91] *Christopher M. Bruner*, Corporate Governance in the Common-Law World: The Political Foundations of Shareholder Power, 2013; *Bruner/Sjåfjell* (fn. NOTEREF _Ref65001886 \h * MERGEFORMAT 7).

[92] The Sustainable Companies Project (2010–2014) and the SMART Project (2016–2020), www.jus.uio.no/companies under Projects, Concluded Projects (last accessed 23.2.2021).

[93] Starting out with *Beate Sjåfjell/Jukka Mähönen*, 'Upgrading the Nordic Corporate Governance Model for Sustainable Companies', European Company Law 11 (2014), 58, and continuing through the SMART Project, culminating in the working paper *Beate Sjåfjell/Jukka Mähönen* et al., 'Securing the Future of European Business: SMART Reform

Danish corporate governance code, a 'considerable driving force in the company's strategy and decision-making processes'.[94] Corporate purpose, as I am using the concept here, should be distinguished against the individual and more detailed purpose expressed by the individual business in instruments of constitution or memorandum and articles of association. A redefined corporate purpose should be expressed on an overarching level in EU legislation, without replacing the specific purposes that the legislation governing the various forms of undertakings in the Member States may set out. Our proposal to redefine the overarching purpose of the undertaking is not one that dramatically changes the nature of European businesses or their specific purposes. It does not take away profit as an intrinsic element of the nature of business or of their value creation. It does not change the differences between various forms of undertakings in the European economy, and how profit is used and distributed in them. For example, cooperatives would still be distinguishable from companies, and multinational enterprises from small- and medium-sized enterprises (SMEs). We do not challenge the distinction between for-profit and not-for-profit, nor do we propose to make all businesses become social enterprises.

What the proposal does do, is to position the value creation of European business, with profit as an intrinsic element, within the context of the transition to sustainability that we all need to undertake, and in such a way that gives European undertakings a level playing field and legal certainty. Drawing on sustainability research, we propose that *sustainable value creation within planetary boundaries* is set as the overarching purpose, outlining the scope within which profit will continue to be made.[95]

Proposals', 7.5.2020. University of Oslo Faculty of Law Research Paper No. 2020–11, available at SSRN: https://ssrn.com/abstract=3595048 (last accessed 3.12.2020); on which the reform proposals in this article draw.

[94] *Danish Committee on Corporate Governance*, 'Danish Recommendations on Corporate Governance', 2020, https://corporategovernance.dk/recommendations-corporate-governance (last accessed 23.2.2021).

[95] *Sjåfjell/Mähönen* et al. (fn. NOTEREF _Ref65004754 \h * MERGEFORMAT 93), section 6.2.1.

Sustainable value creation is an emerging concept in corporate law and corporate governance,[96] which can become a meaningful contribution if interpreted within a research-based concept of sustainability.

Translated into the governance of business, *sustainable value creation* encompasses issues such as fair treatment of employees as well as of workers and local communities across global value chains, with respect for international human rights and core ILO conventions as a minimum, ensuring a 'living wage' and safe working conditions. This further entails supporting democratic political processes and as a minimum not undermining these through engaging in corporate capture of regulatory processes. It also entails contributing to the economic basis of the societies in which the business interacts by not engaging in so-called aggressive tax planning and outright evasion.[97]

Turning to the UNGPs,[98] we see that 'internationally recognised human rights' as the 'benchmarks against which other social actors assess the human rights impacts of business enterprises',[99] refer as a minimum to those expressed in the International Bill of Human Rights. These include the Universal Declaration of Human Rights and its main instruments of codification: the International Covenant on Civil and Political Rights and the International Covenant on Economic, Social and Cultural Rights,[100] as well as the principles concerning fundamental rights set out in the eight ILO core conventions, set

---

[96] Examples include: the 2017 revision of the Australian Council of Superannuation Investors (ACSI) Governance Guidelines, emphasising board oversight of 'sustainable, long-term value creation', see *Victoria Schnure Baumfield*, The Australian paradox: Conservative Corporate Law in a Progressive Culture, in: Beate Sjåfjell/Christopher M. Bruner (ed.), Cambridge Handbook of Corporate Law, Corporate Governance and Sustainability, 2019, p. 161–175; the German Corporate Governance Code on the duty of the Management Board to manage the company 'in the best interests of the company … with the objective of sustainable value creation', see *Rühmkorf* (fn. NOTEREF _Ref65002164 \h * MERGEFORMAT 68); the increased emphasis in the 2016 revision of the Dutch Corporate Governance Code on acting 'in a sustainable manner by focusing on long-term value creation', see *Anne Lafarre/Christoph Van der Elst*, Corporate Sustainability and Shareholder Activism in the Netherlands, in: Beate Sjåfjell/Christopher M. Bruner (ed.), Cambridge Handbook of Corporate Law, Corporate Governance and Sustainability, 2019, p. 260–275; also the new Belgian corporate governance code is based on sustainable value creation, see *Corporate Governance Committee*, 'The 2020 Belgian Code on Corporate Governance', www.corporategovernancecommittee.be/en/over-de-code-2020/2020-belgian-code-corporate-governance, 7.5.2020, (last accessed 23.2.2021).

[97] *Beate Sjåfjell*, 'How Company Law has Failed Human Rights—And What to Do About It', Business and Human Rights Journal 5 (2020), 179.

[98] The Office of the High Commissioner for Human Rights, *Guiding Principles on Business and Human Rights*, (UNGPs), 2011, Principle 12.

[99] UNGPs (fn. NOTEREF _Ref65002318 \h * MERGEFORMAT 98), Principle 12, Commentary.

[100] UNGPs (fn. NOTEREF _Ref65002318 \h * MERGEFORMAT 98), Principle 12, Commentary.

out in the International Labour Organization's Declaration on Fundamental Principles and Rights at Work 1998.[101]

An emphasis on vulnerability arguably resonates with the Commentary to the UNGPs Principle 12, which emphasises that business 'may need to consider additional standards'. These concern, according to the Commentary, the human rights of 'specific groups or populations that require particular attention', elaborated on in United Nations instruments regarding the 'rights of indigenous peoples; women; national or ethnic, religious and linguistic minorities; children; persons with disabilities; and migrant workers and their families'.[102] The elimination of discrimination in respect of employment and occupation, whether on the basis of gender, race, age, disability or migrant status, is crucial.[103]

As an intrinsic element of the transition to sustainable business must be included participatory aspects of the social foundations,[104] of workers, regardless of their labour law status, and of affected communities, including indigenous peoples and ensuring that all affected are fully involved.[105] And yet, we must avoid merely replacing the 'shareholder' in shareholder primacy with 'stakeholder'.[106] While involving affected communities, trade unions and

---

[101] UNGPs (fn. NOTEREF _Ref65002318 \h * MERGEFORMAT 98), 12, Commentary. These eight conventions are: Convention Nos. 87 and 98 on freedom of association and collective bargaining (1948 and 1949); Conventions Nos. 29 and 105 on the elimination of all forms of forced and compulsory labour (1930 and 1957); ILO Convention No. 138 on the minimum age for admission to employment (1973); and ILO Conventions Nos. 100 and 111 on the elimination of discrimination in respect of employment and occupation (1957 and 1958), and the 1999 ILO Convention No. 182 on the worst forms of child labour. These have been criticised as merely promoting 'civil and political rights (and even just a selection of these), while moving away from insistence on broader socio-economic entitlements', Tonia Novitz, 'Past and Future Work at the International Labour Organization', International Organizations Law Review 17 (2020), 10.

[102] UNGPs (fn. NOTEREF _Ref65002318 \h * MERGEFORMAT 98), Principle 12, Commentary. Also, in cases of armed conflict, the Commentary emphasises that business should respect the 'standards of international humanitarian law', ibid.

[103] See the International Labour Organization, ILO Declaration on Fundamental Principles and Rights at Work and its Follow-up, 18 June 1998, art. II(d), SDG 8 targets (including SDG target 8.8 concerning migrant workers); the International Convention on the Protection of Migrant Workers and their Families, New York, 18 December 1990, in force 1 July 2003, A/RES/45/158, and the 2019 ILO Centenary Declaration, including Art. II(A)(xvi) concerning decent work for migrant workers.

[104] Tonia Novitz, 'Engagement with sustainability at the International Labour Organization and wider implications for collective worker voice', International Labour Review 159 (2020), 463. Concerning some of the challenges involved, see also Ian Scoones, 'The Politics of Sustainability and Development', Annual Review of Environment and Resources 41 (2016), 293.

[105] As indeed is envisaged by SDG 16.

[106] See also Marco Ventoruzzo, "On 'Prosperity' by Colin Mayer: Brief Critical Remarks on the (Legal) Relevance of Announcing a Multi-Stakeholders 'Corporate Purpose", Bocconi Legal Studies Research Paper 3546139 (2020), available at SSRN: https://ssrn.com/abstract=3546139 (last accessed 14.1.2021).

civil society is crucial, a mere canvassing of 'stakeholder interests' and giving priority to the ones that make themselves heard the most is insufficient. The backdrop must always be the interconnected complexities and the vulnerability of the often unrepresented groups (whether invisible workers deep in the global value chains, indigenous communities or future generations),[107] and the aim of a sustainable future within planetary boundaries.

Positioning sustainable value creation within *planetary boundaries* has potential significance on three interconnected levels: firstly and most importantly, it brings to the forefront that there are ecological limits (conversely, that being perceived as 'environmentally friendly' while not respecting those limits is totally inadequate). Secondly, it highlights the complex interactions between planet-level environmental processes, recognising for example that climate change, however topical (and difficult to mitigate), is only one aspect of the convergence of crises we are heading towards. Thirdly, it continuously reminds us that state-of-the-art natural science must inform our decisions on a work-in-progress basis, encompassing the uncertainty and complexity of the global challenges.

This entails that a research-based precautionary approach is needed. The conceptual framework for planetary boundaries itself proposes a strongly precautionary approach, by 'setting the discrete boundary value at the lower and more conservative bound of the uncertainty range'.[108] The precautionary principle has a basis in EU law. Article 191(2) TFEU includes the precautionary principle in its stipulation of environmental principles that the EU shall follow. It is recognised that the scope is much wider than environmental issues in practice, and where 'preliminary objective scientific evaluation' indicates that there are 'reasonable grounds for concern that the potentially dangerous effects on the environment, human, animal or plant health may be inconsistent with the high level of protection chosen for the Community', the precautionary principle should be invoked.[109]

The planetary boundaries framework does not offer a finite set of environmental issues that require recognition and protection. Rather, planetary boundaries as a concept forms the rationale by which new boundaries may be identified and better operational quantifications or metrics adopted.[110] This entails that a legislative reform integrating the recognition of planetary

---

[107] *Louis J. Kotzé*, 'The Anthropocene, Earth System Vulnerability and Socio-ecological Injustice in an Age of Human Rights', Journal of Human Rights and the Environment 2019, 62, 73–75.

[108] *Rockström/Steffen* (fn. NOTEREF _Ref65002661 \h * MERGEFORMAT 37), 472–475.

[109] *The European Commission*, Communication from the Commission on the precautionary principle, COM/2000/0001 final, 2.2.2000.

[110] *Sarah Cornell*, 'Planetary Boundaries and Business', working paper on file with current author.

boundaries[111] cannot be satisfied with regulating the protection of the hitherto identified nine boundaries. Instead, the concept of 'planetary boundaries' itself needs to be included, as a general clause, to be interpreted in light of the science as it develops.

With this backdrop, we propose that creating 'sustainable value' should be defined as creating value for the undertaking, while respecting the rights of its members, investors, employees and other contractual parties, and promoting good governance, decent work and equality, and the human rights of its workers and affected communities and peoples. 'Planetary boundaries' should be defined as scientifically recognised processes that regulate the stability and resilience of the Earth system within which humanity can continue to develop and thrive for generations to come.[112]

This overarching purpose of sustainable value creation within planetary boundaries should be operationalised through a redefinition of the duties of the board, outlining in a way that provides legal certainty for undertakings. To clarify a key concept of European company law, which has become somewhat clouded through the influence of the shareholder primacy drive, it should be set out in EU company legislation that the core duty of the board is to promote the interests of the company. This operationalisation would not entail a harmonisation of the definition of the interests of the company in EU company law. Rather, it would draw up the boundaries within which the board shall promote the interests of the company. Developing the understanding of what the interests of the specific company entail in a specific instance should remain with the board to define, within the scope of national legislation, articles of association and existing contracts and commitments.[113]

The duty of the board to promote the interests of the undertaking in a way that contributes to the overarching corporate purpose, should be set out as ensuring that the operations and activities of the business create sustainable value and contributes to global society staying within planetary boundaries. Encompassing respect for human rights, ensuring decent working conditions and promoting good governance should be key aspects of sustainable value creation. Contributing to ensuring that global society stays within planetary

---

[111] Or the 'limits of our planet', as formulated in the Environment Action Programme to 2020, Decision No 1386/2013/EU of the European Parliament and of the Council of 20 November 2013 on a General Union Environment Action Programme to 2020 'Living well, within the limits of our planet', OJ L 354, 28.12.2013, p. 171–200. The EU 7th Environmental Action Programme to 2020 (7th EAP) was adopted in 2013, https://www.eea.europa.eu/policy-documents/7th-environmental-action-programme (last accessed 14.1.2020). The proposed 8th EAP, to guide European environmental policy to 2030, explicitly mentions the planetary boundaries: https://ec.europa.eu/environment/pdf/8EAP/2020/10/8EAP-draft.pdf (last accessed 14.1.2020). See further *Sjåfjell/Häyhä/Cornell* (fn. NOTEREF _Ref65005059 \h * MERGEFORMAT 8), Section 2.

[112] *Sjåfjell/Mähönen* et al. (fn. NOTEREF _Ref65004754 \h * MERGEFORMAT 93), Section 6.2.1.

[113] *Sjåfjell/Mähönen* et al. (fn. NOTEREF _Ref65004754 \h * MERGEFORMAT 93), Section 6.2.1.

boundaries should entail complying with the at any time most ambitious politically adopted targets at the EU level or relevant Member State level, and—within the scope of the business of the undertaking—protecting and regenerating natural resources and processes, and avoiding, or reducing as far as possible, contributions to the transgression of currently identified planetary boundaries.[114]

A legislative reform should specify that the board is to ensure that the business model of the undertaking is in line with the overarching purpose, developing and publishing a strategy that integrates the purpose throughout the business, including in the internal control and risk management systems.[115] Further, it should define clearly the sustainability assessment—including sustainability due diligence—that the board must ensure is undertaken, to identify ongoing negative sustainability impacts and principal risks of future negative sustainability impacts. The due diligence process should be set out so as to encompass consultative processes for engagement with local communities, including indigenous peoples and other groups and persons affected by the operations and activities of the business, encompassing as relevant in the specific case, workers, subcontractors and local or national interest groups and community representatives. Follow-up of the due diligence process should also be stipulated, where identified lack of legal compliance should be rectified immediately. For other identified sustainability impacts and risks, an ambitious continuous improvement process should be drawn up under the leadership of the board. The ambitious continuous improvement plan should include qualitative and quantitative Key Performance Indicators where appropriate. EU rules should also provide for external verification that the due diligence process is undertaken in accordance with the rules, and annual reporting on this should be audited. Together this would provide a good basis for legal certainty for the board that it is following this up as it should and a level playing field in the sense that it would know that other undertakings would be subject to the same rules.[116]

Further, such a process would provide legal certainty for the undertaking as concerns its sustainability impacts, mitigating effectively much of the risks of unsustainability. As Lise Smit and Claire Bright point out, it is important that due diligence does not act as a safe harbour, i.e. that affected parties cannot

---

[114] Biodiversity loss in all ecosystems, including oceans; freshwater pollution and scarcity; land system change, including change in regional vegetation; greenhouse gas emissions; atmospheric aerosol emissions; chemical pollution including synthetic organic pollutants, heavy metal compounds and radioactive materials; and the introduction of novel entities including microplastics and nanomaterials; ozone depletion; nitrogen and phosphorus pollution; and ocean acidification. See further *Sjåfjell/Mähönen* et al. (fn. NOTEREF _Ref65004754 \h * MERGEFORMAT 93), Section 6.2.1.

[115] This resonates with existing requirements in the Non-Financial Reporting Directive as well as with the proposal in the Action Plan for sustainable finance, Action 10.

[116] *Sjåfjell/Mähönen* et al. (fn. NOTEREF _Ref65004754 \h * MERGEFORMAT 93), Section 6.2.1.

file a lawsuit against the undertaking or its board, nor must it devolve into a box-ticking exercise.[117] However, compliance with a thoughtfully formulated mandatory sustainability due diligence regime, would serve as a defence for the undertaking and its board. This will increase the legal certainty for European business, while providing better access to justice for affected workers and communities.

This proposal draws on commonly agreed upon sustainability goals and sustainability science, thereby increasing legal certainty for business in the sense that it would clarify what the boundaries of legitimate business activities are. Followed up by clearly defined rules for sustainability assessment, notably including mandatory sustainability due diligence, risk management would be improved, as would legal certainty in the sense of better knowing the extent of the vulnerabilities of the business. This would give the sustainability-oriented businesses in Europe, of which there undoubtedly are many, the competitive advantage over unsustainable business.

4   Concluding Reflections

The European Commission has through its EU Green Deal signalled an unprecedented broad and ambitious approach towards transitioning to a sustainable future, with a 'just transition' that leaves 'nobody behind', refocusing the coordination of economic policies across the EU to integrate sustainability.[118] Integrating sustainability into corporate purpose and the duty for corporate boards is key to achieving the relevance and reliability of information from businesses. Relevant and reliable information is currently the missing link in the EU's Sustainable Finance initiative.[119] Providing such information will give sustainability-oriented investors and investees the level playing that they are asking for. Integrating sustainability into corporate governance in this

---

[117] Lise Smit/Claire Bright, 'The Concept of a "Safe Harbor" and Mandatory Human Rights Due Diligence', CEDIS Working Papers 2020, https://cedis.fd.unl.pt/wp-content/uploads/2020/12/CEDIS_working-paper_the-concept-of-safe-harbour.pdf (last accessed 23.2.2021).

[118] With reference to the UN Sustainable Development Goals (SDGs); see Ursula von der Leyen, 'A Union That Strives for More. Political Guidelines for the Next European Commission 2019–2024', 16.7.2019, https://op.europa.eu/s/n4FS (last accessed 23.2.2021).

[119] David Monciardini/Jukka Mähönen/Georgina Tsagas, 'Rethinking Non-Financial Reporting: A Blueprint for Structural Regulatory Changes', Accounting, Economics and Law: A Convivium 1 (2020).

way also provides a better basis for Sustainable Public Procurement,[120] and resonates with and strengthens the EU's Circular Economy initiative.[121]

The risks of unsustainability bring home the importance of integrating sustainability throughout the business of any undertaking. If this is not done, the risks will increasingly materialise, as the international trend of lawsuits against European businesses is already showing. Ultimately, the risk of continuing with 'business as usual' is existential. There are a number of scenarios that can lead to societal collapse, and in none of these are steady returns for investors or profitable business likely.

There has never been a stronger case for all partners working together for a better system, and the way out of the Covid-19 pandemic must also be a path towards sustainability. The EU as a global actor and as European policymaker and legislator has a crucial role here to ensure that the international and European regulatory framework for business mitigates the risks of unsustainability as far as possible and secures the contribution of business to a sustainable future.

---

[120] *The European Commission*, 'Green and Sustainable Public Procurement', https://ec.europa.eu/environment/gpp/versus_en.htm (last accessed 19.2.2021); *Marta Andhov/Roberto Caranta* et al., 'Sustainability Through Public Procurement: The Way Forward—Reform Proposals', 23.3.2020, available at SSRN: https://ssrn.com/abstract=3559393 or http://dx.doi.org/10.2139/ssrn.3559393 (last accessed 21.2.2021).

[121] *The European Commission*, 'EU Circular Economy Action Plan', https://ec.europa.eu/environment/circular-economy/ (last accessed 23.2.2021); *Eléonore Maitre-Ekern*, 'Re-Thinking Producer Responsibility for a Sustainable Circular Economy' Journal of Cleaner Production 286 (2021), 125454.

CHAPTER 4

Towards a Smart Regulation of Sustainable Finance

Dirk A. Zetzsche and Linn Anker-Sørensen

1 Nudging or Mandatory Approach to Sustainable Finance?

Regulators worldwide seek to further sustainable investments. For instance, the EU Commission appointed in late 2019 has promised to implement a Green Deal Action Plan[1] and adopted a Strategy for financing the transition into a sustainable economy in two steps on 21 April 2021 and 6 July 2021 (hereafter Sustainable Finance Strategy),[2] with a view to accelerating the efforts which the previous EU Commission had proposed as the Sustainable

---

[1] *See* European Commission, The European Green Deal (11 December 2019), COM/2019/640 final, https://eur-lex.europa.eu/legal-content/EN/TXT/?uri=COM%3A2019%3A640%3AFIN.

[2] For a detailed discussion of the European Commission's work programme announced per 21 April 2021 and 6 July 2021, see infra, at II.2.

D. A. Zetzsche (✉)
Faculty of Law, Economics and Finance, University of Luxembourg, Esch-sur-Alzette, Luxembourg
e-mail: Dirk.zetzsche@uni.lu

L. Anker-Sørensen
University of Oslo, Oslo, Norway

Finance Action Plan in March 2018 (hereafter SFAP 2018).[3] This is not the place to analyse these regulatory efforts in detail.[4] Rather, this chapter seeks to answer one main question of any sustainability-oriented reform agenda: is a nudging or mandatory approach to regulating sustainable finance preferable?

For that purpose, we give a short overview of the sustainability-oriented reforms of EU securities regulation to showcase the shift from a nudging to a mandatory approach (II.), before summarizing what we know about the links between finance and sustainability (III.). On that basis we assess which regulatory approach—nudging or mandatory—is preferable (IV.). Part V. concludes.

2   A Brief Overview of the EU Sustainability-Oriented Securities Regulation

SFAP 2018: A Nudging Approach

The EU's SFAP 2018 aims at a sustainability transformation of the European economy, through essentially three measures: initial state funding shall be leveraged through financial markets, and this leverage shall be facilitated by measures of law, partly nudging and partly forcing EU financial intermediaries to undertake steps that could further the transformation.[5] At the heart stands the Taxonomy Regulation (EU) 2020/852,[6] introducing a joint terminology and standardized approach to 'environmental sustainability'. The taxonomy is cross-sectoral, in that it calls for obedience in all parts of the financial services value chain but also covers issuers of corporate bonds as well as large stock corporations and limited liability companies.[7]

There are also four legislative measures which all aim at enhanced, harmonized and comparable disclosures relating to sustainability.

These measures comprise the following:

> The cross-sectoral Sustainable Finance Disclosure Regulation (EU) 2019/2088,[8] introducing mandatory disclosure for financial market

---

[3] *See* European Commission, Action Plan: Financing Sustainable Growth (3 March 2018), COM/2018/097 final, https://eur-lex.europa.eu/legal-content/EN/TXT/?uri=CELEX:52018DC0097.

[4] See for a more extensive view Zetzsche & Anker-Sørensen, EBOR 2021, ___.

[5] See SFAP 2018, supra note 3, at 2.3.

[6] Regulation (EU) 2020/852 of the European Parliament and of the Council of 18 June 2020 on the establishment of a framework to facilitate sustainable investment, and amending Regulation (EU) 2019/2088, OJ L 198, 22.6.2020, pp. 13–43.

[7] See Article 1(2) Taxonomy Regulation (EU) 2020/852.

[8] Regulation (EU) 2019/2088 of the European Parliament and of the Council of 27 November 2019 on sustainability-related disclosures in the financial services sector, OJ L 317, 9.12.2019, pp. 1–16.

participants and financial advisers on sustainability factors defined by the Taxonomy Regulation (hereafter SFDR) to all EU financial law legislation
The revised Benchmark Regulation (EU) 2019/2089,[9] adding provisions on sustainability benchmarks to the EU rules on benchmark providers
The proposed revisions to EU product distribution rules (in IDD II, MiFID II), demanding that sustainability factors are considered when the suitability of a product for clients is assessed by insurance distributors and investment firms[10]
The proposed revision of Directive 2014/95/EU on non-financial reporting ("NFRD").[11]

Finally, under the SFAP 2018 the European Commission considers legislative measures on the set-up and operational conditions of financial intermediaries, with a view to embedding sustainability risks into financial intermediaries' risk management[12] and combatting undue short-termism.[13]

Methodically speaking, the Taxonomy Regulation aims at answering the question "what is sustainability?", while the disclosure obligations shall help identify "who acts sustainably" or "which product is sustainable"?, respectively. Finally, the review of the set-up and business conduct rules shall ensure that financial intermediaries act sustainably, yet the previous European Commission did not present (draft) legislation on this matter.

While developing and implementing the Taxonomy Regulation will definitely come with challenges, it is nevertheless an ambitious project pursuing a

---

[9] Regulation (EU) 2019/2089 of the European Parliament and of the Council of 27 November 2019 amending Regulation (EU) 2016/1011 as regards EU Climate Transition Benchmarks, EU Paris-aligned Benchmarks and sustainability-related disclosures for benchmarks, OJ L 317, 9.12.2019, pp. 17–27.

[10] See SFAP 2018, supra note 3, at 2.5. The SFAP 2018 resulted in two legislative proposals, yet the proposals have not been adopted by the old European Commission, leaving this work strand for the new European Commission appointed in late 2019.

[11] See SFAP 2018, supra note 3, at 4.1. A consultation preparing the revision was then performed under the new European Commission appointed in late 2019; See also proposed amendments to the NFRD: Proposal for a Directive of the European Parliament and of the Council amending Directive 2013/34/EU, Directive 2004/109/EC, Directive 2006/43/EC and Regulation (EU) No 537/2014, as regards corporate sustainability reporting, COM/2021/189 final.

[12] See SFAP 2018, supra note 3, at 3. The Juncker EU Commission collected feedback in consultations. The implementation was left to the new Commission. See, for instance, *ESMA's technical advice to the European Commission on integrating sustainability risks and factors in MiFID II*, https://www.esma.europa.eu/sites/default/files/library/esma35-43-1737_final_report_on_integrating_sustainability_risks_and_factors_in_the_mifid_ii.pdf; *ESMA's technical advice to the European Commission on integrating sustainability risks and factors in the UCITS Directive and AIFMD*, https://www.esma.europa.eu/sites/default/files/library/esma34-45-688_final_report_on_integrating_sustainability_risks_and_factors_in_the_ucits_directive_and_the_aifmd.pdf.

[13] See SFAP 2018, supra note 3, at 4.2.

clear vision: if a) all issuers would fully disclose data in line with the Taxonomy Regulation, and b) all intermediaries process these data using models that integrate financial and sustainability data, while c) their clients (investor or beneficiaries) prefer taxonomy-compliant financial products over non-compliant products, the Taxonomy Regulation would steer capital flows into environmentally sustainable economic activities, as defined by the Taxonomy Regulation.

While an empirical assessment of the SFAP is not yet available given the early stage of its implementation, we can already identify that the SFAP's main objective is to make sustainable investment "the new normal"[14]: Given that from 2031 to 2050, annual average investments between €1.2 and 1.5 trillion will be necessary to meet the "80% greenhouse gas (GHG) reduction scenarios" contained in the European Commission's long-term vision "A Clean Planet for all"[15] (aiming at a carbon–neutral economy), nothing less will do to achieve these ambitious goals than turning sustainable investments from something niche into the new mainstream. This requires putting EU financial markets in a position where investors have a good understanding of the market's depth and liquidity with regard to products defined as sustainable investments.

To come closer to that objective, the SFAP's core mission is the clarification of terminology so as to ensure that investors can compare sustainable investments and measure their success—by comparing these investments with non-sustainable investments. This addressed one of the main deficiencies in any sustainable finance assessment: the fact that few understood what, exactly, a sustainable investment was,[16] and in turn how profitable truly sustainable investments could be.[17] The legal definition of the term sustainability was

---

[14] The EU Commission uses the term "mainstream investment". For the term used in this paper, see BakerMcKenzie, Sustainable Finance: From Niche to New Normal, 2019.

[15] See European Commission, In-Depth Analysis in Support of The Commission Communication, COM (2018) 773, https://ec.europa.eu/clima/sites/clima/files/docs/pages/com_2018_733_analysis_in_support_en_0.pdf

[16] On the divergence of sustainability ratings, see, e.g., Doni & Johannsdottir 2019 at 440 (arguing on the differences in "scope, coverage and methodology" among different ESG rating providers); Berg et al. 2020 (on the importance of considering original or rewritten data of ESG rating providers: the same providers may change methodology over the years that can significantly change ESG firms ratings impacting empirical research and investment decisions); Dorfleitner et al. 2015, 465 (comparing three of the most used ESG rating approaches, the authors find a clear lack of convergence in ESG measurement); Berg et al. 2020 (comparing six of the most relevant ESG rating providers, the authors find evident divergences in "scope of categories, different measurement of categories, and different weights of categories").

[17] Regarding divergent results of studies on profitability of sustainable investments (independent of the asset class), see, e.g., Cunha et al. (2019), 688–689 (the authors analyse "the performance of sustainable investments in developed and emerging stock markets from 2013 to 2018 " by using "global, regional and country-level sustainability indices as benchmarks" and comparing them with "respective market portfolios", conclude that given the discordant results, conclusions cannot be drawn yet; however, there is increasing

a crucial step towards comparability, yet, not necessarily accuracy. For the purpose of *comparability* of financial products, whether the definition is 100% accurate, and whether the definition leads to truly sustainable investments, is a lesser concern—more important is that the definition is consistently applied throughout the EU financial sector—and potentially beyond.

A second focus point of the SFAP 2018, particularly through the Sustainable Finance Disclosure Regulation (SFDR), is enhancing disclosure. For sure, the SFDR is not lightweight and severely impacts financial market participants and financial advisers which need to review, among others, their risk and remuneration policies as well as amend all product-related disclosures (prospectuses, etc.). Financial market participants need to reveal whether and how they integrate sustainability risks in their investment decisions and remuneration policies, but also whether and how they consider the impact of their decisions on sustainability factors. Further disclosure items relate to the methodologies used for assessing the former, as well as to reliance on indices and the factual basis for relying on terms such as sustainability and carbon reduction throughout the development, investment and marketing of a given financial product.[18]

If sustainability factors are disclosed in a harmonized, comparable way by the product originators (that is financial market participants), regulators can then mandate that these disclosures are read, assessed and used in the remaining parts of the financial services value chain. In line with this, intermediaries involved in the distribution of financial products through investment advice and the provision of life insurance products shall disclose certain sustainability information to end investors.[19] The SFDR implements this requirement with the same catch-all disclosure approach it foresees for financial market participants.

The SFDR certainly comes at a cost for financial intermediaries given disclosure is never without its financial drawbacks and many questions need to be answered to ensure a consistent application.[20] And regardless of whether the definition applied under the SFDR furthers truly sustainable investments, the enhanced disclosure with regard to sustainability factors and methodologies applied by financial intermediaries serves a purpose in itself. These disclosures

---

hope for investors to obtain higher risk-adjusted returns if engaging in sustainable investments in certain geographies); Friede et al. (arguing that 90% of the studies surveyed show a nonnegative correlation between ESG and corporate financial performance); *but see also* Fiskerstrand et al. 2019 (showing no significant relation between ESG and stock returns in the Norwegian stock market); on sustainable investing and higher financial returns, see, e.g., Filbeck et al. 2016 (analyzing socially responsible investing hedge funds compared to conventional hedge funds); on sustainable investing and lower financial performance in mutual funds, see, e.g., El Ghoul & Karoui 2017; Riedl & Smeets 2017.

[18] For details, see view Zetzsche & Anker-Sørensen, EBOR 2021, ___.

[19] See SFAP 2018, supra note 3, at 2.5.

[20] For further details, see Busch 2020; Hooghiemstra 2020.

are "nudging"[21] intermediaries to deal with sustainability as a topic, as well as investors to consider sustainability to a greater extent than previously—whether a product is sustainable or not will thus be an issue confronting the prospective investor. In turn, they can review whether the product is profitable *and* sustainable and have all the means to determine their investment preference with regard to those products.

Beyond additional disclosures, the SFAP 2018 refrains from a "going-all-in" approach. Even if implementing legislation under the SFDR would ask for a quantification of sustainability risks for disclosure purposes, financial intermediaries are not required to integrate the quantification of sustainability risks in their risk models. Moreover, the SFAP 2018 refrains from sanctions in the case of an intermediary tailoring its portfolio in entire disregard of sustainability concerns as long as the intermediary explains why it is doing so. Further, its implementation lacks details: in a financial world each risk requires careful considerations on how to manage it, and unsustainable conduct could create risks; yet regulators do not review whether the risk assessment is in fact correct (from their point of view). So far, each intermediary's own view matters—which will most likely result in a huge variety of risk assessments regarding sustainability factors. When it comes to details, the SFAP 2018 remains silent on what conclusions financial intermediaries could or should draw from *particular* sustainability assessments in terms of risk modelling, investment decisions, capitalization and remuneration.

We do not understand this self-limitation aspect of the SFAP 2018 as a deficiency, but rather as prudent self-limitation: The "nudging" approach stands in contrast to any approach "mandating" sustainability—understood as the forcing of investors to invest into sustainable products.[22]

3   THE SUSTAINABLE FINANCE STRATEGY 2021: TOWARDS MANDATING SUSTAINABILITY

With the EU Green Deal, the sustainability transformation has moved from high importance to super-high importance on the EU's political agenda. In light with this move a fundamental shift is on the horizon: While the SFAP 2018 relied on the idea of nudging, various policy documents issued after the Green Deal was announced examine whether a much stricter strategy is feasible: a mandatory push towards sustainability primarily through the means of increased disclosure.[23]

---

[21] See Thaler & Sunstein2008. *See also* Enriques & Gilotta2015 (discussing the function of market disclosure as a "soft-form substitute of more substantive regulations", dubbed "stealth substantive regulation"). *But see also* Gentzoglanis2019 (arguing that firms preferring a non-regulated or less regulated state of operations will comply with a set of disclosure requirements in order to avoid "substantive" regulation.).

[22] See, e.g., Mancini 2020.

[23] See Zetzsche & Anker-Sørensen, EBOR 2021, ___.

In this context, the European Commission proposed six Delegated Acts[24] which if adopted require sustainability risks to be included in all financial intermediaries' activities and, where required under Article 4 SFDR, to take into account principal adverse impacts on sustainability factors when complying with the due diligence requirements set out in the sectoral legislation. Even more detailed, the Delegated Acts require sustainability risks to be considered as part of the intermediaries' investment processes, their conflicts of interest policy as well as their risk management.[25] These items together are understood by the European Commission as a collective description of the intermediaries' fiduciary duties, broadly understood.[26] The Sustainable Finance Strategy announced in 6 July 2021 complements the measures announced in April 2021 by focusing on the links between financial intermediation and the real economy.[27] In light of these proposals, asset owners and asset managers will have no choice but to include sustainability risks, and when they are larger organizations subject to Article 4 (3) and (4) SFDR, sustainability factors as well, into their overall activities.

While the SFAP 2018 with the SFDR and EU Taxonomy Regulation as key measures, was characterized by enhanced disclosure and "nudging", the measures adopted after the Green Deal—with some details yet to be determined—are harbingers of a heavy-handed mandatory push towards sustainable investment: if the best interest of investors is defined from a sustainability, rather than risk-to-profitability, point of view or at least a combination of the

---

[24] European Commission, Proposal for a Commission Delegated Directive amending (1) Directive 2010/43/EU as regards the sustainability risks and sustainability factors to be taken into account for Undertakings for Collective Investment in Transferable Securities (UCITS); (2) Delegated Regulation (EU) No 231/2013 as regards the sustainability risks and sustainability factors to be taken into account by Alternative Investment Fund Managers; (3) Delegated Regulations (EU) 2017/2358 and (EU) 2017/2359 as regards the integration of sustainability factors, risks and preferences into the product oversight and governance requirements for insurance undertakings and insurance distributors and into the rules on conduct of business and investment advice for insurance-based investment products; (4) Delegated Directive (EU) 2017/593 as regards the integration of sustainability factors into the product governance obligations; (5) Delegated Regulation (EU) 2015/35 as regards the integration of sustainability risks in the governance of insurance and reinsurance undertakings; (6) Delegated Regulation (EU) 2017/565 as regards the integration of sustainability factors, risks and preferences into certain organizational requirements and operating conditions for investment firms (all proposals as of 21 April 2021).

[25] See Commission Delegated Regulation (EU) …/… supplementing Regulation (EU) 2020/852 of the European Parliament and of the Council by establishing the technical screening criteria for determining the conditions under which an economic activity qualifies as contributing substantially to climate change mitigation or climate change adaptation and for determining whether that economic activity causes no significant harm to any of the other environmental objectives, C/2021/2800 final.

[26] European Commission, *EU Taxonomy, Corporate Sustainability Reporting, Sustainability Preferences and Fiduciary Duties: Directing finance towards the European Green Deal*, COM/2021/188 final (21 April 2021), at 13.

[27] See view Zetzsche & Anker-Sørensen, EBOR 2021, ___.

former, we would expect a fundamental change in the discretion exercised by asset managers in investment policies; given that asset management requires procedural guidelines for asset allocation and risk management, intermediaries will need to apply models in which one or the other dimension is prioritized if in conflict. In a similar way, sustainability risks need not only to be assessed from a qualitative perspective, but these risks must be quantified and the quantification must be embedded into risk models. In turn, the investment decision may be influenced as much by sustainability aspects as it is by profitability factors.

In the remainder of this chapter, we will examine whether the time is ripe for such a shift from nudging to mandating sustainable finance in the next section.

4   What Do We Know About the Links Between Finance and Sustainability?

To answer the question whether a nudging or mandatory approach to sustainability in the context of securities regulation is preferable, it is crucial to put the proposed policy steps into the context of sustainable finance research, which is what we know and what we do not know about the link between finance and sustainability.

Lack of Expert Consensus

A close look reveals that research results on some of the most important matters of sustainable finance are inconclusive. The uncertainty extends to the very basics of finance, and thus investment. So, questions unanswered equivocally so far include:

(1) *How* in detail sustainability and which sustainability factors in particular impact on firm and macroeconomic profitability and in which way[28]; the uncertainty even relates to smaller questions such as whether warm weather impacts on firms' productivity.[29]

---

[28] See, e.g., Friede (2020), 1276–1278 ; OECD2020; further, see supra note 17 on results of studies regarding sustainability and financial performance. Some of the more recent studies include Balachandran & Nguyen 2018 (suggesting a causal influence of carbon risk on firm dividend policy); Balvers et al. 2017; (stating that financial market information can provide an objective assessment of losses anticipated from temperature changes if the model considers temperature shocks as a systematic risk factor); Colacito et al. 2019 (finding that seasonal temperature rises have significant and systematic effects on the U.S. economy).

[29] Choi et al. 2020 (finding that stocks of carbon-intensive firms underperform firms with low carbon emissions in abnormally warm weather, since retail investors tend to sell that stock, indicating a premium for low-carbon firms in that environment) *versus* Addoum et al. 2020 (not finding evidence that temperature exposures significantly affect

(2) How investors respond to sustainability risks, i.e., whether there is something like a greenium for sustainable conduct.[30]
(3) Whether investors "are willing to pay" for sustainable investments, that is when they forego profits when investing in ESG products.[31]
(4) How investment decisions impact on sustainability factors,[32] i.e., whether investor preferences actually make a difference with regard to de-browning or greening the planet.

The deficiencies addressed by the Sustainable Finance Action Plan so far—the remarkable variation with regard to sustainability ratings,[33] and the lack of conformity of ESG, and that sustainability indices are far from uniform und

establishment-level sales or productivity, including among industries traditionally classified as "heat sensitive").

[30] *See*, on the one hand, Larcker & Watts 2020 (finding that in real market settings investors appear entirely unwilling to forgo wealth to invest in environmentally sustainable projects. When risk and payoffs are held constant and are known to investors ex-ante, investors view green and non-green securities by the same issuer as almost exact substitutes. Thus, the greenium is essentially zero); Murfin Spiegel 2020 (finding limited price effects to rising sea levels); Baldauf et al. 2020 (stating that house prices reflect heterogeneity in beliefs about long-run climate change risks rather than the severity of the risk itself).
against Eichholtz et al. 2019; (arguing in favour of a premium for corporate environmental (ESG) performance based on commercial real estate investments); Krueger et al. 2020; (arguing that institutional investors believe climate risks have financial implications for their portfolio firms); Alok et al. 2020 (finding that managers within a major disaster region underweight disaster zone stocks to a much greater degree than distant managers, indicating a bias); Painter 2020 (finding that counties more likely to be affected by climate change pay more in underwriting fees and initial yields to issue long-term municipal bonds compared to counties unlikely to be affected by climate change); Bernstein et al. 2019 (finding that homes exposed to sea level rise (SLR) sell for approximately 7% less than observably equivalent unexposed properties equidistant from the beach); Huynh & Xia, *Climate Change News Risk and Corporate Bond Returns*, J. Fin. Quant. (September 2020, in press), https://doi.org/10.1017/S0022109020000757 (finding that investors are willing to pay a premium for better environmental performance); Hartzmark & Sussman 2019; (presenting "causal evidence" from fund inflows that investors market wide value sustainability).

[31] Riedl Smeets 2017 (finding that investors are willing to forgo financial performance in order to invest in accordance with their social preferences); Joliet & Titova 2018 (arguing that SRI funds add some SRI factors to make investment decisions, and thus more than financial fundamentals matter); Rossi et al. 2019; (analysing retail demand for socially responsible products and finding that social investors are willing to pay a price to be socially responsible while individuals who consider themselves financially literate are less interested in SR products than others); Gutsche & Ziegler 2019 (arguing that a left-/green political orientation correlates with the willingness to pay for certified sustainable investments).

[32] See, e.g., Busch et al. (2016), 311 (arguing that the long-term impact of investment strategies may depend on multiple factors and that the consequences of the ESG integration strategies are still uncertain on several aspects) *against* Pedersen et al. 2020 (seeking to model the impact of ESG preferences, trying to define the "ESG-efficient frontier" and showing the costs and benefits of responsible investing.); Bender et al. 2019 (analysing metrics for capturing climate-related investment considerations).

[33] See the references supra note 16.

unambiguous—,[34] complete the picture. In line with these academic insights, the ESAs aire general concerns about the lack of "clear and appropriate taxonomy and labels"[35] on ESG terms.

At the core of this inconclusiveness lies a lack of broadly acknowledged theoretical insights (typically laid down in generally accepted standard models) into the co-relation and causation of sustainability factors with financial data.

Assuming that at least three years of sustainability disclosures need to be assessed leads us to conclude that the first research on the measures adopted under the SFAP 2018 (with the SDFR, benchmark reforms and the Taxonomy Regulation only adopted in 2019 and 2020, and coming into force January 2022 and 2023) will be available in 2030—which coincides with the year when politics has promised to deliver results. We conclude that at least until 2030, the SFAP measures will lack support by standard models and empirical testing. This means the following: until 2030 regulators will lack a scientific basis for drafting rules and standards; while this is often the case, in principle, with regard to minor legislative steps, the extraordinary risk with regard to sustainable finance follows from the fact that regulators seek to transform the financial intermediation function of the whole financial services values chain. The task is enormous, and so are the risks of getting it wrong entirely.[36]

Lack of Data Linking Sustainability and Finance

For achieving consensus among finance experts, the availability of data is an important factor as this allows for the empirical validation of theoretical models.

A closer look reveals that, while data on some sustainability factors such as climate data are abundant, we often miss data that help explain the crucial link

---

[34] See, e.g., Jebe 2019, 685 (arguing on the necessity of merging and harmonizing ESG and financial information disclosure).

[35] See ESAs, Letter to the European Commission, Public consultation on a Renewed Sustainable Finance Strategy (15 July 2020), https://www.esma.europa.eu/sites/default/files/library/2020_07_15_esas_letter_to_evp_dombrovskis_re_sustainable_finance_consultation.pdf.

[36] The Global Financial Crisis of 2007–2009 was evidence of the large impact of unwanted effects stemming from rules relating to interest rates, mortgage credit criteria, derivatives, securitization techniques and accounting rules on society. Compared to the SFAP and Sustainable Finance Strategy 2021, the rules that may have collectively contributed to the Global Financial Crisis were relatively minor in scope. By that comparison, if the Sustainable Finance Strategy 2021 gets it wrong, we would expect value destruction of an enormous size.

between these sustainability factors and financial fundamentals.[37] This assessment, based on research and the state of the law, is confirmed by the July 2020 letter of the three ESAs:

> The current shortage of high-quality data renders it challenging for both firms and investors to identify, assess and measure sustainability risks and opportunities, therefore, to take measures accordingly. (…) Moreover, the comparability and reliability of ESG data will only improve if clear and sufficiently granular taxonomies for "green", "brown" and "social" activities are developed and consistently implemented by the financial sector, together with common and uniformly enforced ESG-related disclosure standards for companies.[38]

We identify four reasons for this data shortage. The first reason is that many entities have not yet reported both types of data in a consistent and periodically reliable fashion. This is partly due to the discretion corporations and financial intermediaries have been granted under the previous non-disclosure regime. Research suggests there is insufficient disclosure as to sustainability factors on the side of intermediaries. For instance, a Frankfurt School UCITS study of 2020[39] showed that 28 out of 101 "green" UCITS did not disclose sufficient data to assess whether they are compliant with the upcoming taxonomy requirements; disclosures on cash flows of the remainder often currently cannot uniformly be classified by the taxonomy standards. Where important sustainable financial intermediaries do not disclose data, data processing does not result in solid results. In such an environment, we do not even need to think about disclosures of traditional financial intermediaries, such as non-green UCITS, AIF and investment firms acting as portfolio managers. The ESAs' proposal to set up a single EU Data Platform[40] covering both financial and ESG information, while reducing the costs of sustainability research, will not provide a fundamental change in the short term given that only data *which are reported by issuers and intermediaries* can be made available via that Platform, and only if that happens in a more or less standardized manner would such data be useful for data-driven analysis.[41]

---

[37] See, e.g., Bender et al. 2019, 191–213 (reviewing data characteristics for metrics such as carbon intensity, green revenue, and fossil fuel reserves, highlighting their coverage and distributional characteristics; even though the data can illuminate risk factors to include in a corporate or investment strategy, we lack a financial adaptation strategy building on such data).

[38] ESAs, July letter, supra note 35.

[39] See Malte Hessenius et al., European Commission. Testing Draft EU Ecolabel Criteria on UCITS equity funds (June 2020), Climate Company and Frankfurt School of Finance & Management, https://op.europa.eu/en/publication-detail/-/publication/91cc2c0b-ba78-11ea-811c-01aa75ed71a1/language-en/format-PDF/source-137198287.

[40] See ESAs, July letter, supra note 35.

[41] This basic insight is supported by research on corporate carbon disclosures. See, for instance Liesen et al. 2017 (arguing that financial markets were inefficient in pricing

Note that the former describes the status quo on the eve of the coming into force of the revised Benchmark Regulation as well as the SFDR in 2021, requiring enhanced disclosures by institutional investors and asset managers on how they integrate sustainability risks in the investment decision or advisory process (see II.3). The situation will indeed improve over time; yet this will take time, and improvement will not be noticeable generally and in all respects. For instance, under the SFDR, consistent with the SFAP's nudging approach, even products non-compliant with the EU Taxonomy may be marketed as sustainable—if only the intermediary explains how it got to this conclusion.[42]

The second reason for inadequate data so far is that any study up to the adoption of the revised Benchmark Regulation and the Taxonomy Regulation lacked authority as to the indices and terms used.[43] That is: where data on the past *are* available, they often do not provide the basis of an expert consensus for lack of a harmonized legal framework in place when these were reported. It will take years until data are generated and reported in the way prescribed by the SFAP measures.

The third reason may be that the SFAP 2018 measures, from a finance perspective, lack additional factors necessary to establish the link between sustainability factors and financial fundamentals. For instance, if the investor type has some influence on the social benefit to be expected, as some research suggests,[44] calculating the impact of sustainable investment faces limitations given that data on investors and their exposures to a given asset are often not transparent.[45]

The fourth and final reason is that European politics started to implement its SFAP at the back end of the financial services value chain, while information flows need to start at the front end, that is the real economy:

(1) the SFDR, adopted first in 2019, covers the financial intermediation chain; while (2) Article 1(2) of the Taxonomy Regulation adopted in June 2020 (hence, after the SFDR), covers beyond financial market participants subject to the SFDR, listed companies that issue "sustainable" (usually green)

publicly available information on carbon disclosure and performance; mandatory and standardized information on carbon performance would consequently not only increase market efficiency but result in better allocation of capital within the real economy).

[42] Article 9 and 10 SFDR do not limit "sustainable" investments and products to products in line with the sustainability definition of the Taxonomy Regulation. While Article 25 of the Taxonomy Regulation inserts some references to the Taxonomy Regulation (in particular, the DNSH principle), financial market participants can still include another explanation on how the sustainability objective is to be attained by other means, as long as this information is accurate, fair, clear, not misleading, simple and concise.

[43] See references supra n 16 and 34.

[44] Chowdhry et al. 2018; (studying joint financing between profit-motivated and socially motivated investors); Barber et al. 2021; (finding that losses due to social commitments vary with investor types, with investors subject to legal restrictions (e.g., Employee Retirement Income Security Act) exhibiting lower losses than publicly and NGO-sponsored vehicles).

[45] See Zetzsche & Anker-Sørensen, EBOR 2021, ___.

bonds as well as large undertakings which are public-interest entities with more than 500 employees during the financial year. Yet, the Taxonomy Regulation will not come fully into force until 2023. In turn, reliable data from the real economy will not be available until 2025 or later. But even then, we should not expect wonders. In the beginning, many issuers will have only partial information at hand, given that the NACE methodology has so far not been the basis of intra-corporate reporting and further, some data may be impossible to get, if part of a group's activities are outside the EU where different laws apply. Under these circumstances we should not expect "complete" reporting, that is truly good disclosures within the next decade. At least until the disclosures triggered by the Taxonomy Regulation have reached the required quality, and been tested and adopted in models, the incomplete disclosures of financial intermediaries at the back end of the financial services chain and deficient information intermediation by benchmark providers will render any market-based transformation of the EU economy a major challenge as markets need decent information.

In this context, it is crucial to speed up the review and expansion of the Non-Financial Reporting Directive (NFRD) as promised in the Sustainable Finance Strategy 2021. Again, after the review firms will need time to implement and adjust their disclosures, so the full impact of taxonomy-based information on markets will not be observed, analysed or *understood*, until some years later, possibly in the 2030s.

Lack of Consistent Application

Third, existing and newly adopted reporting standards are currently applied in an inconsistent way. The inconsistent application is to a lesser extent a result of the unfitness or unwillingness of the financial intermediaries, issuers and services providers subject to the new legislation. Instead, the inconsistent application is evidence of the enormous size of the challenge. A consistent application and reporting under the standards just reported requires nothing less than:

1. The adoption of entirely new scoring and reporting frameworks under both the revised Benchmark Regulation (leading to indices aiming at "de-browning" in pursuit of the Paris accord) and the entirely new Taxonomy Regulation (identifying "Green" investment expenditures)
2. The integration of these standards in generally accepted reporting tools and standards (such as IFRS) and other legal measures to narrow down discretion in sustainability reporting to the extent possible
3. On the side of reporting entities, the necessity to build expertise and to make decisions accordingly, prior to useful disclosures (for instance, management must allocate different parts of the firm to different NACE codes)

- to have software tools in place that collect, aggregate and report the data requested under the new frameworks (for instance, accounting systems must be adapted for sustainability purposes) in a granular manner

4. On the side of information intermediaries (including benchmark providers), the necessity:
   - to build expertise prior to the development and implementation of a useful scoring methodology (including the development and testing for consistency of new scoring models)
   - to have software tools for data aggregation and analysis in place

5. On the side of financial intermediaries (including asset managers), the necessity:
   - to build expertise prior to the integration of the scores into investment decisions and risk management (including the development and testing for consistency of new portfolio and risk management models)
   - to have software tools for data aggregation, analysis and application in place

6. On the side of supervisors of reporting entities, information and financial intermediaries, the necessity:
   - to build expertise prior to issuing supervisory guidelines
   - to develop and implement data-driven supervisory tools, and
   - to have sufficiently qualified and skilled staff for rigid enforcement.

With regard to all of these steps and all the participants listed supra from (1) to (6), the implementing projects have just begun.

In turn, a consistent application of measures just adopted is in fact years away—and equally long will the state in which we know very little about the nexus between finance, sustainability and securities regulation most likely persist.

5 Risks of Regulation in the Dark

Doing Harm to the Sustainability Agenda

In light of the issues just raised, we see challenges for all three dimensions of financial regulation: sustainability-conscious retail investors could get hurt following marketing of apparently sustainable products with an uncertain risk—and profitability-profile to them. Sustainability indices, ratings and other metrics may come to inconsistent results,[46] leading institutional investors

---

[46] See on governance of metrics Chiu 2021 (this volume).

towards less standardized and thus less comparable and potentially less reliable approaches. The former may undermine societal support for the sustainability transformation. Market efficiency may suffer if, due to insufficient data, untested models and inconsistent application of the law, capital flows to less productive uses. Large-scale capital misallocation based on unreliable models may also destabilize the financial system, for instance if the pension portfolios of the future are characterized by large-scale, under-performing asset classes due to the unknown financial effects of sustainability-oriented investments. Depending on how large the issues become, the former may delay the aspired transformation towards a sustainable economy or even bring it to a halt altogether.

This is all the more so since there is a potential domino effect on the horizon. In other words, more and more legislation and disclosure rules build now on the EU Taxonomy Regulation: the International Financial Reporting Standards (IFRS), the Non-Financial Reporting Standards the European Financial Reporting Advisory Group is asked to develop, as well as disclosure and accounting frameworks from accounting firms, law firms and other consultants will draw on the Taxonomy. If the EU Taxonomy turns out to trigger disastrous effects on some parts of the European economy, this could put the whole sustainability transformation project into doubt.

Factoring in Transition Risk

Beyond the externalities that "unmastered" financial regulation can create, further externalities that should keep regulators on their toes could stem from the transition towards the new sustainable financial order: managing the transition in the new world of sustainable finance poses a formidable challenge in itself.

Firstly, understanding, implementing and integrating the Taxonomy Regulation as well as the Benchmark Regulation into disclosures and operating processes, such as investment and risk management models takes time and money. In particular, small and medium financial intermediaries may well wait until software vendors come up with standard approaches on which to rely—which require the former first to develop, programme and market such approaches.

Secondly, the transformation requires the use of new models, many of which have not been tested with abundant data (as most financial models today) simply due to data shortages (see above). To test these models, we need to create data pools sufficient for a five-year modelling span. Even if we apply some backward testing (hoping these data are available), it becomes apparent that regulators face two alternatives: either they ask for the use of models utterly insufficiently tested when put to use (meaning model risk from bad specifications, programming or technical errors, or data or calibration errors, resulting in the potentially large-scale reallocation of funds into the "wrong" assets from an economic perspective); or they need to accept that the transformation to reliable models will take five years *after the disclosure starts*; the latter meaning operational adjustments.

Thirdly, the EU's taxonomy is in itself untested. Some glitches, such as the notion that weapon production could qualify as sustainable business under the Taxonomy, have become apparent—and have been remedied—during the legislative process leading to the adoption of the L2 Delegated Acts. In a piece of legislation as complex as the Taxonomy Regulation, other deficiencies will certainly become more apparent over time.

The fact that any regulatory command has limits, downsides and deficiencies is nothing new to lawyers. The larger the scope of a piece of regulation, the greater the need for thorough debate and analysis ahead of its adoption. In light of this, given that the EU Taxonomy and Sustainable Finance Strategy 2021 cover the whole economic sector while the legislative process was short, particular caution is warranted. Similar to the facts on which the SFAP is grounded, the hidden and potentially unwanted effects of the taxonomy lie in the dark.

The transition risks reinforce the data shortage pointed out above: no data can yet include all the physical, legal or transition risks that come about as a consequence of the new sustainable financial order because these risks are yet to crystalize.

6 How to Regulate Sustainable Finance in the State of Uncertainty?

All in all, as an intermediate result, regulators regulate, and enforce the regulations, to a large extent in the dark. Yet the alternative is not refraining from any sustainability-oriented financial regulation altogether; the legislative train has left the station at high speed and any call for a halt is unrealistic and potentially undesirable, given the factual pressure created by climate change and other indicators of a sustainability crisis.[47]

If regulators regulate in the dark, the best advice for them is to aim at avoiding any harm to the sustainability transformation project by unexpected, if not undesirable and unwanted effects of the newly adopted financial regulation and move forward with care, caution and readiness to adjust adopted regulation quickly instead.[48]

Beyond caution, regulating in the dark requires, from the outset, a focus on illuminating the darkness rather than quack legislation through incorporating experimentation and case-by-case assessments.[49] Further, comply or explain approaches, proportionality clauses and principles in contrast to rules are the preferred style of regulation when regulating in the dark.

[47] Ibid.
[48] Ibid.
[49] See Romano, 1, 28.

Three elements of a regulatory policy are suitable to further the cause: first and foremost, regulators should support all efforts that assist in *creating expertise* on all sides of the sustainable finance value chain, including intermediaries and supervisors alike. Second, regulators shall focus on the *consistent application of existing rules* rather than expanding into new, untested fields. In the context of the SFAP 2018, which means regulators shall focus on supporting consensus and creating comparable datasets through disclosure. On the contrary, any meddling with the set-up and the operating business before sufficient data and models are available risks unwanted effects of regulation. Third, regulators must ensure they *retain the openness of the regulatory framework to innovation*, given that much of what is known on sustainable finance may turn out, with hindsight, to be a myth, while some myth may turn out to be the truth.

The three principles face repercussions as to whether it is desirable to regulate the organization, operations and prudential rules relating to financial intermediaries.

Sustainable Intermediary Set-Up

Fitness & Properness

For now, financial supervisory authorities assess key executives of a financial intermediary in two ways: whether they are experienced in running a financial intermediary (fitness) and whether they are law-abiding, trust-worthy people (properness).

The question is what the sustainability transformation does to the fitness test. At least for now few executives will be sustainability experts. Yet, over time this question will disappear because if sustainable investment is the new normal, then leading any intermediary will come with sustainability matters on a day-to-day basis. The question then is how to accelerate the transformation.

Shall the law require the executives to be trained in sustainability matters or the firm to have a certain number of sustainability experts (similar to accounting experts of today) or shall the law, like the UK Senior Managers Regime, require the board to appoint a sustainability officer? The answer is twofold. On the one hand, financial regulation already requires the training of board members.[50] No doubt, these provisions apply to any new development of relevance to the firm. For instance, for intermediaries where technology is important (as is more or less for all intermediaries) special care shall be taken when training board members and appointing a chief technology officer.[51] The same principle applied to sustainability would then result in sustainability-trained boards and the appointing of a sustainability officer.

The risk of quack governance, however, is real. Other issues of social or economic importance exists that would also warrant attention. For instance, we could envision a gender officer, a globalization officer and so on. If we followed through with this approach, the executive suite would be comprised of a lot of special functions each with separate agendas. This stands in contrast to the principle that—before and after the Sustainability Transformation—the board and the executive suite as a whole shall have the expertise necessary to deal with important matters of the firm.[52] The former is particularly true if sustainable investment is the new mainstream. In other words, the chief operating, investment, risk and technology officers all must understand the implications of sustainability for their business model—*in addition to having solid finance skills*. In creating this expertise, small and large firms face entirely different constraints.

Thus, we propose to encourage creating sustainability expertise on the board as much as the executive level by asking the intermediaries to draft *firm-specific* sustainability development strategies (which may or may not include training and coaching)—yet beyond that, to refrain from details.

Governance

For several years, an expected or perceived short-termism has guided EU policymakers' rulemaking and governance-related rules have been adopted throughout EU securities regulation. Yet, in the absence of clear-cut governance failures in the regulated sector[53] showing deficiency of the fairly new EU

---

[50] See Title IV of the Joint ESMA and EBA Guidelines on the assessment of the suitability of members of the management body and key function holders under Directive 2013/36/EU and Directive 2014/65/EU (CRD IV and MiFID II), at 41 ("Institutions need to provide sufficient resources for induction and training of members of the management body. Receiving induction should make new members familiar with the specificities of the institution's structure, how the institution is embedded in its group structure (where relevant), and business and risk strategy. Ongoing training should aim to improve and keep up to date the qualifications of members of the management body so that at all times the management body collectively meets or exceeds the level that is expected. Ongoing training is a necessity to ensure sufficient knowledge of changes in the relevant legal and regulatory requirements, markets and products, and the institution's structure, business model and risk profile".). Similar provisions requiring induction and training of the governing body can be found in all EU regulations, see for instance Article 21 (d) AIFMD Implementing Regulation (L2),

[51] See Buckley, Arner, Zetzsche et al. 2020.

[52] See Enriques & Zetzsche 2014.

[53] Note that the Wirecard scandal is not evidence to the contrary. Wirecard, under German law, was not regulated at the top level. Further, many important subsidiaries were

governance rules, we encourage the situation to be first assessed prudently *with a focus on understanding the interaction between governance and sustainability factors* prior to regulating in the dark; again.

With regard to sustainability, a further difficulty appears: the link to environmental and supply chain rules[54] all of which are currently under revision by EU institutions. Taking the combination of the various rules—company law, financial law, environmental and other ESG rules—into account, the verdict of "short-termism" cannot be issued without a very careful assessment which cannot start prior to the facts on how legislation just adopted impacts on markets have been gathered and analysed.

Remuneration
Disclosures on the impact of sustainability factors on remuneration policies are required already by the SFDR. In fact, EU financial regulation has a history of tampering with executive pay,[55] following the logic that management will follow financial incentives.

Already in the absence of sustainability concerns, drafting sound remuneration schemes is a (legal) challenge. This challenge does not become easier with sustainability due to a lack of historical data, experience and expertise on all sides concerned, including the board of directors, executives and remuneration consultants.

If management is granted a bonus for a larger stake in sustainable products, we would expect management to shift investments around. Now factor in the uncertainty as to whether sustainable products are profitable and the lack of data which render professional, quantitative investment and risk management a challenge. In extreme cases, if we get it wrong, this could mean unhedged risks and huge losses which if we are unlucky could lead to the failure of the financial institution. These factors together make any cross-sectoral mandatory remuneration requirements regarding sustainability a game too risky to play.

However, we acknowledge one exemption: if an intermediary *organization* (in contrast to a financial product which are already regulated by Article 8 and 9 of the SFDR) frames itself as being sustainable (such as zero carbon) to attract clients, such an organization should penalize managers if in fact the organization does not meet the sustainability factors it has publicly usurped for itself.

---

not regulated. See Langenbucher et al., What are the wider supervisory implications of the Wirecard case? (2020), Study requested by the ECON Committee.

[54] Of course, one could think to rewrite the limited liability rule, a basic principle of company law (or adopt similar radical proposals). Cf. on limited liability in the context of environmental laws Akey & Appel 2020. But the argument against quack legislation aired herein is all the truer for tampering with basic governance features.

[55] See CRD IV, arts. 92–96.

Sustainable Operating Business ?

A sustainability-oriented regulation of the operating business must be handled with even greater care because of the potential impact on the intermediary's operational results, that is profit and cash-flow, and the financial stability issues associated with getting it wrong.

Sustainability-oriented investment and risk models are in their infancy, with multiple questions awaiting answers.[56] In light of the foregoing analysis, investment and risk management is a field where data shortage and a lack of established models come together. With a reliable data trail missing, backward testing is out of the question. Yet, since supervisors who themselves lack data do not know which model is right or wrong, they can hardly impose or assess details of investment portfolio and risk modelling. This renders any demand for a robust sustainability risk assessment and mandatory consideration of these risks[57] less convincing.

At the same time, keeping the status quo does not seem a sensible approach. We hold that a prudent approach would ask financial intermediaries to:

- Consider the risks from unsustainable conduct in their risk management policies *where robust data support the assumption that certain risks could be material*, and
- Develop risk models for these cases *for experimental purposes* for the time being, *to model* the impact of sustainability factors on virtual portfolios.

These two steps would improve sustainability risk management expertise in the financial sector.proportionate to the improvement of the data quality, yet avoid the risk that enhanced risk management requirements enhance model risk in a world where regulators and intermediaries alike know that the data pools are deficient.

Any mandatory requirement, such as the embedding of a sustainability factor into investment strategies and risk models must thus be restricted to cases where the data situation *justifies* such a requirement. For the rest, which at the beginning may form the vast majority, a test-and-learn approach is of the essence, using tools of experimentation and learning rather than mandatory law, with a view to understanding the link between finance and sustainability.

---

[56] See Engle et al. 2020 (researching a model to hedge climate risk, and discussing multiple directions for future research on financial approaches to managing climate risk); Fernando et al. 2017 (distinguishing between environmental risk and "greening" a firm, and arguing that institutional investors shun stocks with high environmental risk exposure, which we show have lower valuations, as predicted by risk management theory. These findings suggest that corporate environmental policies that mitigate environmental risk exposure create shareholder value, while "greening" as such does not).

[57] See ESAs, July letter, supra note 35.

Sustainable Prudential Requirements ?

A crucial part for regulating the operational business concerns the prudential requirements, in particular minimum capital requirements, limitations on financial intermediaries' investment portfolio, liquidity or loan portfolio diversification standards, mandatory insurance for certain risk types, as well as other restrictions intended to limit the type of risks a financial intermediary may undertake.

The complexity of building sophisticated sector-wide prudential rules are well known from the several generations of creating and (re-)shaping the Basel rules (in the EU: the CRR, as well as in a lighter form the IFD). These various editions came with several financial crises in between, so caution when amending capital requirements with a sustainability angle is certainly justified. Consider that these experiences have been gathered in light of near-to-complete financial datasets. The same efforts for sustainability factors where we lack these datasets add up to an insurmountable risk. This is particularly true in light of Mark Carney's concept titled "Tragedy of the Horizon" whereby financial institutions bear the costs of implementation today while benefits accrue to future generations of clients. How to allocate costs and set incentives in such an environment does not come easy.

How to tackle the challenge? (1) Against the learned and well-reasoned opinion of influential commentators,[58] we encourage regulators to refrain from any detailed sector-wide CRR-style rule tied to model results for any type of financial services for now. (2) At the same time, regulators should ask financial intermediaries to develop their own models to the extent that the data quality on certain sectors allows it, and capitalize risks on a case-by-case basis. Such model development should be both supported and scrutinized by standard setters and financial services authorities, so that the limits of these models as well as how one model outcome compares to another is well understood. For instance, an insurance company covering storm risks should slowly but surely increase risk provisioning; a bank engaging with clients whose main business relates to oil shall take into account the effects of environmental legislation and taxation on their clients' business; and an asset manager investing in real estate in the Maldives shall consider the rising sea level. All this factoring in of sustainability risk is necessary and sound, and already provided for in Article 3 SFDR as well as sectoral risk management rules. Regulators seeking to avoid the large-scale impact of sustainability risks on financial institutions[59] have already quite a strong position utilizing existing risk management rules.[60]

Beyond that, regulators shall refrain from tying mandatory capital surcharges to unsustainable products and services for the time being; this is

---

[58] See the discussion in Alexander & Fisher 2019, 15–20 (arguing that sustainability risks, collectively, are of systemic dimensions); see also Alexander 2014, 16–17.

[59] See Alexander & Fisher (2019), 7–34 (arguing that sustainability risks, collectively, are of systemic dimensions).

[60] See Kivisaari (2021), 75, 88–91, 98–100. See also Alexander (2014).

not a sign of disrespect of the importance of sustainability, but rather—in line with a market-based transformation—markets first need time to figure out how to combine sustainability and profitability. As was pointed out above (III.) it will take some years to understand that nexus properly; drafting rules prior to understanding this nexus represents regulatory hazard.

Towards a Smart Regulation of Sustainable Finance

The common conclusion of our view on the three fields—intermediary set-up, operations and prudential rules—is that EU regulators shall, first, implement the taxonomy across sectors, second, ensure reporting based on that taxonomy, third, collect data (and ensure data platforms, comparability, etc.), fourth, assess data with some representative time series, and *finally*, draft rules and standards on the organization, operations and prudential requirements of intermediaries. Until that date, regulators shall not sit idle, but are best advised to further a "test-and-learn" approach across the financial industry with regards to all aspects of sustainability risks and impacts of finance on sustainability factors.

With regard to furthering experimentation, regulating sustainability innovation is not entirely different from regulating other innovations, such as financial and regulatory technologies (FinTech and RegTech). In the context of FinTech and RegTech, the need for "smart regulation" is recognized,[61] with smart regulation being defined as a regulatory approach where regulators retain openness to innovation, engage in learning through constant exchange with innovators, while keeping risks under control through tailor-made, case-by-case decisions following intense scrutiny of new business models in regulatory sandboxes.[62] This is particularly true given that the level of uncertainty and the dynamics of change/progress are similarly pronounced in the area of sustainable investing and fintech. In turn, the regulatory challenges seem, to some extent, to be quite comparable. Waiver programmes, sustainability innovation hubs, regulatory sandboxes and partnerships between financial intermediaries (as experts in finance) and sustainability research centres could work particularly well for certain sustainability matters where regulators and intermediaries lack experience, including risk models, remuneration schemes and portfolio composition.

Such tools could be implemented to benefit early adopters of sustainable finance modelling, under the condition that the elements underlying the models are made available to the public to incentivize the experts' discussion. In a similar vein, regulators could grant some leeway for various modes of

---

[61] Cf. Zetzsche et al. 2017 (coining the term "Smart Regulation").

[62] Cf. Zetzsche et al. 2017 (coining the term "Smart Regulation"); Buckley, Arner, Veidt et al. 2020; Zetzsche et al. 2020; Brummer & Yadav (2019), 248–249 (arguing that innovation poses a challenge for regulators since regulators are expected to warrant financial innovation, simple rules and market integrity at the same time, with limited resources). But see Omarova, JFR 2020 ___ (criticizing a smart regulation approach).

portfolio compositions in terms of investment limits or the provision of new types of sustainability data (previously undisclosed), and even grant prudential benefits for firms that come up with innovative, theoretically grounded models of sustainability risk, as long as these models are being made available to the public to ensure sound discussion among experts.

While many of these approaches will not stand the test of time, the more experts discuss approaches and the more data are being reported which may be included in modelling, the faster we expect the data gap to being closed, risk, governance and remuneration models to be developed, and the consensus to be established that is necessary to make sustainable finance the new mainstream.

7 Conclusion

Sustainable investments are of paramount importance to ensure the sustainability transformation of the European economy. Yet, at the moment we lack in some respects data, in other respects broadly acknowledged theoretical insights (typically laid down in standard models) on the co-relation and causation of sustainability factors with financial data, and in a third respect a consistent application of recently adopted sustainability disclosure rules. The three together hinder as of now a rational, calculated approach to allocating funds with a view to sustainability which we usually associate with "finance". With regard to the nexus of financial and sustainability factors a rational, truly data-driven approach to investing is at its infancy.

While the Taxonomy Regulation's definition of sustainable investments creates legal certainty and can lead to the comparability of sustainability-related disclosures, the implementation of the taxonomy resulting in valuable datasets necessary for empirical assessment and financial modelling will require years. Prior to the availability of these datasets, financial market participants, regulators and investors are subject to transition risk at an enormous scale, given that much of the sustainability agenda within the EU financial markets stands on hollow ground, meaning its regulatory premises are not data-driven, but rather policy-driven. Even in the best of all possible scenarios, the full absorption of the taxonomy in data creation, financial modelling, testing and transposition in lending and investment strategies will take years.

Yet, the same data shortage that has hindered investors to assess and identify sustainable *and* profitable products also prevents financial supervisory authorities from applying a prudent *mandatory* regulatory strategy: If a regulator cannot identify a conduct as "right", that is where regulators effectively fly in the dark, and it is unwise to prohibit certain other conduct by naming it "wrong", as the latter would reduce the options for diversification and increase the risk of unwanted effects.

A truly smart sustainability-oriented securities regulation must aim at generating data and expertise on sustainability factors and sustainable products on the side of both regulators and financial intermediaries, in an effort to prepare the ground for a mature and profitable sustainable investment market. Because a large-scale, long-term *unprofitable* sustainable investment environment is in itself unsustainable.

References

Alexander K (2014) Stability and Sustainability in Banking Reform: Are Environmental Risks Missing in Basel III? Cambridge/UNEP October 2014.

Alexander K & Fisher P (2019) Banking Regulation and Sustainability. In: van den Boezem F.-J. B., C. Jansen & B. Schuijling (eds.) Sustainability and Financial Markets. Wolters Kluwer.

Addoum JM, Ng DT & Ortiz-Bobea A (2020) Temperature Shocks and Establishment Sales. Rev. Fin. St. (33:1331). https://doi.org/10.1093/rfs/hhz126.

Akey P & Appel I (2020) The Limits of Limited Liability: Evidence from Industrial Pollution. J. Finance (early view, September 2020). https://doi.org/10.1111/jofi.12978.

Alok S, Kumar N and Wermers R (2020) Do Fund Managers Misestimate Climatic Disaster Risk. Rev. Fin. St. (33:1146). https://doi.org/10.1093/rfs/hhz143.

Balachandran B & Nguyen JH (2018) Does carbon risk matter in firm dividend policy? Evidence from a quasi-natural experiment in an imputation environment. J. Bank. Fin. (96:249). https://doi.org/10.1016/j.jbankfin.2018.09.015.

Baldauf M, Garlappi L & Yannelis C (2020) Does Climate Change Affect Real Estate Prices? Only If You Believe In It. Rev. Fin. St. (33:1256). https://doi.org/10.1093/rfs/hhz073.

Balvers R, Du D & Zhao Xiaobing (2017) Temperature shocks and the cost of equity capital: Implications for climate change perceptions. J. Bank. Fin. (77:18). https://doi.org/10.1016/j.jbankfin.2016.12.013.

Barber BM, Morse A & Yasuda A (2021) Impact investing. J. Fin. Econ. (139:162). https://doi.org/10.1016/j.jfineco.2020.07.008.

Barnett M, Brock W & Hansen LP (2020) Pricing Uncertainty Induced by Climate Change. Rev. Fin. St. (33:1024). https://doi.org/10.1093/rfs/hhz144

Bender J, Bridges TA & Shah K (2019) Reinventing climate investing: building equity portfolios for climate risk mitigation and adaptation. Journal of Sustainable Finance & Investment (9(3)).

Berg F, Koelbel JF & Rigobon R (2020) Aggregate Confusion: The Divergence of ESG Ratings. MIT Sloan School, Working Paper 5822–19, May 2020. https://papers.ssrn.com/sol3/papers.cfm?abstract_id=3438533.

Berg F et al. (2020) Rewriting History II: The (Un)predictable Past of ESG Ratings. ECGI, Finance Working Paper N° 708/2020, November 2020. http://ssrn.com/abstract_id=3722087.

Bernstein A, Gustafson MT & Lewis R (2019) Disaster on the horizon: The price effect of sea level rise. J. Fin. Econ. (134:253). https://doi.org/10.1016/j.jfineco.2019.03.013.

Brummer C & Yadav Y (2019) Fintech and the Innovation Trilemma. Geo. L. J. (107:235).

Buckley RP, Arner DW, Veidt R & Zetzsche DA (2020) Building FinTech Ecosystems: Regulatory Sandboxes, Innovation Hubs and Beyond. Wash. U. J. L. & Pol'y (61:55).

Buckley RP, Arner DW, Zetzsche DA & Selga E (2020) Techrisk Sing JLS (35).
Busch D (2020) Sustainability Disclosure in the EU Financial Sector. European Banking Institute Working Paper Series 70/2020. https://ssrn.com/abstract=3650407
Busch T et al. (2016) Sustainable Development and Financial Markets: Old Paths and New Avenues. Bus. Soc. (5:303).
Chiu IHY (2021) The EU Sustainable Finance Agenda- Developing Governance for Double Materiality in Sustainability Metrics. __ EBOR __
Choi D, Gao Z & Jiang W (2020) Attention to Global Warming. Rev. Fin. St. (33:1112). https://doi.org/10.1093/rfs/hhz086.
Chowdhry B, Davies SW & Waters B (2018) Investing for Impact. Rev. Fin. St. (32:864). https://doi.org/10.1093/rfs/hhy068.
Colacito R, Hoffmann B & Phan T (2019) Temperature and Growth: A Panel Analysis of the United States. J. Money Credit Bank (51:313). https://doi.org/10.1111/jmcb.12574.
Cunha FAFDS et al. (2019) Can sustainable investments outperform traditional benchmarks? Evidence from global stock markets. Bus Strategy Environ (29:682).
Doni F & Johannsdottir, L (2019) Environmental Social and Governance (ESG) Ratings. In: Filho WL, Azul AM, Brandli L, Özuyar PG & Tony Wall (eds) Climate Action. Encyclopedia of the UN Sustainable Development Goals, Springer, Cham, pp 435–449. https://doi.org/10.1007/978-3-319-95885-9_36.
Dorfleitner G, Halbritter G & Nguyen M (2015) Measuring the level and risk of corporate responsibility – An empirical comparison of different ESG rating approaches. J. Asset Mgmt. (16:450). https://doi.org/10.1057/jam.2015.31.
Eichholtz P, Holtermans R, Kok N & Yönder E (2019) Environmental performance and the cost of debt: Evidence from commercial mortgages and REIT bonds. J. Bank. Financ. (102:19). https://doi.org/10.1016/j.jbankfin.2019.02.015.
El Ghoul S & Karoui A (2017) Does Corporate Social Responsibility Affect Mutual Fund Performance and Flows? J. Bank. Fin. (77:53).
Engle RF et al. (2020) Hedging Climate Change News. Rev. Fin. St. (33:1184). https://doi.org/10.1093/rfs/hhz072.
Enriques L & Gilotta S (2015) Disclosures and Market Regulation. In: Moloney N, Ferran E & Payne J, *Introduction*. In: The Oxford Handbook of Financial Regulation (OUP: 2015, paperback 2017).
Enriques L & Ringe W-G. Bank–fintech partnerships, outsourcing arrangements and the case for a mentorship regime. Cap. Markets L.J. (15:374).
Enriques L & Zetzsche D (2014) Quack Corporate Governance, Round III? Bank Board Regulation Under the New European Capital Requirement Directive. Theoretical Inquiries in Law (16(1):211).
Fernando CS, Sharfman MP & Uysal VB (2017) Corporate Environmental Policy and Shareholder Value: Following the Smart Money. J. Fin. Quant. (52:2023). https://doi.org/10.1017/S0022109017000680.
Filbeck G, Krause TA & Reis L (2016) Socially responsible investing in hedge funds J. Asset. Mgmt. (17:408).
Fiskerstrand SR et al. (2019) Sustainable investments in the Norwegian stock market. J. Sustainable Fin. & Invest. (10:294).
Friede G (2020) Why don't we see more action? A metasynthesis of the investor impediments to integrate environmental, social, and governance factors. Bus. Strategy Environ. (28:1260). https://doi.org/10.1002/bse.2346.

Friede G, Busch T & Bassen A. ESG and financial performance: aggregated evidence from more than 2000 empirical studies. J. Sustainable Fin. & Invest. (5:210).

Gentzoglanis A (2019). Corporate social responsibility and financial networks as a surrogate for regulation. Journal of Sustainable Finance & Investment (9(3):214–225).

Gutsche G & Ziegler A (2019) Which private investors are willing to pay for sustainable investments? Empirical evidence from stated choice experiments. J. Bank. Fin. (102:193). https://doi.org/10.1016/j.jbankfin.2019.03.007.

Hartzmark SM & Sussman AB (2019) Do Investors Value Sustainability? A Natural Experiment Examining Ranking and Fund Flows. J. Fin. (74:2789). https://doi.org/10.1111/jofi.12841.

Hong H, Karolyi GA & Scheinkman JA (2020) Climate Finance. Rev. Fin. St. (33:1011). https://doi.org/10.1093/rfs/hhz146.

Hooghiemstra SN (2020) The ESG Disclosure Regulation – New Duties for Financial Market Participants & Financial Advisers. https://ssrn.com/abstract=3558868.

Huynh TD & Xia Y (2020) Climate Change News Risk and Corporate Bond Returns. J. Fin. Quant. (Sep. 2020, in press). https://doi.org/10.1017/S002210902000757.

Ilhan E, Sautner Z & Vilkov G (2021) Carbon Tail Risk. Rev. Fin. St. (34(3):1540–1571). https://doi.org/10.1093/rfs/hhaa071.

Jebe (2019) The Convergence of Financial and ESG Materiality: Taking Sustainability Mainstream. Am. Bus. L. J. (56:645).

Joliet R & Titova Y (2018) Equity SRI funds vacillate between ethics and money: An analysis of the funds' stock holding decisions. J. Bank. Fin. (97:70). https://doi.org/10.1016/j.jbankfin.2018.09.011.

Kivisaari E (2021) Sustainable Finance and Prudential Regulation of Financial Institutions. In: Fisher G (ed.) Marketing the Financial System Sustainable, CUP.

Krueger P, Sautner Z & Starks LT (2020) The Importance of Climate Risks for Institutional Investors Rev. Fin. St. (33:1067). https://doi.org/10.1093/rfs/hhz137.

Larcker DF & Watts EM (2020) Where's the greenium? J. Acct. & Econ. (69:101312). https://doi.org/10.1016/j.jacceco.2020.101312.

Liesen A et al. (2017) Climate Change and Asset Prices: Are Corporate Carbon Disclosure and Performance Priced Appropriately?. J. Bus. Fin. Acct. (44:35).

Mancini M (2020) Nudging the Financial System: A Network Analysis Approach. UNEP Inquiry and FC4S, April 2020.

Moloney N, Ferran E & Payne J (2015) Introduction. In: Moloney N, Ferran E & Payne J. The Oxford Handbook of Financial Regulation (OUP: 2015, paperback 2017).

Murfin J & Spiegel M (2020) Is the Risk of Sea Level Rise Capitalized in Residential Real Estate? Rev. Fin. St. (33:1217). https://doi.org/10.1093/rfs/hhz134.

OECD (2020) Integrating ESG factors in the investment decision-making process of institutional investors, chapter 4. In OECD Business and Finance Outlook 2020: Sustainable and Resilient Finance, OECD Publishing, Paris. https://doi.org/10.1787/eb61fd29-en.

Painter M (2020) An inconvenient cost: The effects of climate change on municipal bonds. J. Fin. Econ. (135:468). https://doi.org/10.1016/j.jfineco.2019.06.006.

Pedersen LH, Fitzgibbons S & Pomorski L (2020) Responsible investing: The ESG-efficient frontier. J. Financ. Econ. (November 2020, in press). https://doi.org/10.1016/j.jfineco.2020.11.001.

Riedl A & Smeets P (2017) Why Do Investors Hold Socially Responsible Mutual Funds. J. Fin. (72:2505).

Rockström J et al. (2009) Planetary Boundaries: Exploring the Safe Operating Space for Humanity. Ecology and Society (14).

Romano R. Regulating in the Dark and a Postscript Assessment of the Iron Law of Financial Regulation. Hofstra L.R. (43:1).

Rossi M et al. (2019) Household preferences for socially responsible investments. J. Bank. Fin. (105:107). https://doi.org/10.1016/j.jbankfin.2019.05.018.

Shive SA & Forster MM (2020) Corporate Governance and Pollution Externalities of Public and Private Firms. Rev. Fin. St. (33:1296). https://doi.org/10.1093/rfs/hhz079.

Steffen W et al. (2015) Planetary Boundaries: Guiding Human Development on a Changing Planet. Science (347).

Thaler R & Sunstein C (2008). Nudge – Improving Decisions about Health, Wealth and Happiness, Yale University Press, New Haven.

Zetzsche DA, Anker-Sørensen, LC (2021) Regulating Sustainable Finance in the Dark, EBOR ___.

Zetzsche DA, Arner DW, RP Buckley, Kaiser-Yücel A (2020) Fintech Toolkit: Smart Regulatory and Market Approaches to Financial Technology Innovation. Available at SSRN: https://ssrn.com/abstract=3598142.

Zetzsche DA, Buckley RP, Arner DW & Barberis JN (2017) Regulating a Revolution: From Regulatory Sandboxes to Smart Regulation. Ford. J. Corp. & Fin. L. (23:31).

CHAPTER 5

The Duty of Societal Responsibility and Learning Anxiety

Jaap Winter

1 Introduction

In previous publications I have proposed to introduce a duty of societal responsibility of the board of a corporation, to correct the amorality of corporate reality that results from the shareholder primacy doctrine first formulated by Milton Friedman.[1] "In a free-enterprise, private-property system," Friedman wrote in a seminal article in the New York Times Magazine in 1970, "a corporate executive is an employee of the owners of the business. He has direct responsibility to his employers. That responsibility is to conduct the business in accordance with their desires, which generally will be to make as much money as possible while conforming to the basic rules of the society, both those embodied in law and those embodied in ethical custom... there is one and only one social responsibility of business, to use its resources and engage in activities designed to increase its profits so long as it stays within the rules of the game." Although Friedman may not have intended to free the corporation of all moral responsibility (Hess, 2017), nonetheless his theory has been understood, elaborated upon and practiced to generate a corporate reality

[1] See my paper at SSRN, Addressing the Crisis of the Modern Corporation: The Duty of Societal Responsibility, see https://ssrn.com/abstract=3574681 and Towards a Duty of Societal Responsibility of the Board, European Company Law October 2020, vol 17, nr 5, pp. 192–200.

J. Winter (✉)
Vrije Universiteit Amsterdam, Amsterdam, Netherlands
e-mail: Jaap.winter@phyleon.com

© The Author(s), under exclusive license to Springer Nature Switzerland AG 2022
P. Câmara and F. Morais (eds.), *The Palgrave Handbook of ESG and Corporate Governance*, https://doi.org/10.1007/978-3-030-99468-6_5

that the corporation is largely responsibility-free beyond the responsibility to create value to shareholders. The costs of shareholder value creating actions to others, such as employees, customers and wider society (e.g., environmental and climate costs) are of no concern to the corporation itself. In the hands of the capital markets, investment banks, institutional investors, hedge funds and heavily incentivized executives corporations have become, if not immoral, at least amoral. What matters in corporate decision-making is the prospective financial outcome for shareholders, not much else. I oversimplify, I am aware, but I am afraid I am not oversimplifying too much.

But if ever there was a time that society needs business to take responsibility beyond its own immediate financial success, that time is now. The ecological and social crises we are confronted with today require something very different from corporations. We can no longer afford corporations to be amoral. The challenge is to ignite a sense of societal responsibility within corporations. This can only come from the people who make the corporation an actor in society, its directors. I have proposed that a specific duty of societal responsibility should be created for the board of directors of the corporation.[2] This proposal has met fierce resistance from the business community. The plan of the European Commission to specify that the duty of care of directors is to include taking into account sustainability matters, including human rights, climate change and the environment, see art. 25 of the proposed Corporate Sustainability Due Diligence directive, has met a similar fate. Paradoxically, at the same time corporations are becoming more and more vocal about their endeavors to contribute to society. How can we explain that, while more and more corporations seem to be taking their societal responsibility seriously, they nonetheless fiercely resist making this responsibility explicit? In this contribution I will first put forward my version of the analysis of the problem of responsibility of business in society, par. 2. A critical factor is that the amoral corporate context is man-made but often we believe there is little we can do about it. Proposals are made to generate a change in this reality (par. 3) In all solutions that have been put forward so far to address aspects of the problem the role of the board is crucial. But the proposals do not address adequately what is needed for the board to commit the corporation to be a responsible corporate citizen in society. A specific duty for boards to do so could be introduced in corporate law, par. 4. I will then speculate about the source of the resistance against such a duty by introducing the concept of learning anxiety as developed by Edgar Schein, and suggest some ways of dealing with this anxiety, par. 5.

---

[2] See the previous note and for a proposal specifically for Dutch company law Jaap Winter et. al., Naar een zorgplicht voor bestuurders en commissarissen tot verantwoorde deelname aan het maatschappelijk verkeer, Ondernemingsrecht 2020, 86.

2 THE HEART OF THE MATTER: THE MODERN CORPORATION HAS BECOME AMORAL

In a recent paper (Winter, 2019a) I have set out a number of factors that have contributed to a context of corporations that I would describe as dehumanized, stripped of human inspiration, meaning and judgment, disengaged from society, from what the people that make up society believe is important. My analysis is by no means complete. It is focused on the human aspects of the context of the modern corporation, both how this context is created by ourselves and how it affects us. This is essential. The core of the problem is not outside of us, it is in us. It is in how we design our corporate reality and how we conduct ourselves within it. By the combination of design and conduct we shape the context in which corporations exist and act in society. Solutions therefore are also not beyond us, but within our reach, they are about us.

Let me summarize my argument of how the corporation has become amoral. Five factors jointly and interactively cause concerns about the state of the modern corporation: (i) the dominant theory of the firm, (ii) the way capital markets and institutional investors have come to view corporations, (iii) the efficiency-driven organizational practice, (iv) the effects of regulation and (v) the remuneration of executives.

(i) The shareholder primacy theory of the firm that originated from Friedman's work has come to dominate thought and practice on the corporation and inside it. In thought, it was developed into the agency theory: managers are primarily seen as agents who should act in the interest of shareholder as their principals. As managers act rationally in their self-interest, they will not always further the interests of shareholders. Corporate law and corporate governance have the function to ensure that managers do act in the interests of shareholders. In practice, executive remuneration and takeover bids have been developed as mechanisms that discipline managers to do so. Executive remuneration aligns managers' interests with those of shareholders. Takeover bids may force managers out of their jobs if they underperform for shareholders. Other stakeholders who have interests in the corporation and its business, such as employees, customers and the wider community, are no longer seen as relevant principals for whom managers should act as agents. Externalities such as the costs of environmental and climate damage or the social damage caused by restructuring are not factored in this shareholder value model. Put in a different way: in today's dominant corporate context only financial capital is rewarded, at the expense of human capital (well-being of employees), social capital (public goods, social infrastructure, community) and natural capital (resources, environment and climate) (Roche, Jakub, 2017; Mayer, 2018).

(ii) Building on this theory of the firm, capital markets and their main players over time have come to view corporations as bundles of assets and liabilities. They represent separate units of financial value that can be realized at specific prices. Listed corporations do not *have* a balance sheet, they *are* a balance sheet with assets and liabilities that can be realized in individual transactions. The financial value of those individual transactions, the price somebody apparently is willing to pay, is justification for breaking down the corporation as an integrated entity that in myriad ways is connected to society. Derivative financial instruments further remove investors from the reality of the corporation as an organization where real people work and that is connected to society through a multiplicity of parties and relationships. In financial capitalism, investors are only interested in the financial value that emerges from these instruments. Institutional investors who excessively diversify their portfolios are no longer investing in individual corporations but search for mathematically calculated absolute and relative returns. Intermediation of the whole investment process removes investors even further from the corporations they invest in. Financial capital is uncommitted to the corporations in which it invests (Mayer, 2013).

(iii) The world of modern management as taught in business schools and practiced in corporations is characterized by its endless search for efficiency improvement and cost savings as a means of maximizing shareholder value. Employees are a cost factor and staff reductions maximize shareholder value. Employees that stay are directed through key performance indicators, control systems, target setting and measuring and compliance e-tools to ensure they perform and behave according to plan. A plan that is overly based on illusions and scientific pretensions of certainty, while in reality corporations are constantly faced with uncertainty and unpredictability. The dominant organizational pattern is mechanistic, steering towards predetermined objectives and controlling employees through formalistic accountability procedures in bureaucratic processes. The capacity for human judgment of employees is no longer utilized.

(iv) The managerial control bureaucracy is augmented with more and more external regulation defining what corporations can, must and may not do. After every new crisis, new rules give a false sense of control over a reality that was never under control before the crisis hit, with more and more external supervision through regulatory authorities. The paradoxical consequence has been to reduce the sense of responsibility that people have from responsibility for the consequences of their behavior towards others to a responsibility to comply with the rules. We actually feel less responsible and no longer train our moral muscle, a regulatory crowding-out effect. This has become what I call a Perfect System

Syndrome, taken from the words of T.S. Eliot in his poem the Rock from 1934:

> They constantly try to escape,
>   From the darkness outside and within,
>   By dreaming of systems so perfect that no one will need to be good.

(v) Since the mid-1990s, executive remuneration has been developed on the back of the agency theory. Significant short- and long-term incentives seek to align the interests of executives with the interests of shareholders. They produce their own crowding-out effect: executives start to believe they are working hard for the extrinsic motivation of financial reward, rather than intrinsic motivation to do well and act pro-socially. It is no longer in the personal interests of executives to consider any other interests than the shareholders' as relevant for their decision-making.

These factors combined may have devastating results, as we see in the numerous cases of environmental degradation and neglect of employees, customers and wider communities by corporations. A common theme through these factors is an excess of rationalization and system building, in an attempt to control reality, with a view to ensuring one particular outcome, value for shareholders. Such formal rationalization through bureaucracies, systems and processes comes at the cost of human values, as Max Weber already explained some 100 years ago. Weber, one of the founding fathers of sociology, points at one specific factor that is crucial. In a formally rational economy "decisive are the need for competitive survival and the conditions of labor, money and commodity markets; hence matter-of-fact considerations that are simply *nonethical* determine individual behavior and impose *impersonal* forces between the persons involved." Weber believed this formal rationality would come to overtake substantive rationality, in which choices are guided by some larger system of human values. As a result, we were to be left in the modern world with people who simply followed the rules without regard to larger human values (Weber, 1921/1968). Later Erich Fromm would describe the process in different words. Fromm was a German psychoanalyst who moved to the United States in the 1930s. In 1955 he described modern American society from a psychoanalytic perspective in his book The Sane Society. In it, Fromm writes about alienation and conformity. Alienation for Fromm is related to idolatry, man building an idol that he then worships as if it is outside of him. The idol typically is a projection of one partial quality in man himself, which then stands over and above him. Man becomes a servant of a Golem, which his own hands have built. "The very fact that we are governed by laws [of the market, jw] which we do not control, and do not even want to control, is one of the most outstanding manifestations of alienation. *We* are the producers of our economic and social arrangements, and at the same time we decline responsibility.." Authority is no longer overt, but anonymous, invisible, alienated authority. "The mechanism through which anonymous authority operates is conformity. I ought to do what everybody does, hence, I must conform, not be different.. I must not ask whether I am right or wrong, but whether I am adjusted, whether I am not peculiar, not different." (Fromm, 1956).

Weber's and Fromm's descriptions preceded the work of Friedman and the way shareholder primacy has come to dominate the corporate context. But it is striking to see how their insights precisely describe the transformation of the corporate context since the 1970s. In this corporate context, the sole objective is to generate value for shareholders, other considerations are not relevant. As academics, we theorize, research and teach about aspects of the corporate context from this perspective. Within corporations, we steer and organize to maximize efficiency towards this objective, through rational mechanisms that create impersonal forces between the persons involved. Through executive remuneration with substantial short- and long-term incentives executives have a personal interest in maintaining the system. We have made the financial value-maximizing corporate context an idol, something to cherish and worship that we benefit from by worshiping it. In the process, we have started to conform to a corporate context created by ourselves as if this is beyond us, for which we bear no responsibility and which cannot be contested. We have thus successfully alienated ourselves (in the sense of worshiping and not taking responsibility) from the corporate context we have created. The corporation has no responsibility beyond its objective to generate value for shareholders, and neither do we, who can only live with this matter-of-fact, anonymous and impersonal law of the market. And so the corporation and its inhabitants have become amoral. Moral responsibility, in the sense of having responsibility for the consequences of one's behavior towards others, has become an externality itself: it only is meaningful and relevant for and within the corporation to the extent the external world imposes it on the corporation and its inhabitants through laws or other moral codes, as Friedman's quote in the introduction shows. Our ability to conform to this perceived corporate reality as a matter of fact is amazing and even frightening.

3   Proposals for Change

Various proposals and attempts in practice are made to address the corporate context that I have presented above. One direction of these proposals is to reduce shareholder rights so that they cannot effectively discipline directors into actions that favor shareholder wealth. Facilitating takeover defenses (opt-out/opt-in regime of article 12 EU Takeover Bids Directive, enthusiastically implemented by many member states) and restricting shareholder rights in calling shareholders meetings and in corporate litigation are examples of this approach. This direction touches upon a core weakness of corporate governance of the corporation that is listed on a stock exchange. Managers of firms that are not effectively controlled by shareholders (or by others) have wide discretion to act in their own interests and are accountable to none. This is partly what the Friedman paradigm and the agency theory sought to address. Taking away shareholder discipline by reducing their rights and not replacing it with something else, leaves corporations with effectively non-accountable directors. Also, this approach does nothing to actively further including ecological and social matters as part of the corporation's responsibility. Other types of amendments to for example the Takeover Bids Directive would be needed to at least move in the right direction (Winter, 2021).

A second direction is the trend to formulate a wider corporate purpose that includes addressing the needs of society. Benefit Corporations that are being facilitated in a growing number of jurisdictions are part of this trend. The development of B corps is still early days, its future expansion and success remain to be seen (Dorff, Hicks, Solomon, 2020). A strong purpose orientation takes into account not only financial capital provided by shareholders but also human capital provided by employees, social capital in the communities in which the corporation operates and natural capital in the form of natural resources and the impact on environment and climate. Corporations with such a purpose orientation will account for their impact on these various sources of capital (Mayer, 2018). Weaker forms of purpose orientations only seek to mitigate the absoluteness of shareholder primacy, for example by stressing the long term orientation of the corporation (e.g., the Dutch corporate governance code 2016) or by applying an enlightened shareholder approach that allows for serving the interests of other stakeholders if this furthers the creation of shareholder value (sec 172 UK Companies Act 2006). For some this would be enough to strive for (Ferrarini, 2020).

A third direction is through the involvement of other stakeholders. In Europe employees in some member states for decades have enjoyed co-determination rights. The results are mixed and employee co-determination does not address concerns of social capital and natural capital. The multi-stakeholder cooperative may become an altogether new corporate form that includes a wider community of different categories of stakeholders in the governance of the cooperative. Multi-stakeholder cooperatives may become particularly relevant in the digital platform economy. Modern digital information and communication technology, artificial intelligence, 3-D printing and the internet of things allow many to connect to many others, to produce and provide service at ever lower costs and to share assets and resources with others. An economy of collaborative commons may to a large extent replace social-economic activity that is now organized through a classic market economy (Rifkin, 2015; also Zizek, 2017). In such an economy the multi-stakeholder cooperative is a natural legal form to organize and govern the cooperation between multiple parties involved.

These directions for change are positive signs that we start to wrestle ourselves loose from the impersonal forces that hold us, that make us conform to a corporate reality that we no longer want. Signs that we start to take responsibility for the corporate context we have created ourselves. This is all the more so when proposals come from those who inhabit the corporate context themselves, business leaders and others who work with and within corporations and academics who theorize, research and teach about this context.

The Board and the Duty of Societal Responsibility

In each of these three approaches to address the problem of the responsibility of the corporation the role of the board[3] is crucial. Without the commitment of the board to direct the corporation towards achieving a purpose that benefits society and our planet, it is unlikely that the corporation will indeed do so.

This brings us to the core question that is not answered in the proposed approaches: how do we ensure that the board is indeed committed to a broader responsibility of the corporation beyond merely serving the financial interests of shareholders? Mayer stresses the need for commitment but believes this commitment can only be generated when the corporation is owned by large, long-term-oriented shareholders. "United Kingdom, in particular, with its dispersed ownership, is dominated by short-term shareholders rather than long-term shareowners, and is therefore a low commitment economy... Where corporations do not or are unable to offer the level of commitments that society demands of them, then it resorts to prescriptive regulation in place of permissive enabling legislation" (Mayer, 2018).

But why should we stop there? Why should we accept that when a corporation only has uncommitted shareholders, external, prescriptive regulation will have to bring about the level of responsibility that society expects of corporations? In such a view, societal responsibility of the corporation with dispersed owners indeed remains an externality: it only exists to the extent external regulation imposes it on the corporation and its board, as per the Friedman doctrine. We can do better, by turning to ourselves. The core of the problem of the amoral corporation is a human failure of succumbing to self-made formal rationalization and alienation. This human failure can only be conquered by addressing who we as humans are in the corporate context. We need to look at the board of the corporation, the people who make up the board, as its prime agent. It is people who act on behalf of the corporation, who allow for it to participate in economic activity (Rönnegard, Velasquez, 2017). Without people the corporation cannot participate in society.[4] The quality of taking moral responsibility, i.e., responsibility for the consequences

---

[3] I refer to the board in this article as to include both the singular board in the one-tier model as well as both boards in a two-tier model.

[4] More and more corporations' actions can be generated by computers and artificial intelligence. This leads to the human challenge of how to equip computers and artificial intelligence with a sense of human morality, of other-regard (Tegmark, 2017; Russell, 2019). This should press us to pro-actively ensure that corporations assume human responsibility. If we cannot have human morality determine what corporations do when and as long as they still require human agency, we are even worse off when corporate conduct

of one's acts, is a human quality (Comte-Sponville, 2009). It requires empathy, the ability to imagine what others would experience, and more broadly human emotions such as remorse and guilt. In order for societal responsibility to come from within the corporation, the responsibility of the people who are its agents needs to be activated. The board is legally responsible for the actions of the corporation.[5] It takes its core decisions and directs the people and organization of the corporation. The board is the key body that should consider the interests of different stakeholders and the uses of different forms of capital. In the board also rests the authority within the corporation to ensure that others who act on its behalf are guided by the principles and values determined by the board. Corporate law could formulate a duty of the board and the directors to ensure that the corporation acts responsibly with a view to the interests of society and the way it uses financial, human, social and natural capital. I would call this the duty of societal responsibility, a duty to ensure that the corporation acts as a responsible corporate citizen in society. I take these latter words from the King IV Corporate Governance Report of South Africa (2016), principle 3: "the governing body should ensure that the organization is and is seen to be a responsible corporate citizen."[6] If this would not just be an aspirational statement, but a firm duty of the board and directors, this would generate a commitment from within the corporation to drive the corporation towards fulfilling the needs of society.

Based on these thoughts a group of Dutch law professors has proposed to introduce an element of this societal responsibility in the duty of executive and non-executive directors as described under Dutch law. The current duty of executive and non-executive directors is to act in the interests of the corporation and its enterprise, suggesting a broader stakeholder approach. In practice for many listed and private equity-owned corporations this very general statement is implemented in particular to further financial results that are in the interests of shareholders. This is generally accepted under Dutch law, which would only object if in the margins interests of others who are connected to the corporation and its organization are unreasonably prejudiced (Dutch Supreme Court in the Cancun case of 2014).[7] The proposal is to extend the duty of executive and non-executive directors to ensure and monitor that the corporation acts as a responsible corporate citizen. This duty is a general duty for directors of all types of corporations.

---

starts to depend more and more on computers and artificial intelligence, see also Armour, Eidenmueller (2019).

[5] This does not mean that board members therefore face personal liability for everything the corporation is doing. Liability may follow from such responsibility, but typically follows only if certain threshold conditions have been met.

[6] See for the full code https://www.adams/africa/wp-content/uploads/2016/11/King-IV-Report.pdf.

[7] HR April 4, 2014, ECLI:NL:PHR:2013:1826.

The proposal leads to fierce debate in the Netherlands. Some argue that there is no need for this duty as it is already included in the current duty of acting in the interest of the corporation and its enterprise. If this would be the case then directors of a number of listed and private equity-owned corporations in the Netherlands would be breaching their duty, when focusing on financial results that benefit shareholders only. Others argue the responsibility of directors for the conduct of a corporation in society should not be formalized in law, because it could trigger litigation by third parties that would stifle economic risk-taking by corporations. If this were true, it would refute the first claim that the duty of societal responsibility is already included in the current formulation of directors' duties and practiced in reality. As to the risk of litigation, we have suggested that directors under the proposal would have the normal discretion to make judgments as they also have under their current duty. This typically only leads to personal liability of directors if and to the extent a severe personal reproach can be made to them, if no reasonable director in the given circumstances would have taken the contested decision or action. Those who oppose the proposal to explicitly formulate the duty of societal responsibility in corporate law have suggested any such responsibility should be included in the Dutch Corporate Governance Code, which is not deemed to be hard law. However, when proposals for amendment of the Corporate Governance Code were made in the summer of 2021, again they were resolutely rejected by business lobby groups.

The European Commission has indicated to consider as part of the overarching EU Green Deal to introduce a directive that is to further sustainable corporate governance and to include an explication of the duty of care of directors relating to sustainability. The European Parliament has endorsed this and specifically has asked to include in the directive a clear set of rules strengthening the duties of the company board regarding sustainability. In its report the Parliament "[c]alls on the Commission to present a legislative proposal to ensure that directors' duties cannot be misconstrued as amounting solely to the short-term maximization of shareholder value, but must instead include the long-term interest of the company and wider societal interests, as well as that of employees and other relevant stakeholders; believes, in addition, that such a proposal should ensure that members of the administrative, management and supervisory bodies, acting within the competences assigned by them by national law, have the legal duty to define, disclose and monitor a corporate sustainability strategy."[8] In the end the Commission has published a proposal for a directive on Corporate Sustainability Due Diligence in February 2022. Article 25 of the proposal provides that Member States are to ensure that directors when fulfilling their duty to act in the best interests of the company take into account the consequences of their decisions for sustainability matters, including human rights, climate change and the environment, on the short,

---

[8] https://www.europeanparl.europa.eu/doceo/document/TA-9-2020-0372_EN.html.

medium and long term. Member States must also ensure that their laws regulations and administrative provisions providing for a breach of directors' duties apply. The proposal of the Commission has been delayed several times, partly again because of strong resistance from the business community. The resistance is based among others on an academic critique of a report of Ernst & Young commissioned by the Commission, indicating that short-term focus on shareholder value maximization is indeed dominant with listed companies in Europe.[9] My intention here is not to engage in the legal debate about such a duty of societal responsibility and its consequences, or the debate about whether there is evidence for the need to include such a duty or not. Also, it is clear that only introducing a duty of societal responsibility for directors in itself will not generate the transformation that is needed. Such a duty should be elaborated upon in corporate law and corporate governance.[10] Here I want to explore the resistance against these proposals as such.

4   Resistance and Learning Anxiety

The strong resistance against proposals to formulate an explicit duty of societal responsibility puzzles in the light of the growing efforts of corporations and the public statements they make about efforts. It appears that the dominance of the shareholder value maximization dogma is actually waning and that there is a growing and broad acceptance of a wider responsibility of corporations in society, the fulfillment of which directors are to ensure. Let us assume that the corporations and their directors who make explicit statements on their societal responsibility also want to act upon their own sense of that responsibility and seek ways to transform their strategy and business operations (and therefore let us not assume it is all *green-washing* anyway). Why resist against putting in law their responsibility which they actively seem to be fulfilling in practice according to their own statements? Of course, part of this will be genuine concern about the risk of personal liability. Although, as I have said above, this risk in practice is small, directors may have an oversized sense of the risks they are dealing with. On the other hand, if directors are so engaged in transforming the strategy and business operations to fulfill the ecological and social needs of society, they should not have to be so concerned. The fierceness of the resistance also indicates that the resistance may have a deeper source.

Pondering this puzzle, the concept of learning anxiety as developed by Edgar Schein (Schein, 2004) appeared to me as possibly providing an explanation. Could the concern of business leaders about an explicit duty of societal responsibility be driven by the fact that, contrary to their public statements, they actually do not yet know how to transform their business into a successful

---

[9] See for example the critique of the European Company Law Experts (ECLE) https://europeancompanylawexperts.wordpress.com/publications/european-commission-study-on-directors-duties-and-sustainable-corporate-governance/.

[10] I have made suggestions elsewhere, see fn 1 above.

business that is truly sustainable and socially responsible? This would not be a strange concern. For many corporations, it is not at all clear for example how they could sufficiently reduce their CO2 emissions and nonetheless continue to produce and sell their industrial products like steel or chemical products competitively. Or, if they are in the food production chain, how true costing could be implemented and what the consequences for prices, markets and consumer behavior will be. For other businesses the challenges may be much more in the social field, balancing temporary labor with business cycles, dealing with effects in the social community if they shut down a plant to move production to a cheaper place, etc. The examples are numerous and in many cases, corporations need to experiment, try out new ways without being certain of outcomes and success. This is where learning anxiety may come in.

Schein developed the concept of learning anxiety in the context of organizations that need to change and where people need to learn new ways of thinking and behaving. Learning anxiety is a combination of several fears, all of which may be active at any given time as you contemplate having to unlearn something that you have grown comfortable with and to learn something new. Schein distinguishes several fears that may play a role:

- *Fear of temporary incompetence*: during the transition process you will be unable to feel competent because you have given up the old way and have not yet mastered the new way. The fear of looking incompetent in the eyes of others that may lead to rejection adds to this fear.
- *Fear of punishment for incompetence*: if learning takes time, you will fear that you will be punished for lack of productivity or success in the meantime. Having to stay successful during the transition may stop you from spending time on new learning.
- *Fear of loss of personal identity*: your current way of thinking may identify you to yourself and to others and you may not wish to be the kind of person that the new way of working would require you to be.
- *Fear of loss of group membership*: shared assumptions that make up a culture of a group or community also identify who is in and who is out of the group or community. If by developing new ways of thinking you will become a deviant to your group, you may be rejected or even ostracized. To avoid this, you will often resist learning new ways of thinking and behaving.

If learning anxiety is high, resistance to learning new ways of thinking and behaving will be strong. Defensive responses come in the following stages:
- *Denial*: you will convince yourself that the discomforting data are not valid, are temporary, don't really count, are not relevant for you and so on.
- *Scapegoating, passing the buck, dodging*: you will convince yourself that the cause of the problem is somewhere else, that they need to solve it,

that the data do not apply to you, that others need to change first before you do.
- *Maneuvering, bargaining*: you will want special compensation for the effort to make the change; you will want to be convinced that the change really is in your own long-term interest.

Schein continues that organizations will only change if the survival anxiety or guilt is greater than the learning anxiety. If the risk of continuing the old ways becomes very clear, people in the organization are willing to go through the pain of learning and take on the fears that otherwise would stop them from learning. An obvious route for change leaders then appears to be to increase the survival anxiety or guilt. This however may simply increase defensiveness to avoid the threat or pain of the learning process. Schein draws the conclusion that in order to really facilitate change, the change leader must reduce the learning anxiety by increasing the learner's sense of psychological safety, for example by providing a compelling positive vision, formal training, practice fields and feedback, positive role models, support groups and a reward and discipline system consistent with the new way of thinking and behaving.

Learning anxiety may, at least partly, explain the strong resistance from the business community against the introduction of an explicit duty of societal responsibility. The different fears that Schein describes may all play a role as fears of business leaders. Fear of not feeling competent when embarking on an ecologically sustainable strategy and developing fully sustainable business operations. For many business organizations the challenge to become fully sustainable, without CO_2 or other emissions and pollution, may appear to be daunting, with uncertain results. New technologies and business models will need to be developed and explored and it is likely that some will fail. Where the old ways gave a sense of competence, the new sustainable ways of doing business do no yet do so. Fear of being criticized during the transition period, when results are down and the transition is slow. How will the investment community assess results? They may have high ESG ambitions themselves and should support their investee corporations to experiment and learn how to make new technologies and business models work. But they may nonetheless favor immediate results and divest when a corporation underperforms in relation to its peers. Fear of identity may exist if a business leader successful in the old ways of shareholder value maximization needs to let go of a sense of comfort in knowing how to run your business with that objective. The measure of what makes a successful business leader is changing and your old ways of being are not helping you to build new success. Fear of loss of group membership may exist when a business leader in a specific sector explicitly takes on this societal responsibility, which may be seen as disloyal by other leaders in the sector. We also see all the denial, scapegoating and maneuvering that Schein describes as a pattern of resistance against change.

If Schein is right that learning anxiety typically is only overcome when survival anxiety is higher, one could think that in light of the more and more depressing and urgent scientific insights into climate change, survival anxiety is by now much higher than learning anxiety. The recent sixth Assessment Report of the IPPC indicates that climate change is widespread, rapid and intensifying, with some already irreversible changes that have been set in motion.[11] For some, the urgency of having to take decisive measures now may be very clear, but for many apparently it is not. The threat is acknowledged at an abstract level, but it is not yet specific enough for many in order to be able to overcome the fear of learning how to become truly sustainable. In that context, introducing legislation that emphasizes the responsibility of directors to take into account the ecological and social impact of the corporation's activities and to avoid doing harm, only increases the level of defensiveness against learning. It particularly may increase the fear of punishment and rejection if you get it wrong or if you are not performing well in the meantime. Following Schein, it may make more sense to lower the learning anxiety then to increase the survival anxiety.

An obvious step would then be not to introduce a duty of societal responsibility for directors, as this would avoid increasing the learning anxiety. That, however, would be a wrong conclusion. It would be non-sensical to explicitly avoid stating that corporations and their directors have a responsibility to ensure they conduct themselves as responsible corporate citizens, while in fact knowing that is precisely what is needed in order to stand a chance to address the ecological and social challenges we face. Deliberately not stating that this responsibility exists would only create the space for many corporations and business leaders to continue with the old ways and not develop sustainable strategies and business operations. This would probably not affect corporations that are the front runners in the sustainable transition and make efforts to explore and learn how to do it on the basis of their intrinsic motivation. But the laggards would feel justified to remain lagging behind. The large middle group of relatively passive and indifferent corporations would have no indication of what is expected of them and are likely to remain passive and indifferent. For the same reason, I would not be in favor of smoothing the duty of societal responsibility by using diminishing language seeking to provide the comfort that nothing more is needed than some good intentions In my view we can no longer afford a voluntary approach in which corporations start to take their societal responsibility only when they feel up to it.

What we could do instead is to stick to the introduction of an explicit duty of societal responsibility for directors but to take the immediate sting out by providing that for the first period of, say, five years, directors are exempt from personal liability on this ground. The point of introducing the duty of societal responsibility for directors is not so much to be able to hold them personally liable, but to ensure that they actively start to consider what it takes for them to ensure responsible corporate citizenship and to start to act upon this. An

[11] See https://www.ipcc.ch/assessment-report/ar6/.

initial exemption from personal liability would allow directors to experiment, to initiate to learn how to develop a corporate strategy and business operations that are both sustainable, socially responsible and successful, without the immediate risk of punishment through personal liability if you do not get it right in the eyes of others. I would exclude from this exemption liability for damages caused by gross misconduct and negligence, which in all likelihood would already lead to personal liability of directors under current laws, even without any explicit duty of societal responsibility. It would not make sense to introduce a temporary exemption that would give directors a wider discretion to not take their societal responsibility seriously than they currently have.

The introduction of the duty of societal responsibility should be accompanied by disclosure requirements related to the impact of corporations on human, social and natural capital, such as currently envisaged by the EU proposal for a Corporate Sustainability Reporting Directive.[12] On the basis of these disclosure requirements, a robust system for monitoring progress could be set up. This would help to assess what progress corporations make towards becoming ecologically and socially sustainable during this phase of learning. Such a monitoring system could also include or provide the basis for positive incentives to innovate and learn by for example developing rankings of corporations that make most progress towards actual responsible and sustainable strategies and business operations. This could create positive, competitive dynamics to innovate and transform in the direction of sustainable business. After five years, on the basis of the monitoring of progress, it could be reviewed whether there is cause to continue with the exemption from personal liability or not.

Through measures like these, and others that can build on them, the actual learning anxiety of directors in order to make their corporations responsible corporate citizens may be reduced. This would lower their resistance against taking up the societal responsibility that they need to take. We need to start to learn how to do this now and should not allow our fear of learning to continue to hold us back.

References

Armour, John, Eidenmueller, Horst (2019), Self-Driving Corporations? https://ssrn.com/abstract=3442447.

Bebchuk, Lucian and Tallarita, Robert (2020), The Illusory Promise of Stakeholder Governance, https://ssrn.com/abstract=3544978.

Comte-Sponville, André, Le Capitalisme est-il Moral? Albin Michel 2009.

Dorff, Michael, Hicks, James, Solomon, Steven (2020), The Future or Fancy? An Empirical Study of Public Benefit Corporations, https://ssrn.com/abstract=3433772.

---

[12] COM 2021, 189, see https://eur-lex.europa.eu/legal-content/EN/TXT/?uri=CELEX:52021PC0189.

Ferrarini, Guido (2020), An Alternative View of Corporate Purpose: Colin Mayer on Prosperity, https://ssrn.com/abstract=3552156.

Hess, Kendy (2017), The Unrecognized Consensus about Firm Moral Responsibility, in: Eric Orts, N. Craig Smith (eds), The Moral Responsibility of Firms, Oxford University Press 2017, pp. 169–187.

Fromm, Erich (1956), The Sane Society, Routledge 2nd ed 1956.

Hampden-Turner, Charles and Trompenaars, Fons (2015), Nine Visions of Capital. Unlocking the Meanings of Wealth Creation, Infinite Ideas 2015.

Mayer, Colin (2013), Firm Commitment, Oxford University Press 2013.

Mayer, Colin (2018), Prosperity, Better Business Makes Greater Good, Oxford University Press 2018.

Rifkin, Jeffrey (2015), The Zero Marginal Cost Society, St. Martin's Griffin, 2015.

Robé, Jean-Philippe, Delaunay, Bertrand and Fleury, Benoit (2019). French Legislation on Corporate Purpose, https://corpgov.law.harvard.edu/2019/06/08/french-legislation-on-corporate-purpose/.

Roche, Bruno and Jakub, Jay (2017), Completing Capitalism, Berrett-Koehler Publishers 2017.

Rönnegard, David and Velasquez, Manuel (2017), On (Not) Attributing Moral Responsibility to Organizations, in: Eric Orts, N. Craig Smith (eds), The Moral Responsibility of Firms, Oxford University Press 2017, pp. 123–142

Rushkoff, Douglas (2017), Throwing Rocks at the Google Bus, How Growth Became the Enemy of Prosperity, Portfolio Penguin 2017.

Russell, Stuart (2019), Human Compatible, Allen Lane 2019.

Schein, Edgar H. (2004), Organizational Culture and Leadership, Jossey-Bass 2004.

Sinek, Simon (2019), The Infinite Game, Penguin Business 2019.

Van Schilfgaarde-Winter-Wezeman-Schoonbrood (2017), Van de BV en de NV, Wolters Kluwer 2017.

Tegmark, Max (2017), Liefe 3.0, Being Human in the age of Artificial Intelligence, Allen Lane 2017.

Weber, Max (1921/1968), Economy and society (3 vols) Totwa NJ: Bedminster Press (original work published in 1921).

Winter, Jaap (2017), In de Delta van het Rijnland, ESB 2017, 4751, p. 334.

Winter, Jaap (2019a), The Dehumanisation of the Large Corporation, published in Dutch in Ondernemingsrecht 2019a/2, in English available at https://ssrn.com/abstract=3517492.

Winter, Jaap (2019b), The Human Experience of Being-in-the-Board. A Phenomenological Approach, in: Levrau, Abigail and Gobert, Sandra (eds), Governance: the Art of Aligning Interests, Liber Amicorum Lutgart van den Berghe, Intersentia 2019b, pp. 131–148, also available at https://ssrn.com/abstract=3319392.

Winter, Jaap, De Jongh, Matthijs, Hijink, Steven, Timmerman, Vino, Van Solinge, Gerard and 20 co-authors (2020), Naar een zorgplicht voor bestuurders en commissarissen tot verantwoorde deelname aan het maatschappelijk verkeer, Ondernemingsrecht 2020/86.

Winter, Jaap (2021), Sustainable Corporate Governance and the Takeover Bids Directive, in: Liber Amicorum Rolf Skogg, 2021, pp. 1075–1085.

Zizek, Slavoj (2017), The Courage of Hopelessness, Chronicles of a Year of Acting Dangerously, Allen Lane 2017.

CHAPTER 6

ESG and Shareholder Primacy: Why They Can Go Together

Luca Enriques

1 Introduction

Fifty years after the publication of Milton Friedman's essay *The Social Responsibility of Business Is to Increase Its Profits*,[1] the debate on whether directors and managers should only aim to maximize profits (or value) for shareholders rages on.[2]

In a corporate world where institutional shareholders have taken centre stage, this old question must be asked, *mutatis mutandis*, with regard to asset managers and their clients. Where portfolio-value-maximizing "universal owners" dominate the scene, socially responsible corporate behaviour may become more common based on premises that are, on their face, fully consistent with Friedman's framework. To understand that, it is useful first to summarize Friedman's *New York Times* essay.

---

[1] M. Friedman, *The social responsibility of business is to increase its profits*, in The New York Times Magazine 13 September 1970, 32 and 122 ff.

[2] See *e.g.* the numerous contributions by lawyers and economists in *ProMarket* at https://promarket.org/category/friedman50/. See also L. Zingales, *Friedman's legacy: from doctrine to theorem*, 13 October 2020, at https://promarket.org/2020/10/13/milton-friedman-legacy-doctrine-theorem/ (summarizing the ProMarket debate).

L. Enriques (✉)
University of Oxford, Oxford, England
e-mail: luca@enriques.eu

© The Author(s), under exclusive license to Springer Nature Switzerland AG 2022
P. Câmara and F. Morais (eds.), *The Palgrave Handbook of ESG and Corporate Governance*, https://doi.org/10.1007/978-3-030-99468-6_6

2   What Friedman Said

As Alex Edmans has noted,[3] "Friedman's article is widely misquoted and misunderstood. Indeed, thousands of people may have cited it without reading past the title. They think they don't need to, because the title already makes his stance clear: companies should maximize profits by price-gouging customers, underpaying workers, and polluting the environment". That is not, of course, what Friedman wrote. According to Friedman:

A) Talking about the "social responsibility of business" makes no sense because the responsibility lies with people. Public corporations are legal persons and may have their responsibilities, but they act through their directors and managers. Therefore, attention must be focused on the responsibilities of such players.[4]

B) Managers are employees of corporations, which in turn are owned by their shareholders. Therefore, managers must act in accordance with the wishes of the shareholders. Unless the shareholders themselves explicitly determine an altruistic purpose, acting in the interest of the shareholders means "conduct[ing] the business in accordance with [shareholders'] desires, which generally will be to make as much money as possible while conforming to their basic rules of the society, both those embodied in law and those embodied in ethical custom".[5]

C) If managers also had a social responsibility, they would find themselves in the position of having to act against the interests of shareholders, for example by hiring the "hardcore" unemployed to combat poverty instead of hiring the most capable workers. By doing so, they would spend shareholders' money to pursue a general interest. In other words, they would impose a tax on shareholders and also decide how to use its proceeds. That is an eminently political task, which should be the fruit of the democratic process, not the decisions of a private individual chosen by a small circle of individuals (the shareholders themselves) and, in addition, probably lacking the specific skills needed to make political choices.[6]

D) But, it is countered, if there are serious and urgent economic and environmental problems, then it is for managers to face them without waiting for politicians' action, which is always late and imperfect. Friedman replied that it is undemocratic for private individuals using other people's money (and, importantly, exploiting the monopolistic

---

[3] A. Edmans, *What stakeholder capitalism can learn from Milton Friedman*, in ProMarket, 10 September 2020, (https://promarket.org/2020/09/10/what-stakeholder-capitalism-can-learn-from-milton-friedman/).

[4] Friedman (nt. 1), 33.

[5] Ibid.

[6] Friedman (nt. 1), 33 and 122.

rents of the large corporations they lead[7]) to impose on the community their political preferences on how to solve urgent economic and environmental problems, which should instead be addressed through the democratic process.[8]

E) That is a fundamental distinction: the market is based on the unanimity rule; in "an ideal free market", there is no exchange without the consent of those who participate in it. Politics, on the other hand, operates according to the conformity principle, whereby a majority binds the dissenting minority. The intervention of politics is necessary because the market is imperfect. But the social responsibility doctrine would extend the mechanisms of politics to the market sphere since a private subject (enjoying some monopoly power, one may add[9]) would impose its political will on others.[10]

F) Often, the idea of corporate social responsibility (CSR) is just a public relations exercise to justify managerial choices already consistent with the interests of shareholders. Looking after the well-being of employees, devoting resources to the firm's local communities, and so on may well be (and, as a rule, will be) in the long-term interest of corporations. Indeed, cloaking these actions under the label of CSR, as it was fashionable to do in 1970 (and is again today), can in itself contribute to increasing profits.[11]

G) But this game is a risky one: extolling the virtues of CSR and expressing scepticism about the social benefits of profit-making can erode public trust in capitalism and make corrective action by the state more likely if corporations do not live up to the expectations they create with their own rhetoric.[12]

---

[7] Cf. D. Chan Smith, *How Milton Friedman Read His Adam Smith: The Neoliberal Suspicion of Business and the Critique of Corporate Social Responsibility*, working paper, 2020 (abstract available at https://ssrn.com/abstract=3674604): Although Friedman barely mentions monopolies in the essay, it is apparent from his archives that Friedman's aversion to corporate social responsibility was deeply connected with his distrust for monopolies and oligopolies. Without some monopoly power, there would be no room for corporate social responsibility. With monopolists espousing corporate social responsibility, it will be easier for them to obtain favourable treatment from policymakers, i.e. protection from competition, for instance from abroad in the form of tariffs.

[8] Friedman (nt. 1), 124.

[9] See supra note 7.

[10] Friedman (nt. 1), 126.

[11] Friedman (nt. 1), 124.

[12] Friedman (nt. 1), 126.

3 MISSING FROM FRIEDMAN'S PICTURE: THE SHAREHOLDERS

Friedman's essay assigned a totally passive role to what he calls the corporation's "owners" or "the employers"—that is, the shareholders. They are merely the beneficiaries of directors' duty to increase profits, but they have no role to play in pursuing that very goal other than (as he notes in passing) when they elect the board.

That is understandable. When Friedman wrote his piece, the shareholders of US companies were mainly individuals and rarely voted at annual meetings other than to rubber-stamp managers' proposals.[13] Today, a large majority of listed firms' shares are held by institutional investors—that is, managers of other people's money.[14] Institutions have become key players at US (as well as non-US) listed corporations not only as holders of record but also because they regularly vote portfolio shares at shareholder meetings.[15] And their pro-management vote is nowadays anything but certain.[16]

This creates one additional layer of employee/employer relationships, to use Friedman's terminology (today, we would say principal/agent relationships): the one between the institutional investors holding shares, or rather, (as Friedman saw it) their own managers/employees, and the individuals (usually workers and pensioners) whose funds the managers invest. (To be sure, it is often more complicated than that because some institutions, such as pension funds, often delegate management of their assets to asset managers. But this is irrelevant for present purposes).

Friedman's essay raises the question: is there any room for asset managers to assume social responsibility duties in deciding how to invest and how to vote? In Friedman's logic, the answer should be a resounding "no", and it's easy to imagine that he would chastise those fund managers who portray themselves (not always veritably) as socially responsible investors. Like corporate managers, fund managers manage other people's money and should not grant themselves the licence to make political choices, which will inevitably please some of their beneficiaries and not others. Their only goal should be to give their clients the highest returns on the funds invested.

Of course, much like a corporation can be set up with an altruistic (or mixed) purpose, so can asset management products expressly be marketed as socially responsible or ethically invested. Intuitively, investors in such funds expect them to invest and vote in accordance with the socially responsible

---

[13] See J.N. Gordon, *The rise of independent directors in the united states, 1950–2005: Of shareholder value and stock market prices*, in Stanford L. Rev. (2007), 1568; R.C. Clark, *Corporate Law*, 1986, 94.

[14] See *e.g.* OECD, *Owners of the world's listed companies* (2019), *passim* (with data from across the world).

[15] See *e.g.* L. Enriques/ A. Romano, *Institutional investor voting behavior: A network theory perspective*, in Univ. of Illinois L. Rev. (2019), 235.

[16] *Ibid.*

commitments undertaken. But absent a CSR connotation—namely, if the mutual fund has been marketed as a tool for generating financial returns—fund managers have to assume that the fund's investors have a financial objective in mind and do not expect their own political preferences to be promoted by their fund manager, especially if that comes to the detriment of their return. Whether implicitly or explicitly, that's the bargain with each of the mutual fund shares subscribers.

But things are not always as straightforward as that. Universal owners, that is, institutional investors holding the entire market proportionately rather than picking stocks, now hold a considerable share of the stocks listed on exchanges across the world. In the US, indexing or "closet indexing" investment vehicles now represent approximately 30 per cent of the equity market.[17]

As Madison Condon and Jack Coffee have noticed,[18] for investors of that kind, portfolio value maximization may well mean pushing for Environment, Social and Governance (ESG) policies at the individual company level that, while not necessarily profitable *for that company*, will increase *portfolio* returns by making other companies more profitable. Think, for instance, of systemically important financial institutions adopting more conservative risk management policies that significantly reduce the chances of a potentially devastating financial crisis.

Hence, the overlap between socially responsible and profit-maximizing behavior, which Friedman himself acknowledged to be present at the individual company level and criticized only as being politically dangerous, is now even more pervasive at the institutional shareholder level.

In theory, all portfolio value maximizers' decisions on ESG matters should be based on an assessment of the effects that decision would have both on the individual portfolio company's value and on the value of the totality of other portfolio companies. Because ESG policies require widespread adoption to be effective, different scenarios will have to be elaborated and factored in to estimate those effects. Multiple other variables will have to be considered and a number of questionable assumptions made.

Given their rational reticence and misaligned incentives,[19] passive funds' asset managers are unlikely to have the human and financial resources to fully engage with this kind of assessment, let alone reach solid conclusions. And it

---

[17] See V. Suschko and G. Turner, *The implications of passive investing for securities markets*, BIS Quarterly Rev. (March 2018), 115–116. While indexing and universal ownership are different phenomena, the overlap is intuitively significant.

[18] M. Condon, *Externalities and the common owner*, in **Washington L. Rev.** 95 (2020) 1; J.C. Coffee Jr., *The future of disclosure: ESG, common ownership, and systematic risk*, European Corporate Governance Institute—Law Working Paper 541/2020, available at SSRN: https://ssrn.com/abstract=3678197.

[19] R.J. Gilson-J.N. Gordon, *The agency costs of agency capitalism: Activist investors and the revaluation of governance rights*, in **Columbia L. Rev.** (2013), 863; L.A. Bebchuk-A. Cohen-S. Hirst, *The agency problems of institutional investors*, in **Journal of Economic Perspectives** (2017), 89.

would be naïve to assume that political preferences do not affect the simplified analysis they must inevitably resort to in determining how to maximize portfolio value in the presence of externalities.

Owing to shareholder pressure and/or managers' desire to retain their jobs, the ESG preferences of portfolio-value-maximizing institutions may well trickle down to the individual portfolio company level. Under what conditions that is the case will depend on a number of factors, including whether the company is protected from competition, whether undiversified shareholders hold stakes in the company, how politically divisive the socially responsible action is, and so on. Yet, in some cases, and in respect of some of the socially and politically sensitive issues, managers will yield to those preferences. Given Friedman's premise that "increasing profits" must be the only corporate goal because the shareholders are the owners/employers, there is some irony to that.

4 Conclusion

Irony aside, today's corporate world is very different from the one Milton Friedman wrote in. Yet, his essay still provides a workable framework for understanding the implications of managing companies for one purpose or another. And also for answering the reframed question of whether corporate managers *should* cater to the preferences of their portfolio-value-maximizing indexing investors when making decision on behalf of their corporations.

CHAPTER 7

Climate Finance

Miguel A. Ferreira

1 Introduction

It is widely accepted among climate scientists that the global mean temperature is likely to increase by 2 °C relative to the pre-industrial average by the mid- to late twenty-first century.[1] This increase is expected to be associated with more frequent extreme weather events.[2] In 2015, 195 states and the EU adopted the Paris Agreement to promote a global response to limit global temperature increase to less than 2 °C and to attempt to further limit the increase to 1.5 °C.

---

[1] Intergovernmental Panel on Climate Change. 2007. "Climate Change 2007: Synthesis Report." URL: https://www.ipcc.ch/report/ar4/syr/; Id., 2012. "Managing the Risks of Extreme Events and Disasters to Advance Climate Change Adaptation." URL: https://www.ipcc.ch/report/managing-the-risks-of-extreme-events-and-disasters-to-advance-climate-change-adaptation/; Id., 2014. "Climate Change 2014: Synthesis Report." URL: https://www.ipcc.ch/report/ar5/syr/; Id.. 2019. "Special Report on Global Warming of 1.5 Degree Celsius." URL: https://www.ipcc.ch/sr15/download/

[2] David Barriopedro/ Eric M. Fischer/ Jürg Luterbacher/ Ricardo M. Trigo/ Ricardo García-Herrera, *The Hot Summer of 2010: Redrawing the Temperature Record Map Of Europe*. Science, 332(6026), (2011), pp. 220–224.

M. A. Ferreira (✉)
Nova School of Business and Economics, Lisbon, Portugal
e-mail: Miguel.ferreira@novasbe.pt

However, according to the Global Landscape of Climate Finance,[3] the efforts by various economic agents, which led to record levels of climate finance investment, are insufficient to limit global warming to 1.5 °C (Pfeiffer et al., 2018[4]; Tong et al., 2019[5]). In addition, the threat posed by climate change remains underestimated. In March 2017, only 42% of Americans surveyed agreed that the rise in temperatures was a serious risk.[6]

What is the impact of such weather shocks on the real economy? Several studies focus on the direct economic consequences of weather shocks on agricultural outcomes and farmland value (Mendelsohn et al., 1994[7]; Schlenker et al., 2005[8]; Deschênes and Greenstone, 2007[9]; Schlenker and Roberts, 2009[10]; Schlenker and Lobell, 2010[11]; Chevet et al., 2011[12]; and Roberts et al., 2013[13]). There is also growing evidence on the impact of climate

---

[3] Barbara Buchner/ Alex Clark/ Angela Falconer/ Rob Macquarie/ Chavi Meattle/ Rowena Tolentino/ Cooper Wetherbee, *Global Landscape of Climate Finance. A CPI Report* (2019).

[4] Alexander Pfeiffer/ Cameron Hepburn/ Adrien Vogt-Schilb/ Ben Caldecott, *Committed Emissions from Existing and Planned Power Plants and Asset Stranding Required to Meet the Paris Agreement*, Environmental Research Letters, 13(5), (2018), p. 054019.

[5] Dan Tong/ Qyang Zhang/ Yixuan Zheng/ Ken Caldeira/ Christine Shearer/ Chaopeng Hong/ Yue Qin/ Steven Davis, *Committed Emissions from Existing Energy Infrastructure Jeopardize 1.5° C Climate Target*. Nature, 572(7769), (2019), pp. 373–377.

[6] Gallup News Service. 2017. Gallup Poll Social Series: Environment. March 2017. Timberline: 937008, IS: 968, http://www.gallup.com/poll/206030/global-warming-concern-three-decade-high.aspx.

[7] Robert Mendelsohn/ William Nordhaus/ Daigee Shaw, *The Impact of Global Warming on Agriculture: a Ricardian analysis*. The American economic review, (1994) pp.753–771.

[8] Wolfram Schlenker/ W. Hanemann/ A. C. Fisher, *Will US Agriculture Really Benefit from Global Warming? Accounting for Irrigation in the Hedonic Approach*. American Economic Review, 95(1), (2005), pp. 395–406.

[9] Olivier Deschênes/ Michael Greenstone, The Economic Impacts of Climate Change: Evidence Fromagricultural Output and Random Fluctuations in Weather. American Economic Review, 97(1), (2007), pp. 354–385.

[10] Wolfram Schlenker/ Michael J. Roberts, *Nonlinear Temperature Effects Indicate Severe Damages to US Crop Yields Under Climate Change*. Proceedings of the National Academy of sciences, 106(37), (2009), pp. 15594–15598.

[11] Wolfram Schlenker/ D. Lobell, *Robust Negative Impacts of Climate Change on African Agriculture*. Environmental Research Letters, 5(1), (2010), p. 014010.

[12] Jean-Michel Chevet/ Sebastien Lecocq/ Michel Visser, *Climate, Grapevine Phenology, Wine Production, and Prices: Pauillac (1800–2009)*. American Economic Review, 101(3), (2011), pp. 142–146.

[13] Michael Roberts/ Wolfram Schlenker/ Jonathan Eyer, *Agronomic Weather Measures in Econometric Models of Crop Yield with Implications for Climate Change*. American Journal of Agricultural Economics, 95(2), (2013), pp. 236–243.

change on total factor productivity (Graff Zivin and Kahn, 2016[14]; Chen et al., 2018[15]; and Zhang et al., 2018[16]).

It is also relevant to address how climate change influences asset prices. Focusing on residential real estate, Baldauf et al. (2020)[17] show how different beliefs about the threat posed by climate change affects housing prices as homes in climate change "denier" neighborhoods sell for 7% more than in "believer" neighborhoods. Regarding the risk of rising sea levels, Bernstein et al. (2019)[18] reports that houses expected to be affected by the rise in sea level sell at a discount of 7%, and Goldsmith-Pinkham et al. (2019)[19] find evidence of a small price effect in municipal bonds due to the expected rise in sea level, which reflects that the markets do not expect the rise in sea level to cause defaults at least in the short-term. In financial markets, Bolton and Kacperczyk (2021)[20] show that carbon risk is already priced in the U.S. stock market and thus there is a carbon risk premium for companies more exposed to this risk.

Central banks have also started to incorporate the risks of climate change into their frameworks and to evaluate how the physical and transition risks of climate change can impact the macroeconomy and the financial system. Batten et al. (2020)[21] explore the transmission channels through which climate change can impact central banks' monetary policy goals.

However, it is unclear whether and how these shocks affect firm value. In this chapter, we use variation in average temperatures across suppliers of the same client in a year to obtain an estimate of the impact of weather shocks, controlling for firm-specific demand, on firm value due to lost sales. We use

[14] Joshua Graff Zivin/ Matthew Kahn, *Industrial Productivity in a Hotter World: The Aggregate Implications of Heterogeneous Firm Investment in Air Conditioning* (2016). NBER Working Paper, (w22962).

[15] Chen Chen/ Thanh D. Huynh/ Bohui Zhang, *Temperature and Productivity: Evidence from Plant-Level Data*, (2018).

[16] Peng Zhang/ Olivier Deschenes/ Kyle Meng/ Junjie Zhang, *Temperature Effects on Productivity and Factor Reallocation: Evidence from a Half Million Chinese Manufacturing Plants*. Journal of Environmental Economics and Management, 88, (2018), pp. 1–17.

[17] Markus Baldauf/ Lorenzo Garlappi/ Constantine Yannelis, *Does Climate Change Affect Real Estate Prices? Only if You Believe In It*. The Review of Financial Studies, 33(3), (2020), pp. 1256–1295.

[18] Asaf Bernstein/ Matthew Gustafson/ Ryan Lewis, Disaster on the Horizon: The Price Effect of Sea Level Rise. Journal of Financial Economics, 134(2), (2019), pp. 253–272.

[19] Paul Goldsmith-Pinkham/ Matthew Gustafson/ Ryan Lewis/ Michael Schwert, *Sea Level Rise and Municipal Bond Yields*. Rodney L. White Center for Financial Research, (2019).

[20] Patrick Bolton/ Marcin Kacperczyk, *Do Investors Care About Carbon Risk?*. Journal of Financial Economics (2021), ECGI WP 711/2020.

[21] Sandra Batten/ Rhiannon Sowerbutts/ Misa Tanaka. *Climate Change: Macroeconomic Impact and Implications for Monetary Policy*. Ecological, Societal, and Technological Risks and the Financial Sector, (2020), pp. 13–38.

a sample of supplier–client business transactions of firms headquartered in the U.S.

We find that increases in temperature lead to declines in sales. A 1 °C increase in the average daily temperature in a county is associated with a decrease in sales of about 2%. In addition, we find that extreme heat events and extreme cold events can have a disruptive effect on sales at −8 and −36%, respectively. While these results show that weather shocks affect the intensive margin of sales of intermediate goods, we do not find evidence of similar effects on the extensive margin, i.e., we do not find that weather shocks lead to the termination of supply chain relationships.

We show that our results are mostly driven by manufacturing firms and heat-sensitive industries, suggesting that our findings can be explained by a labor supply and productivity channel. We also find that the effect of weather shocks on sales is larger for financially constrained firms and firms with less operational flexibility, suggesting that these firms do not have the resources or the flexibility to adapt and overcome weather shocks without affecting production.

We explore whether input specificity and supplier-client relationship capital can amplify or mitigate the effect of weather shocks on sales. We find that the reduction in sales is more pronounced in industries that sell standardized goods and when the supplier is geographically distant from the client. These findings are consistent with the idea that the supplier-specific economic costs of weather shocks are larger when client switching costs are lower.

This paper contributes to the literature on the indirect costs of climate change on the economy by showing that climate change affects firm sales. Graff Zivin and Kahn (2016),[22] Chen et al. (2018)[23], and Zhang et al. (2018)[24] find that heat affects total factor productivity. We complement these findings by showing that higher temperature affects supplier–client sales via a labor productivity channel. In addition, our results show the role of financial constraints in amplifying the costs of climate change on firm value, with important policy implications as firms emerge from the Covid-19 pandemic with increased levels of leverage.

Lastly, we contribute to the literature on climate change and the supply chain. Pankratz and Schiller (2019)[25] find that heatwaves and flooding at supplier locations lead to the termination of relationships in global supply

---

[22] Joshua Graff Zivin/ Matthew Kahn, *Industrial Productivity in a Hotter World: The Aggregate Implications of Heterogeneous Firm Investment in Air Conditioning* (2016). NBER Working Paper, (w22962).

[23] Chen Chen/ Thanh D. Huynh/ Bohui Zhang, *Temperature and Productivity: Evidence from Plant-Level Data*, (2018).

[24] Peng Zhang/ Olivier Deschenes/ Kyle Meng/ Junjie Zhang, *Temperature Effects on Productivity and Factor Reallocation: Evidence from a Half Million Chinese Manufacturing Plants*. Journal of Environmental Economics and Management, 88, (2018), pp. 1–17.

[25] Nora M. C. Pankratz/ Christoph M. Schiller, *Climate Change and Adaptation in Global Supply-Chain Networks* (2019). Available at SSRN, 3475416.

chain and a reduction in sales. We contribute to this literature by showing that both average weather shocks and extreme weather events lead to changes in supplier–client sales in the intensive margin, but not in the extensive margin.

2   DATA AND METHODOLOGY

Sample and Variables

Our sample consists of supplier–client pairs of firms headquartered in the U.S. This data is available since publicly listed firms in the U.S. must disclose, on a yearly basis, the identity of clients and the sales to clients whose purchases represent more than 10% of total sales. We collect this information from the Compustat Segment files for the period 2000–2015.

We obtain temperature and precipitation data from the PRISM Climate Group (2019),[26] and extreme weather events data from the National Oceanic and Atmospheric Administration Storm Events Database (2019).[27] We map the weather grids in PRISM and extreme weather event locations to counties in the U.S. Census Bureau files. We compute average daily weather variables and the annual number of extreme weather events by event type at the county level for each year. Finally, we match the weather variables in each county to the firms.

Summary Statistics

Our sample consists of 12,439 supplier-client-year observations for 1,856 unique suppliers and 419 unique clients over the period 2000–2015. The annual increase in average temperature is higher than 0.53 °C for 75% of the counties in our sample, and the standard deviation of the change in average temperature is 0.85 °C.

Methodology

Our main objective is to examine whether changes in local temperature affect the firms' economic activity. To investigate this hypothesis, the dependent variable measures the percentage change in the supplier's sales to each client, and the main independent variable is the change in average daily temperature in degrees Celsius in the county where the supplier is headquartered from year $t-1$ to year t. The coefficient of the independent variable estimates the effect of changes in temperature on supplier–client sales, and a negative value

---

[26] PRISM Climate Group. 2019. "PRISM Gridded Climate Data." http://prism.oregonstate.edu.

[27] National Oceanic and Atmospheric Administration, *Storm Events Database*, (with data from January 1950-September 2021).

would indicate that suppliers that observe increases in average daily temperature in their county of location reduce their sales by larger amounts than similar suppliers selling to the same client.

Importantly, our supplier-client data allows us to compare the changes in economic activity across suppliers selling to the same firm. Thus, the estimated difference in sales can be plausibly attributed to supply-side factors, such as changes in labor supply or productivity of suppliers, or an increase in operating costs, both of which potentially leading to lower output. In addition, weather shocks can affect the quality of products or services, or delay deliveries to clients.

3 Results

Main Results

The results show that a 1 °C yearly increase in the average temperature in the supplier county leads to a 1.2% to 1.9% reduction in supplier–client sales. A 1 °C increase in temperature is not uncommon at the local (county) level, where the standard deviation in the annual change in temperature corresponds to 0.85 °C over our sample period. Therefore, our estimates of the average effect of temperature on suppliers' sales are economically meaningful.

Mechanisms

In this section, we exploit the heterogeneity in our data to analyze the channel through which changes in weather might affect firm sales, and which firm characteristics can mitigate or amplify the effect of weather shocks on sales.

Labor Supply and Productivity

We first explore whether the mechanism behind the negative effects on sales might be due to lower labor supply and productivity, consistently with the results in Graff Zivin and Kahn (2016),[28] Chen et al. (2018),[29] and Zhang et al. (2018).[30] If this is the case, we expect that our results are driven by firms whose output is most sensitive to the weather. We consider three measures to test for this mechanism: (1) whether a firm is in heat-sensitive industries; (2) whether a firm is in manufacturing industries; and (3) the ratio of the number of employees to assets as a proxy for labor intensity.

---

[28] Joshua Graff Zivin/ Matthew Kahn, *Industrial Productivity in a Hotter World: The Aggregate Implications of Heterogeneous Firm Investment in Air Conditioning* (2016). NBER Working Paper, (w22962).

[29] Chen Chen/ Thanh D. Huynh/ Bohui Zhang, *Temperature and productivity: Evidence from Plant-Level Data*, (2018).

[30] Peng Zhang/ Olivier Deschenes/ Kyle Meng/ Junjie Zhang, *Temperature effects on productivity and factor Reallocation: Evidence from a Half Million Chinese Manufacturing Plants*. Journal of Environmental Economics and Management, 88, (2018), pp. 1–17.

Firms in industries with predominantly outdoor activities or manufacturing processes are expected to be more sensitive to heat. Following Graff Zivin and Neidell (2014),[31] we identify firms operating in heat-sensitive industries. We find that for firms in heat-sensitive industries a 1 °C yearly increase in the average temperature in the supplier county leads to a 2.2% reduction in sales, while for firms not in heat-sensitive industries we do not find an impact in sales.

Chen et al. (2018)[32] document that higher local temperature reduces total factor productivity. If temperature primarily affects economic performance via a productivity channel, firms in the manufacturing industries are likely to be driving the results. We find that for firms in manufacturing industries a 1 °C yearly increase in the average temperature in the supplier county leads to a reduction in sales that ranges from −2.0 to −2.3%, while for firms outside the manufacturing industries we do not find a significant impact in sales.

Lastly, firms with higher labor intensity are expected to be more sensitive to heat. We find that for firms with high labor intensity (above the median) a 1 °C yearly increase in the average temperature in the supplier county leads to a 2.2% reduction in sales, while for firms with low labor intensity (below the median) we do not find a significant impact in sales.

Financial Constraints and Adaptability
Disruptions to firms' production processes might be particularly severe if suppliers cannot effectively adapt to the changing climate conditions. To measure the ability of firms to adapt to changes in the environment, we consider the following five measures: (1) whether a firm is rated or non-rated; (2) ratio of long-term debt maturing within one year to total long-term debt; (3) total assets as a proxy for firm size; (4) number of employees as a proxy for firm size; and (5) whether a firm is a single-segment firm or a conglomerate.

Firms with a credit rating have access to public debt markets and therefore are less financially constrained. For firms without a credit rating, a 1 °C yearly increase in the average temperature in the supplier county leads to a reduction in sales that ranges from −2.4 to −3.1%. For firms with a credit rating, we find an increase in sales that ranges from 2.4 to 2.7%.

A high ratio of long-term debt maturing within one year to total long-term debt indicates that the firm is more financially constrained. For firms with a higher ratio of debt maturing (above the median) a 1 °C yearly increase in the average temperature in the supplier county leads to a reduction in sales that ranges from −3.8 to −4.2%. For firms with a low ratio of debt maturing (below the median) we do not find a significant effect on sales.

---

[31] Joshua Graff Zivin/ Matthew Neidell, *Temperature and the Allocation of Time: Implications for Climate Change.* Journal of Labor Economics, 32(1) (2014), pp. 1–26.

[32] Chen Chen/ Thanh D. Huynh/ Bohui Zhang, *Temperature and Productivity: Evidence from Plant-Level Data,* (2018).

Firm size can proxy for operational flexibility and financial constraints. Larger firms have more operational flexibility and fewer financial constraints than smaller firms. For firms with lower total assets (below the median) a 1 °C yearly increase in the average temperature in the supplier county leads to a reduction in sales that ranges from −3.0 to −4.2%. For firms with higher total assets (above the median) we do not find an effect in sales. We find similar results when we split the sample by the number of employees. Thus, we find that the negative effects are driven by smaller firms.

The number of business segments can also proxy for operational flexibility and financial constraints. Conglomerates (i.e., multi-segment firms) have more operational flexibility and less financial constraints than smaller firms due to internal capital markets. For single-segment firms, a 1 °C yearly increase in the average temperature in the supplier county leads to a reduction in sales that ranges from −1.7 to −2.1%. For conglomerates, we do not find evidence of an effect on sales. We conclude that the negative effects are driven by single-segment firms.

Overall, we find that the negative effects of climate change are driven by firms with less operational flexibility and more financial constraints as these firms can have more difficulties (or can take more time) to adapt to changes in temperature.

Supplier-Client Relationship
In this subsection, we explore whether input specificity and relationship capital mitigate the negative effects of higher local temperature on sales. We consider three measures for input specificity and the strength of supplier-client relationship: (1) whether a firm is in an industry that sells standardized goods; (2) whether a firm has patents; (3) the geographical distance between supplier-client pairs.

Suppliers selling more standardized goods are likely to have weaker supplier-client relationship, since clients can easily substitute away from a disrupted supplier. Following Burkart, Ellingsen, and Giannetti (2011),[33] we identify industries that are more likely to sell standardized products. For firms in industries that sell standardized goods a 1 °C yearly increase in the average temperature in the supplier county leads to a reduction in sales of -3.6%. For firms that sell less standardized goods, we do not find evidence of an effect in sales.

An alternative measure of input specificity and relationship capital is given by patents. For firms without a patent a 1 °C yearly increase in the average temperature in the supplier county leads to a reduction in sales ranging from −1.4 to −1.9%. For firms with at least one patent, we do not find evidence of an effect in sales.

---

[33] Burkart, M., Ellingsen, T./ Giannetti, M., *What You Sell is What You Lend? Explaining Trade Credit Contracts*. The Review of Financial Studies, 24(4) (2011), pp. 1261–1298.

Supplier–client pairs that are closer to each other geographically are likely to have a stronger relationship. We split the sample into high and low distance according to the median value of its distribution. For pairs that are farther apart a 1 °C yearly increase in the average temperature in the supplier county leads to a reduction in sales ranging from −2.9 to −3.1%. For pairs that are more closely located we do not find evidence of an effect on sales.

Extreme Weather Events

We next examine whether extreme weather events affect firms' economic activity. We test whether excessive heat in supplier counties affects supplier–client sales, and find that an extreme heat event is associated with a further 6.2 to 8.0% reduction in sales, relative to firms with no such event.

We also test whether extreme cold events in supplier counties affect supplier–client sales, and find that firms hit with an extreme cold event suffer an additional reduction in their sales of 31.3 to 35.7%. These results suggest that extreme cold events, even if less often, can have a more disruptive effect on the firm's economic activity.

Extensive Margin

We also evaluate the effect that climate change may have in the extensive margin but do not find evidence that increases in temperature lead to a significant decrease in sales, such that sales to the client falls below the 10% reporting threshold and eventually to zero. This suggests that changes in temperature do not lead to termination of supply chain relationships.

These results contrast with those of Pankratz and Schiller (2019),[34] who find that heatwaves and natural disasters can disrupt the global supply chain in the extensive margin. This may be explained by the fact that our sample is a domestic supply chain network, and client and suppliers may have stronger business relationships, and lower information asymmetries due to their geographical proximity.

4 Conclusion

This chapter studies the economic costs of changes in local temperature. We compare sales of intermediate goods across suppliers that trade with the same client but are exposed to different weather shocks, which allow us to distinguish supply from demand effects.

We show that changes in local temperature can have important effects on supply chain networks activity at the intensive margin. A 1 °C increase in local temperature in supplier counties leads to a reduction in sales of about 2%. We

---

[34] Nora M. C. Pankratz/ Christoph M. Schiller, *Climate Change and Adaptation in Global Supply-Chain Networks* (2019). Available at SSRN, 3475416.

also show that suppliers exposed to episodes of extremely hot and cold weather suffer large reductions in sales.

We examine the channels by which changes in local temperature affect sales. First, the reduction in supplier–client sales is primarily driven by firms in heat-sensitive industries, manufacturing industries, and labor-intensive firms, suggesting that lower labor supply and productivity are driving these effects. Second, larger firms and financially unconstrained firms are better able to deal with the adverse effects of increased local temperature and therefore suffer lower reductions in sales, suggesting that financial constraints play an important role in the ability of firms to adapt to climate change. Finally, input specificity and relationship capital are important drivers of the impact of temperature changes on supplier sales.

Overall, these results suggest that climate change can have important real effects. Suppliers more likely to be affected by climate change can suffer significant decreases in sales, and financial constraints may amplify the effects. Policymakers should consider supply side effects when they design policies to address climate change challenges.

REFERENCES

Baldauf, M., Garlappi, L., and Yannelis, C., 2020. Does climate change affect real estate prices? Only if you believe in it. The Review of Financial Studies, 33(3), pp. 1256–1295.

Barriopedro, D., Fischer, E.M., Luterbacher, J., Trigo, R.M. and García-Herrera, R., 2011. The hot summer of 2010: redrawing the temperature record map of Europe. Science, 332(6026), pp. 220–224.

Batten, S., Sowerbutts, R. and Tanaka, M., 2020. Climate change: Macroeconomic impact and implications for monetary policy. Ecological, Societal, and Technological Risks and the Financial Sector, pp. 13–38.

Bernstein, A., Gustafson, M.T. and Lewis, R., 2019. Disaster on the horizon: The price effect of sea level rise. Journal of Financial Economics, 134(2), pp. 253–272.

Bolton, P. and Kacperczyk, M., 2021. Do investors care about carbon risk?. Journal of Financial Economics.

Buchner, B., Clark, A., Falconer, A., Macquarie, R., Meattle, C., Tolentino, R. and Wetherbee, C., 2019. Global Landscape of Climate Finance 2019.

Burkart, M., Ellingsen, T., and Giannetti, M., 2011. What you sell is what you lend? Explaining trade credit contracts. The Review of Financial Studies, 24(4), pp. 1261–1298.

Chen, C., Huynh, T.D. and Zhang, B., 2018. Temperature and productivity: Evidence from plant-level data.

Chevet, J.M., Lecocq, S. and Visser, M., 2011. Climate, grapevine phenology, wine production, and prices: Pauillac (1800–2009). American Economic Review, 101(3), pp. 142–146.

Deschênes, O., and Greenstone, M., 2007. The economic impacts of climate change: Evidence from agricultural output and random fluctuations in weather. American Economic Review, 97(1), pp. 354–385.

Gallup News Service. 2017. Gallup poll social series: Environment. March 2017. Timberline: 937008, IS: 968, http://www.gallup.com/poll/206030/global-warming-concern-three-decade-high.aspx.

Goldsmith-Pinkham, P., Gustafson, M., Lewis, R.C. and Schwert, M., 2019. Sea level rise and municipal bond yields. Rodney L. White Center for Financial Research.

Graff Zivin, J. and Neidell, M., 2014. Temperature and the allocation of time: Implications for climate change. Journal of Labor Economics, 32(1), pp. 1–26.

Graff Zivin, J. and Kahn, M.E., 2016. Industrial Productivity in a Hotter World: The Aggregate Implications of Heterogeneous Firm Investment in Air Conditioning. NBER Working Paper, (w22962).

Intergovernmental Panel on Climate Change. 2007. "Climate Change 2007: Synthesis Report." URL: https://www.ipcc.ch/report/ar4/syr/.

Intergovernmental Panel on Climate Change. 2012. "Managing the Risks of Extreme Events and Disasters to Advance Climate Change Adaptation." URL: https://www.ipcc.ch/report/managing-the-risks-of-extreme-events-and-disasters-to-advance-climate-change-adaptation/.

Intergovernmental Panel on Climate Change. 2014. "Climate Change 2014: Synthesis Report." URL: https://www.ipcc.ch/report/ar5/syr/.

Intergovernmental Panel on Climate Change. 2019. "Special Report on Global Warming of 1.5 Degree Celsius." URL: https://www.ipcc.ch/sr15/download/.

Mendelsohn, R., Nordhaus, W.D. and Shaw, D., 1994. The impact of global warming on agriculture: A Ricardian analysis. The American economic review, pp. 753–771.

Pankratz, N.M. and Schiller, C.M., 2019. Climate change and adaptation in global supply-chain networks. Available at SSRN, 3475416.

Pfeiffer, A., Hepburn, C., Vogt-Schilb, A. and Caldecott, B., 2018. Committed emissions from existing and planned power plants and asset stranding required to meet the Paris Agreement. Environmental Research Letters, 13(5), p. 054019.

PRISM Climate Group. 2019. "PRISM Gridded Climate Data." http://prism.oregonstate.edu.

Roberts, M.J., Schlenker, W. and Eyer, J., 2013. Agronomic weather measures in econometric models of crop yield with implications for climate change. American Journal of Agricultural Economics, 95(2), pp. 236–243.

Schlenker, W., Hanemann, W.M. and Fisher, A.C., 2005. Will US agriculture really benefit from global warming? Accounting for irrigation in the hedonic approach. American Economic Review, 95(1), pp. 395–406.

Schlenker, W. and Lobell, D.B., 2010. Robust negative impacts of climate change on African agriculture. Environmental Research Letters, 5(1), p. 014010.

Schlenker, W. and Roberts, M.J., 2009. Nonlinear temperature effects indicate severe damages to US crop yields under climate change. Proceedings of the National Academy of sciences, 106(37), pp. 15594–15598.

Tong, D., Zhang, Q., Zheng, Y., Caldeira, K., Shearer, C., Hong, C., Qin, Y. and Davis, S.J., 2019. Committed emissions from existing energy infrastructure jeopardize 1.5 °C climate target. Nature, 572(7769), pp. 373–377.

Zhang, P., Deschenes, O., Meng, K. and Zhang, J., 2018. Temperature effects on productivity and factor reallocation: Evidence from a half million Chinese manufacturing plants. Journal of Environmental Economics and Management, 88, pp. 1-17.

CHAPTER 8

The New ESG Bond Markets

Manuel Requicha Ferreira

1 SGD's, Paris Agreement, and Sustainable Finance

Climate has become one of the greatest governance challenges, no different from the first initial paramount defies posed by remuneration, nominations or audit. Quoting the words of former US President Barack Obama: "*We are the first generation to feel the effect of climate change and the last generation who can do something about it*". The effects will endure for generations and affect disproportionately the poorest and marginalized nations.[1] We are called collectively to a task in which we are the main actors and the Nation-States play a role in (hopefully) helping us make the act. We are facing the global and individual governance ruling of the planet as we know it.

The year 2015 marked two key milestones towards the beginning of that global governance ruling. The first was the adoption of the 2030 Agenda of the United Nations (UN)[2] by all United Nations Member States providing a

---

[1] *See* United Nations Environment Programme, Climate Change and Human Rights, 2015, available at https://web.law.columbia.edu/sites/default/files/microsites/climate-change/climate_change_and_human_rights.pdf.

[2] The declaration of the general assembly of the UN, of 25 September, *Transforming Our World: The 2030 Agenda for Sustainable Development*, available at https://sdgs.un.org/2030agenda.

M. Requicha Ferreira (✉)
Cuatrecasas, Lisbon, Portugal
e-mail: Manuel.requichaferreira@cuatrecasas.com

© The Author(s), under exclusive license to Springer Nature Switzerland AG 2022
P. Câmara and F. Morais (eds.), *The Palgrave Handbook of ESG and Corporate Governance*, https://doi.org/10.1007/978-3-030-99468-6_8

"shared blueprint for peace and prosperity for people and the planet, now and in the future" and establishing 17 Sustainable Development Goals (SDGs) to be achieved by 2030 ranging from equality, climate change, land, and water, poverty.[3]

The second was the twenty-first yearly session of the Conference of the Parties (COP) to the 1992 United Nations Framework Convention on Climate Change (UNFCCC) adopting and ratifying the Paris Agreement (also known as COP21).[4] Several Nations committed to a wide range of collective action measures to slow global warming, in particular (article 2):

a. Holding the increase in the global average temperature to well below 2 °C above pre-industrial levels and to pursue efforts to limit the temperature increase to 1.5 °C above pre-industrial levels;
b. Increasing the ability to adapt to the adverse impacts of climate change and foster climate resilience and low greenhouse gas emissions development, in a manner that does not threaten food production;
c. Making finance flows consistent with a pathway towards low greenhouse gas emissions and climate-resilient development.

However, climate mitigation and adaptation are not for free, they require enormous and expensive investments in low carbon and green infrastructures.[5] The Paris Agreement, acknowledges this financial gap and expressly determines Nations mobilize the private sector in order to finance the investments for the necessary energy transition with a view to carbon emission goals.[6] In fact, the volatility of the Nations-State positions caused by the usual political battle raised in relation to these issues[7] urges the private sector for endured action to overcome any public gridlock.

---

[3] See the SGD's guidelines https://www.un.org/sustainabledevelopment/wp-content/uploads/2019/01/SDG_Guidelines_AUG_2019_Final.pdf.

[4] This Convention is commonly referred as COP21 and was also the eleventh session of the Meeting of the Parties to the 1997 Kyoto Protocol. The Paris Agreement entered into force on November 30, 2016.

[5] See the estimations of OECD expecting $93 trillion in infrastructure investment necessary over the next fifteen years (see OECD green bonds: mobilising the debt capital markets for a low-carbon transition, 2015, available at: https://www.oecd.org/environment/cc/Green%20bonds%20PP%20%5Bf3%5D%20%5Blr%5D.pdf; and the estimation of the United Nations expecting adaptation costs between $280 billion and $500 billion by 2050 (see United Nations Environment Programme, Adaptation Finance Gap Report, 2016, available at http://web.unep.org/adaptationgapreport/sites/unep.org.adaptationgapreport/files/documents/agr2016.pdf). The EC Action Plan refers that Europe has to close a yearly investment gap of almost EUR 180 billion to achieve EU climate and energy targets by 2030.

[6] See Paris Agreement, paragraph 55.

[7] Mainly driven by fear of job-losing, staying behind in the global growth and competitiveness.

In 2018, and following the report issued in January 2018 by the High-Level Expert Group on sustainable finance appointed by the European Commission in 2016, the European Commission approved an *Action Plan: Financing Sustainable Growth*[8] (EC Action Plan), which after recognizing the key role of the financial system in adapting public policies to address climate change concerns, aimed at:

a. reorient capital flows towards sustainable investment in order to achieve sustainable and inclusive growth;
b. manage financial risks stemming from climate change, resource depletion, environmental degradation and social issues; and
c. foster transparency and long-termism in financial and economic activity.

Sustainable finance emerged thus as a unified concept addressing the role of the financial system in supporting and achieving sustainability. Sustainable finance[9] refers to the inclusion in the investment decision-making process of environmental and social considerations that lead to increased investments in longer term and sustainable activities. The environmental considerations refer to climate change mitigation, and adaptation, as well as the environment more broadly and related risks (e.g. natural disasters).[10] Social considerations may refer to issues of inequality, inclusiveness, investment in human capital and communities.[11] The governance of public and private institutions, including management structures, employee relations, and executive remuneration, plays a fundamental role in ensuring the inclusion of social and environmental considerations in the decision-making process. We are introducing the variable of environment and social in the classic factors of financial analysis (i.e. profitability, risk and liquidity).

Sustainable finance and socially responsible investing (SRI), although being part of the broader universe of corporate social responsibility (CSR[12]), are basically taking over CSR and putting in the forefront environmental, gender, diversity and inclusion issues, which were always the weakest element of the

---

[8] Cf. COM/2018/097 final, 8 March, 2018.

[9] Tapia Hermida defines sustainable finance has "the set of mechanisms, people and institutions that intend to facilitate the allocation of savings of the families and companies to the productive investment in a way that considers the limited natural resources of the planet, does not harm the environment and does not affect the future of future generations" (Cf. Tapia Hermida, *Sostenibilidad financiera en la Unión Europea: El Reglamento (UE) 2019/2088 sobre las fianzas sostenibles*, La Ley Unión Europea, 77, 2020).

[10] Paragraph 1 of the EC Action Plan.

[11] See Paragraph 1 of the EC Action Plan.

[12] Historically, sustainability was divided into three pillars (the so-called "triple bottom line"): people, planet, profit and it characterized the CSR movement (see Elkington, Accounting for the Triple Bottom Line, 2 *Measuring Business Excellence*, 1998, 18–22).

CSR equation.[13] The environmental, social and governance (ESG)[14] have emerged as the criteria in the decision-making process and are now fundamental in the definition of the goals of companies, investors, employees, and customers.

2 The Green Bonds: Definition and Origins

The green bonds, the social bonds or the sustainability bonds (and more recently the blue bonds) are the measures (within sustainable finance) intending at introducing an environmental or social elements or both in the way we raise finance and on the projects, assets, and businesses that we finance.

There is no legal definition for green bonds, nor a clear fact pattern that determines what is a green bond and there are bonds that are "green" in substance without being labelled as such. However, there are two clear elements in a green bond: a climate-enhancing (*green*) and a debt instrument (*bond*).

Looking at the OCDE definition of green bonds, they are distinguished from a plain vanilla bond for being "labelled" as green by the issuer as a result of being exclusively earmarked to finance "green" projects, assets, or business activities.[15] Earmarking consists of funds, such as from a bond issuance, which are set aside to pay for a specific project or event.

The International Capital Markets Association (ICMA) established in 2014 a voluntary system of rules, named Green Bond Principles (GBP), that are focused on transparency and publicity as mechanisms of incentivizing the use of green bonds, which has become the most important set of rules in this filed. The GBP's defines green bonds as "any type of bond instrument where the proceeds will be exclusively applied to finance or refinance in part or in full new and/or existing eligible Green Projects",[16] which include development of renewable energy, energy efficiency (e.g. buildings), clean transportation, pollution prevention (e.g. recycling) and control, sustainable management of living natural resources, terrestrial and aquatic biodiversity conservation, sustainable water management, eco-efficient production, processes, technologies, and products.

The first issuances of green bonds were made by multilateral financial institutions. In 2007, the European Investment Bank (EIB) issued the first green

---

[13] See Sjäfjell/Bruner, Corporations and Sustainability, in Beate Sjafjell/Christopher Bruner, *Corporate Law, Corporate Governance and Sustainability*, Cambridge, 2020, 7.

[14] See the approach to sustainability proposed by Sjäfjell/Bruner, as an integrated one with "ecological limits of the planet, putting forward the concept of "Planet Boundaries" (*Corporate* Law note 13, 7).

[15] See OECD green bonds, note 3.

[16] See International Capital Markets Association, The Green Bond Principles 2016: Voluntary Process Guidelines for Issuing Green Bonds, 2017, available at https://www.icmagroup.org/assets/documents/Regulatory/Green-Bonds/GreenBondsBrochure-JUNE2017.pdf.

bond labelling it "climate awareness bond" and was followed, in 2008, by the International Bank for Reconstruction and Development (IBRD) that named it for the first time "green bond".[17] Initially, only multilateral development banks (MDBs) and other public development agencies issued green bonds thus the policies and practices of these issuers influenced strongly the governance of the green bond market. In the first years, green bonds were a small portion of the financing of development agencies but that change rapidly especially since 2013 when green bonds started emerging as one of the keys to sustainable finance.

After the MDBs', the following issuers were State agencies and municipalities, such as California, Goteborg, Connecticut, New York and others.[18] Green bond private issuances were only made in 2013, with Swedish real estate company, Vasakronan, issuing the first corporate green bond. This paved the way for several utilities like Engie (former GDF Suez), EDF, Iberdrola, BBVA or transportation companies such as Toyota to issue green bonds and start the expansion of this market. These issuances have financed the development of renewables and low carbon production as part of the companies' core business models. In 2016, Poland was the first country in the world to issue a green bond and was rapidly followed by France.[19]

Green bonds have emerged as one of the most dynamic and relevant elements of sustainable finance with the demand for purchasing green bonds clearly outstripping the supply leading to what is known as "greenium": i.e. bonds sold with premium vis-à-vis comparable plain vanilla bonds.

In terms of types of bonds, there are four main types of green bonds identified by the GBP's: use of proceeds, green revenue bonds, green project bonds and green securitized bonds.

The use of proceeds bonds consists, as per the definition of the GBP's, on the utilization of the proceeds of the bond for Green Projects and whose debt instrument basically provides the holder recourse against the issuer. The Green Revenue Bond is "a non-recourse-to-the-issuer debt instrument aligned with the GBP in which the credit exposure in the bond is to the pledged cash flows of the revenue streams, fees, taxes etc., and whose use of proceeds go to related or unrelated Green Project.[20] As for the Green Project Bond, they

---

[17] OECD green bonds, note 3.

[18] *See* Office of the Comptroller, City of N.Y., A Green Bond Program for New York City, 2014, available at https://comptroller.nyc.gov/wp-content/uploads/documents/Green_Bond_Program_-September.pdf; Press Release, State of Connecticut Treasurer's Office, Treasurer Nappier Announces Connecticut's Inaugural Issuance of Green Bonds, 2014, available at http://www.ott.ct.gov/pressreleases/press2014/PR102214StateIssuesGreenBonds.pdf.

[19] *See* Helene Durand, *Poland Puts Stake in the Ground for First Sovereign Green Bond*, Reuters, 2016, available at http://www.reuters.com/article/bonds-markets-idUSL5N1E04UD; Anna Hirtenstein, *Green Bond Giant Awakened by Countries Spending to Save Climate*, Bloomberg, 2017, available at https://www.bloomberg.com/news/articles/2017-01-20/green-bond-giant-awakened-by-countries-spending-to-save-climate.

[20] Note 1 to the GBP's.

consist of a project bond for a single or multiple Green Projects for which the investor has direct exposure to the risk of the projects with or without potential recourse to the issuer, and that is aligned with the GBP. Finally, the banks and financial institutions have now started to issue the Green Securitized Bonds: a bond collateralized by one or more specific Green Projects, including but not limited to covered bonds, ABS, MBS, and other structures; and aligned with the GBP.

3 Private Governance of the ESG Bond Markets and Its Limitations

The emergence of the ESG bond markets raises however important issues in terms of governance. Whom determines what a green bond is? Can all bonds be green? Who controls greenwashing[21] and what is done to prevent it?

It is certainly unappropriated for the issuers or the financial intermediaries that underwrite or place the green bonds in the market to determine whether or not the bond is green.[22] The governance of the green bond market has been mainly led by private governance, in particular standards, procedures, and institutions that establish ESG rules and regulations. These private governance rules are filling the lack of public governance rules and their authority does not derive from governments or internal codes of conduct. The ESG rules are often created and enforced by governance clubs and the "membership" in the same is voluntary, requires compliance with certain policies and wants to achieve social and environmental externalities.

One of the most relevant *private governance clubs* is the International Capital Markets Association (ICMA) and its' GBP's in what concerns green bonds regulations.[23] The GBPs set forth four requirements to consider a bond green:

i. *use of proceeds* (allowing for four types of bonds referred above)—which need to be allocated to a Green Project;
ii. *project evaluation and selection*—the issuer needs to communicate to the investors in a clear way the criteria of evaluation and selection of the project so that the investor can assess the effective sustainability nature;

---

[21] A company incurs in greenwashing when it "mislead consumers about its environmental practices or the environmental benefits of a product or service" (see Delmas/Burbano, The Drivers of Greenwashing, 54 *California Management Review*, 2011, 66).

[22] See Tracey M. Roberts, Innovations in Governance: A Functional Typology of Private Governance Institutions, 22 *Duke Environmental Law and Policy Forum*, 2011, 83.

[23] See the requirements of the GBP's on https://www.icmagroup.org/sustainable-finance/the-principles-guidelines-and-handbooks/green-bond-principles-gbp/.

iii. *management of proceeds*—the proceeds need to be allocated to a specific account or portfolio so that the issuer can control effectively the use of the same to the Green Projects; and
iv. *reporting*—the issuers need to provide updated information to the investors on the use of the funds.

The GPB's are focused on transparency and accordingly they recommend external review that can consist of second opinions, verification, certification or ratings third-party opinions. The GBP's are not prescriptive but their overarching mission is for issuers, investors, and other market participants to expand the green bond market through private standards.[24]

Apart from the ICMA, the Climate Bonds Initiative (CBI) is the other *main private governance club* that uses certification as a form of governance. Basically, the CBI issues standards and certification regimes in order for the green bond to receive a certification seal or label, accreditation of the certifier. The CBI has the most developed taxonomy in the green bonds market (the Climate Bonds Taxonomy) and, in order for a green bond or green loan to be certified by CBI, it needs to meet certain standards that have been recently updated[25] and that apply to pre-issuance and post-issuance. It is also possible to obtain labelling or certification after the issuance is made to the extent that the requirements are met.

One of the conditions for the certification is that an independent third-party assurance provider or auditor verifies and ensures that the issuer is complying with the Climate Bonds Standards and prepare and submit to CBI a formal assurance report in accordance with existing auditing and assurance standards, such as ISAE 3000. The CBI's governance regime is more investor-oriented and more inclusive than the GBPs.

The most common form of external verification of compliance/governance[26] is the second opinion. They consist of an "independent review of the framework of rules, regulations, and guidelines used by a green bond issuer". The second opinion focuses on "the analysis on the process by which an issuer selects projects and investments to determine whether the selection criteria contribute to reductions in greenhouse gas emissions" and also "the issuer's broader environmental policies, the role of environmental

---

[24] See Kim Park, Investors as Regulators: Green Bonds and the Governance Challenges of the Sustainable Finance Revolution, 54 *Stanford Journal of International Law*, 2018, 24.

[25] See the Climate Bond Standards Version 3.0, available at https://www.climatebonds.net/files/files/climate-bonds-standard-v3-20191210.pdf.

[26] See Chiu, Standardization in Corporate Social Responsibility Reporting and Universalist Concept of CSR? A Path Paved with Good Intentions?, 22 *Florida Journal of International Law*, 2010, 361.

experts and environmental impact analysis, and reporting frameworks in assessing a green bond issuance".[27]

External assurance is one of the most important governance tools in CSR and, with the dispersion of green bonds' principles and standards, the external assurance gained prominence. CICERO is the largest player but there are others such as Vigeo, DNV GL, Sustainalytics.

Apart from these private governance clubs that issue specific standard or certification rules or external assurance, there are also green bonds indices. The sustainability indices (e.g. Dow Jones Sustainability Indexes; FTSE4Good Index) are a governance tool.[28] Indices are mainly a way of informational regulation, i.e. they disclose company-specific information that can influence decision-making, producers and consumers.[29] The mechanisms of disclosure include labelling, ranking, or reporting.

The indices use different eligibility criteria and methodology and they aim at providing investors the ability to track the performance of green bonds portfolios vis-á-vis the normal bonds.

However, private governance does not solve all concerns around greenwashing, it attempts to take benefit of the financial and commercial advantages associated with a "green culture".[30] Investors may be led to subscribe for green bonds of a certain company based on erroneous information on the sustainability policy of the company and the greenness of the projects it is pursuing. Investors that cannot accurately assess the ESG policy of companies will tend to "over-estimate" and "over-claim from their clients" the green nature projects[31] and private governance does not address these issues. In fact, the dispersion of criteria, the flexibility of the qualifications of green projects, the lack of constant monitoring after the issuance allow for greenwashing strategies. As for scholars point out, investors will then tend to discount the sustainability value and invest on green bonds exclusively on the financial risk, price and other purely financial factors regardless of the green factor.[32] The focus of process rules and policies exclusively on the moment of the issuance

---

[27] CICERO, Framework for Cicero's 'Second Opinions' On Green Bond Investments (April 28, 2016), available at https://cicero.oslo.no/en/posts/single/CICERO-second-opinions.

[28] See Oren Perez, Private Environmental Governance as Ensemble Regulation: A Critical Exploration of Sustainability Indexes and the New Ensemble Politics, 12 *Theoretical Inquiries Law*, 2011, 543.

[29] See Light/Ortz, Parallels in Public and Private Environmental Governance, 5 *Michigan Journal of Environmental and Administrative Law*, 2015, 39.

[30] See Álvaro Gómez Expósito, Finanzas Sostenibles y Bonos Verdes ante la Emergencia Climática, in Revuelta Perez/Alonso Mas, *La regulación de la energía limpia ante la emergencia climática*, Thomson Reuters, Aranzadi, 2020, 442.

[31] See Kim Park, Green Bonds and Beyond. Debt Financing as a Sustainability Driver, in Beate Sjafjell/Christopher Bruner, *Corporate Law, Corporate Governance and Sustainability*, Cambridge, 2020, 606.

[32] See Kim Park, Green Bonds, note 31, 606.

narrows the green analysis to a very specific moment in time without any overarching analysis of the sustainability of the issuer.

Additionally, the degree of transparency is low, which allows the issuers to provide none or very limited information on the effective use of the proceeds. This is mainly a result of the voluntary nature of the governance rules that make green bonds as green as the issuer wants. It's fundamental to ensure that there is constant information regarding the indicators, standards or indexes and restrict the autonomy of the issuers as it is the case of GPB's that manifestly do not intend to stress what indexes, standards or indicators are good.[33]

In the case of social bonds, there are further difficulties in determining what is in fact a social bond, what policies is the issuer required to implement in order to maintain the social labelling and how to monitor adequately the social commitment of the issuer. The ICMA Social Bonds Principles shed some light on the requirements applicable to social bond issuances but are clearly insufficient in what concerns monitoring. The recent EU Sure Social Bond Framework based on such principles is a successful example of the issuance of social bonds but, for being an issuance made by EU entities, it has itself credibility, including on the subsequent monitoring.[34]

4 PUBLIC GOVERNANCE OF THE ESG BOND MARKETS

The lack of response to governance challenges posed by the green bond market calls for the need of public governance rules binding on market players in light of the relevance of green bonds.[35] Such relevance is both qualitative, given their main role in the Sustainable Growth Objectives and also quantitative as a result of the emergent green bond markets.[36]

The first example of public regulation in respect of the green bond market came from China and was followed by India.

In 2015, the People's Bank of China (China's central bank) enacted regulations on the green bond market, which are, in general, aligned with the private governance standards of the most relevant entities referred above.[37]

---

[33] See Álvaro Gómez Expósito, Finanzas, note 30, 441.

[34] The SURE (Support to mitigate Unemployment Risks in an Emergency) instrument "aims at preserving employment in order to sustaining families' income and the economy as a whole, thus targeting the general population impacted by the COVID19 pandemic in the EU" (see https://ec.europa.eu/info/sites/info/files/about_the_european_commission/eu_budget/eu_sure_social_bond_framework.pdf).

[35] See Vogel, Private Global Business Regulation, 11 *Annual Review of Political Science*, 2008, 168–169.

[36] See Freedman, *Financing Green: The Rise of the Green* bond, in *Law360*, 2014, available at http://www.law360.com/articles/552291/financing-green-the-rise-of-the-green-bond?article_related_content1)=1.

[37] See http://english.www.gov.cn/state_council/ministries/2017/03/03/content_281475583659044.htm.

China is the largest green bond market in the world[38] but simultaneously the largest emitter of greenhouse gases. This is the reason why it allows for issuance of green bonds such as "clean coal", not allowed by the majority private governance regimes.

In 2016, the Securities and Exchange Board of India (SEBI) approved the disclosure requirements for issuance and listing of green debt securities, which are in line with GPB's and Climate Bond Initiative standards.[39]

However, the most coherent and expanded response was given by the European Union in 2020 with the approval of Regulation (EU) 2020/852 of the European Parliament and of the Council of 18 June 2020 on the establishment of a framework to facilitate sustainable investment ("**Taxonomy Regulation**").

Taxonomy Regulation sets the criteria for determining whether an economic activity qualifies as environmentally sustainable for the purposes of establishing the degree to which an investment is environmentally sustainable. It's the beginning of a true EU taxonomy on the green bond market with the aim of facilitating funding for environmentally sustainable activities, as the economic activities could be compared against uniform criteria in order to be selected as underlying assets for environmentally sustainable investments. In fact, investors will find it disproportionately burdensome to check and compare different financial products which can discourage them from investing in environmentally sustainable financial products.[40]

Taxonomy Regulation basically establishes four criteria for determining whether an economic activity qualifies as environmentally sustainable:

a. contributes substantially to one or more of the six environmental objectives set out in Article 9 (i.e. climate change mitigation; climate change adaptation; sustainable use and protection of water and marine resources; transition to a circular economy; pollution prevention and control; protection and restoration of biodiversity and ecosystems).
b. does not significantly harm any of the environmental objectives set out in Article 9 and defines in Article 17 certain activities that have this effect (e.g. activity leads to significant greenhouse gas emissions; activity leads to an increased adverse impact on the current climate and the expected future climate, on the activity itself or on people, nature, or assets; activity leads to a significant increase in the emissions of pollutants into air, water, or land, as compared with the situation before the activity started);
c. is carried out in compliance with the minimum safeguards laid down in Article 18 (i.e. alignment with the OECD Guidelines for Multinational

---

[38] See https://www.climatebonds.net/system/tdf/reports/2019_cbi_china_report_en.pdf?file=1&type=node&id=47441&force=0.

[39] See https://www.sebi.gov.in/sebi_data/meetingfiles/1453349548574-a.pdf.

[40] Whereas (12) and (13) of Regulation 2020/852.

Enterprises and the UN Guiding Principles on Business and Human Rights—it's the EU compromise to respect human rights);

d. complies with technical screening criteria established by the Commission, which, in turn, need to comply with several requirements set forth in Article 19 (e.g. identify the most relevant potential contributions to the given environmental objective while respecting the principle of technological neutrality; specify the minimum requirements that need to be met to avoid significant harm to any of the relevant environmental objectives; use sustainability indicators; build upon Union labelling and certification schemes).

In lack of a certification and labelling system of the EU, the existence of an EU taxonomy that allows determining clearly what type of economic activities and projects should be labelled green for being environmentally sustainable brings clarity to the market and increases confidence among investors. Taxonomy Regulation is the first true legal step aimed at channelling private investment to sustainable activities and, by designing a specific taxonomy on sustainable activities, it is creating an incentive scheme to redirect private investment to green projects.[41] The next step, more difficult, is to create a European system of labelling and certification similar, for example to that of the Eco management and Eco audit systems (EMAS) foreseen in Regulation (EC) No 1221/2009 of the European Parliament and of the Council of 25 November 2009 on the voluntary participation by organizations in a Community eco-management and audit scheme (EMAS), as lastly amended by Commission Regulation (EU) 2018/2026.[42]

5   HYBRIDIZATION APPROACH AND CONTRACTUAL APPROACH

Certain authors[43] sustain a different approach to green bond governance, in particular the hybrid public–private governance regimes that have elements of private and public governance to "optimize regulatory outcomes".

The concept of hybridity shows how "private governance regimes and public regulatory systems that regulate a given market, conduct, or sector can work together".[44] The hybridization process is "slow, controversial and

---

[41] It's a phenomenon known as nudging, which defines the method of achieving certain objectives throughout incentives instead of mandates or prohibitions (see Möslein/Mittwoch, *Plan de Acción europeo para financiar el desarollo sostenible*, la Ley mercantile, 61, 2019).

[42] See Möslein/Mittwoch, *Plan de Acción*, note 40.

[43] See the proposal made by Kim Park, Investors as Regulators, note 24, 41–46.

[44] See Kim Park, Investors as Regulators, note 24, 41–46. According to De Búrca and Joanne Scott, there are three forms of hybridity: (i) baseline hybridity, where public regulation defines a baseline of minimum standards that private governance cannot set

uneven" in contrast with the emergence of the green bond market, but it can address legitimacy deficits.

Authors sustaining hybridization of green bonds governance propose the improvement of the use of second opinions as external assurance. They suggest independent third-party audits (which are quite common on other areas of CSR) in order to check compliance with a standard more consistently and rigorously. Another strategy of hybridization is to encourage certification. For example, in France, the labelling of green to investment funds and ESG financial products is managed by the Ministry of Ecology, Sustainable Development and Energy. The labelling could incorporate certification made by private governance entities, such as the CBI. Finally, another hybridization strategy consists of incentivizing the participation in the private governance clubs or regimes, in particular incentives and inducements for the entities to engage more actively with private governance regimes.[45]

Apart from hybridization, the changes to the contractual features of the green bond issuance are also a way to address some of the moral hazard concerns on greenwashing and on green bond issuers monitoring.

The GPB's issued by ICMA now require that the issuers make available annually updated information on the use of the proceeds of the green bonds until they have been fully allocated to the green project. The need for monitoring goes however far beyond this information duty. They require the appointment of independent third parties with broad-ranging powers to request and review confidential information regarding the company's environmental and social data, in particular the impact of its activity. Such third party should not be able to be removed by the issuer and should have own resources to carry out its activity.[46] However, this solution seems quite intrusive, burdensome, and costly for issuers. Additionally, it's fully reliant on the actions carried out by a third party and not allowing for bondholders to act directly. Therefore, the answer to this concern lies more with the regulators and disclosure obligations that are fundamental to ensure full transparency (e.g. EU Regulation 2019/2088).

Another potential solution proposed by scholars is to provide bondholders similar rights to those of the shareholders. This would be achieved throughout the issuance of "contingent convertible bonds" green bonds, which would basically allow green bondholders to convert their bonds into equity in case certain triggers are verified, i.e. breach of information duties or allocation rules

---

aside; (ii) instrumental hybridity, where private governance serves as an instrument to develop the principles defined by public regulation, and (iii) default hybridity, where public regulation as a default case private governance is insufficient or fails (Gráinne De Búrca/Joanne Scott, Introduction: New Governance, Law and Constitutionalism, *Law and New Governance in the EU and the US*, 8).

[45] See Kim Park, Investors as Regulators, note 24, 41–46.

[46] For this proposal, see Kim Park, Green Bonds, note 31, 607–608; Amihud/Garbade/Kahan, A New Governance Structure, 55 *Stanford Law Review*, 1999, 992–995.

regarding green bonds. Authors sustain that conversion does not need to be mandatory, it can just trigger automatically third-party independent audit or a call right to bondholders if the issuer fails to meet certain conditions regarding the use of proceeds of the bonds or its sustainability performance.[47] Again, this is an interesting proposal but, in light of the limited powers granted to shareholders and the difficulty in determining the breach of fiduciary duties based on ESG, it has, in our view, quite limited positive effects on greenwashing. However, the linkage between creditors' rights and ESG performance is an excellent driver for the alignment of incentives and addressing greenwashing concerns. For example, the increase of interest rate, of margins, the call option rights for bondholders or even the need to reallocate other funds of the company linked to ESG performance or use of proceeds are drivers for greenwashing concerns that then need to be completed by a constant monitoring made by an appointed bondholder representative or trustee. This has been used in several bond or loan issuances and should become a recurrent mechanism to protect those that truly want to invest in green and penalize those that want to take advantage of green.

6 INFORMATION DUTIES

One of the most important governance concerns on ESG bond markets is information and transparency. The dispersion of disclosure standards and market-based practices make it difficult to compare financial products, create "an uneven playing field" for such products and for distribution channels allowing for distorting investment decisions. Additionally, they open the door for post-issuance greenwashing that is detrimental to financial market participants and financial advisers.

In 2019, the EU, conscious of this concern and of the lack of adequate response under the existing EU Directives (in particular Directive 2014/95/EU),[48] approved Regulation 2019/2088 on sustainability-related disclosures in the financial services sector ("**Sustainable Finance Disclosure Regulation**" or "**SFDR**"). The aim of this regulation is to reduce "information asymmetries in principal-agent relationships with regard to the integration of sustainability risks, the consideration of adverse sustainability impacts, the promotion of environmental or social characteristics, and sustainable investment, by requiring financial market participants and financial advisers to make pre-contractual and ongoing disclosures to end investors when they act as agents of those end investors (principals)".

---

[47] Making this proposal, see Kim Park, Green Bonds, note 31, 609.

[48] This Directive was based on a voluntary acceptance of the application of the transparency rules defined thereunder with regards to environment and social sustainability, basing itself on the "comply or explain" method (see Tapia Hermida, *Sostenibilidad financiera*, note 9, 9 et seq).

The SFDR was amended by the Taxonomy Regulation, which, following the taxonomy definition on green bonds, created specific disclosure duties that, among others, are applicable to financial products that have sustainably investment as its objective..

Firstly, the environmental objectives need to be listed in the green financial product (i.e. climate change mitigation; climate change adaptation; etc.). Secondly, the way and measure in which the financed projects are considered sustainable investments also needs to be disclosed. Thirdly, financial market participants (and the same applies to financial advisers) need to include pre-contractual disclosures on (i) the manner in which sustainability risk are integrated into their investment decisions and (ii) the results of the assessment of the likely impacts of the sustainability risks on the returns of the financial products they make available. Fourthly, it is necessary to disclose the adverse sustainability impact of a given financial product. Fifthly, where financial product has sustainable investment as its objective and an index has been designated as a reference benchmark, it is necessary to disclose information on how the designated index is aligned with that objective and why and how the designated index aligned with that objective differs from a broad market index. Sixthly, financial market players are required to provide a description of the environmental or social characteristics of the sustainable investment objective and information on the methodologies used to assess, measure, and monitor the environmental or social characteristics or the impact of the sustainable investments selected for the financial product. Finally, financial market participants are required to make available a description in periodic reports: (i) on the extent to which environmental or social characteristics are met in case the financial product promotes, among other characteristics, environmental or social characteristic or (ii) on the overall sustainability-related impact of the financial product by means of relevant sustainability indicators in case of financial products that have sustainable investment as its objective.

Directive 2014/05/EU ("Non-Financial Reporting Directive") did not require the full disclosure regarding environmental sustainability that allowed investors to have the necessary information to introduce sustainability variable into their investment decisions. Regulation 2019/2088 and the Taxonomy Regulation are a clear commitment of the EU to sustainable finance and address some of the most relevant concerns on information and disclosure that failed to be addressed by private governance regimes. This ensures broader transparency in the green bond market but there is still some room for improvement especially in respect of the issuers themselves. This of course needs to be measured vis-á-vis the costs of information duties and the principle of proportionality.[49]

---

[49] See Barnett/Salomon, Beyond Dichotomy: The Curvilinear Relationship Between Social Responsibility and Financial Performance, 28 *Strategic Management Journal*, 2005, which sustain that the potential costs of environmental policies overcome its potential benefits.

7   Blue Bonds

Blue bonds are similar to green bonds because both are based on a debt instrument providing funding to issuers that commit to repay principal with interest. The difference is that the proceeds of blue bonds are used to finance marine projects and ocean-based projects or to safeguard the blue economy, such as promoting biodiversity and supporting economies reliant upon healthy and sustainable fisheries.

In fact, SGD 14 of the UN is concern with the exponential growth of plastic waste added to the oceans and the blue bonds intend to direct investment towards, for example, the goal of reducing plastic waste. Blue bonds are emerging as an instrument to finance solutions at scale.

The first "blue bond" was issued in 2018 by the Republic of Seychelles and consisted on a bond agreement facilitated by the World Bank to offload part of the debt of the country in in exchange for marine protection. This allowed the Republic of Seychelles to stabilize its' credit rating and invest in its economy that is closely connected to the ocean.

In January 2019, Nordic Investment Bank issued a SEK 2 billion Nordic-Baltic Blue Bond under the NIB Environmental Bond Framework, which will fund projects in wastewater treatment, prevention of water pollution and water-related climate change adaptation. Also in 2019, the Asian Development Bank (ADB) committed $5 billion for the next five years for the purposes of promoting sustainable oceans under the Action Plan for Healthy Oceans and Sustainable Blue Economies. Finally, in November 2019, The Nature Conservancy (TNC), committed to raise USD1.6 billion to be allocated to global ocean conservation efforts by mean of a "blue bonds for conservation" scheme.[50] TNC works with countries to refinance a portion of their national debt in a way that reduces their debt burden, secures funding for conservation activities, and enables valuable returns in planning and protection to improve resilience of economies and communities. The debt restructuring creates new financial flows that support governments to reach their protection targets for their ocean areas, including for coral reefs, mangroves and other important habitats, and engage in ongoing conservation work such as improving fisheries management and addressing climate change adaptation.

In 2020, the Bank of China issued a $942.5 million bond with proceeds earmarked for the financing or refinancing of marine-related eligible green projects defined in the issuer's Sustainability Series Bonds Management Statement.

Contrary to green bonds, we are still in the early stages of the blue bond market and the fundamental definitions and concepts are being defined. A recent work from a group of entities and experts on the sector referred that

---

[50] See https://www.nature.org/en-us/what-we-do/our-insights/perspectives/an-audacious-plan-to-save-the-worlds-oceans/.

blue bond project categories might focus on blue natural capital, the sustainable blue economy, conservation, and restoration of coastal areas, as well as the sustainable use of the ocean.

The concept of a "Blue Economy" was first used in the 2012 Rio+20 Conference and emphasizes "conservation and sustainable management, based on the premise that healthy ocean ecosystems are more productive and a must for sustainable ocean-based economies". The core concept is to ensure that socioeconomic marine and costal development do not lead its environmental destruction and aim at outlining the value of the so-called "blue natural capital"[51] in economic activity. Differently, the World Bank defines "Blue Economy" in a broader sense as encompassing the promotion of economic growth, social inclusion and preservation or improvement of livelihoods along with the environmental sustainability. Similarly, the OECD concept also covers all economic sectors that are linked, directly or indirectly to the ocean economy.

The European Commission uses the concept of "Blue Growth" to support sustainable growth in the marine and maritime sectors as a whole with the seas and oceans being drivers for the economy and having great potential for growth and innovation. Regulation 2020/852 inserted, in the EU taxonomy of environmental objectives, the sustainable use and protection of water and marine resources.

In fact, blue bonds are often linked to green bond issuances and are indicated as one of the goals of green bonds. Further to Regulation 2020/852, the GBP's have within its taxonomy items that are relevant for blue bonds, in particular the project criteria for Marine Energy and the Water Infrastructure. For example, CBI Water Infrastructure criteria includes coastal ecosystem conservation/restoration activities. However, and differently from green bonds and social or sustainable bonds, they are not regulated by specified principles issued by private governance regimes, namely the ICMA. There is still a way forward to be done in this regard. In any case, this market has evolved and there are now three types of blue bonds: "blue-sustainable bond", "blue sustainability linked-bond", and "blue-green bond" aligned with the principles developed by ICMA.

Recently, the EU along with the EIB and the Prince of Wales's International Sustainability Unit developed the Sustainable Blue Economy Finance Principles.[52] These principles intend to (i) promote the implementation of the SGD 14 (Life below water), (ii) set out ocean-specific standards while avoiding duplicating existing frameworks for responsible investment and (iii) comply with IFC Performance Standards and EIB Environmental and Social Principles and Standards.

---

[51] Blue natural capital is the natural capital in the coastal and marine environments.
[52] See https://ec.europa.eu/maritimeaffairs/befp_en.

We don't know if the blue bond market will have the same development as the green bond market but it's shaping its way rapidly in the ESG global markets.

8 Conclusion

Green bonds and blue bonds are a strong instrument to mobilize private sector investments for the necessary energy transition with a view to carbon emission goals, reduce plastic waste, and preserve ocean biodiversity. The reorientation of capital flows towards sustainable investment in order to achieve sustainable and inclusive growth is at the heart of a proper governance strategy for sustainability of our lives and our planet.

Green bonds have been modelled by private governance (permissive rules) whose inevitable dispersion following the massive growth of the green bonds markets now requires a stronger intervention from public governance (prescriptive rules) in order not to allow for all bonds to be green. The recent EU legislation is a step in the right direction of the credibility of the market and the subsequent step is to increase monitoring to strengthen the governance of green bonds so that "green remains green".

CHAPTER 9

The Role of Companies in Promoting Human Rights

Ana Rita Campos

1 THE INTERACTION BETWEEN CORPORATE GOVERNANCE AND HUMAN RIGHTS

States have primary duties in relation to human rights, including protecting against harm by companies' business. However, companies also have responsibilities in relation to human rights, since they can significantly affect them in the course of their activities, as showed in the research conducted by the UN[1] when developing the The UN Guiding Principles.[2]

Responsibility to respect human rights should be the baseline global expectation for all companies. This responsibility extends to all internationally recognized human rights including, for example, those set forth on the

---

[1] "Corporations and Human Rights: A Survey of the Scope and Patterns of Alleged Corporate-Related Human Rights Abuse," UN Document A/HRC/8/5/Add. 2 (23 May 2008).

[2] UN Human Rights Council, The Guiding Principles on Business and Human Rights: Implementing the United Nations "Protect, Respect and Remedy" Framework, HR/PUB/11/04 (2011).

A. R. Campos (✉)
Lisbon, Portugal
e-mail: Ana.campos@ecb.europa.eu

© The Author(s), under exclusive license to Springer Nature Switzerland AG 2022
P. Câmara and F. Morais (eds.), *The Palgrave Handbook of ESG and Corporate Governance*, https://doi.org/10.1007/978-3-030-99468-6_9

"The International Bill of Human Rights"[3] and on the International Labour Organization's Declaration on Fundamental Principles and Rights at Work.[4]

Internationally recognized human rights include rights to life and physical security, non-discrimination, rights to freedom of thought, expression and religion, freedom of assembly and movement, rights to education and work, to family life and privacy, to food and water, freedoms from torture, slavery, or forced labour, as well as rights to fair and decent working conditions, freedom of association and the right to bargain collectively, the effective abolition of child labour, and the rights of indigenous people.

Emblematic examples of breaches of human rights involving companies in the 1980s and 1990s include revelations of child labour in Nike's suppliers in SE Asia; the death of thousands following an explosion at Union Carbide's pesticide plant in Bhopal, India; and Yahoo's revealing to Chinese authorities the name of a customer who was then imprisoned for reporting on public unrest. Other impacting examples are the 2013 collapse of the structurally defective Rana Plaza factory in Bangladesh, which killed over 1,100 poorly paid workers and injured about 2,000 others, and more recently, the scandal of goods made in China using forced Uyghur labour.[5]

The laws of individual countries have proven insufficient to regulate the human rights behaviour of global companies with transnational activities, running business in several parts of the world and subject to different legal environments: while some countries have in place binding laws protecting human rights (e.g. provisions covering freedom of association, the right to collective bargaining; the elimination of compulsory labour, abolition of child labour, elimination of discrimination with respect to employment and occupation), others don't. These often fail to comply with international human rights duties and to keep up with the pace of economic changes.

The fact that some companies exercise their activity in countries where human rights are neither respected nor adequately protected increases their exposure to potential breaches of human rights. This in turn demands a higher duty of care and attention in avoiding breaches and in improving conditions to prevent them. In a globalized world, human rights should be high on companies' agendas and ideally embedded in their corporate purpose and seen as part of shareholders' interest. Companies should not overlook them and take advantage of poor human rights frameworks to achieve their results and profits.

---

[3] The International Bill of Human Rights consists of the Universal Declaration of Human Rights, the International Covenant on Economic, Social and Cultural Rights, and the International Covenant on Civil and Political Rights and its two Optional Protocols. For some operating contexts where specific groups may be at special risk (such as indigenous people, women, children) more focused UN instruments should be considered (UN Guiding Principles, Principle 12).

[4] International Labour Organisation, ILO Declaration on Fundamental Principles and Rights at Work, 1998.

[5] *The Economist*, Torment of the Uyghurs, p. 11, October 17th, 2020.

Therefore, human rights and corporations cannot be considered two separate and distinct discussions. The responsibility of companies in the human rights context can and should be framed within their activities, especially when operating in fragile and complex human rights environments. This is expressly recognized in several international instruments, and particularly in the UN Guiding Principles. Notwithstanding, two important questions arise: Is corporate responsibility for human rights mandatory or voluntary? And is there a corporate duty to respect Human Rights beyond what is established by international instruments?

As noted by Barnali Choudhury and Martin Petrin,[6] the wording of the UN Guiding Principles, also when compared with the wording of other international initiatives, points out to a more mandatory than voluntary nature of the corporate responsibility with regard to respecting human rights (references to "should" as opposed to "recommended"). The UN Guiding Principles have also been followed by other international initiatives and notably by many states which are increasingly enacting mandatory and binding instruments in this field.[7]

As for a possible duty beyond corporate responsibility, utilitarian, and non-utilitarian reasons are also put forward to ground the imposition of a mandatory duty for companies to respect human rights, even when there is no legal requirement.

Utilitarian arguments for compliance with human rights are based on companies' self- interest (to avoid negative public opinion and legal claims) and on a duty that serves as an underlying necessity for companies to retain their license to operate. Furthermore, utilitarian arguments include the idea that companies benefit from operating in societies which protect human rights. Voices against these utilitarian arguments argue that this limits companies' responsibility to situations that are not in conflict with their wealth maximization and self-interest.

Non-utilitarian arguments assert the imposition of a mandatory duty to respect human rights. These arguments draw on justifications that impose public duties on companies, such as companies' instrumental power: the power to take decisions which may affect human rights. Furthermore, non-utilitarian arguments also enhance companies' duties to respect human rights arising from their ability to exploit differences in national legal regimes and establish business in the most favourable locations, where protection of human rights may be weaker. In short, corporate's ability to affect human rights and to operate in the above locations justifies a duty of companies to respect human rights, even in the absence of legal obligations. Such a duty is routed in companies' ethical obligations to respect human rights in their interaction

---

[6] Barnali Choudhury and Martin Petrin, *Corporate Duties to the Public*, Cambridge University Press, 2019, pp. 230, 231.

[7] For international and local legislative and regulatory initiatives, see section "Some Domestic Initiatives" below.

with those with whom they engage in their business activities[8] and on the fact that companies have the resources to do so.

Companies' corporate objectives are not incompatible with the respect for human rights. The fact that companies should pursue profit in complying with the best interest of their shareholders, should not hinder a balance that needs to be strike between shareholders and other stakeholders (e.g. employees, community). A balance which should be embedded in the companies' Corporate Social Responsibility (CSR).

Although the duty of companies to respect human rights cannot be compared with the duty of states, both utilitarian and non-utilitarian arguments briefly discussed above, support the idea corporate duties in this area are not merely a voluntary responsibility. Respecting human rights must not be seen as optional for companies. In pursuing corporate objectives, companies have the power, capacity and means to fight against human rights abuses and should be interested in doing so. The scope of such duty comprises their activities and should also be extended to areas in which they can exercise their leverage, inter alia, over those with whom they do business.[9] Fostering a human rights compliance culture, entails proactive, not just reactive, responsibilities and requires companies to actively incorporate respect for human rights in corporate decision-making[10] and culture.

2 REGULATORY POLICY APPROACHES RELATED TO CORPORATE GOVERNANCE—HUMAN RIGHTS INTERACTION

The responsibility and duty of companies to respect human rights have so far been dealt with by states, international institutions, legislators, and companies, in ways that encompass soft law, hard law, and contractual and voluntary schemes.

Although the European Union and other countries outside Europe have adopted and continue to do adopt binding legislative initiatives on corporate responsibility for human rights, there is not yet a specific, overarching international mandatory framework. Therefore, there is a gap in legally enforceable rules on companies' responsibility and duty to respect human rights that can be globally applied and enforced. One of the reasons for this gap lies on the fact that corporate laws and human rights laws tackle different objects and hence have legally developed in separate ways, as unrelated regimes.

This chapter cannot provide an overview of all instruments and their development over time. Nevertheless, the next pages will discuss some of the most important human right instruments. The instruments discussed next include

---

[8] Barnali Choudhury and Martin Petrin, *Corporate Duties to the Public*, Cambridge University Press, 2019, pp. 234, 235.

[9] For further details on the leverage power, see section "Leverage Power" below.

[10] On companies and board's responsibilities, see section "Board Duties (Consideration of Human Rights Issues Ate Board Level)" below.

both hard and soft law and entail rules, standards, principles, and guidelines that should be adopted by companies around the globe in complying with and promoting human rights.

International Initiatives

OECD Initiatives

In response to previous unsuccessful international efforts to establish corporate duties for human rights, in 1976 OECD countries adopted an instrument containing guidelines to regulate multinational companies: the OECD Declaration on International Investment and Multinational Enterprise. Although not containing specific references to human rights, in its non-binding guidelines multinational companies were recommended to make positive contributions to economic and social progress.[11] The OECD's Guidelines, renamed as the OECD Guidelines for Multinational Enterprises,[12] were revised in 2011 to include a specific chapter on human rights and the concept of due diligence, which essentially mirrors Pillar II of the UN Guiding Principles.[13]

United Nations Initiatives

The UN Guiding Principles, adopted in 2011, provided the first global set of rules for preventing and addressing the risk of adverse impacts on human rights linked to business activity. Today they still hold the status of the authoritative global standard on business and human rights and continue to be the largest source of inspiration of public and private initiatives on compliance with human rights around the globe.

The UN Guiding Principles consists of thirty-one Guiding Principles, based on three interdependent pillars, the so called "Protect, Respect, and Remedy" framework. The framework resulted from the combination of three sources of governance: Corporate Social Responsibility (defined as a source of governance based on voluntary conduct and self-regulation that is grounded in a company's long-term self-interest), nation states, and civil society (including stakeholder advocacy groups, trade unions, and investors, etc.).

First, under the UN Guiding Principles states have a duty to protect human rights from abuse by third parties, including business, through appropriate law, policy, regulation, and adjudication (Principles 1–10). Second, companies have a responsibility to respect human rights; this means that they should identify, avoid, and address harm to human rights through their activities and business relationships (Principles 11–24). Third, where individuals and communities have suffered harm, both states and companies have a role to play in providing access to an effective remedy (Principles 25–31).

---

[11] OECD, Guidelines for Multinational Enterprise (1976) 15 I.L.M. 969, para. 12.
[12] reporting.
[13] See section "Corporate Actions Towards Human Rights" below.

In the report underlying the UN Guiding Principles, it is noted that *overall, the corporate responsibility for respecting human rights entailed a requirement not to infringe on the human rights of others*.[14] This responsibility is built upon a twofold perspective: companies should avoid causing or contributing to adverse human rights impacts (preventive, negative nature) and they should address such impacts when they occur (reactive, proactive nature). Additionally, companies should aim *"to prevent or mitigate adverse human rights impacts that are directly linked to...operations, products or services"* through business relationships.[15]

How can companies comply with these responsibilities? They should adopt internal human rights policies including a *"due diligence process to identify, prevent, mitigate and account"* for negative impact on human rights, processes to remediate breach of human rights and the disclosure on such policies and remedies.[16] To support such endeavour the UN published the UN GP Interpretive Guide.[17]

Although it confirms the responsibility of companies with relation to human rights, the UN Guiding Principles are not legally binding and remain a soft law instrument.

Further to the UN Guiding Principles, other isolated UN initiatives took place towards a reflection on possible binding human rights instruments and measures to address companies related human rights abuses.[18] These paved the way to further negotiations and, in June 2014, the UN Human Rights Council took steps to elaborate an *international legally binding instrument to regulate the activities of transnational corporations and other business enterprises:* the business and human rights treaty, the Bill of Human Rights (BHR). At the time of writing, the treaty negotiations and development are ongoing, and its third revised draft has been made public in August 2021.[19]

The BHR would require states to *take appropriate legal and policy measures to ensure that business enterprises, including transnational corporations and other business enterprises that undertake business activities of a transnational character, within their territory, jurisdiction or otherwise under their control, respect internationally recognized human rights and prevent and mitigate*

---

[14] UN Guiding Principles, Principle 11.

[15] UN Guiding Principles, Principle 13.

[16] UN Guiding Principles, Principle 15; Principles 17–21.

[17] The Corporate responsibility to respect human rights—an interpretive guide, 2012.

[18] Statement on behalf of a Group of Countries at the 24rd Session of the Human Rights Council General Debate, "Transnational Corporations and Human Rights" (September 2013); Resolution 26/9 Elaboration of an international legally binding instrument on transnational corporations and other business enterprises with respect to human rights (July 2014).

[19] Legally binding instrument to regulate, in international human rights law, the activities of transnational corporations and other business enterprises.

human rights abuses throughout their business activities and relationships.[20] In short, under the BHR States are required to establish in their domestic law legal liability for human rights abuses, to provide fair, effective, and prompt access to justice and remedies to victims and to require human rights due diligence by companies.[21]

EU Initiatives

At the EU level, the 2014 Non-Financial Reporting Directive[22] requires EU Member States to implement public reporting by large companies of information on the way they operate and manage social and environmental challenges, including respect for human rights. Such disclosure includes a description of relevant risks, policies and outcomes related to the topic and is expected to help investors, consumers, policymakers and other stakeholders to evaluate non-financial performance of the companies in scope and encourage them to develop a responsible approach to business.

In December 2020, the EU Council called for a proposal from the European Commission (EC) for a legal framework including cross-sector corporate due diligence obligations along global supply chains[23] and the EC announced a legislative initiative to ensure EU companies act to prevent and reduce any negative impacts to workers and communities in their operations and supply chains through mandatory human rights and environmental due diligence. This initiative was included in the public online consultation by the EU on its legal framework on sustainable corporate governance, which ended on 8 March 2021.

In response to the EC's announcement, on 10 March 2021, the European Parliament (EP) passed a resolution with recommendations on the envisaged framework,[24] focusing on two main pillars: the obligation of companies to conduct due diligence and the rights of individuals and stakeholders to hold companies liable for non-compliance.[25]

---

[20] Ibid. article 6, nr. 6.2.

[21] Shirin Chua, Freshfields Bruckhaus Deringer LLP, *UN Working Group on a Convention to Regulate Transnational Business and Human Rights—Comments on the 'Zero Draft' Due Soon* (February 2019).

[22] Directive 2014/95/EU of the European Parliament and of the Council, of 22 October 2014, amending Directive 2013/34/EU as regards disclosure of non-financial and diversity information by certain large undertakings and groups.

[23] A due diligence webinar series on this matter was supported by the current Portuguese EU presidency and is available online.

[24] European Parliament resolution of 10 March 2021 with recommendations to the Commission on corporate due diligence and corporate accountability (2020/2129(INL)).

[25] Parliament Press release, MEPs: Companies must no longer cause harm to people and planet with impunity, 11 March 2021 and Recital 26 of the Parliament resolution.

The due diligence obligation requires companies to get to know their supply chains[26] in detail and understand the risks that may be posed to, inter alia, human rights. The scope of application would be large companies operating in the EU, comprising private or state-owned companies operating under the law of a Member State. The obligation would also apply to publicly listed or "high-risk" Small and Medium-sized Enterprises and companies providing financial services and products. Although not binding on the EC, this initiative puts pressure on this topic and gives an indication of the issues which will be addressed by the EC in its legislative proposal.

Other International Initiatives
Other international initiatives also contribute to set and reinforce the scope of corporate responsibilities and duties towards human rights.

Among them it is important to mention the rules and obligations that some international financial institutions—such as the European Investment Bank,[27] the World Bank and the European Bank of Reconstruction and Development[28]—impose to their counterparties, namely borrowing or beneficiary companies. In their contractual frameworks these institutions confirm and recognize companies' responsibility for human rights and embed rules and obligations which include the obligation to comply with human rights and do no harm, to actively promote the support of sustainable development and to address adverse impacts caused or contributed to by business.

An example of such initiatives is the incorporation, by the International Finance Corporation (the private lending arm of the World Bank), of key elements of the UNG Guiding Principles' Pillar II (human rights due diligence), into the performance standards of supported companies.[29] These are tracked by many banks in several countries, covering a relevant stake of project financing in emerging markets.

Some countries take different approaches to companies' responsibilities for human rights, which go beyond the usual due diligence or reporting obligations. An example of this is the Canadian Ombudsperson for Responsible Enterprise (CORE). Created in 2018, the CORE's mandate includes promoting the implementation of the UN Guiding Principles and the OECD Guidelines, advise Canadian companies on their practices and policies with regard to responsible business conduct, receive and review claims of alleged human rights abuses arising from the operations of Canadian companies

---

[26] The Parliament proposes a wide definition of value chains catching all business and investment activities including direct and indirect business relationships upstream and downstream.

[27] European Investment Bank, Environmental and Social Handbook, 2013, p. 2.

[28] European Bank of Reconstruction and Development (EBRD), Environmental and Social Policy (ESP) (7 May 2014), p. 2.

[29] IFC, IFC, Performance Standards on Environmental and Social Sustainability, paragraph 12, (2012).

abroad in the mining, oil and gas, and garment sectors and review, on its own initiative, alleged human rights' abuses.[30]

Some Domestic Initiatives

Legislative initiatives in several countries show that states have been proposing and adopting measures that entail concerns with human rights, along the good standards and practices laid down in international initiatives. However, although the tendency is towards a binding approach to companies' obligations, some countries keep things on the "soft" side, reluctant to adopt legal acts.

The most common initiative remains to require companies to mandatorily disclose the way they are addressing human rights' issues: the main risks faced, policies adopted, the outcomes of any actions undertaken and, information about due diligence processes. In EU countries this is done via the enforcement of the Reporting Directive.[31] However, some countries go beyond general disclosure and impose disclosure on supply chain issues. This is the case of the UK Modern Slavery Act (MSA) enacted in 2015,[32] requiring companies's annual statements to disclose policies and measures taken to prevent slavery and human trafficking from taking place, both within their business and in any of its supply chains. Among other elements, the statement should inform on the nature of supply chains, their policies on slavery and human trafficking, the type of due diligence conducted to gather this information, which aspects of the business may be at risk of slavery and human trafficking and how they are being mitigated.[33] Although these are no legally binding requirements nor entail penalties in case of breach, to induce compliance with disclosure rules the MSA requires the board of directors to approve and sign the annual statement and companies to upload it in their website.[34] In case of non-compliance with disclosure rules themselves an injunction can be brought against the defaulting company.[35]

In 2017 France enacted the Duty of Vigilance Law,[36] which requires companies to take reasonable care in identifying, preventing and addressing

---

[30] Canadian Network on Corporate Accountability, The Global Leadership in Business and Human Rights Act: An act to create an independent human rights ombudsperson for the international extractive sector (Draft model legislation) (2 November 2016).

[31] Directive 2014/95/EU of the European Parliament and of the Council of 22 October 2014 amending Directive 2013/34/EU as regards disclosure of non-financial and diversity information by certain large undertakings and groups.

[32] Modern Slavery Act 2015.

[33] Ibid., s. 54(5).

[34] Ibid., s. 54(6)(7).

[35] Ibid., s. 54(11).

[36] Assemblée Nationale, Proposition de Loi Relative au Devoir de Vigilance des Sociétés Mères et des Entreprises Donneuses D'ordre, in March 2017).

risks to human rights and fundamental freedoms, seriously bodily injury or environmental damage or health risks that result directly or indirectly form the activities of a company. The law applies to companies with more than 50,000 employees and their subsidiaries and to subsidiaries with head office located in foreign countries if the overall group employees exceed 10,000. The activities of the company's subcontractors or suppliers are also in scope. The breach of the rules at stake may be subject to enforceable measures by the competent courts upon a claim by interested parties.

The Netherlands took the lead with respect to binding obligations for companies in respect of child labour, by adopting, in May 2019, the Dutch Child Labour Due Diligence Law (CLDL)[37] which is expected to come into force in 2022 and apply to all companies which sell or supply goods or services to Dutch consumers (whether or not registered in The Netherlands), with no exemptions for legal form or size.

The CLDL obliges companies to investigate whether their goods or services have been produced utilising child labour and to devise a plan to prevent child labour in their supply chains if they find it. There are significant administrative fines and criminal penalties for non-compliance and the law also imposes a reporting obligation: companies in scope must submit a declaration to (a yet-to-be-determined) regulator, confirming that they have exercised an appropriate level of supply chain due diligence in order to prevent child labour. The exact form and content of the statement will be described in forthcoming regulation. Corporate human rights due diligence statements will be made public by the above referred regulator in an online public registry.

Meanwhile, on 11 March 2021 four Dutch political parties submitted a private members' bill which introduces a new corporate duty of care for human rights and the environment. This proposal for a "Responsible and Sustainable International Business Conduct Act" requires companies to identify, prevent, mitigate, and report on the risks and impacts of their activities on human rights, and, increasingly, the environment. The new bill proposes to replace the CLDL, with broad due diligence legislation.

Another relevant Dutch initiative are the multi-stakeholder agreements, aimed at promoting international responsible business conduct, the International Responsible Business Conduct (IRBC) agreements. They cover relevant economy sectors[38] and are agreed by industry associations, government, NGOs, civil society and unions which develop standardized business contract clauses that address improvements and compliance in risk areas, including human rights.

In the IRBC agreements, the parties identify the problems that arise in each sector, describe how they intend to prevent abuses and what each party should do to achieve it. For companies that adhere to IRBC agreements the

---

[37] Dutch Child Labour Due Diligence Law, May 2019.

[38] Garments and textile, banking, gold, true stone, food products, insurance, pension funds, metals, floriculture, sustainable forestry, wind energy and agricultural sector.

main obligation is to conduct due diligence and disclose an annual plan based on the due diligence outcome. Failure to do so may result in them being held to account before the dispute settlement body (usually including NGOs/civil society members and a chair) appointed by the parties to the IRBC agreement. Although being a Dutch initiative, the IRBC agreements can serve as a source of inspiration for companies that wish to ensure their compliance with human rights elsewhere.

At the time of writing, the most recent legislative initiative on corporate and human rights interaction is the German Supply Chain Due Diligence Act, passed by the German Federal Parliament and the German Federal Council in June 2021. Similarly, to the French Duty of Vigilance Law, the GSCA obliges large companies in Germany to carry out due diligence regarding human rights and environmental issues in their supply chains. Under the GSCA there can be imposed fines for violations of due diligence and reporting obligations of up to EUR 8 million depending on the nature and severity of the violation. Companies with an average annual turnover of more than EUR 400 million may be fined up to 2% of their average annual turnover for breaches of the obligation to take remedial action or to implement an appropriate remedial action plan at a direct supplier.

It is also foreseen that the government may temporarily exclude companies from public tenders, for up to three years, when they have been fined above a certain amount for breaching the rules. The GSCA will come into force in stages, applying to companies with more than 3,000 employees from 2023 and to smaller companies from 2014. Companies with fewer than 1,000 employees will be exempt.[39]

3 Integration of Human Rights in the Companies' Agenda

Globalization and the multinational enterprise context, on the one hand, and the fact that companies have a duty to comply with human rights, on the other hand,[40] make it crucial for companies to take a closer look on whether and how human rights are rightly handled and complied within their scope of action in all countries they are active.

More generally, in the 2030 Sustainable Development Agenda, the UN envisioned a larger role for companies, while calling upon all businesses to apply their creativity and innovation in overcoming sustainable development challenges.[41]

---

[39] Overview of the German Supply Chain Due Diligence Act, Taylor Wessing, 28 July 2021.

[40] For further considerations on the duty of companies to respect human rights, see section "The Interaction Between Corporate Governance and Human Rights".

[41] United Nations, Sustainable Development Goals indicators website.

In this context, an effective human rights strategy should be high on the companies' agenda, contribute to accountability towards stakeholders and reinforce companies' social responsibility objectives. Such strategy should embrace applicable rules and duties, both mandatory and anchored in international or national principles, standards and identified best practices. It can also go beyond if companies dare to be creative and innovate.

Board Duties (Consideration of Human Rights Issues Ate Board Level)

In 1970, Friedman's doctrine defended that the social responsibility of business was to increase its profits.[42] In today's world, although the main objective of a company still is to create value generating profit for its shareholders, social and market pressures are increasingly leading to a change in how value is perceived and over which time horizon. There is a growing belief among companies that sustainable business success and shareholder value cannot be achieved solely through maximising short-term profits, but rather through a market-oriented responsible behaviour. Companies are aware that they can contribute to sustainable development by managing their operations in such a way as to enhance economic growth and increase competitiveness while ensuring environmental protection and promoting social responsibility, including the human rights dimension.[43]

Despite the wide spectrum of approaches to CSR, there is large consensus on its main features: a business behaviour over and above legal requirements, voluntarily adopted because businesses deem it to be in their long-term interest; it is also intrinsically linked to the concept of sustainable development: businesses need to integrate the economic, social and environmental impact in their operations; and it is not an optional "add-on" to business core activities, but about the way in which businesses are managed.[44]

Additionally, today more than ever social responsibility may impact companies' wealth and profits. This is stressed by John G. Ruggie and Emily K. Middleton when referring to the fast growing interest on the part of asset owners and asset managers in "ESG" investments, by increasingly taking into account companies' environmental, social, and corporate governance (ESG) performance in making investment decisions".[45] This should in turn reinforce companies interests in considering the impact of these issues and

---

[42] A Friedman doctrine—The Social Responsibility of Business Is to Increase Its Profits, By Milton Friedman, *The New York Times*, September 13, 1970.

[43] Commission of the European Communities, Communication form the Commission concerning Corporate Social Responsibility: A Business Contribution to Sustainable Development, Brussels, 2.7.2002, COM(2002) 347 final, p. 5.

[44] Ibid.

[45] John G. Ruggie and Emily K. Middleton, Money, Millennials and Human Rights: Sustaining 'Sustainable Investing' (2018).

call their boardrooms to action towards improving their ESG standards and performance.[46]

However, well-structured and appealing ESG agendas and public reports on ESG compliance (including human rights),[47] although being an essential starting point to pursue corporate responsibility, are not enough. On substance, they need to reflect and be backed up by effective measures, embedded in shareholders' mandate to the board of directors and hence in its fiduciary duties.

There can be legal, financial and reputational consequences if companies fail to meet their responsibility and duty to respect human rights. Such failure may also hamper companies' ability to recruit and retain staff, to gain permits, investment, new project opportunities or similar benefits essential to a successful, sustainable business. As a result, where a company poses a risk to human rights, it increasingly also poses a risk to its own long-term interests.

The question that emerges is: how can human rights be factored in director's duties to foster a corporate culture compliant with human rights?

Although there is not a specific set of board' duties with respect to human rights, the UN Guiding Principles, supported by the UN GP Interpretative Guide, enable the framing of some key board duties regarding human rights and that can be grouped in three main categories: strategy, human resources and governance.

First, there is a duty to implement a strategy for human rights which comprises knowledge of the applicable legal and regulatory framework, framing of the human rights at risk, budget allocation, identification and interaction with relevant stakeholders, approval of measures (including due diligence and integration of its findings, risk management, remedies, tracking systems and due accountability) to prevent and address human rights abuse, involvement of external expertise (when required), and set of adequate expectations of third parties directly linked to the company's activities.

Second, there is a duty to allocate internal capacity to human rights related issues, which also involves building relevant internal expertise, training of staff, adopting performance indicators linked to human rights and ensuring that responsibility for addressing the impact on human rights is assigned to the appropriate level and function.

Third, there is a duty to have in place sound governance arrangements, enabling the oversight of the strategy for human rights related issues and an effective and timely decision-making to allow appropriate action to address

---

[46] According to the Global Sustainable Investment Alliance (GSIA), *2018 Global Sustainable Investment Review*: within ESG investing, globally, sustainable investing assets in the five major markets stood at $30.7 trillion at the start of 2018, a 34 percent increase in two years. In 2019–2020 environmental, social, and governance-focused indices advanced 40 per cent, according to the Index Industry Association survey, *Financial Times*, ESG Investing—What Does ESG-Friendly Really Mean?, Brooke Fox, 12 April 2021.

[47] Such as the ones required under the Reporting Directive (see above in section "EU Initiatives").

complex challenges. Adequate performance by board members with respect to human rights requires a good and timely information flow which should feed into quality exchange, debate, challenge, and decision-making at board level.

Additionally, sound adequate internal controls and effective reporting lines by managers to the board of directors are also key, as they should always allow for human rights issues to be kept under scrutiny. If working, this interaction should allow directors to report back to shareholders and buy in their support to sustainable and long-term human rights compliance objectives and measures. In large international corporations' contexts, an aligned compliance culture combined with effective policies and reporting lines between parent entities and subsidiaries operating in sensitive human rights environments is particularly relevant.

Board members and senior managers with skills and experience in ESG topics can support the above corporate endeavour and facilitate a more targeted and innovative approach to the issues at hand, while also facilitating the recruitment of internal or external experts who can support companies' human rights strategy.

Finally, in supporting their client companies' objectives towards compliance with human rights, business lawyers and law firms also play an important role and already benefit from specific guidance in this field of expertise, as it is the case of the "IBA Practical Guide on Business and Human Rights for Business Lawyers".[48]

Corporate Actions Towards Human Rights

Under the UN Guiding Principles, responsibility to respect human rights requires that business enterprises *avoid causing or contributing to adverse human rights impacts through their own activities, and address such impacts when they occur; and seek to prevent or mitigate adverse human rights impacts that are directly linked to their operations, products or services by their business relationships, even if they have not contributed to those impacts.*[49]

As referred above, in their strategy to deal with human rights issues, companies must first consider the legal framework applicable to human rights in the countries where they conduct business (irrespective of doing it through their head office, subsidiaries or branches). In cases where such framework does not adequately address human rights concerns, companies should, nevertheless, strive to fully take on board and comply with soft law rules and principles, including those fostered by international organizations applicable to human rights, among which the UN Guiding Principles continue to be the most complete and widely accepted set of rules. Given the fact that, with

---

[48] International Bar Association, IBA Practical Guide on Business and Human Rights for Business Lawyers, 2016.

[49] Guiding Principle 13.

a few exceptions, companies' responsibilities for human rights are still generally addressed through soft law rather than hard law, we next look at some measures and good practices that companies should consider when looking into their compliance with human rights. Among these we focus on due diligence, self-regulation and leverage power, also looking at how companies may also go beyond and undertake additional human rights commitments.

Due Diligence

As mentioned throughout this chapter, due diligence is broadly recognized as one of the most important tools for companies in the human rights' corporate responsibility context, both in soft and hard law frameworks.

The UN Guiding Principles refer to human rights' due diligence as an ongoing process through which a company:

- assesses its risks to human rights, prioritizing the most acute;
- integrates the findings into its decision-making and actions in order to mitigate such risks;
- tracks the effectiveness of these measures;
- communicates its efforts and results internally and externally.

Human rights due diligence is not a single prescriptive formula. Companies of different sizes, in different sectors, with different structures and in different circumstances need to tailor their processes to meet their needs. However, the above elements of human rights due diligence when taken together with remediation processes, should be able to provide companies' management with a framework that enables knowing and showing that they are respecting human rights over time. It therefore usually entails ongoing processes, with recurrent updates overtime, rather than a one-off assignment.[50]

Most of the due diligence related elements are also reflected in the above list of board duties. The way the due diligence is structured, approved, and monitored by the board of directors (and eventually also submitted to shareholder's approval) and cascades down to management and engage staff members is key for a successful implementation. When structuring a due diligence, care should be given, inter alia, in having the right people leading and overseeing it, in involving the right departments, in assessing the need to contract external experts and in timely engaging with directly affected stakeholders. It is also important to ensure that staff responsible for human rights due diligence have the necessary skills and training opportunities and that they have sufficient influence within the organization to adequately perform their tasks.

Self-Regulation

In parallel with international and domestic human rights legal and regulatory initiatives, some companies call upon themselves obligations towards human

---

[50] UN Guiding Principles, Principle 17.

rights through self-regulation. Some of these initiatives may often be driven by due diligence processes, to address identified risks or actual concerns.

The most common self-regulation tools include contractual clauses and codes of conduct.

Contractual clauses may either be clauses bilaterally agreed by companies with third parties (e.g. suppliers) or clauses pre agreed by other stakeholders' groups, to which companies voluntarily adhere to. An example of this commitment is the adherence by companies to the IRBC agreements referred to above.[51] In both cases contractual clauses have the same objective of preventing human rights abuses and promote a responsible human rights business conduct.

The codes of conduct aim at framing and enhancing companies' social responsibility culture, inter alia, in the human rights area. There may also be cases in which companies see the benefit of adopting codes of conduct exclusively dedicated to human rights,[52] for example in countries where protection of human right' abuses are poorly regulated and overseen. When also applying to third parties—such as suppliers—codes of conduct enable an oversight by companies of the impact of such third parties on human rights. While there is not any generally adopted standard or format for codes of conduct, they usually include guidelines, recommendations and rules towards the above objective, with companies having significant discretion to adapt them to their context and specificities.

Barnali Choudhury and Martin Petrin point out the criticisms that codes of conduct have been subject to, ranging from lack of independent monitoring and oversight to mere branding, promotion strategy, aimed at self-promotion, rather than an honest and effective measure to tackle human rights' abuses.[53]

Although in some contexts, such criticisms may prove true, until overarching mandatory rules on the obligations of companies towards human rights are successfully enacted worldwide, soft law principles and standards, complemented by self-regulation tools, such as codes of conduct, play an important role towards compliance with human rights.

Additionally, in a world of real time information, powerful social networks, and investors increasingly interested in sustainable investments, the interaction between companies and human rights is subject to a wide public exposure and scrutiny. As a result, fake or weak corporate self-regulation strategies should increasingly lead to consumer's and investor's boycott and activism.

---

[51] For further details on IRBC agreements, see section "Some Domestic Initiatives" above.

[52] Examples of codes of conduct: Johnson & Johnson "Statement of Human Rights" (December 2012); Fujitsu Global, "Code of Conduct" (undated).

[53] Barnali Choudhury and Martin Petrin, *Corporate Duties to the Public*, Cambridge University Press, 2019, p. 229.

Leverage Power
Companies are the primary and ultimate responsible for ensuring that their activities comply with human rights. While human rights and labour frameworks have been enacted worldwide to ensure such compliance, no general mandatory framework is yet in place and in some countries there are still gaps in the way human rights are ruled and handled. In these countries the importance of corporate human rights due diligence and self-regulation is enhanced, and companies can also leverage on their power to guarantee that their responsibility for human rights is taken seriously when engaging with other entities (e.g. local supply chains, governments and authorities).

Against this backdrop, in the context of the UN Guiding Principles, *leverage is considered to exist where the enterprise has the ability to effect change in the wrongful practices of an entity that causes a harm.*[54]

Leverage of a company over an entity may arise from one or more factors, such as the degree of direct control of the entity by the company, the contract in place between the two, the proportion of business the company represents for the entity, the ability of the company to incentivize the entity to improve human rights compliance (e.g. reputational advantage, capacity-building assistance) and the ability of the company to engage government in requiring improvement in human rights compliance by the entity through the implementation of regulations, monitoring, sanctions, etc.[55]

When dealing with an entity that is adversely impacting on human rights, a company should determine whether it can leverage on any of the above factors to prevent or mitigate such impact. This not being possible, it can seek for ways to increase its leverage, either by offering incentives to the abusing entity to become compliant or collaborating with other players.

If no leverage exists nor it can be increased, the company might have to consider terminating the relationship with the entity, even if the relationship may be "crucial" to the company (e.g. the entity is the sole supplier of a product). Here the severity of the human rights abuse must also be weighted: the more severe the abuse, the faster the company will need to see change before it takes a decision on whether it should end the relationship. In any case, for as long as the abuse continues and the company remains in the relationship, it should be able to demonstrate its ongoing efforts to mitigate the negative impact and be prepared to accept any consequences—reputational, financial, or legal—of the continuing connection.

Beyond Human Rights Rules and Standards
Beyond strict compliance with human rights, enterprises may voluntarily undertake additional commitments, such as the promotion of specific human rights. This management choice may be driven by different reasons, such as

---

[54] UN Guiding Principles, commentary to Guiding Principle 19.
[55] UN GP Interpretive Guide, Box 6 on Guiding Principle 9.

philanthropy, protection and enhancement of reputation or simply development of business opportunities.

Irrespective of the underlying drivers, innovative business initiatives which target companies' objectives and profit goals, while also tackling human rights issues exist, are successfully implemented and should serve as a source of inspiration to boards operating in challenging human rights environments. Examples include companies the operations of which support the expansion of opportunity for low-income individuals, by training their local workers in basic skills, of value to the company's business activity, and paying them a local living wage[56] or companies which provide their employees with health and wellness education, scholarship programs, micro loans and mentorship for those who aspire to be entrepreneurs.[57] These contexts allow for a win–win context in which by investing in the well-being and development of their workforce, companies also ensure the skills, know-how and motivation they need as enablers of performance and productivity.

In this sphere it is also interesting to consider what Michael Porter and Mark Kramer designate as shared value a concept deriving from the efforts by companies to create shared value by reconceiving the intersection between society and corporate performance, by bringing business and society together and connecting business with social progress.[58]

4 COMPANIES AND HUMAN RIGHTS IN COVID-19 TIMES

The UN reports that *"As the COVID-19 crisis brought into stark relief, too many companies still place profit above people and planet – passing the buck at devastating cost to millions of workers, communities and defenders worldwide. Between the onset of COVID-19 (March 2020) and September 2020, we tracked 286 cases of attacks against defenders focused on business-related activities, a 7.5% increase on previous years. From March to May alone, garment workers were deprived of est. US$5.8 billion in wages – yet over half the fashion brands we surveyed last year reported pandemic profits. At the same time, KnowTheChain in its 2020 ICT Benchmark found Europe-based companies scored lower than*

---

[56] The Global Living Wage Coalition defines living wage as the remuneration received for a standard work week by a worker in a particular time and place, sufficient to afford a decent standard of living including food, water, housing, education, healthcare, transport, clothing and other essential needs, including provision for unexpected events.

[57] An example of this kind of business human rights oriented is Sama, a training-data company, focusing on annotating data for artificial intelligence algorithms and driven by the belief that "... connecting people to dignified digital work and paying them living wages can solve some of the world's most pressing challenges - from reducing poverty, to empowering women, and mitigating climate change." While developing its business, Sama also supported people to have meaningful income and move out of poverty.

[58] Michael Porter/Mark Kramer, Creating Shared Value, *Harvard Business Review*, p. 4, 2011.

their North American counterparts when it came to addressing forced labour risks in their supply chains".[59]

Furthermore, news in different parts of the world also point out to increased breaches of human rights in COVID-19 times, often connected with politically unstable and distressed contexts in which individuals in power use the pandemic as an excuse to adopt arbitrary measures and violence against individuals and facilitate corruption. In some countries, officials blame migrant workers, some of whom have been caned and deported. Globally, democracy and human rights are in retreat. Although this began before COVID-19, 80 countries have regressed since the pandemic begun. The list includes both dictatorships that have grown nastier and democracies where standards have slipped. With everyone's attention on COVID-19, autocrats in many countries can do all sorts of bad things, safe in the knowledge that the rest of the world will barely notice, let alone object.[60]

The impact the pandemic is having everywhere, the lessons learned and the effects coming out of it are expected to continue after the pandemic is behind us. Companies should be asked to engaged in a deeper and broader analysis on how they are addressing their human rights responsibilities and conducting human rights due diligence in all countries they operate, with emphasis on those where breaches of human rights may be on the increase. The timing and ways in which companies act—fostering sound and inclusive human rights compliance within their remit—should impact on the way they are perceived and assessed by stakeholders, investors and the public at large.

As stressed by Mark Carney,[61] *"When it's over, companies will be judged by what they did during the war", how they treated their employees, suppliers and customs, by who shared and who hoarded"*.[62]

5   Conclusions

The role and responsibilities of companies with respect to human rights have increased over time. In a globalized and competitive world companies are increasingly doing business in contexts which evidence human rights gaps, while investors are increasingly interested in ESG performance investing opportunities.

Although the interaction between companies and human rights is not generally subject to hard law, international guidelines and standards worldwide confirm responsibility of companies towards human rights. Beyond this responsibility, utilitarian, non-utilitarian, and ethical arguments point out to a duty of companies to respect human rights, even if no law forces them to do so.

---

[59] UN Business & Human Rights Resource Centre.
[60] *The Economist*, Covid-19 and Liberty, No Vaccine to Cruelty, October 17th, 2020.
[61] Mark Carney was Governor of the Bank of Canada and of Bank of England.
[62] *The Economist*, Mark Carney, The World After COVID-19, April 18th, 2020.

Current national and international initiatives show a trend towards an increase of mandatory frameworks governing the human rights–companies interaction. However, they are not enough to ensure a level playing field in the fight against human rights abuses, since compliance with human rights is not yet enforceable everywhere.

Irrespective of operating in a hard or soft law context, the duty of companies to comply with human rights requires them to actively incorporate this objective in corporate decision-making. Companies' duties towards human rights can be addressed through the adoption of regulatory or voluntary measures and policies and should be high on shareholders and boards' agendas.

In COVID-19 times human rights abuses have increased in several vulnerable countries. Companies doing business in those countries should increase their internal control, due diligence and leverage mechanisms to ensure compliance with their responsibilities and duties towards human rights.

References

BIICL, Civic Consulting and LSE, *Study on Due Diligence Requirements Through the Supply Chain*. Final Report (January 2020).

Canadian Network on Corporate Accountability, *The Global Leadership in Business and Human Rights Act: An Act to Create an Independent Human Rights Ombudsperson for the International Extractive Sector* (Draft model legislation) (2 November 2016).

Carney, Mark, The World After COVID-19, *The Economist*, p. 51 (18 April 2020).

Choudhury, Barnali/Martin Petrin, *Corporate Duties to the Public*, Cambridge University Press (2019).

Chua, Shirin, Freshfields Bruckhaus Derringer LLP, *UN Working Group on a Convention to Regulate Transnational Business and Human Rights—Comments on the 'Zero Draft' Due Soon* (February 2019).

The Economist, Torment of the Uyghurs (17 October 2020), 11.

The Economist, Covid-19 and Liberty, No Vaccine to Cruelty, (17 October 2020), 48.

European Bank of Reconstruction and Development (EBRD), Environmental and Social Policy (ESP) (7 May 2014).

European Commission, *Communication from the Commission concerning Corporate Social Responsibility: A Business Contribution to Sustainable Development*, Brussels, 2.7.2002, COM(2002) 347 final.

European Commission, *Communication from the Commission to the Council and the European Parliament, Modernising Company Law and Enhancing Corporate Governance in the European Union—A Plan to Move Forward*, Brussels, 21 May 2003, COM (2003) 284 final.

European Commission, *Commission Recommendation (EU) 2018/1149 of 10 August 2018 on Non-binding Guidelines for the Identification of Conflict-Affected and High-Risk Areas and Other Supply Chain Risks Under Regulation (EU) 2017/821 of the European Parliament and of the Council*.

European Investment Bank, *Environmental and Social Handbook* (2013).

European Parliament, *Corporate Due Diligence and Corporate Accountability*, P9_TA-PROV(2021)0073.

European Parliament, *European Parliament Resolution of 10 March 2021 with Recommendations to the Commission on Corporate Due Diligence and corporate accountability* (2020/2129(INL).

Friedman, Milton, A Friedman doctrine—The Social Responsibility of Business Is to Increase Its Profits, *Financial Times*, September 13 (1970).

Fox, Brooke, ESG Investing—What Does ESG-Friendly Really Mean? *Financial Times* (12 April 2021).

International Bar Association, *IBA Practical Guide on Business and Human Rights for Business Lawyers* (2016).

International Finance Corporation, *IFC Performance Standards on Environmental and Social Sustainability* (1 January 2012).

OECD, *Guidelines for Multinational Enterprises*, OECD Publishing (2011).

Porter, Michael/Mark Kramer, Creating Shared Value, *Harvard Business Review* (2011).

Ruggie, John G./Emily K. Middleton, *Money, Millennials and Human Rights: Sustaining 'Sustainable Investing'* (2018).

Sherman III, John F., Beyond CSR: The Story of the UN Guiding Principles on Business and Human Rights, in Rae Lindsay and Roger Martella (eds), *Corporate Responsibility, Sustainable Business: Environmental, Social and Governance Frameworks for the 21st Century* (2020).

Thomson Reuters Foundation, *Germany to Fine Companies Whose Suppliers Abuse Rights* (12 February 2021).

United Nations, "Corporations and Human Rights: A Survey of the Scope and Patterns of Alleged Corporate-Related Human Rights Abuse," UN Document A/HRC/8/5/Add. 2, 23 May 2008.

United Nations, *Guiding Principles on Business and Human Rights, Implementing the United Nations "Protect, Respect and Remedy" Framework* (2011).

United Nations, *Human Rights, Climate Change and Business, Key Messages* (2021).

United Nations, *Resolution 26/9 Elaboration of an International Legally Binding Instrument on Transnational Corporations and Other Business Enterprises with Respect to Human Rights* (2014).

United Nations, *Statement on Behalf of a Group of Countries at the 24rd Session of the Human Rights Council General Debate—Item 3 "Transnational Corporations and Human Rights" Geneva* (September 2013).

United Nations, *The Corporate Responsibility to Respect Human Rights, an Interpretive Guide* (2012).

United Nations, *The COVID-Driven Humanitarian Crisis of Seafarers: A Call for Action Under the UN Guiding Principles on Business and Human Rights*, Joint Statement by the UN Human Rights Office (OHCHR), the UN Global Compact, and the UN Working Group on Business and Human Rights (5 October 2020).

United Nations, The International Bill of Human Rights (1948).

United Nations, *UN Human Rights "Issues Paper" on Legislative Proposals for Mandatory Human Rights Due Diligence by Companies* (June 2020).

PART II

ESG Regulatory Developments

CHAPTER 10

ESG and EU Law: From the Cradle of Mandatory Disclosure to More Forceful Steps

António Garcia Rolo

1   Introduction: Two Decades of Declarations of Intent

I. The European Union has been trying, since the turn of the century, to determine what role should environmental and social concerns play in corporate governance, first through political statements and then through legislative action. Indeed, in the past decade, EU lawmakers opted to straddle a path that has progressively integrated such concerns in rules directly and indirectly pertaining to corporate governance of some companies.

In the early 2000s, EU lawmakers started talking about "corporate social responsibility", at the time seen as a *purely voluntary* incorporation of social and environmental concerns in business operations (2001 Green Paper

---

The opinions expressed in this text bind the Author exclusively.
The cut-off date for the inclusion of legislation or official documents is 10 October 2021.

A. Garcia Rolo (✉)
Faculty of Law, University of Lisbon, Lisbon, Portugal
e-mail: antoniogarciarolo@fd.ulisboa.pt

© The Author(s), under exclusive license to Springer Nature Switzerland AG 2022
P. Câmara and F. Morais (eds.), *The Palgrave Handbook of ESG and Corporate Governance*, https://doi.org/10.1007/978-3-030-99468-6_10

Promoting a European Framework for Corporate Social Responsibility,[1] the 2002 Communication on Corporate Social Responsibility[2] and in the 2010 "Europe 2020 Strategy"[3]—all non-binding policy documents).

However, since the early 2010s, indications of a more demanding approach emerged, as the Commission reverted its understanding of the voluntary character of corporate social responsibility (in the 2011 Communication on a Renewed Strategy for CSR[4]) and started projecting the construction of a legal framework in which the incorporation of environmental and social concerns in company law and corporate governance became a reality and not a mere option.

II. *Disclosure* of social and environmental information was seen as the point of departure to kick-start a EU-wide regime that would incorporate environmental and social concerns in company law,[5] as we can see in the Commission's Action Plan on European Company Law of 2012,[6]

---

[1] European Commission, *Green Paper: Promoting a European framework for Corporate Social Responsibility*, DOC/01/9, 18.07.2001, pp. 8–18, in which corporate social responsibility is defined as a "*concept whereby companies integrate social and environmental concerns in their business operations and in their interaction with their stakeholders on a voluntary basis*", including concerns with human resources, health and safety, relationship with local communities, NGOs, public authorities and the environment. The document further underlines the intimate link between corporate social responsibility and sustainable development.

[2] Communication from the Commission concerning *Corporate Social Responsibility: A Business Contribution to Sustainable Development*, COM(2002) 347 final, 02.07.2002, in which the Commission analysed the responses to a public consultation on the aforementioned 2001 Green Paper, restating much of the content of the latter.

[3] Communication from the Commission, *Europe 2020: A Strategy for Smart Sustainable and Inclusive Growth*, COM(2010) 2020 final, 03.03.2010, p. 15.

[4] Communication from the Commission to the European Parliament, the Council, the European Economic and Social Committee and the Committee of the Regions, *A Renewed EU strategy 2011-14 for Corporate Social Responsibility*, COM(2011) 681 final, 25.10.2011, p. 6. Such loss of a voluntary character has been deemed a radical change of position and resulted in the shift from private voluntary schemes to public and mandatory systems. For more details on this shift in EU law, cfr. W. Gregory Voss, "The European Union's 2014 Non-Financial Reporting Directive: Mandatory Ex Post Disclosure—But Does It Need Improvement?", in *Extractive Industries and Human Rights in an Era of Global Justice: New Ways of Resolving and Preventing Conflicts* (eds. Amissi Melchiade Manirabona/Yenny Vega Cárdenas), 2019 (359–381), pp. 359 and 362; and Constance Wagner, in "Evolving Norms of Corporate Social Responsibility: Lessons from the European Union Experience with Non-Financial Reporting", St. Louis U. Legal Studies Research Paper no. 2017-6 (08.10.2017), available at SSRN: https://ssrn.com/abstract=3051843, pp. 42–43.

[5] Communication from the Commission to the European Parliament, the Council, the European Economic and Social Committee and the Committee of the Regions, *A Renewed EU strategy 2011-14 for Corporate Social Responsibility*, cit., pp. 7–14.

[6] Communication from the Commission to the European Parliament, the Council, the European Economic and Social Committee and the Committee of the Regions *Action Plan: European Company Law and Corporate Governance—A modern Legal Framework for More Engaged Shareholders and Sustainable Companies*, COM(2012) 740 final, 12.12.2012.

which sought to make companies better and more efficiently run (under the mantra that a well-run company is more sustainable),[7] identifying increased transparency in the corporate world as one of its objectives. By doing so, it indirectly linked such increased transparency with disclosure frameworks pertaining to environmental or social issues, proposing disclosure requirements of board diversity policy and management of non-financial risks.[8]

In 2018, the Commission, while proposing the enhancement of sustainability disclosure in the Sustainable Finance Plan,[9] went further, proposing an "EU classification system for sustainable activities" that eventually led to the Taxonomy Regulation analysed in further detail below.

Finally, a mention is due to the more recent Commission's 2019 Reflection Paper Towards a Sustainable Europe by 2030,[10] 2019 Communication on the European Green Deal[11] and the 2021 Strategy for Financing the Transition to a Sustainable Economy.[12]

III. The countless non-papers, action plans or reflection papers have led to a tangible legislative output, which is going to be addressed in Part 2 of this Chapter (without prejudice to more in-depth analyses in Chapters 13 and 14): (i) early in the game, environmental and social mandatory disclosures were introduced by the Non-Financial Reporting Directive of 2014; (ii) in 2019, the Sustainable Finance Disclosures Regulation (SFDR) brought about a set of disclosure obligations binding certain financial actors; and (iii) the 2020 Taxonomy Regulation completed this legislative trinity through the provision of a conceptual backbone to existing and future legislation. This *disclosure trinity* represents the cradle of a EU-wide ESG regime, with further developments still waiting in the wings.

Other important acts, such as the Low-Carbon Benchmarks Regulation (an important part of the EU's efforts in incorporating environmental concerns in product governance and in the governance of certain companies),[13] disclosure

---

[7] Communication from the Commission *Action Plan: European Company Law and Corporate Governance*, cit., p. 3.

[8] Communication from the Commission *Action Plan: European Company Law and Corporate Governance*, cit., pp. 5–6.

[9] Communication from the Commission, *Action Plan: Financing Sustainable Growth*, COM(2018) 97 final, 08.03.2018.

[10] European Commission, *Reflection Paper: Towards a Sustainable Europe by 2030*, COM(2019) 22, 30.01.2019.

[11] Communication from the Commission to the European Parliament, the Council, the European Economic and Social Committee and the Committee of the Regions, *The European Green Deal*, COM(2019) 640 final, 11.12.2019.

[12] Communication from the Commission to the European Parliament, the Council, the European Economic and Social Committee and the Committee of the Regions, *Strategy for Financing the Transition to a Sustainable Economy*, COM(2021) 390 final, 06.07.2021.

[13] Regulation (EU) 2019/2089 of the European Parliament and of the Council of 27 November 2019 amending Regulation (EU) 2016/1011 as regards EU Climate Transition Benchmarks, EU Paris-aligned Benchmarks and sustainability-related disclosures for benchmarks.

obligations set out for large banks in the Capital Requirements Regulation[14] or for certain pension funds in the IORP Directive[15] or parts of the amendments of the Shareholders' Rights Directive (which took increased transparency in the governance of listed companies to new heights),[16] will not be included in this analysis, which does not preclude their importance in the ESG and corporate transparency frameworks in the EU in general.

IV. In Part 3, it will be shown how EU lawgivers are eager to move past simple disclosure frameworks, as recent preparatory action and concrete legislative proposals indicate potential ground-breaking changes in the basis of EU company law, with EU legislators seemingly willing to adopt a more forceful approach.

2 THE MANDATORY DISCLOSURE TRINITY AS THE CRADLE OF A EUROPEAN ESG FRAMEWORK

Mandatory Disclosure of ESG Information Under the Non-Financial Reporting Directive

I. The first major development in the incorporation of environmental and social concerns in corporate governance was the introduction of mandatory disclosure of non-financial information across the European Union by Directive 2014/95/EU of the European Parliament and of the Council (also known as **"Non-Financial Reporting Directive"** or **"NFRD"**),[17] which added new provisions to Directive 2013/34/EU of the European Parliament and of the Council (also known as the **"Accounting Directive"**).[18]

The amendments introduced by the NFRD in the Accounting Directive on mandatory disclosure of non-financial information (new Articles 19a and 29a of the Accounting Directive)[19] were further complemented by two sets of

---

[14] Regulation (EU) No 575/2013 of the European Parliament and of the Council of 26 June 2013 on prudential requirements for credit institutions and investment firms, as amended.

[15] Directive (EU) 2016/2341 of the European Parliament and of the Council of 15 December 2016 on the activities and supervision of institutions for occupational retirement provisions (IORP), as amended.

[16] Directive 2007/36/EC of the European Parliament and of the Council of 11 July 2007 on the exercise of certain rights of shareholders in listed companies, as amended.

[17] Directive 2014/95/EU of the European Parliament of the European Parliament and of the Council, amending Directive 2013/34/EU as regards disclosure of non-financial and diversity information by certain large undertakings and groups.

[18] Directive 2013/34/EU of the European Parliament and of the Council, of 26 June 2013 on the annual financial statements, consolidated financial statements and related reports of certain types of undertakings, amending Directive 2006/43/EC of the European Parliament and of the Council and repealing Council Directives 78/660/EEC and 83/349/EEC.

[19] In the next pages, when referring to the provisions introduced by the NFRD in the Accounting Directive, such provisions will be referred to in the context of the latter as a

Non-Binding Guidelines adopted by the Commission in 2017[20] and 2019,[21] respectively.

II. What entities are subject to this mandatory disclosure regime? According to Article 19a(1), this framework applies *to large undertakings*[22] that fulfil two cumulative requirements: *(i)* are *public-interest entities*, a term comprising, *inter alia*, listed companies, credit institutions and insurance companies[23]; and *(ii)* have over 500 employees.

A wholly similar regime, laid out in Article 29a(1), applies the very same obligations at a consolidated group level.[24] As such, it will not be autonomously analysed.

III. Article 19a of the Accounting Directive identifies the *main obligation* of this framework—that a certain set of information is reported as a "non-financial statement" in the company's yearly management report,[25] with a hard-to-read and complex text.

To make sense of this complicated provision and the obligations contained therein, and therefore to answer the question "what is to be disclosed?", one must distinguish between the five *aspects* listed in Articles 19a(1)(a)–(e) and

---

whole. When referring specifically to policy goals or other circumstantial information, the NFRD will be referred autonomously.

[20] Communication from the Commission *Guidelines on Non-Financial Reporting (Methodology for Reporting Non-financial Information)*, 2017/C 215/01, 05.07.2017 (**"2017 Non-Binding Guidelines"**).

[21] Communication from the Commission *Guidelines on Non-Financial Reporting: Supplement on Reporting Climate-Related Information*, 2019/C 209/01, 20.06.2019 (**"2019 Non-Binding Guidelines"**).

[22] Please note that the EU legislator uses the economic-oriented term "undertaking" instead of the more legally-oriented "company".

[23] As defined in Article 2(1) of the Accounting Directive.

[24] Being applicable to parent undertakings which are considered public interest entities and which are parents of a group with more than 500 employees (it should be reemphasised, for clarity, that the 500 employees refers to the group and not the parent company). For more detail, cfr. Dániel Szabó/Karsten Sørensen, "Non-Financial Reporting, CSR Frameworks and Groups of Undertakings—Application and Consequences", Nordic & European Company Law Working Paper No. 16-01 (19.06.2016), available at SSRN: https://ssrn.com/abstract=2774170.

[25] According to the Commission's 2017 Non-Binding Guidelines, published to overcome possible differences in reporting format, the non-financial statement should be preceded by a materiality assessment (as seen above) and the information provided should be *"fair, balanced and understandable"*. The statement should be *"comprehensive but concise"* and, *"strategic and forward-looking information should be provided"*, as well as shareholder-oriented and also *"consistent and coherent"*. Furthermore, pursuant to Article 19a(4), Member States may allow the undertakings to disclose information separately from the management report, through a separate report that needs to follow certain requirements. For more details on the difficulties with of this "separate report", cfr. Dániel Szabó/Karsten Sørensen, "New EU Directive on the Disclosure of Non-financial Information (CSR)", cit., pp. 20–21.

the four *matters*[26] described in the first paragraph of Article 19a(1) (environment, social and employee matters, respect for human rights, anti-corruption and anti-bribery matters).

The reader should keep in mind that the undertakings in question must disclose the following information *(the five aspects)* in what concerns the *four matters* further developed below. In compliance with its disclosure obligations, the relevant undertaking must, in the non-financial statement:

a. Describe its *business model* (Article 19a(1)(a))[27];
b. Describe the *policies* it has adopted in what concerns each one of the four matters, including due diligence procedures (Article 19a(1)(b))[28];
c. Indicate the *outcome* of such policies (Article 19a(1)(c)), in order to facilitate monitoring by investors or authorities[29];
d. Outline the *principal risks* pertaining to each one of the four matters linked to the undertaking's operations, including business relationships products or services that are likely to cause adverse impact (Article 19a(1)(d))[30]; and
e. Indicate the *key performance indicators* relevant to the particular business (Article 19a(1)(e)).[31]

The aforementioned information must concern at least the *four matters* identified in Article 19a, i.e. environmental, social and employee matters,

---

[26] This is the word used by the Directive. A fourfold division is used by the Commission's 2017 Non-Binding Guidelines, as well as by Constance Wagner, in "Evolving Norms of Corporate Social Responsibility: Lessons from the European Union Experience with Non-Financial Reporting", cit., p. 28, Other authors, such as Dániel Szabó/Karsten Sørensen, in "New EU Directive on the disclosure of non-financial information (CSR)", cit., pp. 11–12, use a six-fold division—environmental matters, social matters, employment matters, human rights matters, anti-corruption matters and anti-bribery matters.

[27] According to Dániel Szabó/Karsten Sørensen, in "New EU Directive on the Disclosure of Non-Financial Information (CSR)", cit., pp. 13–14, this means that the matters should only be reported on if and only to the extent they are relevant for the undertaking's business, while the Commission's 2017 Non-Binding Guidelines, p. 10, seem to indicate that a mere general description of the company and its activities suffices.

[28] For more details, cfr. Dániel Szabó/Karsten Sørensen, "New EU Directive on the Disclosure of Non-financial Information (CSR)", cit., p. 14. According to the Commission's 2017 Non-Binding Guidelines, pp. 10–11, this includes, for instance, disclosure on who is responsible for setting, implementing and monitoring a specific environmental policy or a description of due diligence processes implemented on suppliers and subcontractors.

[29] Commission's 2017 Non-Binding Guidelines, p. 12.

[30] According to the Commission's 2017 Non-Binding Guidelines, pp. 12–13, this can include the disclosure of information on internally or externally generated risks tied to the four matters, such as a malfunctioning product with possible effects on consumers' safety.

[31] Even though there is no indication in the Directive as to what that really means, the Commission's 2017 Non-Binding Guidelines, pp. 12–13, define them as "*material narratives and indicator-based disclosures*", such as disclosing metrics and targets to assess relevant environmental matters.

respect for human rights, anti-corruption and anti-bribery.[32] Such list of matters does not hinder undertakings from reporting on other topics if wished.[33]

IV. The disclosure of this information is subject to a *materiality test*, foreseen in Article 19(1), as an in-scope undertaking is only bound to disclose information to the "*extent necessary for an understanding of the undertaking's development, performance, position and impact of its activities*".

While the extent and content of this materiality test are unclear in the Directive,[34] the 2017 Non-Binding Guidelines provide some insight, using Article 2(16) of the Accounting Directive as a reference[35] and indicating elements to be included in the assessment,[36] although such indications are mostly vague and unclear in borderline situations where guidance would be useful.[37]

V. Additionally, the Directive provides a major exemption from mandatory disclosure in the fourth subparagraph of Article 19a(1)—information that would otherwise have to be disclosed if such information concerns "*impeding developments of matters in the course of negotiation*", when the disclosure of such information would undermine the "*commercial position of the undertaking*".

---

[32] For developments on the meaning and content of the four matters, please refer to Recital (7) of the Directive and to the examples given in the Commission's 2017 Non-Binding Guidelines, pp. 14–18.

[33] Dániel Szabó/Karsten Sørensen, "New EU Directive on the Disclosure of Non-financial Information (CSR)", cit., pp. 11–12.

[34] For a critical view, cfr. Claire Jeffwitz/Filip Gregor, "Comparing the Implementation of the EU Non-Financial Reporting Directive" (28.11.2017), available at SSRN: https://ssrn.com/abstract=3083368, p. 4; and Koen De Roo, "The Role of the EU Directive on Non-Financial Disclosure in Human Rights Reporting", *European Company Law* 12, no. 6 (2015) 278–285, p. 283, the latter emphasising (before the adoption of the 2017 Non-Binding Guidelines) that the test is too vague and leaves too much of a room for interpretations that neglect indirect relationships across supply chains. Such concern might have been alleviated by the fact that the 2017 Non-Binding Guidelines, in p. 6, clearly state that a company "*may consider that impacts across its upstream supply chain are relevant*", both direct and indirect.

[35] Which defines material information as the "*status of information where its omission of misstatement could reasonably be expected to influence decisions that users make on the basis of the financial statement of the undertaking*".

[36] Commission's 2017 Non-Binding Guidelines, pp. 5–6. Such elements can be topics identified by competitors, customers or suppliers in similar sectors, topics which might be in the interest and expectation of relevant stakeholders, the level of impact that a certain issue might have and the influence of public policies and regulation.

[37] Unfortunately, the Commission's 2017 Non-Binding Guidelines, in pp. 19–20, only provide one enlightening example of what is not material information (i.e., water consumption in offices and branches of a bank) and obvious examples of what is material information (a company producing mineral water may consider disclosing information on the measures it takes to protect the hydric resources it relies upon) and no not dive further into frontier situations. For a critical view, cfr. Georgina Tsagas, "A Proposal for Reform of EU Member States' Corporate Governance Codes in Support of Sustainability", *Sustainability* 12, no. 4328 (2020) 14.

While this *safe-harbour provision* may help undertakings into subjecting the framework introduced by the NFRD to their own commercial interests,[38] it can be pointed out that this provision was drafted in a very restrictive fashion, emphasising its exceptional nature and introducing the possibility of an issuance of an opinion by corporate bodies.

VI. Finally, in what concerns *enforcement,* the framework is mainly built around a comply or explain principle, pursuant to the second subparagraph of Article 19a(1). Further enforcement will depend on the individual Member States and it varies widely, with non-compliance being met with a variety of consequences, such as criminal penalties,[39] administrative monetary penalties[40] or the forced disclosure of withheld information.[41]

VII. Besides the mandatory disclosure of non-financial information that has been analysed up to this point, the NFRD changed the content of another provision from the Accounting Directive—Article 20, which concerns the corporate governance statement of the management report, which now is required to include a description of the undertaking's diversity policy applied in relation to its "*administrative, management and supervisory bodies*", such as diversity of age, gender or educational and professional backgrounds, the objectives of that diversity policy, its implementation and its results (Article 20(1)(g)).[42]

Contrarily to what we have seen up until now, this disclosure obligation laid out in Article 20 only applies to *listed companies* but this time, without any requirement based on number of employees.

This additional disclosure requirement might seem odd in light of the fact that mandatory disclosure on diversity policy could already be subsumed to Articles 19a and 29a, but the Directive seems to make no logical connection

---

[38] Koen De Roo, "The Role of the EU Directive on Non-Financial Disclosure in Human Rights Reporting", cit., p. 284, is particularly critical of the legal solution.

[39] In Germany, §331 of the German Commercial Code (*Handelsgesetzbuch* or HGB) indicates prison time of up to three years and monetary fines for managers and other representatives and members of corporate bodies for the crime of "misrepresentation" of information in financial statements and other mandatory disclosure documents, including the non-financial information statement.

[40] In Italy, pursuant to Article 8 of the Regulation on Reporting of Non-Financial Information (*Regolamento attuativo relativo alla comunicazione di informazioni di carattere non finanziaro*), missing or belated information can lead to various administrative fines, with a maximum of EUR150,000. In Portugal, Article 528(1) of the Companies Code (*Código das Sociedades Comerciais*) provides for an administrative fine between EUR50 and EUR1500 to be applied to the manager who does not submit such mandatory information.

[41] Article L. 225-102-1 (VI) of the French Commercial Code (*Code de commerce*) indicates that, in the absence of the non-financial information, any person with interest or stakeholder ("*toute personne intéressée*") can require the release of the missing information to a judge.

[42] For further details, W. Gregory Voss, "The European Union's 2014 Non-Financial Reporting Directive", cit., pp. 368–369; and Constance Wagner, "Evolving Norms of Corporate Social Responsibility: Lessons from the European Union Experience with Non-Financial Reporting", cit., pp. 27–28.

between the latter and the former, subjecting them to regimes with different scopes, content and format requirements.[43]

In any case, some authors point out that the Directive itself justifies such differences because both disclosure regimes have different objectives.[44]

VIII. It should be noted that Article 8 of the Taxonomy Regulation foresees additional information to be disclosed in the non-financial statement, namely information on how and to what extent the undertaking's activities are associated with environmentally sustainable information in light of the Taxonomy Regulation.[45]

IX. The Commission, after having pledged to do so in its 2019 Communication on the European Green Deal,[46] adopted a proposal for a Corporate Sustainability Reporting Directive[47] on 21 April 2021, which will bring changes to the NFRD provisions currently in force.

The proposed new framework foresees an expanded subjective scope (from 2026 on, even listed small and medium-sized undertakings will be included, though with a simplified regime), an expansion of matters to be disclosed, standardisation of reporting, new rules for external assurance on the information provided and minimum standards for penalties.

If the political process goes smoothly, the new Directive is expected to be adopted by early 2022, with transposition dates in 1 December 2022 and 1 January 2023, though such deadlines are subject to change in the final adopted text.

---

[43] Dániel Szabó/Karsten Sørensen, "New EU Directive on the disclosure of non-financial information (CSR)", cit., pp. 2–3.

[44] As pointed out in Recitals (3) and (18), while the mandatory disclosure regime provided for in Articles 19a and 29a applies to the whole of the company's activity and is more directly aimed at achieving broader sustainability and social goals through the measurement of the company's performance and its impacts on society, the mandatory inclusion of diversity policy in the corporate governance report is aimed at top jobs and seeks to enable a more effective oversight of the management and bring in a wider variety of skills and viewpoints into management. For more details on these differing aims, cfr. Constance Wagner, "Evolving Norms of Corporate Social Responsibility: Lessons from the European Union Experience with Non-Financial Reporting", cit., p. 28; Dániel Szabó/Karsten Sørensen, "New EU Directive on the Disclosure of Non-Financial Information (CSR)", cit., p. 12.

[45] Please note that while the remaining disclosure obligations have their legal basis on national rules transposing the NFRD, the obligation mentioned in the Taxonomy Regulation is directly applicable and binds undertakings without the need for national law provisions. It is applicable from 1 January 2022 and 1 January 2023, depending on the environmental objectives in question.

[46] Communication from the Commission to the European Parliament, the Council, the European Economic and Social Committee and the Committee of the Regions, *The European Green Deal*, COM(2019) 640 final, p. 17.

[47] Proposal for a Directive of the European Parliament and of the Council amending Directive 2013/34/EU, Directive 2004/109/EC, Directive 2006/43/EC and Regulation (EU) No 537/2014, as regards corporate sustainability reporting, COM(2021) 189 final (21.04.2021).

The Sustainable Finance Disclosures Regulation

I. A more recent but equally important piece of legislation aimed at promoting transparency in environmental and social matters within the EU is Regulation (EU) 2019/2088 of the European Parliament and of the Council of 27 November 2019 on sustainability-related disclosures in the financial services sector ("**Sustainable Finance Disclosure Regulation**" or "**SFDR**").[48]

The SFDR's aims include achieving more transparency in how financial market participants and advisers consider sustainability risks in their investment decisions, reducing information asymmetries concerning sustainability information (making it clear to end investors what is sustainable or not) and mitigating greenwashing[49] in the financial industry.[50]

Even though it was published in December 2019 the SFDR has only applied since 10 March 2021.

Regulatory technical standards ("**RTS**") developed jointly by the European Banking Authority, the European Securities and Markets Authority and the European Insurance and Occupational Pensions Authority, on the content, methodologies and presentation of the relevant information to be disclosed were due to be adopted and published in March 2021 with application due 1 January 2022.[51] As of 10 October 2021, their adoption had been officially confirmed as delayed by the European Commission,[52] with a new tentative application date of 1 July 2022.

II. The SFDR shows considerable cross-sectorial application, with the various disclosure obligations listed below being applicable to a wide array

---

[48] Meanwhile amended by Regulation (EU) 2020/852 of the European Parliament and of the Council of 18 June 2020 on the establishment of a framework to facilitate sustainable investment. Regarding the SFDR, see Julien Froumouth/ Joana Frade, Chapter 12 in this book.

[49] For those unfamiliar with this expression, "greenwashing", simply put, is a marketing approach that overemphasises whatever good deeds a company may be doing for the environment in order to appear more virtuous (and benefit commercially therefrom) or to divert attention from any misdeeds in other environmental and social issues. Regarding this terminology see the critical position of Paulo Câmara, Chapter 1 in this book.

[50] Recitals (9) and (10) of the SFDR and Sebastiaan Niels Hooghiemstra, "The ESG Disclosure Regulation—New Duties for Financial Market Participants & Financial Advisers" (22.03.2020), available at SRN: https://ssrn.com/abstract=3558868, p. 1.

[51] For more information, cfr. European Banking Authority's ("**EBA**") page on the public consultation concerning such RTS, https://www.eba.europa.eu/calendar/joint-consultation-paper-esg-disclosures-standards-financial-market-participants.

[52] Letter from the European Commission to the Chair of the Committee on Economic and Monetary Affairs of the European Parliament and to the President of the Ecofin Council within the Council of the European Union, "Information Regarding Regulatory Technical Standards Under the Sustainable Finance Disclosure Regulation 2019/2088", 8 July 2021, FISMA.C.4/LB/mp(2021)4983278.

of companies that fall within the concepts of *(i)* "financial market participants"[53] which include, *inter alia*, several categories of investment firms, fund managers and insurance companies; and *(ii)* "financial advisers",[54] which include credit institutions and investment firms providing investment advice and other entities acting in the same capacity.

This *twofold division between financial market participant and financial adviser* can be seen throughout the regime, as usually the same disclosure obligation applies to both categories but with different legal bases and with slightly tweaked wording. In case an entity pursues activities that can qualify it as both a financial market participant and a financial adviser, such entity shall make the relevant disclosures depending on the capacity in which it acts—for financial market participants when it acts *"in the capacity of manufacturer of financial products"*, including portfolio management, and for financial advisers when it provides an investment of insurance advice.[55]

Some smaller firms are exempt from many of these obligations, in accordance with Article 17(1).[56]

III. Articles 3–11 provide for a wide array of disclosure obligations[57]—most of them applicable to all financial market participants and financial advisers and others having a narrower scope. Taking into account the depth of the disclosure requirements and the limited scope of this text, such obligations will be summarily listed in the following pages.

The information to be disclosed pursuant to Articles 3–11 can be divided into four sets: *(i)* sustainability risk policies; *(ii)* due diligence policies; *(iii)* pre-contractual information; and *(iv)* special disclosure in case of investment and advice on green or sustainable financial products.

The first set of obligations is presented in Article 3, and concerns *sustainability risk policies*. This provision requires financial market participants and

---

[53] Article 2(1) SFDR lists the following entities as *"financial market participants"*: (i) investment firms and credit institutions providing portfolio management; (ii) alternative investment fund manager (**"AIFM"**); (iii) management companies of venture capital funds, social entrepreneurship funds or of an undertaking for collective investment in transferable securities (**"UCITS"**); (iv) insurance undertakings which make available insurance-based investment products; (v) institutions for occupational retirement provisions; (vi) manufacturers of pension products; and (vii) pan-European personal pension product providers (Article 2(1) of the SFDR).

[54] Pursuant to Article 2(11) SFDR, the category of *"financial adviser"* comprises (i) insurance intermediaries and insurance undertakings which provide insurance advice with regard to insurance-based investment products (**"IBIPs"**); (ii) credit institutions and investment firms providing investment advice (under MiFID II rules); (iii) AIFMs providing investment advice; or (iv) UCITS management companies providing investment advice.

[55] See Recital (7) of the SFDR for clarification.

[56] Recital (6) of the SFDR.

[57] These disclosure obligations are well documented in Dirk Zetzsche/Linn Anker-Sørensen, "Regulating Sustainable Finance in the Dark", University of Luxembourg Law Working Paper Series 2021-007; EBI Working Paper 2021, No. 97 (25.08.2021), available at SSRN: https://ssrn.com/abstract=3871677, pp. 10–11.

financial advisers to publish on their website information concerning their policies on integration of sustainability risks in *(i)* their investment decision-making process; and *(ii)* their investment or insurance advice, respectively.[58] It requires those companies to disclose how they address sustainability issues when *they invest* and when *they advise others how and where to invest*.

The obligation provided for in Article 5 will also be included in the first set, with the provision indicating that both financial market participants and financial advisers will include in their remuneration policies information on how those policies are consistent with the integration of sustainability risks and publish that information on their websites. Recital (22) of the SFDR explains that such obligation seeks to promote remuneration structures that do not *"encourage excessive risk-taking with respect to sustainability risks and is linked to risk-adjusted performances"*.

This obligation raises several questions, namely whether it overlaps with the remuneration policy referred to in the Shareholders Rights' Directive[59] and the awkwardness in the compliance with this obligation. It is unclear and only time will tell how the practice will evolve, but one should expect some greenwashing and half-baked explanations.

IV. A second batch of disclosure of obligations, more complicated perhaps, is provided for by Article 4(1), and concerns disclosure and publication of a *statement on due diligence policies,* although it is referred to in other documents as *"adverse sustainability impacts statement"*.[60] The aforementioned provision indicates that financial market participants should publish and keep on their websites:

a. *A statement on due diligence policies with respect to principal adverse impacts of investment decisions on sustainability factors*, if they consider that those adverse impacts exist and apply (Article 4(1)a, (2), (3) and (4)).
b. Where such companies do not identify adverse impacts of investment decisions on sustainability factors, *clarification as to why they do not so* and, if applicable, information as to whether and when they intend to consider such adverse effects (Article 4(1)b)), unless they constitute financial market participants with more than 500 employees (or parent

---

[58] For the sake of clarity: (i) being applicable to financial market participants and (ii) being applicable to financial advisers.

[59] Article 9a, added to Directive 2007/36/EC of the European Parliament and of the Council of 11 July 2007 on the exercise of certain rights of shareholders in listed companies (the Shareholders Rights' Directive or **"SHRD"**) by Directive (EU) 2017/828 of the European Parliament and of the Council of 17 May 2017 (**"SHRD II"**). A possible explanation is the different scopes of application of the two instruments—while the SHRD applies only to listed companies, the SFDR takes into account sectorial criteria—while in many cases there will be an overlap, one is far from guaranteeing that all financial market participants or financial market advisers are listed companies.

[60] For instance, in the RTS.

undertakings of a group exceeding that number of employees), in which case they cannot avoid publication of the statement on due diligence policies on grounds they do not identify adverse effects (Article 4(3) and (4), applicable from 30 June 2021).

For the benefit of clarity, the main difference between Articles 4(3) and (4), applicable to "large" financial market participants, and Article 4(1), of a general nature, lies on the fact that the latter works *on a comply or explain basis* (still applicable, until 30 June 2021, to "small" financial market participants), while the framework provided by the former provides no such possibility—that information *must* be provided.[61]

Article 4(5) lays out a separate regime for *financial advisers*, that will publish and maintain on their website information materially similar to that referred to in Article 4(1) but with respect to the type of financial products they advise on and under a comply or explain the principle.

The draft RTS foresee a yearly report on 30 June each year (Article 4(1)).

V. Article 6 *integrates sustainability factors in pre-contractual information*, such as prospectuses or private placement memoranda, requiring the disclosure of information on: *(i)* how financial market participants integrate sustainability risks into their investment decisions and the results of the assessment of the likely impacts of sustainability risks on the returns of the financial products they make available (Article 6(1)); and *(ii)* how financial advisers include sustainability risks into their investment advice and the results of the assessment of the likely impacts of such sustainability risks on the returns of the financial products they advise on (Article 6(2)).

Article 7 introduces further information to be included in pre-contractual documentation from 30 December 2020.

VI. Finally, there is a set of provisions that aims at mitigating greenwashing through the imposition of strict disclosure requirements on green finance, i.e. financial products that promote themselves as having *environmental or social characteristics and sustainable investment products*,[62] in Articles 8–11. While Articles 8 and 9 mandate the inclusion of pre-determined *pre-contractual information* pertaining to such "green" financial products, Articles 10 and 11 lay out general website publication duties.

---

[61] Sebastiaan Niels Hooghiemstra, "The ESG Disclosure Regulation—New Duties for Financial Market Participants & Financial Advisers", cit., p. 5; Iris Chiu, "Building a Single Market for Sustainable Finance in the EU—Idealism, Policy and Mixed Messages" (20.06.2020), available at SSRN: https://ssrn.com/abstract=3631946, pp. 6–7.

[62] Financial products will qualify as sustainable investments if they pertain to an economic activity that contributes to (i) an environmental objective; or (ii) a social objective, such as tackling inequality or fostering social cohesion, as long as they respect the "do not significant harm" principle (Article 2(17) SFDR). For further details, see Recitals (18) and (19) of the SFDR.

In both cases, the disclosed information should be drafted taking into account Articles 5 and 6 of the Taxonomy Regulation.

The Taxonomy Regulation

I. Regulation (EU) 2020/852 of the European Parliament and of the Council of 18 June 2020 on the establishment of a framework to facilitate sustainable investment (also known as **"Taxonomy Regulation"**) is one of the central legal acts of the EU's ESG framework. Even though it was adopted one year after the SFDR, both share intricate links and cannot be fully understood without one another.[63]

The Taxonomy Regulation has an interesting structure: *(i)* most of its provisions create a framework that identifies criteria according to which economic activities can be considered sustainable, in quite an in-depth fashion; and *(ii)* some provisions complement disclosure obligations already found in the SFDR or foresee new information to be mandatorily disclosed in the non-financial statement issued pursuant to national provisions that transpose the Articles 19a and 29a of the Accounting Directive (or NFRD if only considering the two mentioned provisions).

The Taxonomy Regulation also amended the SFDR, such changes already being included in the analysis above and, being a Regulation, is directly applicable to all Member States.

Among its many objectives are: *(i)* making it easier to identify whether certain activities are environmentally sustainable[64]; *(ii)* avoiding diverging standards among the Member States and consequent internal market and competition distortion[65]; *(iii)* lightening the due diligence burden of potential investors[66]; and *(iv)* mitigating greenwashing through the clear definition of what is considered truly sustainable.[67]

II. First of all, it is difficult to present a straightforward and clear picture of the Taxonomy Regulation's subjective scope because specific obligations such as the disclosure obligations that are laid out in the NFRD and the SFDR are not that predominant in the Taxonomy Regulation, which is more concerned with defining criteria for the determination of sustainable activities. In the context of the Taxonomy Regulation, we have to look at each of the regimes foreseen therein in order to correctly ascertain its subjective scope (granted that the concept of "subjective scope" may not fit an explanation of this Regulation correctly)—the Taxonomy Regulation is *indirectly*

---

[63] Regarding the Taxonomy Regulation, see also Rui Oliveira Neves, Chapter 13 in this book.

[64] Recitals (5) and (6) of the Taxonomy Regulation.

[65] Recital (12) of the Taxonomy Regulation.

[66] Recital (13) of the Taxonomy Regulation.

[67] Recital (11) of the Taxonomy Regulation.

concerned with financial market participants or advisers and large undertakings (the former two in the sense of the SFDR and the latter in the sense provided by the NFRD), even though one could say that the Member States, that must use it as a reference pursuant to Article 4, are directly subject to the Regulation.[68]

III. The crucial question that the Taxonomy Regulation was drafted and crafted to answer to is, in any case, *how to identify environmentally sustainable activities?*

Indeed, after listing relevant definitions and clarifying its subjective scope, the Regulation immediately lists *criteria for environmentally sustainable economic activities* in its Article 3.[69] According to such provision, economic activity is to be qualified as environmentally sustainable if the following conditions are met:

a. if it makes a *substantial contribution to six environmental objectives*, as defined in Article 9,[70] which include climate change mitigation, climate change adaptation, sustainable use and protection of water, transition to a circular economy, pollution prevention and restoration of biodiversity and ecosystems[71] (Articles 3(a) and 9);
b. if it does *no significant harm to any of such objectives* (Articles 3(b) and 17), i.e. it cannot cause more harm to one or more of the six objectives than benefit to other objectives, a principle seemingly crafted to mitigate greenwashing of an activity or a specific financial product by simply saying it substantially contributes to one of the objectives while it might be harming one of the other objectives (for instance, a wind farm operation that is harmful to a local body of water);
c. if such activity is carried out in compliance with minimum social safeguards (Articles 3(c) and 18)), meaning the OECD's Guidelines for Multinational Enterprises and the UN Guiding Principles on Business and Human Rights—this can be understood as a second *do no significant harm* principle, with the fundamental difference of being *social*, not environmental; and

---

[68] Cfr. Marleen Och, "Sustainable Finance and the EU Taxonomy Regulation—Hype or Hope?" (November 15, 2020). Jan Ronse Institute for Company & Financial Law Working Paper No. 2020/05 (November 2020), available at SSRN: https://ssrn.com/abstract=3738255, p. 7.

[69] For a more detailed analysis on the interplay of these criteria, cfr. Dirk Zetzsche/Linn Anker-Sørensen, "Regulating Sustainable Finance in the Dark", cit., p. 7–9.

[70] For a detailed explanation of how to determine the existence of a "substantial contribution" *vis-à-vis* each of the six objectives, cfr. the interesting analysis carried out by Christos Gortsos, "The Taxonomy Regulation: More Important Than Just as an Element of the Capital Markets Union", European Banking Institute Working Paper Series 2020, no. 80 (16.12.2020), available at SSRN: https://ssrn.com/abstract=3750039, pp. 14–18.

[71] For more development on each of these objectives, please refers to Recitals (24)–(26) and the definitions in Article 2 of the Taxonomy Regulation.

d. if it complies with certain technical screening criteria to be developed in delegated legislation (Article 3(d)).[72]

IV. It should be pointed out that, until this point, the Regulation is only defining what it considers environmentally sustainable (its main focus is not disclosure per se, without prejudice to specific disclosure obligations foreseen therein) and that this complex set of criteria can and will be useful to any current or future legislative measures adopted by the EU or Member States setting any rules that concern environmental sustainability.[73]

The Taxonomy Regulation thus has something to add to the set of disclosure obligations already applicable to financial market participants or to other large undertakings, as pointed out above à propos of the NFRD and SFDR. Most of the aforementioned delegated legislation feared to create excessive reporting burden on in-scope entities, was still in development as of 10 October 2021.[74]

The Taxonomy Regulation, having entered into force on 12 July 2020, indicates that the rules relating to climate change mitigation and adaptation objectives apply from 1 January 2022, and the remainder from 1 January 2023, thus presenting a partial application pattern with distinction based on substantive criteria.

Assessment: A Sensible Approach or an Unholy Trinity?

I. The *mandatory disclosure trinity* formed by the NFRD, the SFDR and the Taxonomy Regulation[75] created the first comprehensive framework trying

---

[72] Such criteria must be scientifically determined, respect technological neutrality, build on existing market practices and EU legislation and take into account life cycle impact (Articles 19 and 23 Taxonomy Regulation). For a detailed analysis Christos Gortsos, "The Taxonomy Regulation: More Important Than Just as an Element of the Capital Markets Union", cit., pp. 20–24.

It should be noted that Articles 10(3), 11(3), 12(2), 13(2), 14(2), and 15(2), foresee the adoption of a delegated act on technical screening criteria for each of the six objectives.

[73] Christos Gortsos, "The Taxonomy Regulation: More Important Than Just as an Element of the Capital Markets Union", cit., p. 10.

[74] For more information, cfr. European Banking Authority's ("**EBA**") page on the public consultation concerning such RTS, https://www.eba.europa.eu/esas-consult-taxonomy%E2%80%93related-product-disclosures.

Despite the development of various RTS mandated by the Taxonomy Regulation, the Commission has already adopted, but not yet published in the Official Journal as of 10 October 2021 a Delegated Regulation supplementing the Taxonomy Regulation by specifying the content and presentation of information to be disclosed by undertakings in scope of the NFRD and its methodology. The text is available at: https://ec.europa.eu/finance/docs/level-2-measures/taxonomy-regulation-delegated-act-2021-4987_en.pdf.

[75] As mentioned before, other disclosure obligations exist elsewhere. For instance, Article 449a of the Capital Requirements Regulation includes a requirement for certain credit institutions to disclose "*prudential information on environmental, social and governance*

to incorporate environmental and social concerns into corporate governance *through disclosure obligations*, even though the various instruments had different aims and scopes.

None of these acts target corporate governance *directly*—they do not establish rules on the internal structure of the company, the status of its bodies or the relations established between them.

This is without prejudice to the fact that some disclosure duties are directly concerned with governance issues—for instance, the disclosure of who is responsible for sustainability policies or the mandatory disclosure of diversity in corporate bodies found in the NFRD or the mandatory disclosure of remuneration policies and the integration of sustainability risks therein found in the SFDR.

II. The remainder of the complex web of environmental and social information disclosure duties is expected, at least by EU legislators, to increase *indirect* pressure by investors (essentially institutional investors)[76] on more sustainable approaches to corporate governance and even by the general public.

Indeed, EU lawmakers assumed that, while shareholders and stakeholders did not particularly crave that information, once it was available, they would use it to measure corporate performance on those environmental and social issues and even pressure management to adopt a more proactive role in addressing them.[77]

What was sought was the creation of demand for a type of information for which there was generally little demand, through its mandatory disclosure, such creation taking place not only among environmentally conscious institutional investors but even by an ever-watching public. This can also create be an *incentive for companies to "behave"*—after all, who wants to announce far and wide their doubtful behaviours on environmental or social issues?

Therefore, *mandatory disclosure duties* can be seen as a sensible balance between putting aside the incorporation of environmental and social issues in

---

risks". Implementing Technical Standards on such requirement, applicable from June 2022, are currently under public consultation.

[76] Supporting the idea that nowadays, portfolio value maximisation can mean pushing for ESG policies at individual company level, cfr. Luca Enriques, "Missing in Friedman's Shareholder Value Maximization Credo: The Shareholders" (25.09.2020), available at https://www.law.ox.ac.uk/business-law-blog/blog/2020/09/missing-friedmans-shareholder-value-maximization-credo-shareholders.

[77] Cfr. Tobias Tröger/Sebastian Steuer, "The Role of Disclosure in Green Finance" (10/08/2021). European Corporate Governance Institute—Law Working Paper No. 604/2021, SAFE Working Paper No. 320, LawFin Working Paper No. 24, available at SSRN: https://ssrn.com/abstract=3908617, pp. 4, 6–14, explains in detail the theoretical building blocks of this approach.

Dirk Zetzsche/Linn Anker-Sørensen, "Regulating Sustainable Finance in the Dark", cit., pp. 12–13, when reflecting upon the SFDR, mention that these disclosures are meant to "nudge" intermediaries to deal with sustainability as a topic and investors to consider sustainability more than before.

For a more sceptical view on this assumption, Dániel Szabó/Karsten Sørensen, "New EU Directive on the Disclosure of Non-Financial Information (CSR)", cit., p. 6.

corporate governance and a more authoritarian and interventionist approach that forces such incorporation upon companies, and has been pointed out as a success.[78] In this way, such middle-of-the-road frameworks, though not concerning corporate governance *directly*, do have the ability (and usually the objective) of *indirectly* influencing corporate governance decisions in a direction that promotes the incorporation of environmental and social issues therein.

III. However, the mandatory disclosure trinity was not hailed as a success across the board. Indeed, one can discern two sources of criticism—a policy-oriented scepticism of the assumptions on which this legislation was built on and more technical-level criticism.

IV. On a *policy level*, one can question the basic assumptions underpinning the framework constructed by these three acts—the watchful institutional investor or public who will punish any misbehaviour on an environmental or social issue, whose importance has been seen as overstated.[79]

From a political point of view, mandatory disclosure and imposition of the burden to act proactively on companies can also be seen as a way for national governments or supranational organisations to avoid to act or to issue more in more in-depth regulation[80]—if companies themselves take care of this, why should governments take further action?

Some even argue that some of these rules are nothing more than state-backed greenwashing[81]—not only can simple disclosure incentivise companies to think they have done enough "for the environment" and not act further (as publicise how they care for environmental or social issues) but even an act designed to prevent greenwashing can still allow it in practice. On the latter,

---

[78] Peter Fiechter/Joerg-Markus Hitz/Nico Lehmann, "Real Effects of a Widespread CSR Reporting Mandate: Evidence from the European Union's CSR Directive" (5.12.2020), available at SSRN: https://ssrn.com/abstract=3725603.

[79] Dániel Szabó/Karsten Sørensen, "New EU Directive on the Disclosure of Non-Financial Information (CSR)", cit., p. 6. With the opposite view, Luca Enriques, "Missing in Friedman's Shareholder Value Maximization Credo: The Shareholders", cit.

[80] Cfr. Dániel Szabó/Karsten Sørensen, "New EU Directive on the Disclosure of Non-Financial Information (CSR)", cit., p. 8, indicating that *"by allowing investors to press the business into having a more sustainable profile the EU legislator avoids regulating in detail how best to achieve such a change"*. Thus, disclosure is a *"cheap and non-intrusive way of promoting change towards sustainability"*, also having the benefit of flexibility.
Tobias Tröger/Sebastian Steuer, in "The Role of Disclosure in Green Finance", cit., p. 59, also mention what they call "political opportunity costs", i.e., *"putting bets on a disclosure-centred green finance strategy might exhaust social planners' ambitions to do better (...) inducing them to refrain from fighting for a global carbon tax or emissions trading scheme because they have already shown sufficient problem-solving capacity to their constituents"*.

[81] Koen De Roo, "The Role of the EU Directive on Non-Financial Disclosure in Human Rights Reporting", cit., p. 279; and Marleen Och, "Sustainable Finance and the EU Taxonomy Regulation—Hype or Hope?", cit., pp. 12–13; indirectly, claiming that the Commission's non-binding papers and legislative initiatives have led to the whitewashing of true sustainability initiatives, Georgina Tsagas, "A Proposal for Reform of EU Member States' Corporate Governance Codes in Support of Sustainability", cit., p. 14.

some argue, for instance, that the Taxonomy Regulation allows the marketing of financial products as "sustainable" even if their underlying activity is slightly harmful to any ESG factor.[82]

Keep in mind that these sceptical, policy-focused, positions can either be espoused by those who oppose or distrust further environmental and social regulation in general or by those who think mere disclosure does not go far enough.

Furthermore, there are some fears that, with the full entry into force of the SFDR and Taxonomy Regulation, one might be promoting the predictable outsourcing, by financial actors, of their data gathering and assessment to sustainability rating agencies with no significant supervision or scrutiny[83]—if sustainability assessment is outsourced, how will businesses genuinely reorient themselves towards such aims?

Other non-surprising criticisms concern high costs and burdens on companies to gather and assess all data.[84] Linking criticism based on high costs to the idea of a state-sponsored greenwashing, one author points out that such costs might end up being incentives for companies to just disclose that a product is non-sustainable, rather than going through the costly in-depth assessment of all activities—maybe this will turn sustainable finance into a costly niche product, rather than making it mainstream,[85] more or less what happens with many mainstream retail products marketed as sustainable or biological. Additionally, other authors cite a lack of evidence that investors reward sustainable or green investments.[86]

V. From a more technical point of view, these acts have been criticised for an excessive *ex post* focus that transforms disclosure of environmental and social issues into a mere box-ticking exercise, without bringing about genuine

---

[82] Marleen Och, "Sustainable Finance and the EU Taxonomy Regulation—Hype or Hope?", cit., pp. 12–13 argues that, under the Taxonomy Regulation, if a product is, let's say, 80% compliant with the Regulation and 20% harmful, it could be marketed as compliant.

[83] Marleen Och, "Sustainable Finance and the EU Taxonomy Regulation—Hype or Hope?", cit., p. 10.

[84] Marleen Och, "Sustainable Finance and the EU Taxonomy Regulation—Hype or Hope?", cit., p. 11, argues that such high costs can have palpable economic implications for companies or other financial actors (it is still unknown to whom will these costs be passed on to, to end investors and therefore make a sustainable financial product less competitive and less attractive). In a similar vein, Tobias Tröger / Sebastian Steuer, "The Role of Disclosure in Green Finance", cit., pp. 58–59.

[85] Marleen Och, "Sustainable Finance and the EU Taxonomy Regulation—Hype or Hope?", cit., pp. 11–12.

[86] Dirk Zetzsche/Linn Anker-Sørensen, "Regulating Sustainable Finance in the Dark", cit., pp. 21–22.

concern in corporate boards,[87] limited subjective scopes and enforcement[88] and inconsistencies between the acts themselves.[89]

3 OUT OF THE CRADLE OF DISCLOSURE AND NEXT (FIRST STEPS)

I. Despite or *because of* some of the criticism mentioned above, the EU seems to be ready to move in a different direction and to truly legislate *directly* on governance matters. Fundamental changes seem to be on the way in what concerns the legal framework governing corporate boards, directors' duties of care and even corporate purpose.[90]

Indeed, environmental and social concerns in company and financial law seem to be leaving their *cradle of disclosure* and taking their *first steps* in a more forceful direction. At the time of the conclusion of this text (10 October 2021), two important lines of action[91] were developing and were soon to materialise as legislative proposals creating: (i) rules on directors' duty of care and other corporate board matters; and (ii) rules on due diligence across supply chains.

---

[87] Koen De Roo, "The Role of the EU Directive on Non-Financial Disclosure in Human Rights Reporting", cit., pp. 283–285; W. Gregory Voss, "The European Union's 2014 Non-Financial Reporting Directive", cit., pp. 376–377; and Constance Wagner, "Evolving Norms of Corporate Social Responsibility: Lessons from the European Union Experience with Non-Financial Reporting", cit., pp. 76.

[88] Koen De Roo, "The Role of the EU Directive on Non-Financial Disclosure in Human Rights Reporting", cit., pp. 283–285; W. Gregory Voss, "The European Union's 2014 Non-Financial Reporting Directive", cit., pp. 376–377; and Constance Wagner, "Evolving Norms of Corporate Social Responsibility: Lessons from the European Union Experience with Non-Financial Reporting", cit., pp. 76.

[89] Cfr. EBA's, European Securities and Markets Authority's and the European Insurance and Occupational Pensions Authority's, *Final Report on draft Regulatory Technical Standards with regard to the content, methodologies and presentation of disclosures pursuant to Article 2a(3), Article 4(6) and (7), Article 8(3), Article 9(5), Article 10(2) and Article 11(4) of Regulation (EU) 2019/2088*, JC 2021 03, 02.02.2021, pp. 144, 147, 161.

[90] Please note that certain national legislations have already started tweaking their classic provisions on corporate purpose—Article 1833 of the French *Code Civil* indicates that a company is to be managed in its own self-interest but must consider the social and environmental impact of its operations. For more details on the new French paradigm, cfr. Blanche Segrestin, "When the Law Distinguishes Between the Enterprise and the Corporation: the Case of the New French Law on Corporate Purpose", *Journal of Business Ethics* (30.01.2020). More specifically on the ongoing debate, cfr. *inter alia*, Thomas Lee Hazen, "Corporate and Securities Law Impact on Social Responsibility and Corporate Purpose", *Boston College Law Review* 62, no. 3 (2021) 18–25.

[91] Statements of intent can be found in the Communication from the Commission, *Action Plan: Financing Sustainable Growth*, cit., pp. 6 and 11, where Action 10 recommended the *"fostering of sustainable corporate governance and attenuating short-termism in capital markets"* through clarification of directors' duties vis-à-vis the company and by requiring corporate boards to develop and disclose a sustainability strategy, including appropriate due diligence through the supply chain; and in the *The European Green Deal*, cit., p. 17.

In parallel to these developments, the aforementioned 2021 Sustainable Finance Strategy hints at a more forceful approach to sustainable investment.[92] In the following pages, we will focus on the prospective new rules on directors' duties and due diligence.

II. In what concerns changes to the content of directors' duty or care and to rules on corporate boards, it all started with a controversial study on director's duties and sustainable corporate governance published by the Commission in July 2020.[93] Such study identified as its core problem a perceived excessive focus of EU publicly listed companies on short-term maximisation of shareholder value rather than on the long-term interests of the company, which led directors to prioritise shareholder value and short-termism, investors to pressure directors for short-term gains and director remuneration structures that in turn incentivised more short-termism.[94] The study alleged that such short-termism also led to the sidelining of sustainability as a strategic priority and identified that board composition did not fully support a shift towards sustainability.[95]

Many authors expressed several doubts on the premises of the study, namely on two different but intrinsically connected grounds: (i) the study departed from a highly biased viewpoint, whereby its authors seem to have an exaggerated distrust of European companies, painting a grim picture of widespread short-termism and a greedy and unscrupulous business that does not correspond to the truth[96]; and (ii) serious methodological shortcomings, such as

---

[92] For a critical view, Dirk Zetzsche/Linn Anker-Sørensen, "Regulating Sustainable Finance in the Dark", cit., pp. 19–29.

[93] Ernst/Young (for the European Commission), "Study on Directors' Duties and Sustainable Corporate Governance: Final Report" (July 2020), available at https://op.europa.eu/en/publication-detail/-/publication/e47928a2-d20b-11ea-adf7-01aa75ed71a1/language-en.

[94] "Study on Directors' Duties and Sustainable Corporate Governance: Final Report", pp. 9–40.

[95] "Study on Directors' Duties and Sustainable Corporate Governance: Final Report", pp. 9–40.

[96] Claus Richter/Steen Thomsen/Lars Ohnemus, "Consultation on 'Study on Directors' Duties and Sustainable Corporate Governance" (08.10.2020), available at https://ec.europa.eu/info/law/better-regulation/have-your-say/initiatives/12548-Sustainable-corporate-governance/F594592, pp. 2, 4–9, refer that there is no empirical evidence that European companies have no regard for sustainability, society or the environment, on the contrary; Mark Roe/Holger Spamann/Jesse Fried/Charles Wang, "The European Commission's Sustainable Corporate Governance Report: A Critique" (14.10.2020), available at https://www.hbs.edu/ris/Publication%20Files/21-056_51410b50-5488-477a-9aa3-df8f81138e53.pdf, pp. 3–8; Alex Edmans, "Response to the EU Commission Study on Sustainable Corporate Governance" (01.10.2020), available at https://ec.europa.eu/info/law/better-regulation/have-your-say/initiatives/12548-Sustainable-corporate-governance/F556360, pp. 4–7, 9–11.

unsubstantiated assumptions, limited sample of countries and companies and random, biased and limited literature selection.[97]

III. From the conclusions of the study, the Commission moved on to the publication of an impact assessment document in July 2020[98] and the launching of a public consultation in October 2020.[99] From the conclusions of the study on directors' duties and the intents laid out in these two more recent documents, one can expect that future legislation on this matter, will possibly be a Directive and in the form of a modification of the Codified Company Law Directive[100] and/or amendments to the Shareholders' Rights Directive, will include:

a. *Changes to the content of directors' duty of care*—the rules on directors' duty of care should cease to give primacy to shareholder interest and should take into account the interests of all stakeholders which are relevant for the long term sustainability of the firm or the interests affected by them, with possible tweaking of corporate purpose principles;
b. *Rules on enforcement of directors' duties of care*—strengthening enforcement mechanisms outside internal board structures and shareholders' meetings, probably including other stakeholders in such enforcement[101];
c. *Changes in remuneration practices*—through rules that incentivise long-term value creation instead of short-term value creation. An example would be rules tying remuneration (or a part thereof) to non-financial performance; or
d. *Changes on rules on board composition*—potential requirements to have a certain number or percentage of board members with environmental, social and/or human rights expertise or regular training.

---

[97] Claus Richter/Steen Thomsen/Lars Ohnemus, "Consultation on 'Study on Directors' Duties and Sustainable Corporate Governance", cit., pp. 2-4; Alexander Bassen/Kerstin Lopatta/Wolf-Georg Ringe, "Feedback Statement on the Sustainable Corporate Governance Initiative of the Directorate-General for Justice and Consumers" (08.10.2020), available at https://ec.europa.eu/info/law/better-regulation/have-your-say/initiatives/12548-Sustainable-corporate-governance/F594615, p. 2; Alex Edmans, "Response to the EU Commission Study on Sustainable Corporate Governance", cit., pp. 1-2, 7-8; Mark Roe/Holger Spamann/Jesse Fried/Charles Wang, "The European Commission's Sustainable Corporate Governance Report: A Critique", cit., pp. 8-9. See also Abel Ferreira in "ESG and Listed Companies", Chapter 17 in this book.

[98] *Proposal for Legislation Fostering More Sustainable Corporate Governance in Companies*, Ref. Ares(2020)4034032 (30.07.2020), available at https://eur-lex.europa.eu/legal-content/EN/ALL/?uri=PI_COM:Ares(2020)4034032.

[99] Such public consultation, which ran up to February 2021, can be found here https://ec.europa.eu/info/law/better-regulation/have-your-say/initiatives/12548-Sustainable-corporate-governance/public-consultation.

[100] Directive (EU) 2017/1132 of the European Parliament and of the Council of 14 June 2017, relating to certain aspects of company law (codification).

[101] The public consultation mentions employees, the environment or people affected by the operations of the company as represented by civil society organisations.

There is a clear intent of *leaving the cradle of disclosure and taking steps to directly legislate on governance*. The public consultation document clearly states that while disclosure frameworks only presents incentives to report, the new initiatives aims at introducing duties "*to do*".[102]

These possible changes have not been without their critics. Some have pointed out that a radical shift in company purpose and interest would lead companies to ignore their owners' interests and become "self-driving autonomous entities", undermining the entire governance system as we have known it.[103] Other authors have pointed out—and rightly so—that, by expanding the range of stakeholders and concurrent liability allocation that accountability to an immense number of stakeholders means accountability to no one.[104] Proposals to tie remuneration to sustainability metrics were also seen as of questionable efficacy.[105]

IV. Another interesting set of changes concerns *new due diligence requirements*. The gigantic study on due diligence requirements through the supply chain released by the Commission in January 2020,[106] after reviewing market practices and legal and regulatory frameworks of twelve Member States, tried to conclude whether there was a need for any action and what form should that action take. The study opted to defend new regulation that would require companies to carry out due diligence to identify, prevent, mitigate and account for actual or potential human rights and environmental impacts in their operations and supply or value chain, as a legal duty or standard of care.[107] Issues within the scope of such due diligence would be human rights, climate change

---

[102] Public consultation, p. 2.

[103] Claus Richter/Steen Thomsen/Lars Ohnemus, "Consultation on 'Study on Directors' Duties and Sustainable Corporate Governance'", cit., p. 9.

[104] Claus Richter/Steen Thomsen/Lars Ohnemus, "Consultation on 'Study on Directors' Duties and Sustainable Corporate Governance'", cit., pp. 9–10; Mark Roe/Holger Spamann/Jesse Fried/Charles Wang, "The European Commission's Sustainable Corporate Governance Report: A Critique", cit., pp. 10–11, the latter putting into question the efficacy of "doubling down" on increased director accountability to stakeholders.

[105] Alex Edmans, "Response to the EU Commission Study on Sustainable Corporate Governance", cit., pp. 16–17; Mark Roe/Holger Spamann/Jesse Fried/Charles Wang, "The European Commission's Sustainable Corporate Governance Report: A Critique", cit., pp. 14–15, with both texts, but especially the former, emphasising that it transforms sustainability in a compliance issue, rather than making an economic case of it.

[106] British Institute of International and Comparative Law, Civic Consulting, London School of Economics (for the European Commission), "Study on Due Diligence Requirements Through the Supply Chain: Final Report" (January 2020), available at https://op.europa.eu/en/publication-detail/-/publication/8ba0a8fd-4c83-11ea-b8b7-01aa75ed71a1.

[107] "Study on Due Diligence Requirements Through the Supply Chain: Final Report", pp. 290 ff.

mitigation, natural capital, land degradation, ecosystems degradation, pollution, efficient use of resources and raw materials, hazardous substances and waste.[108]

There have been some doubts on how wide should the subjective scope of this mandatory due diligence requirement be (if it should only include large undertakings in the sense of the NFRD or if it should be applicable to all companies), on the objective scope (sector-specific or thematic, principles-based approach or minimum process approach) and what kind of enforcement mechanism should be conceived—if civil liability and/or enforcement through designated authorities in the Member States.

V. On both of these possible upcoming changes—directors' duties and due diligence requirements, at the time this text was closed (10 October 2021), no proposal had been published, but there have been, in any case, indications that a proposal is likely to be put forth by the Commission somewhere in the fourth quarter of 2021.[109]

The legislative process should be swift *vis-à-vis* the European Parliament, as there is wide support among its members for new rules on both these issues.[110] It is likely that more in-depth objections come from the Council.

4 Concluding Remarks

I. As we have seen, much has changed since the non-binding documents on a voluntary "corporate social responsibility" of the early 2000s. Even though a policy shift led to the first mandatory rules on incorporation of environmental and social concerns in company law—the aforementioned *mandatory disclosure trinity*—such rules, for the most part, did not directly concern *governance*, though one could argue that they exerted some sort of *indirect* influence thereon.

In any case, the disclosure obligations laid out in those three acts—the NFRD, the SFDR and the Taxonomy Regulation—were applicable to a limited

---

[108] Public consultation, p. 19.

[109] As indicated in the roadmap available at https://ec.europa.eu/info/law/better-regulation/have-your-say/initiatives/12548-Sustainable-corporate-governance.
In particular concerning the new due diligence requirements, cfr. https://www.csreurope.org/newsbundle-articles/eu-due-diligence-law-proposal-delayed.

[110] The European Parliament, in the resolution, *Sustainable Corporate Governance: European Parliament Resolution of 17 December 2020 on Sustainable Corporate Governance (2020/2137(INI))*, available at https://www.europarl.europa.eu/doceo/document/TA-9-2020-0372_EN.pdf, stressed the importance of providing a framework for a sustainability-oriented directors' duty of care and additional measures to make corporate governance more sustainability-oriented and called for the Commission to present a legislative proposal to ensure that directors' duties cannot be limited to short-term maximisation of shareholder value, but must instead include the long-term interest of the company and wider societal interests, as well as of employees and relevant stakeholders, at least to large undertakings as defined in Article 3(4) of the Accounting Directive. For an in-depth analysis of this resolution, see Guido Ferrarini, Chapter 2 in this book.

number of companies, either using the "large undertaking" concept, which included most publicly listed companies, credit institutions or banks (NFRD) or using a functional approach to subjective scope, as the SFDR did when bringing financial actors or advisors into its scope. Most of these rules lack enforcement mechanisms or leave that to the Member States, resulting in varying approaches to enforcement and a mostly comply-or-explain regime.

II. However, things seem to be changing and the first steps of a truly EU-wide ESG framework are being taken as we speak. Even though the mandatory disclosure trinity was only completed in 2020 (with some delegated legislation still pending), the EU seems to be ready to move on into a more forceful direction and *directly* legislate on the incorporation of environmental and social concerns in corporate governance. The scope of these new rules, which might initially be limited to a smaller group of large companies or companies that pursue certain activities, will likely expand with time, and such expansion will likely be very contentious.

Companies and stakeholders alike would do well in taking heed of future legislation that will reconfigure the content of directors' duties (and even corporate purpose) and impose new requirements in board composition. What will be more interesting to watch is how EU lawmakers plan to involve other stakeholders in the enforcement of these renewed duties.

Regardless of one's position in the debate on the pertinence or measure of introducing such concerns in EU law, one must be ready to welcome, anticipate or withstand a seismic change in corporate governance in Europe.

CHAPTER 11

Sustainability and Sustainable Finance: A Regulator's Perspective and Beyond

Gabriela Figueiredo Dias

1 THE TRANSITION FRAMEWORK

The integration of environmental, social, and corporate governance factors in the business models and conduct and more generally in the financial activity has gained increasing importance in recent times.

In fact, we are facing dramatic environmental problems calling for an urgent response from governments and the public sector, as well as from business and the financial sector.

It is well known that public funds are insufficient to finance the climate and environmental pledges, particularly the increased pledges to reach net-zero carbon emissions by 2050. Therefore, the role of the financial system in channeling private sector savings into sustainable projects is crucial. The urgency in dealing with climate and environmental problems for the preservation of our planet has justified the paramount importance that has been given to the environmental ("E") factor, both in terms of EU regulation (taxonomy, benchmarks, labels), and in the supply of green financial products, in particular green bonds.

However, social and government factors should not be relegated to a second plan. In fact, all factors are very interconnected, as an example of the link between environmental and social issues, poor populations are much more affected by the climate changes strong impacts, such as droughts, floods,

G. F. Dias (✉)
International Ethics Standards Board for Accountants, Munich, Germany
e-mail: figueiredodiasgabriela@gmail.com

© The Author(s), under exclusive license to Springer Nature Switzerland AG 2022
P. Câmara and F. Morais (eds.), *The Palgrave Handbook of ESG and Corporate Governance*, https://doi.org/10.1007/978-3-030-99468-6_11

and gigantic fires. Issues such as human rights, the need to ensure decent working conditions, ethical tax approaches and the fight against hunger and social inequalities—in other words, preserving the people's wellbeing, dignity, and minimum living standards—are eventually as important as preserving the planet where we live. Furthermore, the relevance of the "S" and the "G" factors have become even more evident in the context of the Covid 19 crisis.

The 'S' fired with the crisis, calling for effective measures to respond to the challenges the pandemic has posed. Important projects with social impact, such as the issuance of bonds to fund projects with positive social impact, were already rising even before the Covid 19 crisis. But the timid emergence of "*social bonds*" in previous years has been replaced by an exponential increase in the issuance of these products during the Covid 19 crisis and in its aftermath to support employment and respond to social inequalities and vulnerabilities (according to the S&P Global Ratings, the issuance of "social bonds" will have quadrupled in the first half of 2021). Between October 2020 and May 2021, the European Commission (EC) issued a total of €89.64 billion of social bonds in seven issuances under the EU SURE Bonds program of up to €100 billion, to financial support the 19 Member States that asked to benefit from the scheme aimed for employment protection. SURE has thus become the world's largest social bond scheme. This instrument calls for private parties to contribute in a very innovative approach (even revolutionary, if one thinks about the disruptive effects such measure will probably have in the traditional European institutional framework), and there was very strong investor interest in this highly rated instrument—resulting in favorable pricing terms for such bonds.

The "G" factor is also an increasingly essential part of the smooth functioning and success of companies and of the entire financial system, including issues such as the remuneration of managers not fostering short-term bias, boards' diversity, gender balance and equality, or the debate on the fiduciary duties of managers, among many other aspects, such as the distribution of dividends that crisis recently highlighted.

Governance has already been among the concerns of managers and economists for many years, in the logic of maximizing the efficiency of companies. What is somehow new, particularly since the great financial crisis, is the way we approach the idea of ethics and sense of purpose in business, and the possible (yet polemic) transition from shareholder capitalism to stakeholder capitalism, which is already making its way.

2 Sustainable Finance—Is There a Role for Financial Regulation and Supervision?

The financial system has a key role to play in the transition to a more sustainable planet, society, and economy. It is critical to ensure that the financial system, including financial products and financial institutions, make the necessary progress and adjustments to be prone to channel sustainable funding into sustainable projects. In other words, it is not only necessary that the financial system is able to carry over the necessary funds to support sustainability and sustainable projects; but it is also imperative that such funds are gathered in a sustainable manner and from sustainable entities and businesses. In an even simpler formulation, it is fundamental to ensure sustainability across the whole value chain.

The strong and crucial involvement of the financial system and capital markets sector, as well as their positive impact in the sustainable transition, implies a similar involvement and contribution from regulators and supervisors. The credibility of sustainable finance and the trust and confidence of those who decide whether to put their money in sustainable projects, which is the most central and critical cornerstone of the market, strongly rely upon the existence of appropriate regulation and supervision.

It goes without saying that for financial regulators and supervisors, this has been proving to be one of the most relevant challenges they have ever faced. Regulating and supervising sustainability is *per se* a giddying project. But trying to make sustainability—this subjective, slippery, and made of perceptions (and not rarely, made of illusions) issue—fit into the traditional financial regulatory world and addressing it with the same prudential and transparency concepts and instruments and the same market approach usually used in traditional financial regulation is probably the greater challenge ever for financial regulators.

But this may not stop regulators. The sustainability path is a one-way road also for regulators, and no reverse gear is possible anymore. On the contrary, regulation and supervision must speed up in identifying the gaps in the existing frameworks and in providing investors and the industry with appropriate, bespoke concepts and instruments adjusted to address the sustainable finance challenges, risks, and opportunities.

The regulators' approach, however, must go very well beyond the strict classical regulatory approach. Even if conceiving and enforcing appropriate regulation is an important task for regulators—the central one—, the sustainable finance subject has spotlighted the need to progress toward an advanced regulatory vision and mission of full commitment to a purpose.

In line with this advanced regulatory approach, the role of regulators must move beyond the traditional and rigid regulatory, supervisory, and enforcement activity to make its contribution to the global sustainability purpose, making much more use of the complementary tool of moral suasion, shaping business models and conducts according to the higher ethical standards, progressing towards a molder regulator role.

This integrated regulatory approach is crosscutting and should be applied in all areas of financial supervision and regulation. But it becomes particularly decisive when sustainable finance is at stake. And this is so, not only because sustainable finance is in itself a transversal issue, but mainly because it is a different issue in nature, hardly measurable, difficult to capture, and strongly linked to perceptions and values, making it impossible to fit into the classical dogmatic and conceptual financial and supervisory framework and mindset without important adjustments and changes of paradigms.

Being impossible and unwise to address all the relevant regulatory and supervisory dimensions of sustainable finance, I will only address in more detail two of the ESG dimensions which for different reasons deserve a specific mention.

One important dimension is about the role of regulated information and institutional communication on ESG matters.

Another one is about ethics and culture in financial business and the transition to a new paradigm for the company and the capital market arising from a sustainable approach to business.

3   Information and Communication in/for Sustainable Finance

Regulated information and communication are different but complementary regulatory tools in sustainable finance.

While regulated information aims to prevent and combat greenwashing, favoring informed and responsible investment decisions and promoting equal opportunities for market participants, institutional communication is growingly key as a critical factor of dissemination of knowledge, awareness creation, and a motivation driver.

The big challenge in respect to communication on sustainable finance is to know how to deal effectively with shareholders, but also with workers, potential investors, creditors, suppliers, customers, and other stakeholders, who have very different motivations. Being successful space requires an ability to build messages that resonate with the addresses' motivational drivers. It is necessary to decode the way people think, are motivated and are likely to behave in various situations. If we want to get investors and the industry to move forward in a sustainable direction, we must pick them up where they are and not just wait for them to join.

The Covid 19 crisis, with the impacting ingredients of isolation, remoteness, distancing, and detachment, but also and at the same time, of acceleration of digitalization, web surfing, and social media, has exponentially increased the importance of good communication.

In July 2021, the European Commission launched its Renewed Sustainable Finance Strategy, an ambitious and comprehensive package of measures to help improve the flow of money toward financing the transition to a sustainable economy, highlights the need and usefulness of communication to boost the attractiveness of the company to the various stakeholders and consequently, to its growth.

Together with regulated information, institutional communication comes as an imperative to address and minimize the reputational risks of greenwashing, by creating investors' awareness about the need to distinguish mere and sometimes elusive marketing communications from concrete and impactful sustainable action from financial operators. It must contribute above all to explain ESG to existing investors shareholders while winning future ones.

But sustainable finance is already subject to relevant regulated information and reporting duties, at least in Europe, where the European Commission is pushing to speed up the setting of an extensive and demanding regulatory framework in this respect. The Non-Financial Reporting Directive (NFRD), in force since 2017, or the Sustainable Finance Disclosure Regulation (SFDR), which was introduced in 2019 and came into effect in March 2021 and the Taxonomy Regulation, gradually applicable as from January 2022, are part of a new wave of European regulation aimed at building a sustainable economy, focusing on transparency and reporting duties on ESG matters. Adding to those, the ambitious proposal from the European Parliament of a Sustainable Governance Directive, requiring companies to undertake due diligence exercises along the whole supply chain, is a very significant signal of the approach that policy makers are adopting to sustainability and business.

These legal requirements, at European level, have been incentivizing companies to reflect (positively) on its ESG performance, while at the same time capturing the attention of investors on the importance of scrutinizing such performance and integrating it in their investment decisions.

However, the information report requirements on ESG matters pose several challenges. The lack of quality and reliability of the information disclosed by the companies has been highlighted, implying that (institutional) investors must resort (and bear the costs) to ESG data providers.

As said above, a major challenge for supervision is to learn how to adapt the supervisory mindset to a totally new reality. But supervision has also an important role to play in helping identify the weaknesses of a regulatory framework being built in a speed mode. Market participants, in particular asset managers, were also requested by the EU to play a key role in leading the transition to a more sustainable finance environment, since their clients will increasingly require responsible and sustainable funds, since at least institutional investors will have to observe sustainability requirements themselves. Asset managers are therefore at the center of an important system which renders them a particular position of facilitators of sustainable investment.

Auditors, will play a key role as well—actually, they already play it. Their duties of skeptical observation of reality, questioning, requirement of reasoning, and documentation are already fundamental in financial information (and to this extent a fundamental contribution to the relationship between the ESG and performance). But the first line of responsibility with respect to non-financial information lies with preparers of such information and with the Chief Financial Officers (CFO) of companies. While auditors, who in any case come after, are subject to hard regulation and rules, namely in what concerns the aforementioned duties and independency request, the prepares and CFO are generally not subject to any specific regulatory frameworks. That raises the importance of international ethics standards by which preparers of financial and non-financial information must abide, such as the ones in the IESBA

Code of Ethics, which provide strong and ethical-based guidance to those professionals and a significant convergence of practices, principles, and values.

Europe has been leading the regulatory response to the sustainable finance perplexities, progressing very fast in the conception of new regulatory pieces aiming at capturing the ESG challenges and opportunities in a comprehensive set of rules. This will most probably take Europe to pave the way and set the pace of the sustainable finance journey. But this path also encompasses significant risks of dissociation and detachment from and of other regions and regulatory and standard setting initiatives, resulting in a regulatory gap between Europe and the rest of the world in case the level of regulatory complexity moves other regions and markets away from the European model.

On the other hand, the existence of a multiplicity of very important and (individually) useful standards for the sustainability, non-financial information reporting with which companies are being confronted, risks to negatively impact the consistency and comparability of non-financial information.

It is however irrefutable that international reporting and assurance standard setters which have been working in this field play a precious and irreplaceable role in providing boards, auditors, and accountants with innovative, tailormade technical and ethical standards, material and guidance on how and what to report and assure with respect to sustainability information. To preserve and take full advantage of this work and deliverables from the standard setters, it is therefore also evident that regulators and decision makers at all levels must strive to ensure globally consistent and convergent regulatory and supervisory responses.

It is critical that the global regulatory and standard setting community joins efforts to build a common standard of reporting at international level, which makes the IFRS work materialized in the creation of International Sustainability Standards Board, in developing a global standard base built on the prototype that resulted of the alliance of the main standard setters, of the utmost importance.

4   Sustainable Finance Reporting and Disclosure: A Burden or an Opportunity?

In any case, we need to fight the simplistic misconception that ESG is a burden.

It is true that the requirements to report on ESG matters put an additional burden on companies, which have to comply with additional reporting duties, to learn how to address its technicalities, to deal with ESG metrics, etc. It is also true that there is a risk that this regulatory trend, given its complexity and specificity, may widen the gap between more and less developed markets and firms.

But sustainable finance information requirements may also work as a lever to induct a change of mentality and culture of companies, shareholders and investors, as the integration of ESG factors in the conduct of business activity is essential for the long-term sustainability of companies and for the transition to a greener, inclusive, and resilient economy, which is now an inevitability.

The regulatory informative requirements on sustainability are therefore not only a lever for achieving collective ESG objectives, but eventually for the success, growth, profitability and sustainability of the company itself. Investors, in particular institutional ones, have increasingly taken an active role in the sustainability of companies, directly engaging and debating their ESG approach with management and intervening in general meetings in the discussion and approval of sustainability policies.

On the other hand, there is an increase in the demand and supply of ESG investment funds, so the selection of assets for the portfolios of funds must be based on reliable information. Some of the larger global asset managers have already announced their intention to integrate only investment funds with sustainability criteria in their portfolios soon.

The integration of ESG elements is increasingly becoming a demand from investors and consumers in general. In fact, investors have the perception that the integration of ESG factors in the conduct of business activity can be a driver for its reputation and financial performance and that a long-term business sustainability management approach is a good indicator of good governance. It is an opportunity to use 'finance for good'.

But as conscious, purpose-driven consumerism seems to be gaining momentum, attention to risks of greenwashing and social washing must double—and so must the regulators' attention to the issue, as part of their mission to ensure the integrity and stability of the financial market, investor protection, and market promoting.

In particular, they play an important role in informing and creating knowledge in this matter, modeling trends and behaviors, raising market standards and incentivizing the adoption of best practices, while combating greenwashing and social washing and keeping alive a constant dialogue with market players in adapting to the new regulatory environment.

5 Sustainability and Firm's Culture in Capital Markets

Alongside their traditional supervisory responsibilities, financial supervisors, and particularly capital markets supervisors, play a very important role in the urgent process of changing business cultures toward more ethical, values-based and sustainable development models, given its impact on investor protection, on the integrity of the markets and on the stability of the financial system as a whole.

The securities regulatory system has two basic goals—the protection of investors and the promotion of market efficiency—which are, in fact, two sides of the same coin.

We know, however, how much investors' confidence and their willingness to channel savings to the market has been affected over the past two decades by the devastating effect of some financial events on their trust in firms.

The problem lies not only on the financial losses they incurred—that, for some, meant their lifetime savings. The main problem arises also or mainly from the widespread and persistent inadequate market practices and conducts, revealing deep corporate culture failures, that started on bad risk management and short termism and went all the way up to pure fraud.

Investors and the society more broadly have lost confidence in the financial sector for well-known reasons, and not enough has been made to restore it, namely in what regards assuring that significant changes were achieved in conduct and professionalism of market agents.

At the same time, the Capital Markets Union, a project designed to boost the market, seems to have failed, in its initial configuration, to achieve the objectives set out. One of the reasons for this failure is the way this European project disregarded the investor's core role in it. This has been more clearly addressed in CMU 2.0., but for its success to happen, corporate culture and sustainability need to be high in the agenda of the development of fair, efficient, and competitive financial markets in the European Union.

A new regulatory approach focused on culture, sustainability, and purpose of supervised entities is hence fundamental.

The classical regulatory approach toward the promotion of investor protection rests upon an increasingly complex set of rules, the majority of which are market conduct rules on disclosure of information, reduction of information asymmetries, marketing, trading, etc. The hotspot of market regulation and supervision still focus on the conduct of issuers, intermediaries, and asset managers in their direct relationship with clients or investors, materialized in a regulatory system built upon this main topic of concern.

But this limited vision on the role of capital markets supervision, having financial and marketing transparency as the main goals of market regulation, is no longer adjusted to the current context and was, so far, unable to promote effective investor protection and to reverse the mistrust trend. The importance of a regulatory approach focused on purpose and culture of supervised entities is becoming more and more evident.

We must look at other critical dimensions of market conduct besides transparency: organization, governance, decision-making processes, conflicts of interest and mitigation mechanisms, fiduciary duties, fraud, ethical concerns in meeting business objectives, the definition of 'sustainable'. In short, it is mostly about the vision and suitability of firms' leaders, about the tone coming from the top, about purpose in business models, and about integrating sustainable values in the strategy of value creation and growth—in a word, it is about culture and sustainability. Restoring investors' confidence in the market depends on changing corporates' conduct by changing corporates' culture, caring about the negative impacts of the business activity on the planet and the people, and focusing on long-term, sustainable goals, more than any other thing. Alongside with the culture of the organization and the ethical robustness of leaders, professionalism is a critical element for market recovery and investor confidence.

The classical regulatory approach to what is deemed to be sound market conduct practices must thus be complemented with sustainable and purposeful corporate culture, which determine such practices. Acting on marketing, on pre and post contractual information, disclosures, market integrity, and other dimensions is of course important to prevent the poor outcomes arising from misconduct. But effectively preventing such practices also critically depends on the determinant influence of firms' culture and adherence to strong professional standards on firms' conduct.

In this context, the question is what should be the regulators' role and how can they promote better culture in financial firms and in their leaders, or how they can be effective in promoting such results.

It is essential that regulators set clear priorities among the various dimensions that I've just mentioned, using their powers and tools on the genetic causes of problems, so as to induce structural improvements regarding the robustness of the companies' culture and their strict adherence to professional standards.

Having that in mind, there are at least three critical issues regarding corporate conduct and culture which deserve our attention and probably a more assertive action: governance, business models, and sustainability.

Board and Oversight Functions

One key governance issue is board and oversight functions in supervised entities. Looking at board composition, functioning and effectiveness in supervised entities, be they issuers, financial intermediaries, or asset managers, it is critical to promote the firms' culture and a fair treatment of investors and stakeholders.

If we really want supervision to make a meaningful contribution for that purpose, then it should focus on board nomination processes, board diversity (including, but not only, gender diversity), reputable, independent, and skilled directors, sound and long-term oriented remuneration policies and say-on-pay, robust and independent oversight structures and procedures (preparers of financial and non-financial information, CFO, audit committees, supervisory boards and external auditors) and appropriate management of conflicts of interests at the board level.

The powers given to supervisors by the European regulatory framework already allow them to assess and act on boards' culture, effectiveness, and integrity. It's time to use such powers for the good and intervene at the earlier stages of the dramatic governance gaps and vulnerabilities aiming at avoiding their devastating impacts and consequences.

Fitness and Propriety | Suitability

Fitness and propriety, or more widely, suitability of executives is probably the most relevant and effective supervision path to be tread when we aim at strengthening firms' culture and make them sustainable. It is there, in executives' experience, skills, professionalism and in the ethics and values of those who design and manage business models, governance structures, market practices, inducements and, at the end of the day, decide about the long-term, sustainable goals to achieve and about how to achieve them, that the big difference can be made.

It is not an easy task for supervisors to assess and decide about fit & proper issues. Even if we set up some objective criteria, like the number of years of experience, or the professional standards to abide by, or the relevant judicial decisions against board members, it mostly relies on prognosis' judgment on the future behavior regarding investor protection and market integrity of those who are subject to a fit and proper assessment. The inherent difficulty of this prognosis judgment (for preventive purposes), which is per se a delicate exercise, poses additional difficulties in terms of legal enforcement, that are particularly relevant in some countries and jurisdictions, for cultural and legal reasons. But we should not rest our case—on the contrary, the regulators' action in this field must be significantly improved, by preventing those who prove to be unfit and unproper from leading financial firms and any public interest entities, but also by using moral suasion to embed values and ethics in the leaders' approach to management.

Fiduciary Duty

Working on firms' culture about pursuing long-term, sustainable objectives also requires rethinking the board's fiduciary duty. The classical approach to the fiduciary duty as the ultimate board's duty to pursue (only) profit in the (only) interest of their investors/shareholders/clients is being consistently challenged and does not reflect the new mainstream proposals of transition from "shareholder capitalism" to the "stakeholder capitalism." More than an ideologic approach, the integration of other interests and values, like sustainability, social responsibility, equality, alongside with the profit goal, and considering all stakeholders, namely workers or the community itself, is becoming imperative. There is growing evidence that this is a more robust, profitable, and sustainable approach in the long run, including for shareholders.

At the same time, the fiduciary duty in financial regulation, traditionally understood as a business conduct rule, is growingly absorbing prudential concerns. The number and intensity of public interests that financial entities' boards are required to respect and which, according to the existing regulation, already prevail over the individual, private interests of the clients, lead to the conclusion that the fiduciary duty has already undergone a deep transformation in consequence of financial prudential regulation, triggered by underlying public and private interests. Thus, the compliance of this duty must be addressed both in a conduct and prudential supervision perspective.

Supervision of fiduciary duty performance by board members of issuers, financial intermediaries, and asset managers, already assessed according to the classical approach, has hence not only to be reinforced but also revisited to take into account such new features, such as the integration of long-term concerns, and the consideration of sustainable and collective values in the firms' culture and business model.

Sustainability

Coming to the core and wrapping topic about firms' culture and long-term viability and profitability of the firm in the interests of all its stakeholders, sustainability is one of the most relevant issues impacting firms' culture and the community's perception about firms' culture; and hence, as already made clear above, it is already a supervisory priority.

Environmental and inequality challenges have contributed to a growing awareness of the need to adopt sustainable development models, that meet the needs of the present generations without compromising the ability of future generations to meet their own needs and create long-term value. The industry is responding to that demand by adapting business practices and offering green investment opportunities, issuing green bonds and setting up sustainable investment funds.

There is therefore a fundamental need to induce business cultures to take sustainability issues seriously into account, ensuring that investors have appropriate information about how market agents are pursuing the sustainability goal.

The supervision of mandatory disclosure of non-financial information by listed companies requires capital markets regulators to contribute to the sustainable finance movement, even if, as highlighted above, this duty is proving hard to comply with by most issuers.

But there are other matters of concern: distribution and pre and post contractual information on "green" products and green investment policies is already a supervisory issue to be addressed not only with the new but also with the existing regulatory tools, such as MiFID, UCITS, AIFMD, and PRIIPs rules. Likewise, the mismatch between what asset managers proclaim doing on sustainability ('greening' their investment options) and what they are actually doing (often exercising their voting rights in invested entities against green options) should be assessed by supervisors, to ensure that market agents do walk the talk and are not just selling dreams.

The Business Models' Analysis (BMA)

The business models' analysis (BMA) of supervised entities is an essential vector of market prudential supervision, not to be neglected by markets supervisors. It aims at assessing the viability and or sustainability of the entities' activity, by assessing the source and sufficiency of their profitability, adjusted to the risk incurred in their activity and its long-term sustainability. This should be compatible with the risk appetite defined at the strategic level and which is part of the entity's own culture, defining its positioning in the market. The level and type of indebtedness and leverage exposure that the entity is prone to accept is indeed one of the most relevant supervisory issues of business models.

Doing BMA requires a forward-looking and critical approach from the supervisor, as well as its capacity to challenge the entities' strategic business plans and the assumptions set out therein, in order to identify possible vulnerabilities that may put their own viability at stake and take action on them.

Organizational Issues

Finally, there are some organizational issues related to a sustainability culture that fully depend on the specific governance and business models set up by the firm and where regulators have significant responsibilities.

Risk management, internal control, compliance function, inducements, costs and performance fees structures, and all those elements of a business model that may significantly exacerbate risks, impair the fair treatment of customers and undermine the integrity of financial markets and confidence in the financial system, namely when they are not strong enough to prevent money laundering, bribery, or fraud. Sometimes, regulators have intervened too late. It is critical to dive deep into the organization of supervised entities, adopt a look-through approach, instead of focusing only on conduct, and checking the robustness of their processes, organization and internal governance, the appropriateness of their human capital, the independence of supervisory functions, their goals and what they are ready to do to achieve them.

6   CONCLUSION

Investors' confidence depends vertically on a critical cultural change in companies and on fierce adherence to the highest quality standards of each profession.

In this regard, market regulators have a role to play in supervising certain dimensions of firms' organization and activity, thus contributing significantly to strengthen the sustainable culture of supervised entities and executives, with a firm view to restore investors' confidence. The traditional separation between conduct and prudential supervision does no longer make sense in capital markets: just like market integrity and market stability, or investor protection and financial stability, they are two sides of the same coin.

CHAPTER 12

ESG Reporting

Joana Frade and Julien Froumouth

1 Calls for Consistency, Quality and Reliability in Reporting and Provision of Comparable Information

A Buzzword Does not Suffice

In the aftermath of COP21,[1] policy makers—backed by a growing attention from the civil society—have indicated that they will increasingly dedicate specific attention to transparency, accountability and compliance with Environmental, Social and Governance ("ESG") related topics. Trustworthy information is indeed pivotal in channelling capital towards low-carbon and sustainable activities. Accurate, timely and reliable information is expected to allow investors to make informed decisions on their capital allocations. In relation to transparency, strong voluntary practices would need to be defined with further granularity and based on common and widely-accepted definitions, in

---

[1] 21st Conference of the Parties that signed up to the 1992 United Nations Framework Convention on Climate Change, held in Paris.

J. Frade (✉)
Fundação Oriente, Lisbon, Portugal
e-mail: Frade.j@gmail.com

J. Froumouth
Luxembourg Bankers' Association, Luxembourg City, Luxembourg
e-mail: Julien.froumouth@abbl.lu

© The Author(s), under exclusive license to Springer Nature Switzerland AG 2022
P. Câmara and F. Morais (eds.), *The Palgrave Handbook of ESG and Corporate Governance*, https://doi.org/10.1007/978-3-030-99468-6_12

order for these to serve as a baseline against which ESG performance and comparability can be measured and tracked.

The sustainability reporting topic was until recently likely to remain self-regulated, with the advantage of allowing for an open and flexible sectorial-approach. However, it has been found that such flexibility should be carefully balanced against investors' trust, which strongly relies on the consistency, the quality, the reliability and the comparability of information disclosed.

Not to meet more stringent reporting objectives is currently deemed to jeopardise the credibility of companies and institutions, even those who happen to be active in the field of sustainable activities and/or finance.

Appropriate ESG disclosures and the related sustainability reports will therefore ultimately need to be matched by further convergence of minimum common standards to allow, among other aspects, for product, service and companies performance's comparability, notably in the financial sector. Hence, relevant mandatory sustainability reporting is expected to become a powerful tool to enhance the efficiency of capital markets and a risk-based allocation of financing channelled in economic activities contributing to environmental, social and governance-related objectives.

With a view to enabling trust in sustainability reporting and disclosures, it is necessary that the basis of such reporting is clearly defined and the reported figures and information are comparable across countries and industries.

This means that ESG-related disclosures and transparency are here to stay—it appears now of essence that legislators and regulators worldwide enforce their enshrinement within legal frameworks by setting more stringent mandatory milestones to be complied with, in order to provoke a clear awareness that it is no longer an option not to consider such sustainability aspects when conducting business.

Policy makers and other internal and external stakeholders recognise the growing importance of holding companies accountable for their impacts on climate, environmental and social factors, increasing the need for adequate disclosure on their strategies, the associated risks and their action plan to manage, monitor, track, measure and finally report on the impacts of their activities as well as the resilience of their business models with regards to sustainability.

It has in fact been found that, additionally to engaging in sustainable activities, the establishment of reporting obligations on such business standards contribute to the creation of a proper environment for fostering existing initiatives. In fact, to describe such activities and to provide information about them to others is key for changing pre-existing patterns and to persuade market players to walk the talk on the ESG agenda.[2]

---

[2] Refer inter alia to recent papers such as Adoption of CSR and Sustainability Reporting Standards: Economic Analysis and Review, Hans B. Christensen, Luzi Hail and Chsti-tian Leuz, 2019, ECGI; ESG Performance and Disclosure: A Cross-Country Analysis, Florencio Lopez-de-Silanes, Joseph A. McCahery, Paul C. Pudschedl, 2019, ECGI; The

The objectives of standardised reporting include providing information, enabling comparability, allowing the implementation and development of internal control systems, providing for compliance and establishing proper supervision.

The multitude of activities entered into by the broad range of very diverse market players and the inexistence of harmonised reporting standards do not add value to the process and prevent stakeholders (such as clients or supervisors) from ascertaining if the reporting entities are properly embedding ESG principles in their governance. This behaviour has been found to enable practices such as greenwashing.

In "Four Things No One Will Tell You About ESG Data",[3] George Serafeim and Sakis Kotsantonis highlighted "the sheer variety, and inconsistency, of the data and measures, and of how companies report them".

In this paper, a multitude of forms that companies resort to for reporting employee health and safety data was listed, which led its authors to argue that "such inconsistencies lead to significantly different results when looking at the same group of companies".

To date, the pre-existing framework of voluntary and scattered reporting was in fact allowing market players to prepare multidimensional and incomplete reporting, jeopardising resources, enabling the continuation of information asymmetries and contributing to the non-assimilation of the moral hazard by the recipients of such information.[4]

Enhancements to a Scattered Framework

Fortunately, there has been some encouraging improvements. The normative context in which sustainability reporting exists has been constantly developing, diversifying and becoming more specific. By relying on pre-defined standards, norms and labels, a more proactive approach can be adopted to respond to changes in transparency requirements and market expectations—an approach that must be structured by reference to an operational framework drawn up by recognised organisations. Recent years have seen the development of a wide range of national, European and other international norms and standards. Different and yet at the same time complementary, those norms and standards encompass varying characteristics which need to be understood, as

---

Future of Disclosure: ESG, Common Ownership, and Systhematic Risk, John C. Coffee, 2020, ECGI.

[3] George Serafeim/ Sakis Kotsantonis, Four Things No One Will Tell You About ESG Data, 2019, available at https://www.hks.harvard.edu/centers/mrcbg/programs/growthpolicy/four-things-no-one-will-tell-you-about-esg-data.

[4] Such reporting was mainly the result of the voluntary application of Stewardship Codes and Responsibility Investment Principles by some early bird market players, which created the need for auditing such reports. Auditing methods have also been built up on a case by case and non-harmonised basis (vg Sustainalytics EGS, Bloomberg).

well as the added value that they respectively offer, in order to enable companies to determine which set of rules are best suited and relevant to meet their specific needs and objectives.

Given the diversity of those norms and standards, it may appear difficult to select the right frame of reference against which to assess an organisation's sustainability performance.

For the purpose of sustainability reporting, norms and standards most commonly encountered may be classified not only according to the sustainability objectives they tend to reflect but also to the extent to which they are recognised and accepted (global, European or national influence).

The number of issued texts and initiatives on the subject has built up, step by step, a structure enabling the normative framework of sustainability reporting to emerge and take shape, going back to the Organisation for Economic Co-operation and Development Guidelines for Multinational Enterprises published in 1976, which provided a comprehensive overview of the main instruments and methods available to enterprises intending to conduct business in a responsible manner.[5]

In 2000, the United Nations launched the Global Compact[6] initiative, a non-binding act whereby undertakings, non-governmental organisations and associations covenant to respect ten universally defined principles concerning human rights, labour standards, the environment and measures to combat climate change and publish each year a report on the progress made in implementing such principles.

The International Organisation for Standardization's ("ISO") 26000 standard issued in 2010[7] and the adoption by the United Nations of the 2030 Agenda programme comprising seventeen Sustainable Development Goals ("SDGs") in 2015[8] have finally empowered an international consensus around the "responsibility of an organisation for the impacts of its decisions and activities on society and environment, through transparent and ethical behaviour that contributes to sustainable development [...]" and allowed for the global community to "acknowledge the importance of corporate sustainability reporting and encourage companies, where appropriate, especially publicly listed and large companies, to consider integrating sustainability information into their reporting cycle". Among those SDGs, a specific goal[9] encourages "companies, especially large and transnational companies, to adopt

---

[5] Available at https://www.oecd.org/daf/inv/mne/oecdguidelinesformultinationalenterprises.htm.

[6] Vide https://www.unglobalcompact.org/what-is-gc/mission/principles.

[7] Available at https://www.iso.org/iso-26000-social-responsibility.html; vide also https://www.iso.org/files/live/sites/isoorg/files/archive/pdf/en/iso26000_sr.pdf, p. 1.

[8] Vg https://sdgs.un.org/2030agenda.

[9] SGD 12.6 Live Tracker available at https://sustainabledevelopment.un.org/partnership/?p=9851.

sustainable practices and to integrate sustainability information into their reporting cycle".

The Sustainability Accounting Standards Board (SASB),[10] founded in 2011, issued a set of globally applicable sectorial standards designed to assist organisations to report on the impacts that they have on the environment, on the economy and governance, and on society as a whole. Those standards are aimed, in essence, at investors since they are oriented towards financial aspects of sustainable development.[11]

In 2015, the Financial Stability Board,[12] international body that monitors and makes recommendations about the global financial system, recognised that climate change embeds a financial risk to the economy and established the Task Force on Climate-Related Financial Disclosures (TCFD), which recently provided a framework for reporting on climate risk,[13] allowing organisations to better understand, consider and report on such risk.

The Global Reporting Initiative[14] ("GRI") constitutes to date an internationally recognised frame of reference for sustainability reporting. The GRI's standards aim to enable all undertakings and organisations, in particular financial institutions, to account for their performance across four dimensions, namely the economic, the environmental, the social and governance aspects, by applying indicators and guidelines specific to each activity and sector. The GRI published an internationally recognised standard for non-financial reporting, whereby an organisation draws up a public report on its economic, environmental and/or social impacts and consequently, on its positive or negative contributions to the attainment of the objective of sustainable development. In light of the issuance in 2014 of the European Directive on non-financial and diversity disclosure ("NFRD"),[15] the GRI issued a document to inform users on how the GRI Standards can be used to comply with all aspects of the European Directive.[16] This linkage initiative sheds light on the multitude of existing standards to report on the sustainability topic and moreover on the need to establish harmonised and comparable reporting standards.

---

[10] Available at https://www.sasb.org/.

[11] Vide https://www.sasb.org/company-use/.

[12] Information available at https://www.fsb.org/.

[13] Available at https://assets.bbhub.io/company/sites/60/2020/10/FINAL-2017-TCFD-Report-11052018.pdf.

[14] Founded in 1997, with the objective of creating the first accountability mechanism to ensure companies to adhere to responsible environmental conduct principles, eg https://www.globalreporting.org/.

[15] Directive 2014/95/EU of the European Parliament and of the Council of 22 October 2014 amending Directive 2013/34/EU as regards disclosure of non-financial and diversity information by certain large undertakings and groups.

[16] Refer to https://www.globalreporting.org/media/mwydx52n/linking-gri-standards-and-european-directive-on-non-financial-and-diversity-disclosure.pdf.

The International Integrated Reporting Council ("IIRC")[17] also issued general guidelines and formulated recommendations to assist undertakings and other organisations wishing to prepare an integrated report which aims at enabling decision-useful reporting by integrating and communicating a holistic range of factors that materially enhance the organisation's ability to create value.

Finally, global consulting undertakings as AccountAbility have also been disclosing standards to enable users to report on sustainability—AA1000 AccountAbility Principles ("AA1000 AP")[18] is a set of internationally recognised general guidelines whereby organisations evaluate, manage, improve and communicate their responsibility and performance in terms of sustainable development. The guidelines are based on the principles of inclusivity of stakeholders, of the materiality in identifying relevant issues, of the responsiveness to actions carried out and of the impact of actions undertaken.

State of Play—Materiality Seems to Fit All

As we believe to have demonstrated, there is no single set of metrics and indicators that properly cover all ESG aspects for all companies globally. Moreover, the landscape of ESG criteria has been rapidly evolving and some issues that were overlooked are becoming of greater importance.

Therefore, criteria for a balanced sustainability report should be grouped together according to the following fundamental principles:

1. Relevance: the information provided must have a connection with the relevance of the analysis of the issues and priority impacts involved in companies' activities;
2. Balance: the information provided must show not only the positive but also the negative/adverse impacts on the social and societal, environmental and economic factors;
3. Inclusion of stakeholders: the information must provide the organisation's responses to all relevant stakeholders' expectations and interests;
4. Quality of information: the information must be reliable, comparable, clear, balanced (according to its relevance), verifiable and linked to a given period.

In addition, another dimension that has been increasingly noted is the materiality of information that needs to be included on a sustainability reporting. In light of the tremendous amount of ESG data to be considered as a basis for meaningful transparency and reporting of business sustainability strategies, materiality is becoming an essential filter and criterion for determining

---

[17] Global not-for-profit organization founded in 2015—vide https://integratedreporting.org/.

[18] Available at https://www.accountability.org/standards/.

what information will be truly relevant to fit the communication objective of companies' specific reports.

It is worth detailing to what extent the concept of materiality can be applied to all ESG-related matters and shall not be limited to financial information.

Materiality has been generally defined both within the European Union ("EU")[19] and national legislations as information which, if not (accurately) provided, is susceptible of influencing its users' decisions.

Moreover, for non-financial information reporting purposes, it has been found that the interests of the widest range of stakeholders possible should be considered, for different needs and perspectives to be envisaged when determining that said information is material.

The EU has been emphasising the concept of "double materiality" to enhance the qualitative assessment to be considered when disclosing information and to require the reporting not only on the impact of sustainability risks on business models—outside-in risks—but as well on the impact of businesses on the sustainability factors—inside-out risks.[20]

The various dimensions described above—that appear to be relevant to build a reliable and useful sustainability reporting—have been progressively considered within the significant acceleration of the European (and international) regulatory agenda.

2 How the EU Agenda is Accelerating the Move from Voluntary Approaches to Mandatory Regulatory Regimes

A Challenging European Policy Issuing Process

When it comes to shaping, directing and ultimately triggering financial institutions, corporates and investors' incentives, an adequate policy landscape is key.

As such, Europe's regulatory agenda emerged as a reaction to the status quo and has been setting the pace by forcing market players to phase out from a voluntary and non-standardised scenario of ESG reporting into an increasingly mandatory harmonised environment.[21]

---

[19] See namely article 2 (16) of Directive 2013/34/EU of the European Parliament and of the Council of 26 June 2013 on the annual financial statements, consolidated financial statements and related reports of certain types of undertakings, amending Directive 2006/43/EC of the European Parliament and of the Council ("Audit Directive") and repealing Council Directives 78/660/EEC and 83/349/EEC.

[20] Albeit having intended to consider such concept in the NFRD, it has been argued that "the directive does not include an adequate definition of the concept of materiality"—vg https://www.europarl.europa.eu/RegData/etudes/BRIE/2021/654 213/EPRS_BRI(2021)654213_EN.pdf.

[21] Listed below are some of the most relevant EU's regulatory initiatives on the sustainability agenda:

In the financial sector, a considerable number of EU regulations currently converge on the need for establishing an appropriate flow of information on ESG factors to all stakeholders benefiting from financial institutions' activities (e.g. investment, lending, insurance, asset management) to allow each of such undertakings to comply with specific disclosure requirements. Timely availability of appropriate and relevant sustainability data is one of the current most important challenges for the financial sector to meet disclosure requirements and to measure the real impact of their activities on the economy and the society.

Undertakings are therefore called upon to take part in a movement designed to integrate sustainability into their strategy and reporting. Nevertheless, such undertakings remain faced with a dilemma where the most responsible decision does not necessarily appear to be the most profitable one. This obstacle undeniably reinforces the need to impose certain (but not limited to) transparency and disclosures obligations on undertakings. Those obligations have gradually taken shape within the upcoming framework of European law.

Several key actions from the EU Action Plan on Financing Sustainable Growth clearly suggest that ESG disclosures and transparency cannot be considered a passing trend.

The Sustainable Finance Disclosure Regulation

The present paragraph focuses on current ESG reporting obligations within the EU, mainly those foreseen in the SFDR. This regulation constitutes, together with the Taxonomy Regulation and Benchmark Regulation, the cornerstone of the 2018 EU Sustainable Finance Action Plan, which embeds a

i. Directive 2013/34/EU of the European Parliament and of the Council of 26 June 2013 on the annual financial statements, consolidated financial statements and related reports of certain types of undertakings, amending Directive 2006/43/EC of the European Parliament and of the Council and repealing Council Directives 78/660/EEC and 83/349/EEC, as amended by Directive 2014/95/EU of the European Parliament and of the Council of 22 October 2014 amending Directive 2013/34/EU as regards disclosure of non-financial and diversity information by certain large undertakings ("NFRD");
ii. Directive 2007/36/EC of the European Parliament and of the Council of 11 July 2007 on the exercise of certain rights of shareholders in listed companies;
iii. Regulation (EU) 2019/2089 of the European Parliament and of the Council of 27 November 2019 amending Regulation (EU) 2016/1011 as regards EU Climate Transition Benchmarks, EU Paris-aligned Benchmarks and sustainability-related disclosures for benchmarks ("Benchmark Regulation");
iv. Regulation (EU) 2019/2088 of the European Parliament and of the Council of 27 November 2019 on sustainability-related disclosures in the financial services sector ("SFDR");
v. Regulation (EU) 2020/852 of the European Parliament and of the Council of 18 June 2020 on the establishment of a framework to facilitate sustainable investment, and amending Regulation (EU) 2019/2088 (Taxonomy Regulation").

strong political ambition to redirect capital flows towards sustainable activities and foster greater transparency and long-termism in financial and economic activities.

The SFDR applies for the most part from 10 March 2021 and lays down harmonised rules for financial market participants and financial advisers on transparency with regard to the integration of sustainability risks and the consideration of adverse sustainability impacts in their processes, as well as the provision of sustainability-related information with respect to financial products.

Its objective is to provide investors/clients (both professional and retail) with more transparent information and to guide them in considering ESG contributions on targeted investments, in addition to strict financial return. After having assessed the information describing the products' ESG characteristics and/or sustainable objectives, and how markets participants and financial advisers manage sustainability risks, investors are expected to be able to make better-informed investment decisions as regards the sustainability of such financial products.

The new rules introduced by the SFDR are complex and they are having a considerable impact on the obliged financial entities. The SFDR foresees obligations for financial market players (i) at entity level, by imposing general ESG disclosure duties; (ii) at product level, by foreseeing specific ESG disclosure duties; and (iii) policy amendments in order to incorporate ESG principles.

These new disclosure rules require the preparation of new information to be added to existing pre-contractual documentation, websites and periodic reports. These rules complement the existing information requirements on ESG aspects of investment strategies, policies or products/services, which are frequently already being reported by concerned financial institutions.[22]

The complexity, the scope of rules, the amount of information that needs to be collected to comply with the rules and the challenging timelines require significant effort from in-scope firms and joint efforts from experts in several departments.

Thus, the SFDR brings substantial changes to the current mandatory disclosure requirements for financial institutions by adding a completely new category of sustainability-related disclosures, with the frequently mentioned argument that even mere disclosures are supposed to create incentives for boosting financial products with sustainability-related credentials.

However, both regulators and market players have been arguing that the SFDR is to some extent unclear as to what obliged entities are expected

---

[22] Notably refer to regulations such as the NFRD, the Directive 2014/65/EU of the European Parliament and of the Council of 15 May 2014 on markets in financial instruments and amending Directive 2002/92/EC and Directive 2011/61/EU or the Regulation (EU) 2017/1129 of the European Parliament and of the Council of 14 June 2017 on the prospectus to be published when securities are offered to the public or admitted to trading on a regulated market, and repealing Directive 2003/71/EC.

to disclose, and that said regulation needs to be complemented with more granular rules.

To provide additional clarity on the construction of the approved new set of rules, the European Supervisory Authorities ESMA, EBA and EIOPA ("ESAs") raised, in a letter dated 7 January 2021,[23] a set of questions addressed to the European Commission ("EC") on priority issues relating to the application of the SFDR.

This action led to the issuance, on 6 July 2021, of the much-awaited decision C(2021) 4858 final, which provides further guidance to all market players.[24] The EC answers do not extend the obligations already contained in the applicable legislation but clarify certain provisions, notably confirming the regulatory neutrality in terms of financial products design and contributing to the concept of "promotion" of ESG characteristics.

On February 2, 2021, the ESAs additionally issued, through their Joint Committee, their final report on draft Regulatory Technical Standards ("RTS") with regard to the content, methodologies and presentation of sustainability-related disclosures under the SFDR.[25]

Having considered the feedback received from stakeholders to the public consultation which preceded the mentioned draft report, the ESAs has updated (i.e. reduced) the core set of mandatory indicators for principal adverse impacts, which is supplemented by an extended list of opt-in indicators. The ESAs have also decided to develop specific indicators for investments in sovereigns and real estate assets.

Steven Maijor, Chair of the ESAs Joint Committee, has recently stressed that the issued set of rules *"strike a careful balance between achieving common disclosures across the range of financial products covered by the SFDR and recognising that they will be included in documents that are very diverse in length and complexity"*.[26]

On February 25, 2021, the ESAs further issued a joint supervisory statement on their report recommending that impacted stakeholders refer to the draft RTS when applying the SFDR during the interim period within which the final RTS are not in force.[27] This would serve as guidance for the impacted market players, in light of the goal of harmonisation and would also allow supervisors to properly prepare for the effective and consistent application

---

[23] Available here: https://www.esma.europa.eu/sites/default/files/library/jc_2021_02_letter_to_eu_commission_on_priority_issues_relating_to_sfdr_application.pdf.

[24] Vide https://www.esma.europa.eu/sites/default/files/library/sfdr_ec_qa_1313978.pdf.

[25] Available here: https://www.eiopa.europa.eu/content/final-report-draft-regulatory-technical-standards_en; the final RTS may differ from the draft.

[26] Vide https://www.eba.europa.eu/three-european-supervisory-authorities-publish-final-report-and-draft-rts-disclosures-under-sfdr.

[27] Available here: https://www.esma.europa.eu/sites/default/files/library/jc_2021_06_joint_esas_supervisory_statement_-_sfdr.pdf.

and national supervision of the SFDR, promoting a level playing field and protecting investors.

Considering the complexity of the new set of rules, in a letter dated 8 July 2021, the EC deferred application of regulatory technical standards under the Sustainable Finance Disclosure Regulation 2019/2088 (SFDR level 2 measures) to 1 July 2022.[28]

On 22 October 2021, the ESAs jointly released a Final Report on draft RTS regarding taxonomy-related disclosures under the SFDR[29] which foresees templates for pre-contractual and periodic product disclosures. These new RTS will be incorporated with the original ones, submitted to the Commission in February 2021, in one instrument.

In light of the length and complexity of the issued RTS and in order to facilitate the implementation of the delegated act by product manufacturers, financial advisers and supervisors, the date of application of the single ruleset to be issued was postponed from1 July 2022 to 1 January 2023.[30]

At a national level, legislators and regulators are complementarily issuing sets of rules to comply with EU's regulations on ESG reporting.

In Portugal, the Comissão do Mercado de Valores Mobiliários published on February 2, 2021 its Model Report for disclosure of non-financial information by listed companies[31] and, on March 5, 2021, adopted the ESA's recommendations on the application of the SFDR, urging market participants to prepare for the entry in force of the RTS by implementing the ESAs draft during the interim period of year 2021.[32]

In March 2021, the Autorité des Marchés Financiers ("AMF") has published guidance on the implementation of the SFDR and its articulation with the AMF position-recommendation ("AMF doctrine")[33] published in March 2020 (and updated in July 2020) and which applies to French undertaking for collective investment in transferable securities ("UCITS") and alternative investment funds, as well as non-French UCITS that consider ESG criteria and that are authorised to be marketed to French retail investors. The AMF doctrine aims to prevent the risk of greenwashing by requiring that information provided to non-professional investors regarding fund's consideration

---

[28] Vide https://www.esma.europa.eu/sites/default/files/library/com_letter_to_ep_and_council_sfdr_rts.pdf.

[29] Available here: https://www.esma.europa.eu/sites/default/files/library/jc_2021_50_-_final_report_on_taxonomy-related_product_disclosure_rts.pdf.

[30] Vide letter dated 25 November 2021 to European Parliament and Council, available here: https://www.esma.europa.eu/sites/default/files/library/com_letter_to_ep_and_council_sfdr_rts-j.berrigan.pdf.

[31] Available here: https://www.cmvm.pt/pt/Legislacao/ConsultasPublicas/CMVM/Documents/Modelo%20de%20Informa%c3%a7%c3%a3o%20N%c3%a3o%20Financeira.pdf.

[32] Vide https://www.cmvm.pt/pt/Comunicados/comunicados_mercado/Pages/20210305a.aspx.

[33] Available at https://www.amf-france.org/en/regulation/policy/doc-2020-03.

of non-financial characteristics is proportionate to the actual consideration of these factors.

In Luxembourg, the Commission de Surveillance du Secteur Financier has implemented in December 2020 a fast-track procedure to facilitate the submission of the funds' prospectus and issuing document updates limited to reflect changes required under the SFDR.[34]

National doctrines therefore complement the SFDR, with the objective to making each country's approach converge to the maximum extent with the EU regulatory framework.

Other Regulatory Initiatives

The EU had however been active within non-financial disclosure topics even before the SFDR. In 2014, the NFRD[35] had been issued, which lays down the rules on disclosures of non-financial information and diversity disclosures for certain large undertakings and groups with more than 500 employees. For public interest entities concerned, it foresees, on a consolidated basis, the issuance of a non-financial statement containing information relating to, as a minimum, environmental, social and employee matters, respect for human rights, anti-corruption and bribery matters.

In their annual report, in-scope companies must publish information in accordance with the five areas as referred to in the NFRD: business model, policies pursued, due diligence processes implemented, the policies' outcomes and the principal risks and how they are managed, including key performance indicators ("KPIs").

Since 2014, the European authorities issued complementary guidelines to the NFRD, in order to enlighten markets players as to how to better meet the NFRD's objectives.

In June 2017, the EC published non-binding guidelines on the NRFD,[36] which set out key principles for providing useful, relevant and comparable information: (1) disclosure of material information, (2) fair, balanced and understandable information, (3) comprehensive but concise information, (4) strategic and looking-forward information, (5) stakeholder-oriented information, (6) consistent and coherent information.

---

[34] Vg https://www.cssf.lu/en/2020/12/communication-on-regulatory-requirements-and-fast-track-procedure-in-relation-to-regulation-eu-2019-2088-on-the-sustainability-related-disclosures-in-the-financial-services-sector/.

[35] As specified in Whereas (25) of the SFDR, the form and presentation required by NFDR was found not always to be suitable for direct use by financial market participants and financial advisers when dealing with end investors, which should have the option to use information in management reports and non-financial statements for the purposes of SFDR in accordance with NFDR, where appropriate.

[36] Available at https://eur-lex.europa.eu/legal-content/EN/TXT/PDF/?uri=CELEX:52017XC0705(01)&from=EN.

On June 18, 2019, the EC published guidelines on corporate climate-related information reporting, which in practice consist of a new supplement to the previous guidelines.[37] This supplement provides companies with practical recommendations on how to better report the impact that their activities are having on the climate as well as the impact of climate change on their business.

In January 2020, the EC launched a consultation on the review of the Non-Financial Reporting Directive ("NFRD"), seeking to collect feedback from stakeholders in order to help to standardise and simplify companies' reporting at the European level, including through the introduction of EU reporting standards as well as to give effect to the changes required by the SFDR and the Taxonomy Regulation.

To this end, on June 25, 2020, the EC has mandated the European Financial Reporting Advisory Group (EFRAG)[38] to issue a report setting out recommendations on the development of EU sustainability reporting standards, which were issued on March 8, 2021 and embed a roadmap for the development of a comprehensive set of EU sustainability reporting standards.[39]

However, in parallel, the EU Platform on Sustainable Finance ("Platform")[40] advocates reforms to EFRAG's governance structure and funding (if it were to become the EU sustainability reporting standard setter) to ensure that future EU sustainability reporting standards would be developed resorting to an inclusive and rigorous process.

The expert group of which the Platform is composed aims to "have a single, coherent view on the relationship of SFDR, NFRD and Taxonomy reporting obligations to double materiality concepts"[41] and is therefore advising on how to define reporting requirements to enable companies in communicating how and to what extent their activities are aligned with the EU taxonomy as well on their transition plans.

The Platform relies on six reporting principles to guide companies on sustainability reporting requirements:

---

[37] Available here: https://ec.europa.eu/finance/docs/policy/190618-climate-related-information-reporting-guidelines_en.pdf.

[38] General information available at https://www.efrag.org/Assets/Download?assetUrl=/sites/webpublishing/SiteAssets/Letter%2520EVP%2520annexNFRD%2520%2520technical%2520mandate%25202020.pdf.

[39] Roadmap available at https://www.efrag.org/Assets/Download?assetUrl=%2Fsites%2Fwebpublishing%2FSiteAssets%2FEFRAG%2520PTF-NFRS_MAIN_REPORT.pdf.

[40] Vide https://ec.europa.eu/info/business-economy-euro/banking-and-finance/sustainable-finance/overview-sustainable-finance/platform-sustainable-finance_en.

[41] Vg https://ec.europa.eu/info/sites/info/files/business_economy_euro/banking_and_finance/documents/210319-eu-platform-transition-finance-report_en.pdf, p. 19.

1. Proportionality
2. Integrity
3. Relevance
4. Consistency
5. Predictability
6. International application.

On April 21, 2021, the EC has released the EU Sustainable Finance—April package,[42] which comprises an impressive number of legislative initiatives that form part of the European Green Deal[43] and intend to further orientate investors towards more sustainable technologies and businesses and is expected to be in force as from 2022 onwards.

Within such package, a proposal was drafted of a Corporate Sustainability Reporting Directive ("CSRD"),[44] which, if adopted, will consist in:

i. Amendments to the NFRD, extending its scope of application to all large companies and all companies listed on regulated markets (except listed micro-enterprises), setting additional guidance on the mentioned principle of double materiality, and providing clarification to the maximum extent on several ambiguous reporting obligations, notably imposing Member States to approve legislation stating that sustainability information is to be reported as part of the management report, in a "single electronic reporting format", and foreseeing statutory auditing requirements (e.g. "limited assurance engagement") on companies sustainability reporting,[45]

ii. Amendments to the Transparency Obligations Directive,[46] introducing the concept of "sustainability", imposing statements to be issued by companies' representatives and referring to auditing requirements on sustainability,

iii. Amendments to the Audit Directive, envisaging the mentioned "assurance for sustainability reporting", and setting the rules and procedures which will govern such auditing activity,

---

[42] Please refer to https://ec.europa.eu/info/files/sustainable-finance-communication-factsheet_en.

[43] Please refer to the webpage https://ec.europa.eu/info/strategy/priorities-2019-2024/european-green-deal/actions-being-taken-eu_en.

[44] Available here: https://ec.europa.eu/finance/docs/law/210421-proposal-corporate-sustainability-reporting_en.pdf.

[45] The EC is expected to adopt sustainability reporting standards by means of Delegated Acts.

[46] Directive 2004/109/EC of the European Parliament and of the Council of 15 December 2004 on the harmonisation of transparency requirements in relation to information about issuers whose securities are admitted to trading on a regulated market and amending Directive 2001/34/EC.

iv. Amendments to the Audit Regulation,[47] containing detailed governance rules to be implemented in order for a sound assurance to be issued (e.g. on conflict of interests).

In addition to the ambitious CSRD proposal, a new Taxonomy Climate Delegated Act was approved[48] to allow classification of which activities will best contribute to mitigate and adapt to effects of climate change, and six other Delegated Acts were put forward to amend other sectoral legislations, such as the Solvency II Directive[49] and the Insurance Distribution Directive,[50] as well as to other delegated EU acts (e.g. on MiFID II Directive[51] related topics as the UCITS, Alternative Investment Fund Managers, insurance-based investment products, investment firms), in order for sustainability to be transversely considered by financial firms, such as advisers, asset managers or insurers, in their procedures and their investment advice to clients, both at entity and product levels.

More recently, on 6 July 2021, the EC published its Renewed Sustainable Finance Strategy ("RSFS") with various legislative and non-legislative proposals aimed at supplementing and enhancing the EU Sustainable Finance Action Plan.[52]

The RSFS is built around 4 main pillars: (1) extend the existing EU Taxonomy and toolbox to enable all economic actors to adequately finance their transition plans, (2) improve inclusiveness to further access of citizens and small and medium-sized enterprises to sustainable finance, (3) improve the financial sector's resilience and combat greenwashing, (4) foster global ambition through deepened cooperation and convergence of goals and standards.

The RSFS has been accompanied by a legislative proposal for an EU Green Bond Standard and an updated delegated act on article 8 disclosures of

---

[47] Regulation (EU) No 537/2014 of the European Parliament and of the Council of 16 April 2014 on specific requirements regarding statutory audit of public-interest entities and repealing Commission Decision 2005/909/EC.

[48] Provisional texts of the Act and Annexes available at https://ec.europa.eu/finance/docs/level-2-measures/taxonomy-regulation-delegated-act-2021-2800_en.pdf, https://ec.europa.eu/finance/docs/level-2-measures/taxonomy-regulation-delegated-act-2021-2800-annex-1_en.pdf and https://ec.europa.eu/finance/docs/level-2-measures/taxonomy-regulation-delegated-act-2021-2800-annex-2_en.pdf.

[49] Directive 2009/138/EC of the European Parliament and of the Council of 25 November 2009 on the taking-up and pursuit of the business of Insurance and Reinsurance (Solvency II).

[50] Directive (EU) 2016/97 of the European Parliament and of the Council of 20 January 2016 on insurance distribution.

[51] Directive 2014/65/EU of the European Parliament and of the Council of 15 May 2014 on markets in financial instruments and amending Directive 2002/92/EC and Directive 2011/61/EU.

[52] Documents available here: https://ec.europa.eu/info/publications/210706-sustainable-finance-strategy_en.

Taxonomy Regulation by undertakings in scope of the NRFD whereby those companies must publish information on how and to what extent their activities are associated with economic activities that qualify as "environmentally sustainable" under the Taxonomy Regulation.

The implementation by EU market players of this newly disclosed set of rules—corresponding to a clear priority of the EU agenda—, some of which are still to be approved and further regulated, constitute an increased challenge to the financial industry and will no doubt involve a continued investment on the sustainability topic.

En Route to a Global Standardised Sustainability Reporting

EU regulations are also impacting players outside the European territory, such as the United States of America ("US"), as rules on product distribution within the EU apply regardless of the home country of the distributor. As such, US asset managers also have to disclose, among other, climate, diversity and governance data for investments by funds to be marketed in the EU, and are forced to comply with European rules on sustainability-related disclosures under SFDR, notably by disclosing the potential harm their investments could do to the environment and society.

In light of growing demand and regulatory pressure for climate change information and ESG data as well as considering questions about whether current disclosures adequately inform investors, the US Securities and Exchange Commission ("SEC" or "Commission")[53] has been reassessing its regulation of climate change disclosures. Since 2010, guidance was provided to issuers as to how existing disclosure requirements apply to climate change matters, investor demand and political push for such review have grown significantly.

In March 2021, SEC published a dedicated statement[54] on ESG disclosures, which clearly suggests that such Commission is also shifting towards promoting increasingly mandatory and voluntary ESG information disclosure, as well as producing an international framework on sustainability reporting standards drafted by the International Financial Reporting Standards Foundation ("IFRS Foundation").[55]

More recently, in May 2020, SEC's Advisory Committee approved recommendations advocating for the Commission's efforts in updating reporting requirements which should request issuers to include material, decision-useful

---

[53] General information available at: https://www.sec.gov/.

[54] Available here: https://www.efrag.org/Assets/Download?assetUrl=/sites/webpublishing/SiteAssets/Letter%2520EVP%2520annexNFRD%2520%2520technical%2520mandate%25202020.pdf.

[55] Vide https://www.ifrs.org/.

ESG factors.[56] In December 2020, the ESG Sub-Committee of the SEC Asset Management Advisory Committee issued a preliminary recommendation that the Commission require the adoption of standards by which corporate issuers disclose material ESG risks.[57]

3 FUTURE OF SUSTAINABILITY REPORTING(S)

Considering the above, it seems unequivocal that sustainability reporting has gone far beyond being a pure marketing ploy.

Looking ahead, what is to be expected?

Available indications and increasing regulatory pressure suggest that we can foresee further evolution of standards and practices to report on sustainability risks and factors and on disclosure of ESG performance at both entity and product levels.

There are growing signs that seem to indicate that regulation and standardisation is on the horizon for extending the scope of reporting to themes beyond climate change and environmental issues.

In fact, human rights, nature and/or biodiversity are still under-developed in this respect, and further development on reporting on such matters is anticipated, in line with what seems to constitute the EU agenda's direction: stringent standards, growing credibility, increased scope of action.

Biodiversity for instance has been more and more under the spotlight. The UN Principles for Responsible Investment[58] have recently published a discussion paper on investor action on biodiversity[59] and issued recommendations to investors, urging market players to collaborate with peers and stakeholders to enhance nature-related financial disclosures.

The United Nations also adopted in March 2021 a new framework[60] to integrate natural capital in economic reporting. This System of Environmental-Economic Accounting—Economic Accounting aims to ensure that the contributions of nature—forests, wetlands and other ecosystems—are properly recognised and valued as benefitting people and the economy. More than thirty countries are compiling ecosystem accounts on an experimental basis as a reaction to SEEA-EA's call for action.

As with climate and environment-related risks, nature-related risks need to be better integrated and disclosed. With biodiversity loss moving up the agenda of governments, civil society and financial institutions, efforts are

---

[56] Recommendations available at https://www.sec.gov/spotlight/investor-advisory-committee-2012/recommendation-of-the-investor-as-owner-subcommittee-on-esg-disclosure.pdf.

[57] Vg https://www.sec.gov/files/potential-recommendations-of-the-esg-subcommittee-12012020.pdf.

[58] Vide https://www.unpri.org/.

[59] Available here: https://www.unpri.org/download?ac=11357.

[60] https://seea.un.org/.

being intensified to increase knowledge, explain terminology, map and disclose how nature loss poses risks to companies. In this context, the University of Cambridge Institute for Sustainability Leadership[61] has recently produced a handbook for identifying nature-related financial risks.[62]

With companies having to disclose more and more with regards to sustainability reports they should already be prospectively assessing those additional requirements in light of existing disclosure obligations, in order to prevent "over transparency" from being materially misleading.

As previously stated, ESG issues need to be addressed globally through global and harmonised reporting solutions and market players must be aware that one size does not fit all, as a framework pertaining to the same risk category faced by the similar companies within the same activity sector may not equally apply. Undoubtedly, the proportionality principle will have to be taken into account when imposing disclosure obligations.

In this regard, the work of the IFRS Foundation to establish a sustainability standards board, combined with the progress made by the EFRAG and the TCFD appears promising.

However, the task of establishing a global and internationally recognised sustainability reporting framework is complex and not without challenges.

The current market and regulatory evolution raises a number of questions and considerations and these will need to be managed carefully. As the EU action plan as well as many international initiatives open up to more and more ESG criteria and data to be included in sustainability reporting, it will be critical to ensure that there is a "chorus approach" and that all players are properly equipped and financially capable to comply with the new rules and provisions. A failure to do so risks undermining the entire credibility of such disclosures. This will require a rigorous, inclusive and transparent process for developing global standards, including all relevant stakeholders within the process. The EU is no doubt playing a leading role in such task, setting the pace for other jurisdictions which will be able to build upon the European framework to address their own needs and targets.

---

[61] https://www.cisl.cam.ac.uk/.
[62] https://www.cisl.cam.ac.uk/resources/sustainable-finance-publications/handbook-nature-related-financial-risks.

CHAPTER 13

The EU Taxonomy Regulation and Its Implications for Companies

Rui de Oliveira Neves

1 Introduction

Sustainability is the key driver of change in this decade. Never as today there has been such a social consensus around the major concern of our civilization—climate change. Both physical persons and institutions have become growingly conscious of the need to promote measures to reduce their carbon footprint. Corporations will have to be in the frontline of this effort. Transition to a green economy depends mostly on private investment and on aligning the type of investments required from companies with the carbon-neutrality goal.

Aligning investment flows with sustainability needs and reforms is by far the largest challenge towards carbon neutrality. The European Commission reflected its concerns in this respect in its Communication about the EU Taxonomy and sustainability reporting: *"estimates and early testing of the climate taxonomy criteria showed a low overall Taxonomy alignment today in companies' activities and investment portfolios (between one and five percent, with many companies and investment portfolios standing at zero). While this figure is expected to rise significantly with the implementation of the Green Deal,*

R. de Oliveira Neves (✉)
Universidade Católica Portuguesa, Law School, Lisbon, Portugal
e-mail: ron@mlgts.pt

Universidade de Lisboa, Law School, Lisbon, Portugal

it highlights the extent of the transition still required towards climate neutrality by 2050".[1]

The discussions at COP26 and the outcomes from the summit, in particular the Glasgow Climate Pact, also reflect the extreme gap between national pledges and effective results.

In spite of this, sustainability is at the core of today's concerns and public narratives for both private and public organizations. There is a huge pressure around the carbon footprint of businesses and people and therefore a wide recognition on the need to decrease and to offset emissions to prevent further harm to the environment. And for this to occur there is the need for a paradigm shift both at the economic as well as at the financial level to address the negative externality effects derived from climate change.

Since the middle of the last decade, there has been a rapidly growing awareness and focus on the importance of the Environmental, Sustainability and Governance topics, the so-called ESG. Investors and companies have begun to incorporate ESG indicators into their decision patterns, as well as in their corporate reporting by publicly disclosing performance on several ESG topics, which, one after another, are starting to impact the way to do business and corporate organizations.[2,3]

But there is also scepticism on whether climate change and ESG are actually at the centre of the new investment strategies of companies and even ESG concerned funds. Financials continue to drive investment decisions and the benefits of sustainability tend to be considered as non-financial. Market analysts and investment houses continue to issue their investment reports on the basis of the financial performance of equity and debt issuers while ESG commitments and their corresponding delivery are not yet considered as a material factor. Moreover, the pandemic crisis has in some extent brought up some short-termism investing behaviour allied to high-frequency trading, which both seems to be neglective of the long term and sustainable approach that frames corporate strategies and fiduciary duties.

---

[1] Communication from the Commission on EU Taxonomy, Corporate Sustainability Reporting, Sustainability Preferences and Fiduciary Duties: Directing finance towards the European Green Deal of April 21, 2021, see https://ec.europa.eu/finance/docs/law/210421-sustainable-finance-communication_en.pdf.

[2] One of the key trends that has developed over the last decade refers to integrated reporting under the *International Integrated Reporting Council* coalition, which was formed in 2010 and has in January 2021 published a new version (first version dates back to December 2013) of an *International Integrated Reporting Framework* that encompasses financial, environmental, social and governance disclosure.

[3] *TCFD—Task Force on Climate-Related Financial Disclosures*, currently with more than 2000 supporters of its 4 recommendations, *PRI—Principles on Responsible Investment*, currently with nearly 2700 signatories of its 6 principles or *CDP—Carbon Disclosure Project*, which promotes a global environmental disclosure system, benchmarking more than 2500 companies, are some of the main institutions that have been encouraging ESG topics to be integrated into the strategy of corporations and their respective disclosures.

Additionally, ESG benchmarks fail to provide a consistent assessment of sustainability behaviours for two main reasons: first, the elements considered are of different nature, making their measurement and correlation difficult to attain; secondly, there are distinct metrics to assess ESG performance, which necessarily impact the ability to compare companies resourcing to different ratings.

While sustainability-based investments are not an active contributor to operational results, cash flow generation and shareholder value creation it is likely that sceptics are not incentivised to promote them on the basis of the importance of ethical investment for our society's development goals.

Effectively, recent science-based ESG research demonstrated that currently ESG strategies do not add significant outperformance to listed companies in the US and other developed markets nor offer significant downside risk protection. Nonetheless, it is recognized the intrinsic value of ESG strategies for the specific benefits they can provide, *"such as hedging climate or litigation risk, aligning investments with norms and making a positive impact for society"*.[4]

Responsible investment has been a designation used to express this new conceptual approach to the ESG compliant investments. Green bonds and other sustainability-linked financing instruments are one of its main tools, as investors capital allocation decisions get driven by ethical reasons, in particular by climate change mitigation.[5]

More importantly, the expression *responsible investment* also conveys an idea of sustainability in a broad sense encompassing three essential dimensions. The first, environmental sustainability addressing the mitigation of business impacts on environment, by adapting processes and using technology that can contribute to prevent and reduce pollution and water consumption or to protect biodiversity. On the economical perspective, sustainability reflects the corporation's resilience to generate results from its businesses on the long term. Finally, on the social dimension, sustainability evinces the companies' contribution to society through its activities, including by creating jobs or sharing the benefits of its activity with the communities, by respecting human rights of the populations impacted by those activities or by promoting equality or integrating minorities.

---

[4] "Honey, I shrunk the ESG Alpha": Risk-Adjusting ESG Portfolio Returns, April 2021, Scientific Beta, available at https://www.bankingexchange.com/recent-articles/item/8684-honey-i-shrunk-the-esg-alpha-new-research-questions-outperformance-data.

[5] The main framework on voluntary principles for issuing green bonds is set out by the International Capital Market Association, which was updated in June 2021 to reinforce transparency and encourage disclosure about alignment of projects with official or market-based taxonomies—https://www.icmagroup.org/assets/documents/Sustainable-finance/2021-updates/Green-Bond-Principles-June-2021-100621.pdf.

2　The Taxonomy and Its Functions

Concept

There is a fundamental mismatch between the priorities of economy and the institutional and social goals in terms of sustainability. As a society, Europe recognizes the importance of developing a sustainable economy that can contribute to mitigate the impacts of climate change. To achieve such a result the level of investment required is enormous in order to enable transformative changes in economic activities.

The concept underlying the taxonomy established in the Regulation (EU) 2020/852 of the European Parliament and of the Council of 18 June 2020 on the establishment of a framework to facilitate sustainable investment, and amending Regulation (EU) 2019/2088[6] (the "Taxonomy" or the "Regulation") is that of creating a classification methodology about environmentally friendly economic activities to guide the capital markets on making informed investment decisions on businesses that can be qualified as sustainable, based on scientific and industry experience.

For this purpose, the Regulation lists those economic activities and the respective performance criteria based on their contribution to six sustainability-related objectives: (i) climate change mitigation, (ii) climate change adaptation, (iii) sustainable use and protection of water and marine resources, (iv) pollution prevention and control, (v) protection and restoration of biodiversity and ecosystems and (vi) transition to a circular economy.

The Regulation is therefore purpose-oriented to pursue certain goals related with climate change, environmental protection and circular economy. This brings to evidence not only that the Taxonomy is one of the tools to achieve certain economic targets within the European Union, but also that the ends elected in the Regulation are those intended to contribute to a new economic environment in the European Union driven by sustainability concerns.

In spite of being addressed at financial markets institutions, the Regulation does not measure or set criteria on financial performance, nor does it envisage to discriminate between business sectors. Its main aim is that an economic activity can only qualify as environmentally sustainable across the European Union if any of the established environmental objectives is pursued.

The Taxonomy evidences a clear reaction to certain movements in the financial markets oriented to associate investments to green goals irrespective of an actual or substantial contribution being achieved and of an effective full disclosure of positive and negative information on sustainability; this is particularly highlighted in recital 11 of the Regulation: "*(...) Requirements for marketing financial products or corporate bonds as environmentally sustainable investments,*

---

[6] There have been prior initiatives to incentivize climate-related funding, such as Decision No 1386/2013/EU of the European Parliament and of the Council of 20 November 2013 on a General Union Environment Action Programme to 2020 "Living well, within the limits of our planet".

including requirements set by Member States and the Union to allow financial market participants and issuers to use national labels, aim to enhance investor confidence and awareness of the environmental impact of those financial products or corporate bonds, to create visibility and to address concerns about 'greenwashing'. In the context of this Regulation, greenwashing refers to the practice of gaining an unfair competitive advantage by marketing a financial product as environmentally friendly, when in fact basic environmental standards have not been met".

By the opposite, if a business is not part of the Taxonomy it does not mean that such activity lacks environmental sustainability. There is no intention of discrimination of economic activities, but rather to create a clear and predictable set of activities that can properly qualify as sustainable investment opportunities. This is the whole objective of introducing a transparency tool: each investor will take its investment decisions on the basis of the best available information in the market. And if it decides to invest in a green financial product, such investor will have access to information that evidences the underlying environmentally sustainable economic activities.

The option to relate the Taxonomy to financing and financial products directly relates to the European Union investment environment, where most SME funding come from banking financing.[7] It is also consistent with the latest action plan for the Capital Markets Union, which, among other goals, envisages to support economic recovery to develop a green, digital, inclusive and resilient economy, by facilitating the access to financing by European companies.

The foregoing shows that the Regulation is focused on the E-pillar, i.e. the environmental requirements, instead of encompassing a full range ESG vision for European companies. The social and governance aspects have been kept in the shadows by the Regulation and the only (and very limited) progress effectively made in this respect was the inclusion of certain human rights safeguards in article 18 of the Regulation, as we will analyse further on. However, governance topics have been left completely aside by the Regulation, failing to promote an integrated approach to the ESG topics.

By setting sustainability purpose-driven standards to economic activities, the Regulation must necessarily be accompanied by new perspectives on corporate governance. There is a necessary impact in terms of the governance of a company if the requirements applicable to its business change. To manage and steer sustainable businesses, it is necessary that sustainability becomes embedded in the governance structure, rules and practices. There needs to be an alignment between governance drivers and the actual economic activities that companies are engaged in so as to ensure consistency and actual delivery of results.

---

[7] According to the European Commission, 75% of SME's financing is supported through banking financing.

Regulation will surely play an important role in creating the right incentives for this alignment to be achievable. And within this, governance rules will be essential to steer this transformation. The European Commission clearly shares this understanding and continues to take measures to create a sustainable finance framework in which the Taxonomy is becoming a main pillar.

Effectively, the Taxonomy leans on and deepens the path initiated with the Sustainable Finance Disclosure Regulation,[8] the latter requiring financial participants offering financial products in the EU to publish on their reports and websites qualitative and quantitative information about sustainability impacts of the investments associated to financial products offered in the EU.

And this sustainable finance framework, which results ultimately from the Sustainable Finance Action Plan, is being continuously developed, as it is evidenced by the latest amendment to the MiFID Regulation that requires financial participants offering financial products to include the sustainability preferences of each client as part of their suitability test,[9] as well as by similar requirement in relation to product governance of financial products.[10]

Additionally, the European Commission has approved a proposal of Corporate Sustainability Reporting Directive, which intends to set out reporting duties to large companies and listed companies, from January 1, 2023, and to SME's, from January 1, 2026, on their sustainability risks and impacts, both on the perspective of the impact of the companies' businesses on the climate and of the impact that climate change has on their respective business, as well as on how those risk are managed by those companies.

The role of governance to steer the management of sustainability risks and challenges is also clearly underpinned in the proposed Corporate Sustainability Reporting Directive. Companies within the European Union will be required to disclose, among others, (i) the role that its administrative, management and supervisory bodies have in relation to sustainability matters, (ii) its business ethics and corporate culture, including anti-corruption and anti-bribery, (iii) its political engagements, including its lobbying activities, (iv) the management and quality of its relationships with business partners, including payment practices and (v) its internal control and risk management systems, including in relation to the reporting process. Accordingly, the governance environment

---

[8] Regulation (EU) 2019/2088 of the European Parliament and of the Council of 27 November 2019 ('SFDR') on sustainability-related disclosures in the financial services sector.

[9] Commission Delegated Regulation (EU) 2021/1253 of 21 April 2021 amending Delegated Regulation (EU) 2017/565 as regards the integration of sustainability factors, risks and preferences into certain organizational requirements and operating conditions for investment firms.

[10] Commission Delegated Directive (EU) 2021/1269 of 21 April 2021 amending Delegated Directive (EU) 2017/593 as regards the integration of sustainability factors into the product governance obligations.

will necessarily need to evolve and develop so that practices meet these new regulatory requirements.

Still, this continues to be a limited approach. While on the environmental side there are actual material requirements already in place, on the social and governance dimensions there is just a reflex approach to the problem: setting disclosure duties does not reach the broader effects resulting from the definition of an integrated ESG regulatory policy.

In any case, both the Taxonomy Regulation and the Corporate Sustainability Reporting Directive shall have inevitable corporate governance impacts since they will influence and constrain the activity and the decision-making process of both directors and shareholders.

On the shareholders side, decisions will need to be taken first in relation to the investments to be included in their portfolios based on sustainability impact criteria. Secondly, on those companies included in their investment portfolio decisions on the strategic alignment of public interest companies with investments that are Taxonomy compliant will be required.

Consequently, directors will be conditioned on their decision-making depending on the strategic decisions and routes taken by public interest companies' shareholders regarding Taxonomy-alignment.

Economic Activities

The key criterion used by the Regulation refers to environmentally sustainable economic activities, which are those underlying sustainable investments. For this end, an economic activity must provide a substantial contribution to an environmental objective either through its own performance or as ancillary and enabling other activities to make such contribution.

Accordingly, the Regulation sets out three objective requirements for economic activities to be qualified as environmentally sustainable:

a. make a substantive contribution to one of the following six environmental objectives: (i) climate change mitigation, (ii) climate change adaptation, (iii) sustainable use and protection of water and marine resources, (iv) pollution prevention and control, (v) protection and restoration of biodiversity and ecosystems and (vi) transition to a circular economy,
b. do no significant harm to any of the other environmental objectives and
c. meet minimum safeguards on governance and ethics.

In order to assess the first two requirements, it is necessary that the economic activities comply with technical screening criteria set by the European Commission through delegated acts in accordance with Articles 10/3, 11/3, 12/2, 13/2, 14/2 or 15/2 of the Regulation. In other words, the technical screening criteria are the operational tools to verify if specific economic activities qualify as contributing substantially to climate change mitigation and climate change adaptation and for determining if such economic activities cause significant harm to any of the other environmental objectives.

There are already level 2 measures in force, since the EU Taxonomy Climate Delegated Act[11] has set out the technical screening criteria, which include a description of the economic activities and their classification under the NACE code for statistical classification of economic activities established by Regulation (EC) No 1893/2006, as well as the substantive requirements that must be met and evidenced regarding both the substantial contribution for climate change mitigation and adaptation, the first two of the six environmental objectives of the Regulation, and the "do no significant harm" to any of the other environmental objectives.

Economic sectors and economic activities included in the Regulation have the potential to make a substantial contribution to climate change mitigation or climate change adaptation. The approach differs for each of these objectives, reflecting their nature. Sectors selected for climate change mitigation are those that have a large emissions footprint (responsible for 93.5% of direct greenhouse gas emissions in the European Union). The rationale is that activities making a substantial contribution to climate change mitigation are more likely to have a large impact in these sectors.

Segregating by category, the Regulation comprises economic activities which are already low carbon and, as such, compatible with a 2050 net zero carbon economy, as well as economic activities contributing to a transition to a zero net emissions economy by 2050, while currently not operating at that level and enabling activities destined to facilitate emissions reduction.

This does not mean that there is no room for high emitting activities to be considered as making a substantial contribution to climate change, but under significantly demanding conditions: net carbon neutrality being achieved by 2050 or being a top performer of the EU-ETS benchmark are mandatory. This

---

[11] Delegated Regulation (EU) 2021/2139 supplementing Regulation (EU) 2020/852 of the European Parliament and of the Council by establishing the technical screening criteria for determining the conditions under which an economic activity qualifies as contributing substantially to climate change mitigation or climate change adaptation and for determining whether that economic activity causes no significant harm to any of the other environmental objectives. The EU Commission has also adopted on 6 July 2021 a delegated act specifying the content, methodology and presentation of the information to be disclosed by both non-financial and financial undertakings required to report about the alignment of their activities with the EU Taxonomy: Commission Delegated Regulation (EU) 2021/2178 of 6.7.2021 supplementing Regulation (EU) 2020/852 of the European Parliament and of the Council by specifying the content and presentation of information to be disclosed by undertakings subject to Articles 19a or 29a of Directive 2013/34/EU concerning environmentally sustainable economic activities, and specifying the methodology to comply with that disclosure obligation. On 9 March 2022, the EU Commission formally adopted a complementary Delegated Regulation to amend the Taxonomy Climate Delegated Act by adding technical screening criteria for certain economic activities in the natural gas and nuclear energy sectors that have not been included in that Delegated Act, which shall come into effect with its official publication.

will be quite demanding for the energy industry in particular since traditional economic activities *per se* will not be eligible and initiatives aligned with climate change mitigation, such as carbon capture, must be implemented for achieving eligibility.

The "do no significant harm" test relies mostly on existing EU regulations to assess whether an economic activity that has a substantial contribution to an environmental objective causes significant harm to any other environmental objective. The purpose of this criterion is to avoid economic activities being exclusively focused in pursuing one environmental objective, but that actually cause harm to the environment to an extent that outweighs their contribution to such specific environmental objective.

This is essentially a consistency test that is reasonable in the context of a science-based regulation. Nonetheless, such criteria should take into account the life cycle of the products and services provided by that economic activity in addition to the environmental impact of the economic activity itself, including taking into account evidence from existing life cycle assessments, in particular by considering their production, use and end of life.

Finally, the ethics and governance requirements push into the spotlight the importance of governance as part of sustainability, but the Regulation has approached it in a shy manner. Sustainability is not a unidimensional feature that can be achieved on the basis of an ecological approach. It is a much wider and broader challenge that requires good corporate governance practices to be implemented and an ethical behaviour across economic sectors. The Regulation already lifts the veil on the social objectives as being material for sustainability assessment, by requiring the Commission to describe the provisions that would be necessary to extend the scope of the Regulation beyond environmental topics (article 26, paragraph 2b), but it is still missing the governance perspective as a fundamental part of sustainability.

In fact, under article 18 of the Regulation, an economic activity already cannot qualify as environmentally sustainable if it is not compliant with human and labour rights. The Regulation selects certain fundamental conventions or guidelines on human and labour rights (in particular, the OECD Guidelines for Multinational Enterprises and UN Guiding Principles on Business and Human Rights, including the declaration on Fundamental Principles and Rights at Work of the International Labour Organisation (ILO), the eight fundamental conventions of the ILO and the International Bill of Human Rights) as sustainability requirements, but this should not rely on an environmental sustainability assessment and rather a fully integrated ESG assessment on sustainability.

This assertion finds a strong support on the fact that, as recognized in recital 35 of the Regulation, should there be more stringent requirements related to the environment, health, safety and social sustainability set out in Union law, those shall apply. So, the absence of social and governance requirements in the Regulation does not prevent their application, as a matter of Union law compliance.

Scope of Application

Apart from the EU and its Member States, the Regulation applies to financial market participants offering financial products in the European Union and to large companies required to provide a non-financial statement under the Non-Financial Reporting Directive.

The European Commission took a prescriptive approach in respect to defining financial market participants, which on the one hand reduces uncertainty on the application of the norms, but on the other hand leaves a large non-regulated area to which the Regulation rules do not apply.

Although credit institutions and investment firms are the entities covered in the broadest sense by this definition, they only become subject to the Regulation to the extent that portfolio management services are provided. The remainder of the defined entities are dedicated to specific purposes in the financial markets' environment, such as an institution for occupational retirement provision (IORP), a management company of an undertaking for collective investment in transferable securities (UCITS management company), a manufacturer of a pension product, an alternative investment fund manager (AIFM), a pan-European personal pension product (PEPP) provider, a manager of a registered qualifying venture capital fund or a manager of a registered qualifying social entrepreneurship fund. Insurance undertakings are also subject to the Regulation, but only to the extent that they make available insurance-based investment products (IBIP).[12]

The other entities subject to the Regulation are public interest entities, as defined in Directive 2013/34/EU of the European Parliament and of the Council of 26 June 2013, which include listed companies, credit institutions, insurance undertakings and other undertakings designated at each Member State level, with a minimum average number of 500 employees on a given financial year.[13]

This double criterion to select entities that are subject to the Regulation evidences a double concern. On the one side, the Regulation intends to cover virtually every market participant irrespective of their place of origin to the extent that they offer green financial products to EU investors, as a result of their portfolio management or financial products issuance activities. The main concern in this case is to spark discipline in market agents and to protect EU investors, ensuring that financial products purchased for their sustainability contribution actually have the merits of achieving such contribution. On the other side, by covering public interest entities the Regulation reaches not only the most substantial volume of EU investors, but also those investors that

---

[12] Article 1, paragraph 2b) and article 2, paragraph 3 of the Regulation and article 2, paragraph 12 of Regulation (EU) 2019/2088 of the European Parliament and of the Council of 27 November 2019 on sustainability-related disclosures in the financial services sector.

[13] Article 1, paragraph 2c) of the Regulation.

can promote or influence more investment being allocated to environmentally sustainable economic activities.

The merit of this approach is that the potential impact can be effectively significant to change the course of history by focusing investment on the green economy through incentives that reach across society.

Functions

There is an obvious concern on the ability of this regulatory measure to create sufficient confidence to investors in the capital markets by removing ambiguity on the labelling of environmentally friendly economic activities. But our analysis deserves to go further. This harmonization is also supposed to generate a reduction on transaction costs, envisaging that "green" labelled financial products can become more competitive in the market thereby incentivizing the allocation of resources to these products and hence to environmentally sustainable economic activities.

The key function of the Taxonomy is to create a level playing field to support and allow access for the financing of sustainable activities. It has therefore to become a tool to help implement the European Green Deal on its financial dimension. By defining harmonized criteria for determining whether an economic activity qualifies as environmentally sustainable, investors and corporations may resource to the Taxonomy to express the alignment between their investments and/or activities and the transition efforts to a sustainable economy.

The setting up of the Taxonomy also serves a purpose for the finance industry. By elevating the standards applicable to the underlying businesses, the Taxonomy harmonizes financial products that are targeting investors on the grounds of GHG emissions reduction. In fact, there has been a substantial increase in the green financing market in the last few years, although it continues to represent a very limited part of corporate finance market.

Another function that the Taxonomy serves is to generate information on the sustainability of the economic activities developed by corporations and on the investment products available in the European Union financial markets in order to reinforce investors' confidence in issuers and accordingly to contribute for the protection of the interests of investors. By resourcing to science-based technical criteria for setting out a regulatory framework for recognizing environmentally sustainable economic activities, the Regulation becomes a creditworthy reference for investors, strengthening protection vis-à-vis greenwashing financial products and other initiatives.

Furthermore, this allows investors to set out standards and criteria to compare companies regarding information that reflect the sustainability of their businesses in a fast-changing world. This ability to compare corporations will also play a relevant role on assessing the resilience of their businesses and consequently their value. Although the Regulation is structured by reference to the qualitative nature of businesses and refers to non-financial disclosure, the fact is that it focuses on the sustainability of businesses. And this has inevitably financial and operational materiality for investors on a long-term perspective.

Alongside, the Taxonomy will be useful for companies to plan and report the transition to an economy that is aligned with those environmental objectives. The disclosure obligations contained in the Regulation will stir environmentally friendly measures since companies should report their progress regarding the achievement or the initiatives developed towards achieving the green criteria set out therein.

Each of these effects that we are identifying are also being screened and assessed by companies, which are not indifferent to the changes they are identifying in the markets, but rather shape their strategy, their capital allocation and their businesses on the basis of the developments occurring in areas so different as regulation, technology or consumer preferences.

If companies' constituents, such as capital market investors and consumers in particular, are increasingly valuing ESG topics, this will necessarily impact companies. On the one hand, ESG creates pressure to reinforce the resilience of their business portfolios to put into evidence the sustainability of their activity. On the other hand, it influences shareholders' and management decisions to align their governance with these new standards and to give effect to the underlying business choices.

3 Corporate Disclosure—The Comply or Explain

Disclosure ensures transparency and brings the discipline that is required for efficient markets to function. For this reason, disclosure requirements are fundamental mostly to level information among agents in the market, to support adequate decision-making by investors and to instil appropriate and interest-aligned behaviours from agents towards principals.

In the case of periodic disclosure, it catalyses an appropriate management of expectations from both the side of the principals and prospective principals as from the side of the agents. Companies and their investors have levelled expectations in relation to the timing and the content of the disclosure that is presupposed to manifest. These are called the "known unknowns". It is known to happen at a certain moment, but its content is not yet known. On this account, either the absence of disclosure or the quality of the disclosure can have a substantial impact and prompt principals to take action.

Likewise, we cannot neglect that the quality of ESG-based disclosure influences governance, since it enables principals to conduct an effective monitoring of agents based on public information. Conversely, it can be significant for the public reputation of managers, namely on their commitments and contributions to the development of the company's business.

This is precisely the type of disclosure elected in the Regulation to deal with the exposure of public interest entities to the green economy. Pursuant to article 8 of the Regulation, public interest entities must annually provide disclosure to the market on the extent of their association to environmentally sustainable economic activities, starting from January 1, 2022 in respect of economic activities contributing substantially to climate change mitigation and to climate change adaptation. In relation to the remaining environmental objectives—sustainable use and protection of water and marine resources, pollution prevention and control, protection and restoration of biodiversity and ecosystems and transition to a circular economy—disclosure duties shall only apply from January 1, 2023. This timeline takes in consideration the period of time required for the development of technical screening criteria by the European Commission for the other four environmental objectives.

Compliance with this periodic disclosure will bring to light the effective commitment of companies across the European Union and of their managers and shareholders towards environmentally sustainable investments and the influence deriving therefrom on the decisions that investors and prospective investors will take on the investment criteria to be followed.

In any case, the positive approach adopted by the Regulation is fundamental to avoid legal discrimination of activities based on their environmental impact. This is to ensure compliance with constitutional rules, under the EU Treaty, namely non-discrimination and the right of establishment. In other words, the new disclosure rules operate as a means to incentivizing corporations to progressively make investments in green businesses, but it is not intended to discriminate those that do invest in other businesses.

The disclosure legal command determines identifying the specific environmentally sustainable economic activities that are pursued by public interest entities, as well as the level of commitment put to that end. This transparency obligation is currently focused on three economic indicators: (i) capital expenditure, (ii) operational expenditure and (iii) turnover. It is these indicators that will determine the quality of the information that is disclosed to the market and the positive or negative signals that will derive therefrom, namely from the side of investors and prospective investors.

In our opinion the first two indicators are those that currently can better serve the purpose of promoting environmentally sustainable investments. The level of maturity of businesses and technologies around green economy is quite variable and many of those require substantial levels of investment to reach to a market level that can bring them into competition with conventional economy.

Capital allocation will put into evidence the proportion of investment that public interest entities commit to environmentally sustainable economic activities. The value of this indicator is showing the potential for business development and growth resulting from the investment that corporations will make.

Many environmentally friendly activities do not generate results for several years. An example is the investments on renewable energy[14] that require two to five years of development and construction to reach commercial operations. During such period, investment is the key factor and no significant OPEX or turnover will be available to be disclosed to the market. By the opposite, low capital activities will be comparable on the basis of their operational expenditure and potentially turnover.

Looking attentively to these criteria it is hard to say that these relate to non-financial disclosure. In spite of associating this new disclosure requirement to the transparency of non-financial statements, the Regulation actually resources to financial criteria to determine the contents of the disclosure. This passes a powerful message to the market since it actually permits to mensurate the exposure of corporations to environmentally friendly businesses and not only to have access to qualitative disclosure in this regard.

Considering the different nature and performance of environmentally sustainable economic activities it is fundamental that the weight of the different indicators disclosed by public interest entities is considered by reference to the concrete activities promoted by those corporations. Intensive capital activities should not be measured on the basis of turnover for the appropriate period of time for revenues to start being obtained.

The European Commission delegated regulations would be a good opportunity to clarify these differentiations. But even if that is not the case, investors surely will be attentive to them and benchmark comparable activities on the basis of each of those indicators.

However, these requirements can easily lead to the creation of benchmarks and taxonomy profiling that use taxonomy performance indicators to evaluate companies. From there to ranking taxonomy compliant companies could be a step.

We consider that these results would be contrary to the legal framework created by the Regulation and accordingly would not be compliant neither with the text nor the purpose of the Regulation. Mechanisms of this type could have a fundamental impact on companies and would be discriminatory by nature.

Although these disclosure duties have only come into force in 2022, it is important to consider that the European Commission also defined, pursuant to Article 2 of Directive 2014/95/EU of the European Parliament and of the Council of 15 November, guidelines to assist companies to disclose climate-related information as part of their non-financial reporting.

From a governance perspective, the Regulation and the approach that is generally being pursued is essentially supported on disclosure duties. For companies to be able to adapt to these duties, a new governance framework

---

[14] According to the European Commission, "*the energy sector accounts for approximately 75% of greenhouse gas emissions in the Union and thus plays a key role in climate change mitigation*" (recital 14 of the Commission Delegated Regulation).

needs to be put in place to address the effective business challenges associated to sustainability. ESG cannot stay at a disclosure level; it needs to be embedded in the organizational structure and in the functional model of operating a company.

Specific governance measures are required for the Taxonomy application to be effective. Boards of directors need to reflect ESG expertise within their composition. Discharge of directors' duties of care should include actions on ESG matters. Remuneration structures should also reflect ESG metrics and key performance indicators to create the right incentives for managers.

These are just some examples of how the Taxonomy's impact on companies will necessarily go beyond transparency. Transparency suffices only when proper governance is in place to secure that reporting is the translation of effective business management.

4 Conclusions

The clarification of a Taxonomy for qualifying economic activities as environmentally sustainable will have a reach far beyond from that expressed in the Regulation. Further from designing green financial products, foreseeable uses by investors can include expressing investment preferences aligned with ESG concerns and relying on the taxonomy as an *affidavit* of environmental sustainability. Likewise, it can allow a specialization and selection of holdings by investors by reference to environmentally sustainable economic activities, which can incentivize an increase on the allocation of investments to these activities supporting the private financing required to pursue EU's climate change mitigation targets.

From the angle of corporate governance improvement, the Taxonomy Regulation can play an additional function and support the development of benchmarks that measure the environmental performance of a financial product or of a security. Moreover, it can even support the measurement of the ethical performance of financial market participants and companies.

There are also pitfalls that can result from the Taxonomy's application. Transparency requirements are different between EU and non-EU countries, which impacts companies and other entities. EU Member States are subject to the Regulation, but there is no common framework at a wider international level, which creates an unbalanced regulatory environment namely among the parties to the Paris Agreement. This may lead to a risk of an exodus of investment and investors that are not committed to an environmentally friendly investment strategy from the European Union. It will be the exit or voice decision on the part of institutional investors.[15]

---

[15] Pacces, Alessio M., Will the EU Taxonomy Regulation Foster a Sustainable Corporate Governance? Amsterdam Law School Legal Studies Research Paper No. 2021-32 (November 2021), p. 13. Available at SSRN: https://ssrn.com/abstract=3940375.

On the companies' side, effects can be even deeper as listed companies, financial institutions and other investors from the European Union will be subject to the stringent transparency obligations from European law while corporations from other countries will not be bound to the same levels of disclosure on environmentally sustainable investments. This will lead at least to a replacement of investments from public interest entities that will tend to promote divestment from economic activities that are not compliant with the screening criteria. But new investors that are not public interest entities dedicated to activities that do not fall within the scope of the Regulation will be exempt of any disclosure.

The highly regulated approach taken by the European Union, as results from the level 2 regulations, may also prove to be counter-productive to effectively promote the engagement of financial markets participants and investors. Effectively, the green financial products market is still quite small and may have difficulties to grow in an excessively regulated market. While science-based criteria are fundamental to avoid ideological or political approaches to the green finance topic, there must be a pragmatic balance between highly detailed rules and effective technical screening. Excessive regulation tends to induce high externalities and high costs, which more likely work against adherence to the regulated standards.

A potential future enlargement of the scope of these transparency rules could be considered by applying the same disclosure to State-owned entities. However, apart from this, we do not advocate the extension of the Taxonomy's scale.[16] Extending the Taxonomy to all economic activities does not only proves to be needless since, in practical terms, it would represent an exclusion of activities by means of a regulatory statement, but more importantly it would raise serious concerns on legal discrimination of those activities unable to meet science-based criteria to qualify as environmentally sustainable.

Given the substantial difference of regulatory context at an international level, the recent transparency requirements in the European Union may actually lead to a paradoxical confusion and obscurity regarding non-environmentally friendly economic activities that will continue to be pursued wherever legal frameworks allow.

Governance rules and practices can help to mitigate the detrimental effects of inconsistent sustainability regulation across the world. Incorporating disclosure principles equivalent to those of the Taxonomy in the next review of the OECD Principles of Corporate Governance could widen the effect of the Regulation and promote transparency at a wider level. Similarly including those disclosure contents within the recommendations from the Task Force

---

[16] Och, Marleen, Sustainable Finance and the EU Taxonomy RegulationHype or Hope? (November 15, 2020). Jan Ronse Institute for Company & Financial Law Working Paper No. 2020/05 (November 2020), pp. 13–14. Available at SSRN: https://ssrn.com/abstract=3738255.

on Climate-related Financial Disclosures could open transparency to more countries and entities.

While governance and financial products-related measures play an important part to support the way for the European Union to develop a green economy, we cannot forget that other mechanisms will be determinant for the success of the economic transition envisaged with the Green Deal and the related tools, such as the Taxonomy Regulation. On the same day that the European Commission was disclosing the Climate Delegated Regulation and other governance measures and reaching an agreement on the Climate Law, in the United States of America the Senate Finance Committee presented a proposal to consolidate current energy tax incentives into emissions-based provisions designed to incentivize energy efficiency, clean transportation and clean electricity.[17] Taxpayers' incentives are probably one of the most powerful measures to align economic behaviours.

This is still the dawn of the green era for the World's economy. There is however a sense of urgency to deal with climate change threats that requires all players to move in a sensible manner to support or implement a new economic order that is driven by sustainability objectives. Creating the appropriate governance frameworks both at governmental and companies' levels will be determinant to manage appropriately the required institutional transformation and the new economic paradigms.

The Taxonomy Regulation is just a piece of the puzzle (albeit important) and, as discussed, must be used as a positive element to promote such transformation on the economy and on the institutions. A fully integrated ESG approach that provides the environmental, social and governance frameworks to implement the green economy standards is fundamental to effectively achieving balanced and sustainable solutions for companies to perform their role in this change.

---

[17] See also the US Clean Energy for America Act (2021). Available at https://www.finance.senate.gov.

CHAPTER 14

Business Judgement Rule as a Safeguard for ESG Minded Directors and a Warning for Others

Bruno Ferreira and Manuel Sequeira

1 INTRODUCTION

There is a long-standing and recurring debate as to which interest the companies' directors and managers should pursue while managing the company.[1] Some believe that the directors are required to act with the exclusive interest of the shareholder in mind.[2] Others share the view that directors have a fiduciary duty not only towards the shareholders, but also in relation to the

---

[1] Caetano Nunes, *O Dever de Gestão dos Administradores de Sociedades Anónimas*, Almedina, 2012, pp. 256–270, with several references to American legal commentators and case law. This is also an old debate in Germany (Caetano Nunes, *ob cit*..., pp. 331 and ss.) and in Portugal, where the debate has been going on since the period before the Portuguese Companies Code ("PCC") was enacted, and it is still subject to discussions (Caetano Nunes, *ob cit*... pp. 454 and ss.).

[2] Adolf A. Berle, "Corporate Powers as Powers in Trust", in *Hard Law Review*, no. 44, 1930–1931, pp. 1049–1050 and "For Whom Corporate Managers Are Trustees — A Note", in *Hard Law Review*, no. 44, 1932, pp. 1365–1372. Michael Jensen and William Meckling, "Theory of the Firm: Managerial Behavior Agency Costs and Ownership Structures", *Journal of Financial Economics* 3, 1976, pp. 305–360.

B. Ferreira (✉)
Lisbon, Portugal
e-mail: Bruno.ferreira@plmj.pt

M. Sequeira
PLMJ Advogados, Lisbon, Portugal
e-mail: Manuel.sequeira@plmj.pt

company itself, as a permanent organisation in which several constituencies are involved, including employees, consumers, suppliers and other stakeholders, or society in general, with a consequential sacrifice of the shareholders' exclusive interest.[3] In other words, this second approach argues that, although directors have a discretionary margin for decision, other factors have to be considered.

Companies are currently facing challenges stemming from the pandemic, technological disruption, globalisation, social media and political instability. The idea that companies should engage in socially responsible business practices or initiatives relating to environmental, social and governance (ESG) matters began to take shape mostly by the 1970s and its prominence has been growing ever since, even if not uncontested.[4] Although ESG comprises a wide range of factors, the definitions differ and there is still no consensus on what socially responsible activity actually means or comprises or on the rationale for its pursuit, in particular, because the concept is evolving in parallel with the society's needs.[5]

There are three main ideas, one for each letter.[6] Firstly, the "E" means environment, including energy and natural resources and the consequences for living beings as a result. Energy and resources are used by every company in carrying on its business. This means each company affects and is affected by the environment. Secondly, the "S" stands for social, addressing the company's relationships with its partners, employees, customers and suppliers, its reputation in the market and society in general, and the defence of values for social change (such as racial and gender equity or intergenerational mobility[7]). Lastly, the "G" stands for governance, and it relates to the internal system of practices, controls and procedures for the governance of companies, which are essential to adopt efficient decisions, comply with the law and benefit stakeholders. ESG is also used to refer to "all nonfinancial fundamentals that can impact firms' financial performance, such as corporate governance".[8]

The consideration of ESG factors has been one of the main market drivers in recent years, and it has shifted the way agents evaluate a company's performance and influenced investor decisions. Although there is no legal concept

---

[3] E. Merrick dodd, "For Whom Are Corporate Managers Trustees?", *in Harvard Law Review*, no. 44, 1932, pp. 1145–1163. Milton Friedman, "A Friedman Doctrine—The Social Responsibility of Business Is to Increase Its Profits," *New York Times Magazine*, September 13, 1970, available at https://www.nytimes.com/1970/09/13/archives/a-friedman-doctrine-the-social-responsibility-of-business-is-to.html.

[4] Elizabeth Pollman, "Corporate Social Responsibility, ESG, and Compliance", Draft of November, 2019, p. 9, available at www.ssrn.com.

[5] Elizabeth Pollman, "Corporate Social… cit., p. 2.

[6] Witold Henisz, Tim Koller, and Robin Nuttall, "Five Ways that ESG Creates Value", in *McKinsey Quarterly*, November 2019, available at www.mckinsey.com.

[7] World Economic Forum, The Future of the Corporation Moving from balance sheet to value sheet, in collaboration with Baker McKenzie, White Paper, January 2021, p. 6, available at www.weforum.org.

[8] Elizabeth Pollman, "Corporate Social… cit., p. 5, available at www.ssrn.com.

of ESG, it is unanimously recognised that it covers the interests of constituencies other than just shareholders. There has been an effort to introduce those factors into the daily activities of companies and to create new standards of conduct for directors. Bearing in mind the limitations the size and scope of this study, we cannot address all risks, problems and drivers of ESG, but all these matters are interconnected and have a mutual overlapping impact. We will therefore focus on "G" and "S".

Considering the broad implications of ESG, this study's objective is to provide a brief overview of the director's duties and liability under Portuguese law and to assess the impact that recent ESG factors may have on their activity, in particular in the duty of care. Following a brief presentation of the duty of care and its correlation with the business judgement rule ("BJR")—which was transplanted from common law jurisdictions—we will analyse the impact of ESG on the activity of directors of Portuguese companies.

This study will not address specific duties of care in the context of control transactions (e.g. mergers or takeovers), as these have a different legal framework specific to the country in question and this causes directors to have different roles in this context. Furthermore, there are usually country-specific provisions addressing individual duties for establishing a balance within the companies' corporate bodies. Conversely, considering the limited space, we will also not address the particularities resulting from executive or non-executive offices and from one-tier/two-tiers governance models.

2 Brief Overview of Directors' Duties

The general duty of any director is to represent and to manage the company.[9] This duty of management consists of the "obligation to promote the establishment's success by doing a large number of acts and directors are, therefore, granted a wide margin of discretion in their actions. It is up to them to decide, according to discretionary criteria, what the most appropriate or opportune acts are to pursue the corporate interest".[10] Risk management, usually associated with innovation and creativity, has the potential of benefiting the company and the shareholders.[11]

Although the duty of management grants directors a wide margin of discretion, such duty is intensified by undetermined and general fiduciary duties,

---

[9] Caetano Nunes, *O Dever de Gestão... cit.*, pp. 469 and ss.

[10] Bruno Ferreira, "Os deveres de cuidado dos administradores e gerentes (Análise dos deveres de cuidado em Portugal e nos Estados Unidos da América fora das situações de disputa de controlo societário)", in *Revista de Direito das Sociedades*, I year, no. 3, Almedina, 2009, p. 709.

[11] Pais de Vasconcelos, "Business Judgement Rule — Deveres de cuidado e de lealdade, ilicitude e culpa e o artigo 64.º do Código das Sociedades Comerciais", in *DSR*, no. 2, 2009, p. 12, Coutinho de Abreu, *Responsabilidade Civil dos Administradores de Sociedades*, IDET, no. 5, Almedina, 2007, pp. 22. Ricardo Costa and Figueiredo Dias, *Código das Sociedades Comerciais em Comentário*, vol. I, Almedina, 2010, p. 728.

consisting in standards of conduct. In other words, when the standard of conduct expected from the duty is not followed, it is considered that directors are in breach. Usually, directors are subject to two main fundamental duties: the duty of care and the duty of loyalty, which shape all directors' activity and decision-making. These duties are based on the fiduciary relationship stemming from managing assets and interests of third parties. They impose standards of conduct in the context of each act and decision by the director. Each of the said duties outlines specific and instrumental duties which jointly define the scope of activity of directors,[12] considering the particular circumstances of each company. Furthermore, some specific provisions set out specific duties of the directors in a certain context and, in these cases, the discretion is narrower. Directors are liable to the company, creditors and third parties for wrongful mismanagement if they fail to perform their duties.

Despite the director's duties being a central matter in the day-to-day management of European companies, no European-wide regulations on this matter have been enacted to this date. According to the European Commission, the Member States have a "variety of solutions as regards corporate and board duties",[13] which is not surprising, considering the well-known existence of both Member States with Civil Law and Common-law legal frameworks.

Between October 2020 and February 2021, the European Commission launched a consultation on "sustainable corporate governance",[14] in which the Commission expected to receive inputs on the needs and objectives for EU intervention, costs and benefits of different policy options and knowledge in specific issues (in particular on national frameworks, enforcement mechanisms and case law). Therefore, and according to the information publicly available, it is expected an EU Directive proposal in the last quarter of 2021.

Portuguese legal framework is a paradigmatic example to be considered within the EU level initiative, considering that, although being generally described as Civil Law, it has some recent influences of Common-Law in specific matters (as is the case of those under analysis here). In Portugal, these fundamental duties are set out in article 64(1) of the Portuguese Companies Code ("PCC"). This provision was last amended in 2006, by Decree-Law 76-A/2006 of 29 march ("Decree-Law 76-A/2006"), which introduced the duty of loyalty and the BJR into the Portuguese legal framework.

Before 2006, this Portuguese provision was mainly inspired by the German legal framework. Article 64 basically incorporated, with some amendments, the wording of Decree-Law 49.831 of 15 November 1969, which was inspired by § 93/I of the German *Aktiengesetz* of 1965 (as amended in 2005[15]) stating that "managers, directors or officers of a company must act with the diligence

---

[12] Ricardo Costa and Figueiredo Dias, *Código... cit.*, p. 729.

[13] European Comission, "Inception Impact Assessment", Ref. Ares(2020)4034032 - 30/07/2020, available at https://ec.europa.eu.

[14] Available at https://ec.europa.eu.

[15] An English translation is available at www.gesetze-im-internet.de.

of a careful and orderly manager, in the interest of the company, taking into account the interests of shareholders and employees". The last part of the article was added at the last minute based on the proposal for the 5th Directive on companies,[16] which was never approved and severely criticised.

As a result of Decree-Law 76-A/2006, the wording of article 64 (1) of the PCC is currently the following:

Article 64

Fundamental Duties

1 – The company's managers or directors must comply with:

a) Duties of care, evidencing availability, technical expertise and understanding of the company's business appropriate to their role, and executing their duties with the diligence of a careful and orderly manager; and

b) Duties of loyalty, acting in the company's interests, bearing in mind the shareholders' long-term interests and considering other parties' interests relevant to the company's sustainability such as employees, clients and creditors.

Roman-Germanic law countries have always inspired the Portuguese legal framework on these matters—in line the with civil legal framework on obligations and civil liability—and the introduction of concepts from common law countries in 2006 generated a wide debate about directors' duties and liability. This constituted a major shift in the law-making approach to these matters, particularly with respect to the introduction of the BJR.

The Portuguese courts have only sporadically analysed the merits of the decisions,[17] as directors decide between alternative decisions at their disposal under the management duty provided by law. However, they have always been somewhat reticent about finding directors in breach of their duties, especially on the grounds of non-compliance with the duty of care. In fact, even after 2006, it was rare to come across court decisions in which directors were sentenced by courts on this basis (and none has addressed the shareholders' v. stakeholders' interests), as courts often decide on director's accountability linked with the duty of loyalty,[18] for example, on conflicts of interests.

---

[16] Menezes Cordeiro, "Os deveres fundamentais dos administradores das sociedades", in ROA, no. 66, vol. II, 2006, available at www.oa.pt.

[17] Ferreira Gomes, *Da administração à fiscalização das sociedades. A obrigação de vigilância dos órgãos da Sociedade Antónima*, Almedina, 2015, p. 817, with references to Portuguese court decisions.

[18] Although the majority of the courts refer to breach of both the duties of care and loyalty, the actions reviewed concern the potential violation on the duty of loyalty, on which the courts typically concentrate their analysis. There is, however, an important STJ decision of 16.05.2000 (case no. 259/2000), available at *Boletim do Ministério da Justiça*, no. 497, 2000, pp. 396–405, regarding the analysis of the "diligence of a careful and orderly manager" and, in particular, concerning the duty to obtain information, prior to the 2006 amendments to the PCC.

3 DUTY OF CARE AND THE BUSINESS JUDGEMENT RULE

General Notes on the Duty of Care

The duty of care stems from US case law dating back to the nineteenth century and it was then developed in the context of tort law, as a special application resulting from the negligence law.[19] However, it is also frequently characterised, in parallel with the duty of loyalty, as a fiduciary duty.[20] In particular, the duty of care provides standards of conduct intended to ensure the quality of the company's management by establishing the minimum criteria for management actions and decisions, so as to enable a lawful analysis by the courts of directors' acts in specific situations.[21]

In common law countries, this duty was established by case law. As is widely known remarkable court decisions have contributed to the consolidation and stability of this duty. These include Litwin v. Allen from the Supreme Court of New York in 1940, Selheimer v. Manganese Corp. of America from the Supreme Court of Pennsylvania in 1966, and Smith v. Van Gorkom from the Supreme Court of Delaware in 1985. The Turquand v. Marshall case in United Kingdom, decided in 1869, is also commonly mentioned by scholars.

In article 64 of the PCC, the legislature opted for a general duty of care but also identified some instrumental specific duties of care impending over directors. As some legal commentators highlight, this is a mere exemplification, to be complemented by scholars and case law.[22] For a complete overview of the specific duties of care, this provision should be read together with article 72 of the PCC (excerpt in the next section). On the other hand, in 2009, one of us[23] argued for the existence of at least five instrumental and specific duties of care, which are the:

---

[19] Melvin Aron Eisenberg, "The Duty of Care of Corporate Directors and Officers", in *University of Pittsburgh Law Review*, vol. 51, 1989, pp. 945–949, available at https://lawcat.berkeley.edu, Dr Yoram Danziger and Omri Rachum-Twaig, "Re-Evaluating the Justifications for the Existence of an Independent Duty of Care", *in The Company Lawyer*, no. 35, Issue 9, Thomson Reuters (Professional), p. 266, available at www.ssrn.com.

[20] Paulo Câmara, "O Governo das Sociedades e os Deveres Fiduciários dos Administradores", in *AAVV, Jornadas: Sociedades Abertas, Valores Mobiliários e Intermediação Financeira*, Almedina, 2007, p. 167 and Pais de Vasconcelos, "Business Judgement Rule, p. 30.

[21] Bruno Ferreira, "Os deveres de cuidado… cit., p. 710.

[22] Menezes Cordeiro, "Os deveres fundamentais… cit., Paulo Câmara, "O Governo das Sociedades… cit. p. 166, and Bruno Ferreira, "Os deveres de cuidado… cit., p. 711.

[23] Bruno Ferreira, "Os deveres de cuidado… cit., p. 711, in which several references to national and foreign legal commentators were made. With a different classification, and analysing the duties of vigilance, inquiry and to be available all together, Coutinho de Abreu, *Responsabilidade Civil…cit.*, p. 20.

i. **Duty of vigilance**[24] over the company's activity (including of other directors and individuals carrying out management duties) and **duty of investigation**, arising from the American "duty to monitor" and "duty to enquire". These duties impose a duty on directors to know about their company's activity and to enquire when there is any alert or risk of there being an irregular event connected with the company. These duties also require the creation of "information and reporting systems" that can allow the board to assess corporate compliance with all applicable laws;

ii. **Duty to be available** to manage, by investing "time and energy" in carrying on the company's activity,[25] despite the absence of an exclusivity duty and with the objective of avoiding the appointment of directors for purely formal purposes;

iii. **Duty to properly prepare management decisions**, which requires directors to collect and analyse the information on which the decision will be based, thus avoiding reckless decisions. The level of effort required and the procedures adopted for collecting information depends on the context in which the decision is taken[26];

iv. **Duty to take rational management decisions**, which means that any decision considered incomprehensible, without any coherent explanation[27] or without any meaning or wisdom[28] is unlawful;

v. **Duty to take reasonable management decisions**, considering that directors have wide discretion during the selection of decisions that they deem more convenient or appropriate to pursue the company's purpose. Thus, the planned decision must be reasonable when compared with the excluded alternative decisions (considering the risk) which could have been adopted by a careful and orderly director.[29]

Legal references to the "technical expertise", "company's sustainability" and "diligence of a careful and orderly manager" should be interpreted as elements that help in the assessment of the level of care by which the directors

---

[24] The existence of this duty was already highlighted by the STJ, in a decision of 19 November 1987, in *BMJ* 371 (1987), pp. 473–489.

[25] Perestrelo de Oliveira, *Manual de Governo das Sociedades*, Almedina, 2017, p. 233.

[26] Coutinho de Abreu, *Responsabilidade Civil...cit.*, p. 21, highlighting, in particular, the importance of the decision, its urgency, costs of obtaining the information, and if the decision relates to the ordinary course of business or to the scope of extraordinary management.

[27] Ricardo Costa, "Responsabilidade dos Administradores e Business Judgment Rule", in *AA. VV., Reformas do Código das Sociedades,* Colóquios no. 3/IDET, Almedina, Coimbra, março de 2007, p. 84. Coutinho de Abreu, *Responsabilidade Civil...cit.*, p. 46.

[28] Calvão da Silva, "'Corporate Governance – Responsabilidade civil dos administradores não executivos, da Comissão de Auditoria e do Conselho Geral e de Supervisão", in *RLJ*, no. 136 (September–October), 2006, p. 55.

[29] Bruno Ferreira, "Os deveres de cuidado... cit, p. 729.

are bound.[30] All directors are required to get to know and learn[31] any matters relating to their duties and at least be able to assess whether they need further assistance from expert professionals. Therefore, if appointed as director of a company which does not have the required general competence, this director should refuse and resign when he or she cannot update and maintain the required knowledge and skills for the position.[32]

Director's Accountability and Business Judgement Rule

As explained above, under Portuguese law, directors are liable towards the company (article 72 of the PCC), creditors (article 78 of the PCC) as well as shareholders and third parties (article 79 of the PCC) for wrongful mismanagement. The BJR was introduced in Portugal in 2006 by the abovementioned Decree-Law 76-A/2006 of 29 March, in article 72(2) of the PCC, which provides as follows:

> "Article 72
>
> Board members' accountability towards the company
>
> 1 – Managers or directors will be liable towards the company for losses caused by acts or omissions in breach of their legal or contractual duties, unless they prove that they acted without fault.
>
> 2 – Liability will be excluded if any of the persons mentioned in the previous paragraph proves that he/she acted in an informed manner, free of any personal interest and according to corporate rationality criteria".

Before 2006, there was a discussion among legal commentators on whether BJR would be applicable, despite it not being expressly provided for in the legislation. Some argued that directors would only be liable for breaches of duties established by law or in the articles of association or agreements (or, in case of gross negligence, under the bona fides principle). Others assumed that management errors, including those not covered by gross negligence, would only be relevant to assessing accountability if specific processes and/or

---

[30] Some Portuguese legal commentators consider this to be a level of diligence more intense than the civil law *bonus pater familias*, Luís Brito Correia, *Os Administradores de Sociedades Anónimas*, Almedina, 1993, p. 600, Gomes Ramos, *Responsabilidade Civil dos Administradores e Diretores de Sociedades Anónimas Perante os Credores Sociais*, Coimbra, 1997, p. 95, Coutinho de Abreu, *Responsabilidade Civil...cit.*, p. 24 and 25 (note 36), Ricardo Costa, "Responsabilidade dos Administradores... cit., p. 78. Ferreira Gomes, *Da administração... cit.*, pp. 730–731. Also, the Portuguese Supreme Court of Justice, in decisions of 28.02.2013 (case no. 189/11.3TBCBR.C1.S1) and of 30.09.2014 (case no. 1195/08.0TYLSB,L1.S1), available at www.dgsi.pt. Others consider that the civil law criterion should be adopted to the specific case, this being a concretization. Calvão da Silva, "'Corporate Governance'... cit., pp. 51–52.

[31] Perestrelo de Oliveira, *Manual de Governo... cit*, p. 233.

[32] Gomes Ramos, *Responsabilidade Civil... cit.*, p. 92, Coutinho de Abreu, *Responsabilidade Civil...cit.*, p. 24.

care standards (similar to those under BJR) were not followed. These resulted from the "diligence of a careful and orderly manager" required by law. In this context, there is only one published decision from a first instance court, dated 27 October 2003, in which the BJR was used as a limit on the court's *ex-post* analysis of management decisions.[33]

As is widely known, the BJR was created by the US courts, based on the tort of negligence, as a limitation on judicial review of management decisions. In other words, it was created as a recognition of the management prerogatives of directors. It is argued that, while the duty of care is a standard of conduct, the BJR is a standard of review.[34] However, other authors reject this view, considering the difficulties of distinguishing between a standard of conduct and a standard of review.[35]

It is also argued that the courts should not review the merits of a director's decisions, because (i) judges do not have the means (including the technical expertise) to do this[36]; (ii) there are no objective rules for management[37]; and (iii) due to the absence of clear standards, hindsight bias (with information collected *subsequently*) "can make even the most reasonable managerial decision seem reckless *ex-post*".[38]

These justifications are based on the notion that risk-taking is crucial to the process of business decisions and to the wealth of the company and that any *ex-post* review of decisions would severely limit business activity and prevent talented people from accepting positions as directors. Taking all of this into consideration, some scholars hold that the BJR is a compromise between authority and directors' accountability, and it saves the effectiveness of management decisions from judicial review.

Nowadays, the BJR is itself subject to a different interpretation from country to country (and sometimes, even within the same country, as is the case of the US[39]). Nonetheless, while in the US it is mainly applicable to exclude accountability, in Portugal there is a discussion about the nature of

---

[33] Published at Caetano Nunes, *Corporate Governance*, Almedina, 2006, p. 37.

[34] Melvin Aron Eisenberg, "The Divergence of Standards of Conduct and Standards of Review in Corporate Law", in *Fordham Law Review*, no. 62, 1993, pp. 462–464.

[35] D. Gordon Smith, "A Proposal to Eliminate Director Standards from the Model Business Corporation Act", *University of Cincinnati Law Review*, no. 67, 1999, pp. 1203–1209.

[36] Luca Enriques, Henry Hansmann, and Reunier Kraakman, "The Basic Governance Structure: The Interests of Shareholders as a Class", in *Anatomy of Corporate Law—A Comparative and Functional Approach*, Second Edition, Oxford University Press, 2009, p. 79. Although this argument is very criticized, since there are other specific areas (e.g. medicine) where this does not prevent judicial review, as explained by Ferreira Gomes, *Da administração... cit.*, p. 838, with reference to the thoughts of Gevurtz.

[37] Coutinho de Abreu, *Responsabilidade Civil... cit.*, p. 39.

[38] Luca Enriques, Henry Hansmann, and Reunier Kraakman, "The Basic Governance ...cit., p. 79.

[39] According to the Principles of Corporate Governance of the American Law Institute.

the BJR.[40] In particular, the question is whether it is only an exclusion of fault[41] (since it establishes the level of diligence required from a director), an exclusion of the unlawful nature of the act (as it provides the standard of conduct that directors must follow),[42] both,[43] or a manifestation of the duty of management which, in practical terms, result in a general exclusion of accountability.[44]

There is also a discussion on whether the BJR is only applicable in cases of breaches of duty of care.[45] This would mean that the BJR does not provide guidance on defining the scope of directors' fundamental duties. Instead, would define the scope of their duty to indemnify, in case of a breach.

Contrary to some US States (e.g. in Delaware), in Portugal the BJR is not a presumption that the directors acted correctly, as was immediately recognised by CMVM (the Portuguese securities markets authority)[46] in the documents for public consultation regarding the amendments introduced by Decree-Law 76-A/2006 to PCC. Once this was enacted, legal commentators immediately highlighted the fact that article 72(2) should be interpreted restrictively, because it was not to apply to breaches of duties provided in the law, in articles of association, or in agreements when there is no scope for judgement by the directors.[47] They argued that it only concerns judicial review in the context of management decisions.

---

[40] The same discussion exists regarding §93/I of the German Aktiengesets of 1965. Caetano Nunes, *O Dever de Gestão... cit.*, pp. 322 and ss.

[41] Calvão da Silva, "'Corporate Governance'...cit., pp. 53–57 and Menezes Cordeiro, "Os Deveres Fundamentais... cit, available at www.oa.pt.

[42] Coutinho de Abreu, *Responsabilidade Civil...cit.*, pp. 42–43, and Paulo Câmara, "Governo das Sociedades e a Reforma do Código das Sociedades Comerciais, *in Código das Sociedades Comerciais e Governo das Sociedades*", Almedina, 2008, pp. 50–53. Pais de Vasconcelos, "Business Judgement Rule", p. 30 states that, in these cases, the law deems the acts done to be lawful.

[43] Ricardo Costa, "Responsabilidade... cit, pp. 64 and 73–79."

[44] Carneiro da Frada, "A business judgement rule no quadro dos deveres gerais dos administradores", *in Jornadas Sociedades Abertas, Valores Mobiliários e Intermediação Financeira*, Almedina, 2007, pp. 223 and ss. and 230 and ss. Ferreira Gomes, *Da administração...cit.*, p. 887. Caetano Nunes, *O Dever de Gestão... cit.*, pp. 515 and ss., deciding not to enter in this discussion, states that it is a clause of "accountability exclusion". Caetano Nunes, *O Dever de Gestão... cit.*, pp. 515 and ss., deciding not to enter into this discussion, states that it is an "exclusion of accountability" clause.

[45] See Caetano Nunes, *O Dever de Gestão... cit.*, pp. 462 and ss. and 517, as well as Ferreira Gomes, *Da administração... cit.*, pp. 887 and ss., both with several references.

[46] "Governo das Sociedades Anónimas: proposta de Alteração ao Código das Sociedades Comerciais — Processo de Consulta Pública", no. 1/2006, p. 18, available at www.cmvm.pt.

[47] Calvão da Silva, "'Corporate Governance'... cit., p. 57, Carneiro da Frada, "A Business Judgement Rule... cit., pp. 222, and Ricardo Costa, "Responsabilidade...cit, pp. 65 and 67 and ss. Others argue that the courts can judge the gross error, which are excluded from BJR's scope. Menezes Cordeiro and Barreto Menezes Cordeiro, *Código das Sociedades Comerciais Anotado*, 3rd edition, Almedina, 2020, p. 356.

Considering the difficulties in the review of the merits of management decisions by the courts, the legislature intended that, if there was evidence that the directors' actions were informed, free from personal interests and rational, as set out in article 72(2), the court would not analyse the merits and appropriateness of the management decision. Instead, they would consider such behaviour to be sufficient to comply with the duty of care and exclusion of accountability.[48] Therefore, management errors will be relevant if the management decisions were not reasonable and when it is not proved that they were rational, preceded by adequate information and free from any personal interest—the minimum standard for action by directors.[49]

However, as previously argued by one of us,[50] failure to establish that these instrumental duties of adequate preparation and rational decision (and the fact of being free of personal interest) were fully observed, does not automatically trigger a breach of the duty to make substantially reasonable decisions.[51] Such cases only enable courts to review the director's actions (and management options) and, therefore, establish whether the director has complied with the duty of care.

In addition, while breaching the duty to prepare for the decision does not automatically mean that the principal duty of care was breached (and compliance with this main duty does not necessarily mean that the duty to prepare decisions was fully observed), a breach of the duty to take rational decisions will inevitably result in the breach of the main duty of care, since irrationality of the management decision will necessarily result in unreasonableness.

4 ESG's Impact on the director's Duty of Care

Shareholders have intense essential powers over the company's destiny, since some of these powers may only be exercised by the shareholders' general meeting, in particular, appointing and dismissing directors. According to some

---

[48] Lisbon's Court of Appeal, decision of 11.11.2004 (case no. 5314/06.3TVLSB.L1-7). The Portuguese Supreme Court of Justice ("STJ"), in a decision of 28.04.2009 (case no. 09A0346) also considered that directors should comply with the fundamental duties within the decision-making process. Both decisions are available at www.dgsi.pt.

[49] Ricardo Costa, "Responsabilidade ...cit, p. 70 and Bruno Ferreira, "Os deveres de cuidado ...cit, p. 725.

[50] Bruno Ferreira, "Os deveres de cuidado... cit., p. 726.

[51] It was mentioned, obtiter dictum, in the Aronso v. Lewis case by the Supreme Court of Delaware, that for the directors to benefit from the BJR, they would have to comply with the duty of obtaining information within the decision-making process. This was formally decided in Smith v. Van Gorkom case by the Supreme Court of Delaware, in 1985. Moreover, in Portugal, some legal commentators refer to an "adequate decision-making proceedings". Menezes Cordeiro and Barreto Menezes Cordeiro, *Código das Sociedades* Barreto Menezes Cordeiro, *Código das Sociedades... cit*, p. 356.

authors[52] and studies, this encourages shareholder activism based on short-term strategies to collect profits, rather than on long-term sustainable value creation. This is especially so in countries where the ownership of the shares is more concentrated (creating individual or groups of controlling shareholders).

Thinking proactively and acting with ESG values in mind has lately become even more pressing for companies. A weight of accumulated research has concluded that companies that pay attention to ESG concerns do not experience a drag on value creation (rather the contrary).[53] As a result, the mindset of investors has been shifting from an individual perspective to a new outlook, where social involvement is also considered. Moreover, there has been an unprecedented wave of initiatives concerning these matters. We have seen an overwhelming number of groups, associations or informal public commitments[54] in defence of ESG or Corporate Social Responsibility ("CSR") values, as well as the increase of legislative initiatives, some of hard law,[55] others of soft law[56] (and also corporate governance mechanisms of self-regulation).

---

[52] The dangers of short-termism have been highlighted for more than 40 year. Recently, Tim Koller, James Manyika, and Sree Ramaswamy, "The case against corporate short termism", in Milken Institute Review 2017, available at www.mckinsey.com and Nicolas Grabar and Fernando Martinez, "The Short-Termism Debate" February 2021, available at https://corpgov.law.harvard.edu/. Additional references on Lucian Bebchuk, "Don't Let the Short-Termism Bogeyman Scare You", 2021, which part is available at https://corpgov.law.harvard.edu/, which has been trying to point out that short-termism is not necessarily dangerous.

[53] As emphasised, based on several studies, by Witold Henisz, Tim Koller, and Robin Nuttall, "Five Ways that ESG Creates Value", in *McKinsey Quarterly*, November 2019, available at www.mckinsey.com. Susan N. Gary, "Best Interests in The Long Term: Fiduciary duties and ESG integration", in *University of Colorado Law Review*, 731 (90), 2019, pp. 747 and ss.

[54] In August 2019, the Business Roundtable—a group of 181 prominent companies' CEOs, including JPMorgan Chase, Amazon, Apple, and Walmart, among others—released a statement declaring that the purpose of the corporation no longer gives shareholders special consideration, but rather that companies should pursue the interests of customers, employees, suppliers, communities in which they work, and those of shareholders. The statement, in favour of the stakeholder's capitalism, is available at https://www.businessroundtable.org.

[55] The European Commission has issued (i) Directive 2014/95/EU of the European Parliament and of the Council, of 22 October 2014, amending Directive 2013/34/EU, requiring large companies to report on their social and environmental impacts; (ii) Directive (EU) 2017/828 of the European Parliament and of the Council of 17 May 2017 amending Directive 2007/36/EC as regards the encouragement of long-term shareholder engagement. More recent and also very relevant, (a) Regulation (EU) 2019/2088 on sustainability-related disclosures in the financial services sector (Disclosure Regulation); (b) Regulation (EU) 2019/2089 amending Regulation (EU) 2016/1011 (BMR) as regards EU climate transition benchmarks, EU Paris-aligned benchmarks and sustainability-related disclosures for benchmarks (Low Carbon Benchmark Regulation) and (c) Regulation (EU) 2020/852 on the establishment of a framework to facilitate sustainable investment (Taxonomy Regulation).

[56] Elizabeth Pollman, "Corporate Social... cit., p. 12, indicating as examples, UN Global Compact (UNGC), the Global Reporting Initiative (GRI) Standards, and the Organization

For example, the G20 Finance Ministers and Central Bank Governors asked the Financial Stability Board (FSB) to review how the financial sector can take account of climate-related issues. The FSB established a Task Force on Climate-related Financial Disclosures (TCFD)[57] to issue recommendations (released in 2017[58]) for more effective climate-related disclosures that could "promote more informed investment, credit, and insurance underwriting decisions" and, in turn, "would enable stakeholders to understand better the concentrations of carbon-related assets in the financial sector and the financial system's exposures to climate-related risks". The TCFD identified several categories of climate-related risks and opportunities, including investors and companies considering longer-term strategies and most efficient allocation of capital in light of the potential economic impacts of climate change. The duty of care is evolving in a way that requires investors to take account of ESG issues in their investment processes.

Additional European legislation on ESG matters is expected to come. In 2018, the European Commission launched an action plan on financing sustainable growth. This plan encourages transparency, and long-termism in financial and economic activity is featured as one of the main goals. It also establishes a clear and detailed EU taxonomy, a classification system for sustainable activities, creating an EU green bond standard and labels for green financial products, among others. In this respect, on 1 February 2019 the Commission also requested advice from the ESMA, EBA and EIOPA on undue short-term pressure from the financial sector on corporations[59] and, later, from Black-Rock Financial Markets Advisory, which disclosed on 27 August 2021 a final study on the development of tools and mechanisms for the integration of ESG factors into the EU banking prudential framework and into banks' business strategies and investment policies.[60]

Although the negative effects of "shareholder primacy" are disputed,[61] the rise of ESG is contributing to fighting it, and disregard of ESG matters may inclusively have a negative impact on public opinion. It seems that the biggest

---

for Economic Co-Operation and Development's (OECD) Guidelines for Multinational Enterprises.

[57] The Task Force's 32 international members, led by Michael Bloomberg, include providers of capital, insurers, large non-financial companies, accounting and consulting firms, and credit rating agencies.

[58] The 2017 TCFD recommendations report is available at fsb-tcfd.org/publications/.

[59] The European Authorities published their reports in December 2019 and they recommended strengthening disclosure of ESG factors to facilitate institutional investor engagement. They are available at https://ec.europa.eu/.

[60] Also available at https://ec.europa.eu/.

[61] As explained by Mark J. Roe, Holger Spamann, Jesse M. Fried and Charles C.Y. Wang, "The European Commission's Sustainable Corporate Governance Report: A Critique", in *ECGI Working Paper Series in Law*, Working Paper N° 553/2020 November 2020, available at www.ssrn.com, with several references to legal and financial studies.

challenge is how to give greater weight to stakeholders' interests without undermining the morality of free markets and fiduciary duties.[62]

In Portugal, the amendment to the PCC in 2006 had, in part, placed ESG matters under the directors' "radar", as it is clear that the central duty of directors is to satisfy the interests of the company (as the common interests of the shareholders). These are the "reference interests", the ones that guide (and should guide) the directors' decisions and acts. However, the law currently requires directors to consider "the long-term interests of the shareholders" and also to "considering other parties' interests relevant for the company's sustainability such as employees, clients and creditors", despite this obligation being subordinated to the "reference interests".[63] Directors have a wide scope of discretion in the decision-making process, especially in respect of the more convenient or appropriate route for the company. Therefore, the introduction of the stakeholders' interests is an "exhortation" to directors to take into account, as much as possible, the interests of stakeholders (and therefore, ESG). Considering, however, the lack of enforcement of directors' duties in Portugal to date, in practical terms, the amendment to the PCC is interpreted narrowly and is therefore a mere appeal to directors.

Despite the change in the rules of the game provoked by ESG, societies in general tend to resist embracing new models. Some others are trying to avoid a shift to stakeholder primacy, but it seems the dichotomy of shareholder v. stakeholder primacy does not necessarily exist, as there is vast space between the two.[64] Shareholders have an important say over company's performance through their role in the election of directors, through shareholder activism, and by means of exit. Shareholders' rights are protected by corporate law and securities regulations, and directors have fiduciary duties of care, loyalty and good faith. Moreover, shareholders can bring claims against the directors if those duties have not been met. But none of this logically implies that any move beyond shareholder primacy *ipso facto* is a move towards stakeholder primacy. It represents a desire to move beyond the narrow confinement of the of principal–agent construction, beyond the "owners" and "employees" conception of the body corporate.[65]

Therefore, despite these ESG advancements, directors are still strictly conditioned to act in accordance with the controlling shareholders' interests, with

---

[62] Commissioner Hester M. Peirce, "Markets, Morality, and Mobsters: Remarks at the 18th Annual Corporate Governance Conference", speech on 27.08.2020, available at https://www.sec.gov/news/speech/peirce-markets-morality-mobsters-2020-08-27.

[63] As explained by Catarina Serra, "The New Company Law: Towards a Responsible Corporate Governance", in *Scientia Iuris*, Londrina, v. 14, November 2010, pp. 162–169.

[64] Paul Barnett, "The Shareholder v Stakeholder False Dichotomy", August 2016, available at https://www.linkedin.com/pulse/shareholder-v-stakeholder-false-dichotomy-paul-barnett/ and John Gerard Ruggie, "Corporate Purpose in Play: The role of ESG Investing" Draft chapter for Sustainable Investing: A Path to a New Horizon Edited by Andreas Rasche, Herman Bril & Georg Kell, p. 12, available at www.ssrn.com.

[65] Gerard Ruggie, "Corporate Purpose… cit., p. 12.

disregard of stakeholders' interests as set out by law. This was discussed in the European Commission's research into sustainable corporate governance[66] (2020), which argued that one of the key drivers of short-termism in publicly traded companies is a tendency for directors' duties and companies' interests to be interpreted narrowly, with a tendency to favour the short-term maximisation of shareholder value. In this study, it was also discussed whether directors' duties of care should be more clearly defined in legislation. In particular, this could include providing a duty for directors at the European level, since one of the alleged key drivers for short-termism is a tendency for directors' duties and companies' interests to be interpreted as to consider a wider range of stakeholders and to balance the interests of all stakeholders, rather than giving primacy to shareholders. An obligation to identify the relevant stakeholders and to implement procedures to ensure that adverse environmental, human rights and social impacts on stakeholders are addressed is also being considered. Also, following the European Commission's aforementioned consultation on "sustainable corporate governance", an EU level intervention is expected to come soon.

The rules on ESG introduced additional standards of conduct[67] in the Portuguese legal framework, to be observed by the directors, in particular in the context of risk analysis, organisation of governance models and adoption of reasonable management decisions. Complementary to laws already enacted (such as the 2006 amendment to the PCC),[68] expected ESG rules

---

[66] See "Study on directors' duties and sustainable corporate governance — Final report", prepared by Ernst & Young (EY) for the European Commission DG Justice and Consumers, available at www.europa.eu. This report identified the following seven key problem drivers, including (i) directors' duties and company's interest tending to favour the short-term, (ii) companies failing to identify and manage relevant sustainability risks and impacts, (iii) long-term interests of stakeholders not being incentivised by corporate governance and (iv) limited enforcement of the directors' duties. The report then proposes that the EU should act and proposes three ways of doing so (two of soft law and one of hard law).

Strong critics of this report are Mark J. Roe, Holger Spamann, Jesse M. Fried and Charles C.Y. Wang, "The European Commission's...cit., who conclude that "The Report fails on every important dimension. It does not define the problem properly, presents inapposite evidence, fails to address, or even cite the relevant academic research, and neglects elementary problems with its policy proposals. No EU policymaker should rely on this Report". They also state that "(...) changing the jurisdictional status here is unlikely to have a discernible impact. The better way to deal with this recognition is to accept this and other evidence—not cited in the Report—that such formulations are unlikely to make a meaningful difference".

[67] Against this, Jr. Leo E. Strine, Kirby M. Smith and Reilly S. Steel, "Caremark and ESG, perfect together: A practical approach to implementing an integrated, efficient, and effective Caremark and EESG strategy", in Harvard Law School, Discussion Paper no. 1037, Cambridge, 07/2020, available at www.ssrn.com, who, instead of adding a new component to the traditional fiduciary duties of loyalty and care, situate ESG within the established legal framework and propose a way for boards to address the demands of ESG and compliance in an integrated, efficient, and effective way.

[68] Other national legislation was already enacted regarding ESG factors, but not specifically regarding directors' duties: (i) Decree-Law no. 89/2017 of 28 July, on the disclosure

(in particular from European law[69]) will intensify the reshaping of governance structures[70] and directors' work, in particular, by making compliance with the duty of care more challenging. Directors should consider material ESG factors, particularly those that have a financial impact.

Especially concerning the instrumental duties of care mentioned above, ESG has already had impact (which will then be analysed in accordance with the specific circumstances) as follows[71]:

i. On **duty of vigilance** and **duty of investigation**: if there is any alert or risk of an irregular event occurring in connection with ESG matters (e.g. racial or gender discrimination), directors must ensure they are resolved and, when necessary, address them directly. These duties also require the creation of "information and reporting systems" concerning ESG matters[72] and to "triage ESG-related information".[73] Depending on the company's activity and specific circumstances, this may even require the creation of specific departments and contracting of staff for this purpose, in order to allow the board to monitor corporate compliance with all applicable laws. It is essential that directors understand scope, challenges and opportunities of ESG, so that they are able to comply with this duty;

ii. On **duty to properly prepare management decisions**: in the process of collecting and analysing the information that management decisions will be based on, ESG metrics, risks and information should be collected by the board to avoid reckless decisions. Especially in the coming decade,

---

of non-financial information and information on diversity by large companies and groups, which transposes the Directive 2014/95/EU; and (ii) Law no. 62/2017 of 1 August, on the balanced gender representation in the management and supervisory bodies of the public sector entities and listed companies, providing certain thresholds for the appointment of members of each gender for those corporate bodies.

[69] On 10 March 2021, the above-mentioned Regulation (EU) 2019/2088 on sustainability-related disclosures in the financial services sector became, in its majority, directly applicable in the legal systems across the European Union, including in Portugal. This Regulation establishes certain harmonised transparency rules applicable to financial market participants and financial advisers in relation to the management of the financial risks arising from climate change, resource depletion, environmental degradation and social issues. It thus imposes the consideration of ESG factors in their investment decisions and sets out certain transparency duties in the way financial market players take into account sustainability risks in their investment decisions or in their investment advisory activities.

[70] Martin Lipton, "It's Time to Adopt the New Paradigm", in *Harvard Law School Forum on Corporate Governance*, February 2019, available at https://corpgov.law.havard.edu.

[71] No impact is expected on the duty to be available.

[72] As explained by Martin Lipton, "It's time… cit., these mechanisms to be placed must aid employees to "seek guidance and alert management (…) about potential or actual misconduct without fear of retribution".

[73] John W. White, Matthew Morreale and Michael Arnold, *Responding to the ESG Paradigm Shift: Practical Steps for Boards and Management*, Cravath, Swaine & Moore LLP, March 2021, p. 3, available at https://www.cravath.com.

directors will be required to invest in knowledge regarding these matters, so that they can try to predict developments, assess material issues and, therefore, define the areas of the company that need improvement and plan the best way to manage it. However, considering the volume of information produced in recent months, "it can be overwhelming for an officer or director to track and consider all relevant developments". This highlights the importance of creating the information system mentioned above.[74] Moreover, the impact of the decision and alternative decisions on shareholders' and stakeholders' interests (especially when these are not aligned) must be considered. Of course, the level of effort required and procedures adopted for collecting information and foreseeing the decision's effects still depends on the context in which the decision is taken;

iii. On **duty to take rational management decisions**: when possible, ESG matters should also be taken into account in the coherent explanation on which a decision is based. This means that directors will have to become ESG experts in the near future, because taking decisions will be more complex. Therefore, discussing them within the board (and explaining them to shareholders) will become more and more demanding;

iv. On **duty to take reasonable management decisions**: considering that directors have a wide scope of discretion during the selection of decisions that they deem more convenient or appropriate to pursuing the company's object, the introduction of the stakeholders' interests constitutes a limitation on directors pursuing shareholders' interests beyond what is reasonable or ethically admissible. ESG stimulates directors to make decisions and perform socially responsible acts, in short, to promote CSR,[75] especially those that have a positive impact or do not drag on value creation. This should therefore be analysed in a context of a judicial review of whether a management decision is reasonable when compared with the excluded alternative decisions (more or less risky) which could have been adopted by a careful and orderly director.

The ESG factors should therefore be considered by directors, who are expected to pursue credible, long-term business strategies and adopt rational and reasonable decisions.

---

[74] John W. White, Matthew Morreale and Michael Arnold, *Responding... cit.*, p. 3.

[75] Also, Coutinho de Abreu, "Deveres de cuidado e de lealdade dos administradores e interesse social", in *Reformas do Código das Sociedades*, Almedina, 2007, pp. 46–47, Carneiro da Frada, "A Business Judgement Rule... cit., pp. 216–217), and Catarina Serra, "The new Company Law... cit., p. 168.

5 Conclusions—The Business Judgement Rule as a Safeguard for ESG Minded Directors and a Warning for Others

The BJR prevents courts from stepping into the shoes of directors and reviewing the merits of management decisions. In the past, one type of criticism was raised by commentators who thought that the BJR overshadowed the duty of care to the extent that it no longer existed, and that there was a good reason to change corporate governance rules in order to bring back a deterrent for directors in their process of business decision-making.[76]

The BJR balances conflicts between directors' discretionary powers and judicial review in the context of the duty of care. Meanwhile, the rules on ESG matters balance the conflicts between shareholders' and stakeholders' interests. They do this by introducing into the legal framework additional standards of conduct to be observed by the directors in their work, in particular in risk analysis, organisation of governance models and adoption of reasonable management decisions. Although the law still provides for and legitimates shareholder capitalism,[77] ESG rules require directors to protect other constituencies. This confirms that, when faced with alternative decisions, directors may sometimes take a decision that also pursues the stakeholder's interests (even if they are not fully aligned with the shareholders' interests). However, and despite not being fully aligned with the shareholders, they will still be pursuing the company's best interests, such as investing in capital expenditures and in employee development and innovation.

The BJR will apply in case of a breach of duties provided for in the law, the articles of association or agreements, in the context of management decisions by directors of Portuguese companies. Consequently, courts will be prevented from analysing the merits and opportunity of a management decision when, in preparing the decision, directors have complied with the instrumental duties of care set out in article 72(2) of PCC. This means that directors can manage companies in accordance with some standards of "stakeholder capitalism", as long as they comply with the minimum standards of the duty of care above described, as they are protected by the BJR. By requiring a responsible decision-making process for application, the BJR is the safeguard of directors' discretionary management decisions. As such, management errors will only be relevant if the decisions were not (i) preceded by adequate information (ii) rational, and (iii) free from any personal interest.

However, if there is a lack of evidence that the instrumental duties of adequate preparation and rational decision (or evidence that the director acted free of personal interest), the courts should therefore review the director's

---

[76] Dr Yoram Danziger and Omri Rachum-Twaig, "Re-evaluating the Justifications for the Existence of an Independent Duty of Care", in *The Company Lawyer*, no. 35, Issue 9, Thomson Reuters (Professional), p. 267, available at www.ssrn.com.

[77] Caetano Nunes, *O Dever de Gestão... cit.*, p. 491.

management decision. Considering the wide scope of directors' discretion, and for all the reasons set out above, the court should limit the review to compliance with the minimum standard of the duty of care established by law. In other words, it should assess whether the decision was reasonable. In this context, the rising importance of ESG factors introduces additional standards that will have a significant impact on claims brought by shareholders and stakeholders against directors (e.g. judicial removal from office or accountability actions). Despite the supremacy of shareholders' interests, directors' decisions must not result in an intolerable, unnecessary, or disproportionate sacrifice of any of the interests of stakeholders, in favour of the shareholders' interests.[78]

This analysis will be subject to a proportionality method,[79] based on the general principle of proportionality existing in Portuguese law, in particular, subject to adequacy and necessity tests, as well as to a benefits-*versus*-costs analysis. Therefore, directors must act in the company's best interest, considering the critical role of the shareholders, but also taking seriously the idea that companies are independent entities serving multiple purposes. The fact that shareholders do not agree with the directors' decisions no longer necessarily means that directors have breached its duties, notably the duty of care.

The less ESG minded directors must consider their situation carefully. A director that is currently ignoring ESG factors is, depending on the circumstances, most probably not complying with the instrumental duties of care set out in article 72(2) of PCC. Is such a director taking an adequately informed decision when ignoring the ESG advancements? Has she made a sufficient effort to consider the best alternatives in the *leges artis* of management? The risk here is that directors will not be afforded the BJR's protection in case they are not considering ESG progress, and this is a clear warning sign for them.

Notwithstanding, bearing in mind the lack of enforcement by the courts, and with the majority of judgements only discussing a potential breach by directors of the duty of loyalty, it is expected that future amendments to law will not have very practical results. The Portuguese experience regarding BJR is a paradigmatic example to be considered by the EU legislature, when drafting EU legislation regarding the directors' duties, since it indicates that effective consideration by directors of the stakeholders' interests and following ESG principles will depend, in particular, on market (and tax) incentives, self-regulation and investors' incentives. In fact, the current incentives are still for directors to act in their shareholders' interests, on which their future (re-appointment) as directors relies. Sometimes, it is already difficult for directors to serve different "masters", as the company's shareholders may have totally opposing profiles—for example, long-term investor v. hedge fund.

---

[78] Catarina Serra, "The new Company Law… cit., p. 168.

[79] Karl Larenz, "Metodologia da Ciência do Direito", translated by José Lamego, 6th edition, Gulbenkian, 2012 (1991), p. 586 and André Figueiredo, "O princípio da proporcionalidade e a sua expansão para o Direito Privado", in *Estudos Comemorativos dos 10 anos da Faculdade de Direito da Universidade Nova de Lisboa*, vol. II, Almedina, p. 2008, pp. 25 e ss.

In the end, the degree of success of ESG will depend on shareholders selecting directors based on their profile (for example, directors who take ESG matters seriously), and replacing them when they do not meet the shareholders' expectations.

CHAPTER 15

ESG and Executive Remuneration

Inês Serrano de Matos

1 Introduction

In this chapter we will seek to assess how executive remuneration is an essential mechanism to align the interests of executive directors with the promotion of ESG factors and the corporate purpose. In view of the complex and extensive body of legislation governing remuneration, namely in Europe—which differs depending on the type of company in question— our analysis will focus on the alignment potential of the variable remuneration of the executive directors of banks.[1]

---

[1] New legislation (and still evolving) regarding remuneration matters in the context of banks derives from several sources, with particular emphasis on: (i) Reg. (EU) 2019/2088 of 27 November 2019 on sustainability-related disclosures in the financial services sector (SFDR), (ii) Reg. (EU) 2020/852 of 18 June 2020 on the establishment of a framework to facilitate sustainable investment, and amending SFDR (Taxonomy) and (iii) Directive (EU) 2019/878 (CRD V) of 20 May 2019 amending Directive 2013/36/EU (CRD IV), as regards several matters, including remuneration.

I. S. de Matos (✉)
Faculty Law, Universidade de Coimbra, Coimbra, Portugal
e-mail: ines.serrano.matos@fd.uc.pt

© The Author(s), under exclusive license to Springer Nature Switzerland AG 2022
P. Câmara and F. Morais (eds.), *The Palgrave Handbook of ESG and Corporate Governance*, https://doi.org/10.1007/978-3-030-99468-6_15

Among the various reasons that led us to choose banks as the focus of this chapter,[2] one stands out: due to their activity, banks are in a privileged position to carry out the aforementioned promotion and, in doing so, to act as a catalyst of ESG adoption throughout the economy enormous impact.

2 Banking as a Promoter of ESG Factors

It is in the banks' interest, as leading figures in the financial market and in economic development, to promote best practice, mitigate potential risks and provide sustainable policies that do not neglect the future. The news that have come to light on the incorporation of ESG factors in banks show some progress although the trend is still to privilege financial metrics; in fact, in October 2020, it was reported that banks "did not shine in terms of sustainability", although a progressive trend was noted.[3]

The same conclusion can be drawn from the analysis of the public governance reports of some major banks. Although the bank's intent is to take those factors into account in their business culture in terms of principles, in practice—i.e., in the strategic priorities and corporate governance—there is not yet a widespread understanding of how to implement them (i.e., objectives, deadlines, concrete metrics). It is in this context of growing importance of ESG factors in banking that it becomes essential to relate executive directors' financial incentives (and indeed other banks' employees) to ESG adoption and performance.[4] Without such integration of ESG factors into

---

[2] For developments on what motivates the differences between the corporate governance of banks and other companies, see for all, Klaus J. Hopt, "Corporate Governance of Banks and Other Financial Institutions After the Financial Crisis: Regulation in the Light of Empiry and Theory", *in: Journal of Corporate Law Studies*, Volume 13, Part 2, 2013, [219–253], available at: ssrn.com.

In recent years we have witnessed the trend to turn the analysis of *corporate governance* to the reality of banks, essentially due to a wave of financial scandals and their aftermath at the beginning of this century. However, the year 2020 also marked a change in the way the community in general views banks, forcing their recognition as an ally in the pandemic crisis: their influence, the circle of potentially reachable subjects and the extremely relevant credit intermediation activity, place them in a unique position (i.e. the widely discussed legal moratorium on loans is a good example). Moreover, banking is especially relevant in this study due to the prominent role it plays in financial markets and the fact that it is increasingly alert to the need to incorporate ESG factors into its core business; For example, in Portugal, many banks are now signatories of the *Principles for Responsible Investment* and of the 2019 *Carta de Compromisso para o Financiamento Sustentável em Portugal* (Commitment Letter for Sustainable Financing in Portugal), which for some banks represents a deepening of sustainability strategies already in place.

[3] In this regard see the news story "Bancos brilham pouco na sustentabilidade", available at: jornaldenegocios.pt.

[4] Although currently only one Portuguese bank issues shares admitted to trading on a regulated market (*Millenium BCP*) and therefore this approach does not pay particular attention to *soft law*, we should stress that there are several recommendations aimed at implementing practices and policies aimed at long-term, sustainability and remuneration

executive compensation arrangements, ESG will never go beyond a set of good principles without much practical application.

The bank's proactive attitudes towards sustainability are embodied, in particular, in the financing decisions[5] and in the type of products and services that it offers. Those decisions and services are grounded both in business strategy and in risk management decisions, that should take into account ESG risks.[6]

The pursuit of socio-environmental and governance best practices by banks occurs on two levels: internal and external.[7]

At an internal level, a call for the adoption of the best environmental, social and governance practices is evident as we witness the promotion of one or some of these factors internally (even though this may also have external consequences). For example, a bank may promote important social practices internally allowing equal access to employment for people with disabilities. For another example, *banks may also champion* good governance practices, becoming more reliable and less prone to corruption, for example in the composition of the board of directors, favouring the independence and diversity of its members and promoting the structuring of remuneration policies fairly aligned with the consideration of ESG factors.

as a way to promote them—cfr. recommendations IV.3., V.2.7. and V.2.8. of IPGC's Corporate Governance Code of 2018 (revised in 2020).

[5] This particular emphasis is related to banks' credit intermediation and capital-raising activities; they are notably active in attracting deposits from customers and analysing the targets for those amounts, particularly by granting loans. Today, non-financial values—in their environmental, social and governance dimensions—have also weighed on banks' actions and decision-making processes. By channelling private and corporate savings to the destination that best ensures the promotion of those values, banks are taking part in their implementation. Moreover, given their de facto stronger position—as they are the capital providers par excellence—they are also in a position to make demands to ensure that the financing they provide promotes those factors—cfr. "Environmental, Social and Governance Integration for Banks: A Guide to Starting Implementation", 2014, available at: wwfint.awsassets.panda.org and «Banca e Seguros, Ambiente e Sociedade – Desafiar Mentalidades, Definir Novas Oportunidades de Negócio (Guia para a inclusão dos riscos ambientais e sociais na concessão de crédito dos Bancos Portugueses)», 2007, available at: sustentare.pt.

The Portuguese example of Caixa Geral de Depósitos is elucidative of a good practice, as this bank has adopted the so-called "CGD Principles of Sector Exclusion and Limitation", according to which the institution restricts or excludes from credit policy activities or projects that may negatively impact sustainability; the 2018 list is available at: cgd.pt.

[6] The Portuguese Securities Market Commission (CMVM) published the «Risk Outlook» for 2021, in which the three most significant risks were highlighted, among which the ESG and the impact of negative externalities resulting from the disregard of these factors – cfr. «Risk Outlook | 2021» by CMVM, 2021, available at: cmvm.pt, especially pp. 12 et seq. and 76 et seq.

[7] Despite the interdependence of these levels, when we refer to the internal and external levels, we aim to distinguish the measures adopted by the bank, the impact of which is more internal (on its organizational structure, for instance), or more external (on stakeholders such as customers, the community at large, etc.). This distinction is without prejudice to the evident interdependence between them.

From the examples briefly listed we can state that the actions carried out are benefiting *stakeholders'* interests (even the *community at large*, especially when environmental protection is at stake). The shareholders themselves may also benefit from this action, by obtaining a higher financial return: share price may increase as a result of the action of investors who specifically value the integration of ESG factors in the bank's culture and policies.

It may also happen that, jointly or separately, the bank (also) wishes to promote the triad of factors in an external plan, going beyond the *corporate universe* aiming for the *general welfare* of the community, which brings us back to an external approach, namely to sustainable financing,[8] a theme sufficiently wide-ranging to encompass several actions that are increasingly popular, namely (i) the placement of securities such as *green bonds*[9]—allocating the capital to the financing of certain environmental purposes—or *social bonds*—allocating the capital to the financing of projects with a social impact, such as fighting poverty; (ii) making the remuneration of deposits dependent on the performance of social or environmental criteria by companies; or (iii) the promotion of lines of credit for companies created by the unemployed to pursue their own business.

These are some of the possibilities for action in the field of sustainable finance, where the satisfaction of interests that may also go beyond the *confines of the company* and meet the *common good* are at stake.

However, one should not be naïve enough to think that the bank may become a philanthropic institution: it may be that the reasons behind this behavioural change and awareness-raising have selfish motivations, such as the need to reduce risks associated with environmental, social and governance reasons that will certainly impact the banking activity (on its reputational component) or as a strategy for economic recovery, all contributing, after all, to the leverage of numbers and the bank's financial targets.

---

[8] For further developments on sustainable financing, see, for all, *Sustainable Finance in Europe—Corporate Governance, Financial Stability and Financial Markets* (edited by Danny Busch / Guido Ferrarini / Seraina Grünewald), EBI Studies in Banking and Capital Markets Law, Palgrave Macmillan, 2021 and Rui Pereira Dias / Mafalda de Sá, «Deveres dos Administradores e Sustentabilidade», *in: Administração e Governação das Sociedades*, Coimbra: Almedina, 2020, [33–85], pp. 68 et seq.

[9] See the Proposal for a Regulation of the European Parliament and of the Council on European green bonds of 6th of July 2021 available at: europa.eu. The Proposal aims to establish the regulatory framework applicable to that means of financing and establishes rules concerning (i) European green bond issuers and (ii) external reviewers. This Proposal is anchored to the Taxonomy Regulation.

On *green bonds*, also see Mafalda Miranda Barbosa, «Green Bonds: Riscos e Responsabilidade», *in: Revista de Direito da Responsabilidade*, Year 1, 2019, [834–861], pp. 838 et seq. and Stephen Park, "Green Bonds and Beyond: Debt Financing as a Sustainability Driver", *in: Cambridge Handbook of Corporate Law, Corporate Governance and Sustainability*, Chapter 42, Beate Sjåfjell / Christopher M. Bruner (eds.), 2019, available at: ssrn.com.

3 THE RENEWED INTEREST IN THE ISSUE OF REMUNERATION

The centrality of the remuneration issue in the context of *corporate governance*[10] is justified for both positive and negative reasons. On the one hand, remuneration policies have been perceived as compromising sound and effective risk management by encouraging excessive risk-taking in order to pursue short-term interests, all of which may have led to the collapse of financial institutions in 2007/2008; on the other hand, and traditionally seen as good corporate governance, attention has been turned to management remuneration as a positive way of aligning the interests of directors with those of the shareholders.[11]

It so happens that the renewed interest in the remuneration issue is now particularly related to the increasingly decisive weight of the ESG criteria (and the necessary articulation between both[12]), with the global pandemic crisis that ravages the planet and that gave rise to the (current) economic, financial and social crisis (which highlighted the relevance of such criteria)[13] and, lastly, with the very redefinition of the corporate purpose.[14] These three aspects

---

[10] Among other Authors, Diane K. Denis identifies executive compensation as a controversial issue of the last 20 years—"Twenty-Five Years of Corporate Governance Research ... and Counting", *in: Review of Financial Economics* 10, [191–212], p. 201; in the same vein, see Lucian Arye Bebchuk / Jesse M. Fried, "Executive Compensation as an Agency Problem", *in: Journal of Economic Perspectives*, vol. 17, no. 3, 2003, [71–92], p. 71.

[11] Despite these different perceptions, risk taking is not independent of the alignment of management with shareholder's interests: the first is intertwined with the second, since risk taking is of the essence of successful business performance, thus contributing to maximize results (for the shareholders).

[12] See for all, Stefania Sylos Labini / Antonia Patrizia Iannuzzi / Elisabetta D'Apolito, "'Responsible' Remuneration Policies in Banks: A Review of Best Practices in Europe", *in: Socially Responsible Investments—The Crossroads Between Institutional and Retail Investors*, Chapter 2, M. La Torre / H. Chiappini (eds), Palgrave Studies in Impact Finance. Palgrave Pivot, Cham, 2019, [5–36].

[13] On the impact of the ongoing pandemic crisis on remuneration matters and on the distribution of dividends, see Paulo Câmara, «COVID-19, administração e governação das sociedades», *in: Administração e Governação das Sociedades*, Coimbra: Almedina, 2020, [15–32], pp. 26 et seq.; European Banking Authority, "Statement on Dividends Distribution, Share Buybacks and Variable Remuneration", 31.03.2020, available at: eba.europa.eu; Banco de Portugal, «Comunicado do Banco de Portugal sobre recomendação de não distribuição de dividendos», 01.04.2020, available at: bportugal.pt.

[14] Although we will (briefly) return to the topic of *corporate purpose* in this chapter, we should stress the inexistence of a definition for it; we consider as sound the one according to which "companies should be governed according to the purpose (corporate purpose) of caring for all those who are affected by their activity. In this scenario, the corporate purpose overcomes the antithetical logic (*stakeholders vs shareholders*) and becomes a central concept in corporate governance"—see Paulo Câmara, «COVID-19, administração...», cit., p. 31. This definition is in line with the *Davos Manifesto 2020*, according to which "[t]he purpose of a company is to engage all its stakeholders in shared and sustained value creation. In creating such value, a company serves not only its shareholders, but all its stakeholders

have a common denominator: a greater permeability of the society to *external interests*.

Even before the current crisis, the *three pillars of sustainability*—as well as the very issue of sustainability and corporate social responsibility—were ascending into the media spotlight, and Covid-19 (among many other non-pathological effects) contributed to make the remuneration of directors a salient issue to a number of actors, among which institutional investors stand out.[15] In fact, a catastrophe such as this pandemic, so unpredictable and

– employees, customers, suppliers, local communities and society at large", available at: weforum.org.

About the redefinition of corporate purpose, see Guido Ferrarini, "Redefining Corporate Purpose: Sustainability as a Game Changer", *in*: *Sustainable Finance in Europe*..., cit., pp. 85 et seq.

[15] Due to the extensive documentation and non-financial reporting duties imposed on most *large* companies—provided for in the SFDR and Taxonomy Regulations (in force and, to date, already in effect, save for the exceptions provided for), in the *Final Report on draft Regulatory Technical Standards* (JC 2021 03, 02.02.2021), in Directive 2014/95/EU of the European Parliament and of the Council of 22 October 2014, which is at an advanced stage of revision—abundant and formal information reaches the market to which investors resort as (dis)investment criteria.

Both at national and European level, the reporting of non-financial information within the scope of sustainability has been subject to legislative amendments and regulatory contributions. Without dwelling on this topic, we note, based on Articles 66-B and 508-G of the CSC, that these rules are nothing more than good intentions; in fact, paragraph 3 of both provisions is revealing: if the company does not apply ESG policies, the non-financial statement should only provide a (clear and reasoned) explanation, referring to *soft law* and the respective *comply or explain*. But if the director is aware—because the remuneration policy so determines—that an incentive (remuneration) depends on the practical implementation of these values, this becomes an enticement. This is because, disregarding the highly subjective ethical and moral factors, there are two ways to guarantee, with a higher degree of probability, the fulfilment of objectives: through the institute of responsibility or through an increase in remuneration. It is more likely that a company linking the implementation of social, environmental and governance objectives with the remuneration of directors will effectively seek to achieve them.

The information provided will help in the sorting task, separating sustainable societies from those that are not, always bearing in mind the *greenwashing* phenomenon, as "the practice of gaining an unfair competitive advantage by marketing a financial product as environmentally friendly, when in fact basic environmental standards have not been met"—cfr. recital (11) of the aforementioned Taxonomy Regulation.

See also, a study, by Julie Segal, that associates hedge funds to the detection of companies that practice greenwashing—cfr., from the said Author, «Activist Hedge Funds Can Smell Greenwashing, Study Finds», Institutional Investor, 2020, available at: institutionalinvestor.com.

Regarding ESG reporting, we cannot ignore the relevant role of rating agencies; for an overview of their performance, cfr. Bcsd Portugal – Conselho Empresarial para o Desenvolvimento Sustentável, «Guia para apoiar as empresas a reportar os indicadores ESG (Environmental, Social and Governance)», available at: sustainablefinance.pt, especially pp. 17 et seq. However, these agencies are the target of criticism/hesitation as demonstrated by some studies "that there is a lack of a commonality in the definition of ESG (i) characteristics, (ii) attributes and (iii) standards in defining E, S and G components. We provide evidence that heterogeneity in rating criteria can lead agencies to have opposite opinions on the same evaluated companies and that agreement across those providers is substantially low"—cfr. Monica Billio / Michele Costola / Iva Hristova /

lethal, allowed to focus on non-financial values and to understand that there is a myriad of interests that go far beyond the creation of value for shareholders and that can greatly contribute to the adaptability of companies at various levels. Although it may be a common sensation—not least because climate change, the depletion of resources, etc., are not recent events—there is nothing like experiencing a global and highly restrictive situation to redefine horizons and priorities: what is the point of paying directors well and ensuring the distribution of good dividends when the company has no concern for the future and our planet is being destroyed with each passing day? What is the use of a company generating extraordinary profits if, when faced with an overwhelming situation, it does not have the capacity to react and, as a result, lives and jobs are lost?

In view of the above, remuneration represents, on the one hand, an attractive topic for investors who are at the central stage to oversee its fairness and structure due to the economic consequences of the pandemic—do the current circumstances imply a moral conscience in the sense of reducing or renouncing the variable component of remuneration?; on the other hand, it is important to note how remuneration is placed at the service of an end goal: that of more effectively ensuring the pursuit of non-financial values—as it is not possible to require from all directors equal ethical and moral principles and concern for their surroundings, can the increase in remuneration represent a strong incentive to act in line with the adoption of the triad of factors?

The aim is not to assess the value of the remuneration—checking whether it is excessive or not—but to determine whether it is suitable to fulfil its purpose of aligning the directors' interests with the pursuit of environmental, social and governance objectives and the corporate purpose.

Taking a bank as a reference point, due to its activity and the set of rules and recommendations, both prudential and behavioural, that govern it, we cannot expect it to *reinvent* itself as so many other non-financial companies do. In particular, we expect it to have a purpose: to take an interest in its surroundings, which goes far beyond the maximisation of results for its shareholders (without prejudice to recognising the importance of financial return); to take an interest in the markets to which it directs its financing; in selling particular types of products, in adopting certain types of policies and favouring markets that promote ESG factors. This necessity to attend interests that go beyond the *bank's walls* meets the mentioned redefinition of the corporate

---

Carmelo Latino / Loriana Pelizzon, «Inside the ESG Ratings: (Dis)agreement and Performance», University Ca' Foscari of Venice, Dept. of Economics Research Paper Series No. 17/WP/2020, 2020, [1-39], available at: ssrn.com. On the other hand, "a deep analysis of the criteria also shows that ESG rating agencies do not fully integrate sustainability principles into the corporate sustainability assessment process"—cfr. Elena Escrig-Olmedo / María Ángeles Fernández-Izquierdo / Idoya Ferrero-Ferrero / Juana María Rivera-Lirio / María Jesús Muñoz-Torre, "Rating the Raters: Evaluating how ESG Rating Agencies Integrate Sustainability Principles", *in: Sustainability*, 11(3), 2019, [1-16], available at: mdpi.com.

purpose. This subject gained renewed impetus in August 2019 at the *Business Roundtable*,[16] where the perspective of conducting corporate governance for the benefit of all *stakeholders* and not only *shareholders*[17] was defended, an idea that is particularly fitting, and has already been advocated, in relation to banking management.[18] The possible antithetical relationship that has been opposing the interests of shareholders to those of *stakeholders* is less and less plausible, a reality that is evidenced by the current period in which the assumed conviction is that in a company everyone counts, from shareholders to the community in general.

The management body is particularly in the sights of the markets and investors, and high expectations are placed on the performance of this body as a promoter and implementer of sustainability[19]; in this chapter we seek

---

[16] See Business Roundtable, "Statement on the Purpose of a Corporation", available at: system.businessroundtable.org. This declaration differs from those issued since 1997 in that "[e]ach version of the document issued since 1997 has endorsed principles of shareholder primacy – that corporations exist principally to serve shareholders. [...] the new Statement supersedes previous statements and outlines a modern standard for corporate responsibility"—see "Business Roundtable Redefines the Purpose of a Corporation to Promote 'An Economy That Serves All Americans'", available at: businessroundtable.org.

On the reaction of investors to the redefinition of the corporate purpose, see Karen Firestone, "How Investors Have Reacted to the Business Roundtable Statement", *in: Harvard Business Review*, 2019, available at: hbr.org and Ken Bertsch, "Council of Institutional Investors Responds to Business Roundtable Statement on Corporate Purpose", Council of Institutional Investors, 2019, available at: cii.org.

While not surprised, we note that the issue of adopting a remuneration policy as promoting corporate purpose and with the ESG matter was not discussed, as CEOs do not wish to address a *sensitive* issue such as their own remuneration.

[17] See the study by Ira T. Kay / Chris Brindisi / Blaine Martin / Soren Meischeid / Gagan Singh, "The Stakeholder Model and ESG—Assessing Readiness and Design Implications for Executive Incentive Metrics—A Conceptual Approach", 2020, available at: paygovernance.com. The reflections of this study date from September 2020 and stem from the Business Roundtable; one of the conclusions reached by its AAs is "(...) companies may look to include some stakeholder metrics in their compensation programs to emphasize these priorities. As companies and Compensation Committees discuss stakeholder and ESG-focused incentive metrics, each organization must consider its unique industry environment, business model, and cultural context".

[18] In fact, as the doctrine rightly states "instead of an orientation towards *shareholder value* it has been understood that bank management should be oriented towards *stakeholders*, which is more – and somewhat different – from merely taking into account the interests of these subjects, in a logic of *enlighted shareholder value*. That is, management should be done not only (and not primarily) in the interests of the shareholder but of the *debtholders*"—cfr. Ana Perestrelo de Oliveira, «Governo dos bancos públicos: autonomia de gestão e limites da influência do Estado», *in: RDS* IX, n.º 4, 2017, [743–761], p. 759.

[19] Ana Perestrelo de Oliveira studies the fundamental role and responsibility of the management body of banking institutions—cfr. of the referred Author, «Governo dos bancos públicos...», cit., pp. 752 a 755.

Sustainability should be one of the focuses of the management body and in this reflection, in particular, it is sought that this focus, to achieve good levels of sustainability, focuses on the adoption of remuneration policies in line with the implementation of ESG factors and the establishment of the respective metrics; all the more evident in the pandemic crisis we are going through—cfr., namely, Gigi Dawe / Carola van Lamoen /

to assess the ability to adopt a remuneration policy that aligns the interests of directors with the implementation of ESG factors and the pursuit of the corporate purpose, i.e., with the conduct of corporate governance also for the benefit of all *stakeholders*, focusing on sustainability and the long term. The remuneration, as emphasized, has been understood as an enticement to achieve this purpose and, therefore, a relevant mechanism to achieve sustainability.

Therefore, in the expectation that certain types of companies, due to their size or activity—as is the case of banks—, incorporate environmental, social and good governance factors in their culture, strategy and risk definition, it is increasingly common to set goals/objectives—particularly of a non-financial nature—which, if achieved, determine an increase in the remuneration of executive directors. By indexing the attribution of increased financial benefits to their performance in the attainment of measures that safeguard those criteria or by decreasing them in the case of exposing the bank to excessive risks (for example, through *malus* clauses), the bank seeks to guarantee that the executive director aligns his/her interests (especially the economic ones) with the attainment of a *corporate purpose*.

Remuneration, particularly in its variable component, may work as a way to implement the set of principles and practices—which are often nothing more than high-sounding phrases in documents with public visibility, such as sustainability reports—aimed at safeguarding social, environmental, governance and stakeholder interests in general. The reinforcement of these policies is often ensured by an internal committee of the bank, namely a Corporate Governance Committee/Sustainability Committee which, in articulation with the management body, carries out the promotion of social and environmental sustainability. Moreover, these principles should be at the basis of the respective remuneration policy and be effectively reflected in the remuneration of executives thereby incentivizing the achievement of sustainability goals and more sustainable business decisions.

George Dallas, "COVID-19 and Executive Remuneration", *in: International Corporate Governance Network*, available at: icgn.org and Robert G. Eccles / Mary Johnstone-Louis / Colin Mayer / Judith C. Stroehle, "The Board's Role in Sustainability", *in: Harvard Business Review*, 2020, available at: pbs.up.pt.
Some legal scholars even discuss the possibility of the Directors being responsible for *sustainability duties*—see Rui Pereira Dias / Mafalda de Sá, «Deveres dos Administradores e Sustentabilidade», cit., pp. 79 et seq.
The prominent role of the board of directors is also the result of a study carried out by ERNST & YOUNG (EY) in July 2020, which concluded that the European Commission needs to legislate in order to find *corporate governance* mechanisms to be implemented in companies to ensure sustainability. In this sense, according to the auditing firm, the role of management should be strengthened at several levels, including the promotion of mechanisms in the scope of remuneration with a view to contributing to greater sustainability – cfr. «Study on directors' duties and sustainable corporate Governance (Final Report)», conducted by EY for the European Commission, available at: op.europa.eu.

4 THINKING AHEAD: REMUNERATION POLICY AND GOOD PRACTICES ON ESG

The Remuneration Policy: A Strong Incentive?

The remuneration policy is a document intended for multi-annual cycles, establishing the periodicity, structure and payment conditions of the remuneration due to the members of the corporate bodies and other employees; it represents the "statement of the remuneration benefits objectives and clarification of their structure, particularly in their variable component and the relation between it and the management performance, without, however, interfering in the concrete establishment of the remuneration benefit".[20],[21]

The criteria for setting the variable component of remuneration should be based on the *sustainable and risk-adjusted performance of the credit institution*, as well as on the *extraordinary performance* of the executive directors' duties. In order for social and environmental factors to be reflected in executive directors' remuneration, a prior step must be taken; it is decisive for each company to define which factors it wishes to invest in and how it wishes to do so—from the outset, whether it wishes to focus on sustainability on an internal level and/or on the bank's activity (external level)—, by incorporating them in the company's culture and defining strategic and risk policies accordingly.[22]

After such reflection, and once the company has understood *if* and *how* it wishes to implement these factors and avoid the risks arising thereafter, it must find performance indicators and metrics that are appropriate for its business and the purposes they are intended to safeguard; the management body must ensure that the individual performance assessment process—including criteria of a financial and non-financial, quantitative and qualitative natures—is in line with the remuneration policy.[23]

In short, it is necessary for the company to find metrics and indicators, financial and non-financial, individual and corporate, based on well-defined criteria, which include also qualitative information, according to which the executive director's performance is assessed and, consequently, remunerated.

While remuneration policy seeks to align interests and encourage behavioural inducement, in the case of banks this task may prove to be difficult

---

[20] See Paulo Câmara, «A comissão de remunerações», *in*: *RDS*, Ano III, no. 1, Coimbra: Almedina, 2011, [9–52], p. 37.

[21] According to Article 5 of the above-mentioned SFDR Regulation, information on how these policies correspond to the integration of sustainability risks should be included in banks' remuneration policy, with a view to ensuring transparency and information.

[22] A public consultation launched by the European Commission on how sustainability issues should be integrated into the *corporate governance* of companies was open until 8 February 2021, a consultation which was mainly based on the above-mentioned EY study.

[23] See, in this sense, Article 42 of BdP Notice no. 3/2020; on this *new* Notice, see the comments of Luís Costa Ferreira/ Benedita Magalhães da Cunha, «Breve comentário sobre o Aviso do Banco de Portugal n.º 3/2020 em matéria de conduta e cultura organizacional e sistemas de governo e controlo interno», *in*: *RDS* XI, 3–4, 2019, [737–749].

for a number of reasons: (i) there are many more (and more sensitive) interests at stake; (ii) the awareness of the sustainability issue is recent; (iii) financial targets are (still) the most revealing metric to chalk out the remuneration policy and, due to the combination of these obstacles, (iv) the identification of non-financial metrics in the performance objectives proves to be complex.[24]

Remuneration Committee Role
The soundness of the solution of placing remuneration at the service of the pursuit of environmental, social and governance values also implies placing the weight of this mission on the remuneration committee, which has the crucial role of designing remuneration policies[25] and carrying out a benchmark analysis.

The remuneration committee is a relevant governance structure[26] that should be set up in *significant credit institutions*, composed of non-executive directors or members of the supervisory body, with a majority of independent members.

The committee—and if it does not exist, the management body—submits the remuneration policy for members of the management and supervisory bodies to the annual general meeting for approval. This duty of appraisal of the remuneration policy evidences the *say on pay*[27] which represents the ideal *moment* for shareholders to have a *say* on the composition of remuneration, increasingly aware of the need for remuneration to influence sustainability. Shareholders will be taking sides in the discussion on linking executive remuneration to the achievement of sustainable goals, representing a vote in favour of encouraging best practices. However, and despite the undeniable relevance of sustainability, this is also a decisive moment for shareholders in disagreement with the remuneration policy for example, because of a perception that the preponderance attributed to the rights and interests of stakeholders is endangering their own) to vote against it, being ideal to identify the aspects that

---

[24] Complexity still remains a factor despite the fact that the necessity to take into account qualitative criteria and non-financial indicators in the assessment of the management performance (in addition to the financial ones) is not a recent narrative.

[25] On the two fundamental vectors that should guide the definition of an adequate remuneration structure and composition in relation to the board of directors—the sound and prudent management of risks and the management of conflict of interests—reference is made to Diogo Costa Gonçalves, «A remuneração dos administradores das instituições de crédito: o comité de remunerações», in: *A Governação de Bancos nos Sistemas Jurídicos Lusófonos*, Coimbra: Almedina, 2016, [225–248], pp. 226 et seq. and our own, «O Governo dos Bancos: reflexões em torno da remuneração dos administradores executivos», in: *Direito das Sociedades em Revista*, Year 11, vol. 22, 2019, [181–216], pp. 193 et seq.

[26] The *committee* is provided for in Article 115-H of the RGICSF (corresponding to Article 131 of the CAB's Draft) and in Article 45 of the BdP Notice no. 3/2020.

[27] On this relevant principle, see Paulo Câmara, « 'Say on Pay': O Dever de Apreciação da Política Remuneratória pela Assembleia Geral», in: *Revista de Concorrência e Regulação*, Year 1, n.º 2, 2010, Almedina, [321–344] and Ana Perestrelo de Oliveira, *Manual de Governo das Sociedades*, Coimbra: Almedina, 2017, pp. 206 et seq.

motivated their non-approval. In order to envisage remuneration as a mechanism at the service of sustainability, everyone is responsible to contribute, especially the board of directors (executive and non-executive) and the general meeting.

The Variable Component of Remuneration, the Consideration of ESG Factors, Key Performance Indicators (KPIs) and Key Purpose Indicators
The consideration of ESG factors is still at an embryonic stage and in a recommendatory and voluntary domain, whereby disregarding these social and environmental factors does not entail immediate[28] consequences from the point of view of the director's liability,[29] nor does it lead to asset loss—or rather, a *non-gain*—, as financial targets and metrics still prevail as the preferred criteria in the performance assessment and definition of the remuneration structure. It so happens that, for some years now, other (non-financial) variables have started to weigh on the remuneration package of directors and to dictate a possible equity incentive if these are achieved; it is true that it still represents—in general and for those banks that have already taken *the step*—a low percentage of the variable component (especially reflected in the long-term component), which nonetheless shows some level of commitment.[30]

In any event, encouraging news have recently come to light about banks that see the current year as a milestone for change regarding the link between the remuneration of their executive directors and sustainability goals. The doctrine has already highlighted the need for each company to find its own key purpose indicators based on the corporate purpose—which implies its definition a priori—which, in turn and increasingly, should be linked to the achievement of results in environmental, social and governance matters. The consideration of key purpose indicators does not invalidate the consideration

---

[28] The consequences may not be immediate, but there are risks to be expected in the near future and which have already been advanced by the CMVM; see note 6 above.

[29] In fact, we have already had the opportunity to examine, in a study regarding the *controlling creditor*, that the disregard of the interests foreseen in Article 64(1) second part of the CSC does not entail the responsibility of the director—see Inês Serrano de Matos, "Debt Governance – O papel do credor activista", *Direito das Sociedades em Revista*, Year 7, vol. 14, 2015, [161–198], pp.189 and 190 as well as the references to reference doctrine identified therein.

[30] See, by way of example, the 2019 Annual Report and Accounts of the Millennium BCP bank, which explains that "[the] calculation of the amount of the AVR is based on the results of the performance assessment for the AVR assessment period in question, and results from the sum of two autonomous and independent components:

- 80% of the amount derives from the assessment of the degree of achievement of quantitative objectives (corporate KPIs);

- 20% of the amount derives from the performance assessment of each director in relation to the qualitative objectives" – see ind.millenniumbcp.pt, p. 57.

of classic KPIs; in fact, the inadequacy of the trade-off between the choice of the best ESG practices and the focus on financial performance and shareholder value creation is increasingly evident. In a simplistic way, we would say that key purpose indicators are KPIs that consider in their respective performance non-financial objectives and sustainability criteria that privilege the creation of long-term value.[31]

The combination of fixed remuneration (which tends to be attractive[32]) with a variable component may be a suitable mean of aligning the directors' interests with consideration for ESG factors and, more broadly, the corporate purpose, all of which imply the pursuit of a sustainable, long-term interest. However, in order to achieve the intended effect, the design of remuneration packages must be adjusted to the desired aligner purpose. It is not enough to *pay a lot and well*, it is necessary to design remuneration structures adjusted and in accordance with well-defined criteria. In fact, it is of no use setting up remuneration policies which encourage a short-term vision (*short-termism*) and which exacerbate the impassioned search for maximum and immediate benefits (reflected, in particular, in the variable component of remuneration), contributing only to excessive risk taking and a mindset of indifference towards the future of the society.[33]

The variable component of remuneration has therefore been perceived as a way of aligning interests[34] given the widespread conviction that *money* is a good incentive for the pursuit of objectives, notwithstanding the fact that

---

[31] On some difficulties perceived in relation to those non-financial factors, cfr. Alexander Bassen / Ana Maria Masha Kovacs, "Environmental, Social and Governance Key Performance Indicators from a Capital Market Perspective", *in: Zeitschrift für Wirtschafts- und Unternehmensethik*, no. 9/2, 2008, [182–192], available at: ssrn.com—"[t]hese factors are however of a qualitative nature and therefore difficult to express in numerical figures. Consequently, disclosure and the relevancy thereof to investors are problematic".

[32] On the inexistence of a correlation between company performance and the amounts paid to directors, see Nuno Fernandes, "Board Compensation and Firm Performance: The Role of 'Independent' Board Members", *in: ECGI—Finance*, Working Paper No. 104/2005, 2005, available at: ssrn.com; on the link between higher remuneration and the assumption of greater risk, see Miguel Ferreira, «Política de remuneração e risco», *in: A Governação de Bancos nos Sistemas Jurídicos Lusófonos*, Coimbra: Almedina, 2016, [211–224], p. 218 and the Authors identified therein.

[33] Legal scholars have been warning that "remuneration schemes based on variable components do not guarantee the intended goals and may constitute a perverse incentive for directors to adopt short-term strategies, distort results and manipulate the price of shares in order to increase the value of options, so as to obtain a higher remuneration, but endangering the sustainability of the company and the long-term return on shareholders' investments"—see Ana Raquel Frada, «A Remuneração dos Administradores das Sociedades Anónimas – Tutela Preventiva e Medidas *ex post*», *in: Questões de Tutela de Credores e de Sócios das Sociedades Comerciais* (coordination by Maria de Fátima Ribeiro), Coimbra: Almedina, 2013, [321–360], p. 328.

[34] See Paulo Câmara, «COVID-19, administração e governação das sociedades», cit., p. 27. In this sense, the EBA's position in "Statement on dividends distribution, share buybacks and variable remuneration", cit.

the prestige of the position of director may also act as an incentive to good practices[35] as well as personal characteristics such as pride.

Once the principles, criteria and metrics that guide the attribution of a variable component have been defined—which may contemplate short and long-term incentives, always adequate to each company—, there are strategies as to the definition of the variable component that seek to encourage executives to practice conduct tending towards the sustainability of the company and the long term[36]:

i. a substantial part of the variable remuneration is *deferred*[37];
ii. the payment being made in *instruments* provided for in Article 94(1) (l) of the Directive 2013/36/EU (amended by CRD V),[38] while stipulating clauses that *limit the transferability of* the instruments; and

---

[35] Observing economic incentives as a good behaviour inducer, see Janice Koors / Pearl Meyer "Executive Compensation and ESG", available at: corpgov.law.harvard.edu.

Despite globally seeing the variable component as a desirable feature of remuneration policy design, Miguel Ferreira prospects three difficulties, namely: (i) "[the ease with which] managers can influence the results of the Key Performance Indicators (KPIs) by which they are being assessed"; (ii) "differences in the profiles of managers and shareholders due not only to information asymmetries but also to the fact that managers tend to have more experience, which enables them to perceive risks much better, it is possible that their incentives remain misaligned with those of shareholders, even if the remuneration of both is based on the same indicator"; and (iii) "the decision on the KPIs to be associated with the variable component of remuneration is highly complex"—see of the mentioned Author, «Política de remuneração e risco», cit., pp. 216 and 217.

[36] For a study that seeks to answer the following questions: "(a) To analyse the adoption of these metrics by the most important European banks through the elaboration of an "ad hoc" governance score; (b) To verify the qualitative and quantitative diversification of these non-financial (or sustainability) indicators; and (c) To identify and examine some best practices adopted by European banks"; see Stefania Sylos Labini / Antonia Patrizia Iannuzzi / Elisabetta D'Apolito, "'Responsible' Remuneration Policies in Banks...", cit., in particular pp. 27 et seq.

[37] The CRD V, already imbued with a *spirit of sustainability*, finds in the deferral one of the main changes—extending the minimum deferral period and imposing a higher limit for members of the board of directors and senior management of significant institutions—, without prejudice to the principle of proportionality also being a relevant feature of the Directive by restricting the application of the rules on deferral to *small companies* and members of the corporate bodies whose variable remuneration does not exceed 50 thousand euros and does not represent more than ¼ of the total remuneration.

[38] Regarding the specific instruments for the remuneration policy, see Miguel Ferreira's study on: (a) the allocation of shares; (b) the allocation of options; (c) the allocation of preferential shares subject to non-transferability; and (d) the alternative instruments: debt and risk measures; the author identifies the advantages and how they promote an alignment of interests and, on the other hand, to what extent one can foresee disadvantages to those instruments– see of the mentioned Author, «Política de remuneração e risco», cit., pp. 218-223.

iii. subjecting the variable component to *"ex post adjustment mechanisms"*[39]—specifically, reduction («malus»)[40] and reversion («clawback»)[41]—being the bank's responsibility to define the exact criteria of its application, taking into special consideration the behaviour of the *employee*.[42]

In this context, the identification of Key Performance Indicators for ESG (also referred to as Key purpose indicators or Non-financial Key Performance Indicators) is essential; this involves the establishment of non-financial KPIs, underpinned by non-financial/qualitative criteria, such as "deployment of renewable energy or waste" (in the context of the environmental factor), "diversity or credit loans, undergone ESG screening" (in the context of the social factor). Based on these criteria, the bank can determine KPIs which, taking into account the examples provided, can be chalked out as follows: "% of energy in kwh from renewable energy sources as of total energy consumed/ % of energy in kwh from combined heat and power generation as of total energy consumed"; "waste by unit produced/ % of waste recycled"; "percentage of female employees as of total/ percentage of female managers as of total"; "percentage of credit loans undergone ESG screening".[43]

5 Prospective Reflections: The Difficulties Ahead

Having reached this point and despite the undeniable importance that sustainability and ESG factors have gradually been assuming, we cannot end this

---

[39] "These mechanisms correspond to posterior (*ex post*) adjustment mechanisms, based on risk, through which the institution has the possibility to adjust the remuneration of its Identified Employees, depending on the materialisation of risks arising from the performance of these same employees"—Benedita Magalhães da Cunha, «Políticas e práticas remuneratórias nas instituições de crédito», *in*: *Direito dos Valores Mobiliários II*, AAFDL Editora, 2018, [248–285], p. 280.

[40] Cfr. Article 115-E(10) paragraph (a) of the RGICSF, corresponding to Article 181(3) paragraph (c) of the CAB Draft.

[41] Under Article 115-E(10) paragraph (b) of the RGICSF, corresponding to Article 181(3) paragraph (d) of the CAB Draft. Thus, "if the reversion mechanism is triggered, the employee is obliged to return to the institution the amounts that he/she has already received"—see Benedita Magalhães da Cunha, «Políticas e práticas remuneratórias...», cit., p. 281.

[42] On the mechanisms of reduction and reversal see, among others, Tiago Guerreiro, «A efectividade das cláusulas de redução e reversão de remunerações em instituições de crédito», Working Paper no. 3/2016, Governance Lab, available at: governancelab.org.

[43] All advanced examples can be found at DVFA Society of Investment Professionals in Germany / EFFAS European Federation of Financial Analysts Societies, "Key Performance Indicators for Environmental, Social & Governance Issues Financial Analysis and Corporate Valuation. Version 1.2", available at: ec.europa.eu, version 3.0 can already be accessed at: dvfa.de.

reflection without highlighting the difficulties ahead—at least the most immediate and foreseeable—in the task of conceiving the remuneration of executive directors as a means capable of aligning interests and, therefore, of operating as a corporate governance mechanism placed at the service of sustainability. These are presented in three levels: (i) that of the alignment of interests and the possibility of extending directors' duties; (ii) that of sustainability and ESG factors; and (iii) that of remuneration.

In what concerns the *alignment of interests*, the interests looming at the core of a bank's activity are many. There are several categories of stakeholders and, within the same category, there may be dissonant interests, which may result in several conflicts of interests. It should be noted that the line of approach that has been taken—in face of a *broad social interest* understanding, which does not make the shareholders' interest prevail, but encompasses the interest of the several interested parties, shareholders and other stakeholders—makes the task even more difficult. Such difficulties are associated with the idea of the essentiality of the compatibility of several interests at stake. It is as if there is a *common interest*, superior to the sum of the interests of all interested parties. However, there is a need for a practical articulation of the coexistence of these purposes with the maximization of the creation of shareholder value.[44] The fear is the need for management to choose, which may fall into undesirable extremes: between the pursuit of shareholders' interests, for which they are primarily accountable[45] and the excessive attention to the interests of stakeholders, neglecting the rights and interests of shareholders.

The difficulty is, therefore, to harmonize interests, which may not be simple, since the administration will have to identify the stakeholders, their interests, and the associated risks and opportunities. It follows that a new set of duties related to the decision-making process and compatibility of interests is emerging in the administration sphere, alongside the need to define the corporate purpose, set targets in ESG matters, adopt a remuneration policy that relates those factors with the performance of executives and assess the achievement of targets. All of this adds to another hesitation: the possible direct accountability of the administration for the breach of those duties and, if so, in what way.

---

[44] For a positive view on this topic see the following examples, Daniela Salvioni / Francesca Gennari, "CSR, Sustainable Value Creation and Shareholder Relations", Symphonya. Emerging Issues in Management (symphonya.unimib.it), 1, 36–49, 2017, available at: ssrn.com and Robert G. Eccles / Ioannis Ioannou / George Serafeim, "The Impact of Corporate Sustainability on Organizational Processes and Performance", Working Paper 17,950, 2012, available at: nber.org.

[45] The concerns in this respect are not new and we have already had the opportunity to share them when analysing the controlling creditor which, moreover, warrants the addition of clauses to safeguard its credit position—see Inês Serrano de Matos, "Debt Governance – O papel do credor activista", cit., pp. 168 et seq.

More generally, in what concerns sustainability and ESG factors there remain several challenges which will continue in the next few years, especially related to (a) the novelty of the issue of sustainability and ESG factors, (b) the denialism as to its relevance within businesses, often arguing that the concerns with the environmental or social dimension should not be the responsibility of the private sector, and they contribute to a slowdown in business growth and (c) difficulties in changing the business strategy to a more sustainable one.[46] Additionally, (d) the *large* and public interest companies are increasingly subject to strong documentation and reporting duties in the field of sustainability, since an informative approach is intrinsically associated with these matters. However, if, on one hand, this information is welcomed as it seeks to ensure that the pursuit of the triple value does not become empty words in the reports and to fully inform the market in general, on the other hand, these duties represent high costs for the company that are borne by shareholders—to which must be added those that could result from the implementation of a different remuneration policy—, who will only bear them if they expect an increase in earnings for themselves. Therefore, the trend will be to seek to achieve financial metrics or to set low percentages for the variable component related to the qualitative criteria to be attributed to the executive directors.

Finally, the last difficulty is related to *the* remuneration itself[47]; not only with the establishment of the metrics and key purpose indicators that lead to its attribution—explaining how executives are remunerated for their performance in favour of achieving the three objectives—, but also the composition of the remuneration package, in order to affirm it as a potential aligner of interests.

Despite these difficulties, relevant steps have been reported in the field of banking with a view to considering these three pillars of sustainability, in order to contribute to an environmentally sound, socially inclusive and well governed world. Therefore—and we have already studied this issue elsewhere[48]—, it is widely acknowledged that the introduction of a variable component in the remuneration of executive directors may lead to good practices, but nothing better than the passage of time to assess the appropriateness of these considerations.

---

[46] In the same vein, see Filipe Manuel Morais, «Conselhos de Administração e Fatores ESG», available at: governancelab.org.

[47] Dirk A. Zetzsche / Linn Anker-Sørensen stress that "[A]lready in the absence of sustainability concerns, drafting sound remuneration schemes is a (legal) challenge. This challenge does not become easier with sustainability due to a lack of historical data, experience and expertise on all sides concerned, including the board of directors, executives and remuneration consultants"—for further developments, see, from the quoted Authors, "Regulating Sustainable Finance in the Dark" (August 25, 2021), *in*: *University of Luxembourg Law Working Paper Series 2021–007*, mainly pp. 37 et seq., available at: ssrn.com.

[48] See Inês Serrano Matos, «O Governo dos Bancos...», cit., pp. 186 et seq.

CHAPTER 16

How Can Compliance Steer Companies to Deliver on ESG Goals?

Lara Reis

1   Introduction

The grand challenge of our generation and of the ones to come is to stop the clock to the full consumption of our resources and the irreparable degradation of the world's social foundation.[1] We live in a fragile balance between meeting the needs of humanity and avoid pushing the planet beyond what scientists understand as the boundaries of the life supporting systems, and this requires equity—within and between countries—in the use of natural resources, as well as efficiency in transforming those resources to meet human needs.

---

[1] The concept of world's "social foundation" as a minimum for certain dimensions (water, food, health, education, income & work, peace & justice, political voice, social equity, housing, networks, and energy), below which the humanity is in deprivation, is explored in the "Doughnut Theory", created by K. Raworth. For more information about this see, "A Safe and Just Space for Humanity: Can We Live within the Doughnut?" (2012) Oxfam Discussion Papers.

L. Reis (✉)
Solicitor, Lisbon, Portugal
e-mail: Larareis81@gmail.com

The debate about sustainability[2] and the consciousness that we are dangerously heading to a precipice has decades, but the sense of urgency that we feel today results from the attention paid by all public and private actors to the topic.

The Paris Agreement[3] and the UN 2030 Agenda for Sustainable Development, including the Sustainable Development Goals ("SDG"),[4] marked the beginning of a new paradigm for the world economy and of a worldwide crusade for the protection of the planet. Building on the lead from governments around the world, the European Commission committed to the objective of making Europe carbon–neutral by 2050 in the European Green Deal[5] and upgrading Europe's social market economy in the Communication for a Strong Social Europe for Just Transitions.[6] The Circular Economy Action Plan,[7] the Biodiversity Strategy for 2030[8] and the Farm to Fork Strategy,[9] followed on in 2020, aiming at accelerating the transformative change required by the European Green Deal and foreseeing, amongst other deliverables, several legislative initiatives planned for 2021.[10] In what regards the finance front, the Commission Action Plan for the Financing of a Sustainable Growth[11] laid down the foundations and the EU strategy for sustainable finance at the EU level. One of the key principles behind

---

[2] The terms "sustainability", "environmental, social and governance" (ESG), "non-financial" or "corporate social responsibility" (CSR) are all used interchangeably in this chapter.

[3] See conclusions agreed in December 2015 by 195 countries at http://unfccc.int/paris_agreement/items/9485.php.

[4] Available at https://www.un.org/sustainabledevelopment/sustainable-development-goals/. The Sustainable Development Goals are part of the Resolution adopted by the General Assembly on 25 September 2015, available at https://www.un.org/ga/search/view_doc.asp?symbol=A/RES/70/1&Lang=E.

[5] COM(2019) 640 final, available at https://eur-lex.europa.eu/resource.html?uri=cellar:b828d165-1c22-11ea-8c1f-01aa75ed71a1.0002.02/DOC_1&format=PDF.

[6] COM(2020) 14 final, available at https://eur-lex.europa.eu/resource.html?uri=cellar:e8c76c67-37a0-11eaba6e-01aa75ed71a1.0003.02/DOC_1&format=PDF.

[7] See https://ec.europa.eu/environment/circular-economy/pdf/new_circular_economy_action_plan.pdf.

[8] COM(2020) 380 final, available at https://eur-lex.europa.eu/resource.html?uri=cellar:a3c806a6-9ab3-11ea-9d2d-01aa75ed71a1.0001.02/DOC_1&format=PDF.

[9] COM/2020/381 final, available at https://eur-lex.europa.eu/resource.html?uri=cellar:ea0f9f73-9ab2-11ea-9d2d-01aa75ed71a1.0001.02/DOC_1&format=PDF.

[10] The European Commission launched a consultation on sustainable corporate governance on 26 October 2020. The consultation is a preliminary step towards a future policy intervention, seeking feedback on potential changes to several aspects of corporate governance. The consultation closed on 8 February 2021 and a formal proposal is expected to be published in Q2 2021, but further detail on the direction of the new initiative may be included when the Commission publishes its Renewed Sustainable Finance Strategy (now expected to be in the first half of 2021).

[11] COM(2018) 97 final, available at https://eur-lex.europa.eu/legal-content/EN/TXT/PDF/?uri=CELEX:52018DC0097&from=EN.

these initiatives is to encourage companies to consider long-term objectives in their decision-making process, instead of focusing on short-term financial performance.

The proliferation of goals and principles issued by international organizations, the multiplication of laws enacted by supranational and national bodies; the innumerous initiatives lead by companies and the impact on individual consumer behaviour; evidence that we are on a path to integrated sustainability at all levels. Companies that did not want to be left behind the sustainability trend already jumped on the bandwagon, but the ambitious environmental, social and economic transformation that humankind needs, will require an impactful response.

This chapter aims to unveil the influence of corporate compliance vis-à-vis the laws that demand obedience and the remaining principled-based rules that make up the sustainability framework. We start by looking at the main features of the sustainability legal framework in Europe, including some paradigmatic laws at the national level, with a focus on their strengths and shortcomings from companies' perspective as law-takers. Subsequently, we discuss how companies respond to sustainability rules and to what extent the response is linked to their compliance maturity. In conclusion, we aim to respond to the instigating question of this chapter by unleashing possible compliance mechanisms and interventions that may help companies navigate the sustainability shifting waters.

2 VERTICAL COMPLIANCE

The Debate Between a Harder or Softer Stance in the Pursue of ESG Goals

The system of laws and regulations and soft-law instruments such as action plans, treaties, or declarations (for this purpose jointly referred to as "rules") is intertwined with the corporate world. Rules provide the legal concepts that enable, restrict and shape companies; provide the templates for interactions amongst them, and fora where the legal tools are employed. Simultaneously, companies are the field where the rules are interpreted, structured and embedded in society.

As discussed by Edelman and Suchman[12] the nature of the interconnectedness of rules and companies may vary depending on the harder or softer stance of such rules and be characterized in three fashions: in a *facilitative environment*, rules provide the tools and fora that companies employ to accomplish internal goals and hence this environment is more prone to litigation; a *regulatory environment* relies on a system of edicts as a means of society directly influencing corporate life; and a *constitutive environment* is

---

[12] L. Edelman and M. Suchman (1997). The Legal Environments of Organizations. *Annual Review of Sociology*, Vol. 23 (1997), pp. 479–515. Retrieved January 27, 2021, from http://www.jstor.org/stable/2952561.

based on definitional and conceptual rules that work as building blocks shaping the relationship between corporate actors.

The legal framework addressing ESG matters has different development speeds around the world and does not have a universal approach across the board.[13] There are examples of rules that follow a coercive model (*hard law*) and others fall under the normative model (*soft law*), additionally, in some instances, both models co-exist in regulating certain ESG topics. The pros and cons of each stance are assessed in the ongoing debate about the best approach to achieve sustainability purposes, as briefly described below.

The supporters of *soft-law* argue that this model has a higher potential for an "extraterritorial reach", due to the flexibility ensured by a framework based on principles and values. Those building blocks can then be interpreted and applied in light of the national context and absorbed by MNEs according to their standards. Naturally, *soft-law* instruments are limited in their impact due to the absence of legal effects, but they equally provide useful guidance and prepare the ground for *hard-law* to be developed.

Conversely, others point out that the voluntary adherence[14] that is the underlying principle of the normative model, may not achieve the desirable corporate response. Being left to the companies' discretion, the internalization of rules is driven by the preference between a structural or superficial response, and, in the case of the latter, the role of law in corporate life risks being marginalized as "window dressing".

Although offering, in theory, more credibility and mechanisms for control and enforcement, *hard-law* initiatives have also raised practical difficulties resulting from being fragmented internationally or, in certain aspects, incoherent or presenting gaps.[15] Critic voices of this model argue that compelling compliance under the threat of sanctions and penalties may have the perverse effect of incentivizing search for the legal loopholes and circumvention of rules.

There are also examples of legal initiatives that only go as far as establishing frameworks and goals, leaving to the companies the freedom to decide on how to pursue those goals, and there are cases of legal instruments that impose specific obligations but lack prescribed consequences for breach or non-compliance.

In the particular context of global supply chains, there is a notable shift from a voluntary approach to *hard-law* solutions, especially by the increasing

---

[13] Muzaffer Eroglu, "How to Achieve Sustainable Companies: Soft Law (Corporate Social Responsibility and Sustainable Investment) or Hard Law (Company Law)" (January 1, 2015). Kadir Has University Law School Journal, 2014, Available at https://ssrn.com/abstract=2736867.

[14] Voluntary compliance results from the conviction that the law is fair and it is fairly applied.

[15] Jingchen Zhao, "Extraterritorial Attempts at Addressing Challenges to Corporate Sustainability", in "*The Cambridge Handbook of Corporate Law, Corporate Governance and Sustainability*" [Edited by] Beate Sjåfjell, Christopher M. Bruner, pp. 29–42.

number of disclosure and due diligence requirements.[16] We now look into more detail to the compliance challenges arising from such reporting and due diligence legal initiatives.

Reporting Initiatives

The introduction of reporting obligations is a mechanism that ensures minimum intrusion in corporate life.[17] It focuses on external monitoring of corporate behaviour rather than directly imposing actions or omissions. The principle behind this approach is the duty of vigilance of the society in general and reliance on public disclosure as an incentive to behavioural change.

The disclosure regulations covering ESG matters seek a dual purpose: fostering leadership by example and allowing interested recipients of information to act accordingly.

Disclosing companies lead by example by sharing achievements, experiences, goals and lessons learned. They set standards and best practices that put their peers under pressure to catch up.

On the other hand, companies are mindful that the quality of the information made public is critical for investment decision-making by market participants and an important marketing tool to attract talent.[18] Additionally, financiers will check disclosure against internal criteria as a condition for financing. The public scrutiny of any misleading or inaccurate data may also catch the attention of regulators or authorities and result in unexpected inspections. This reflects a shift from the traditional reporting to serve shareholders' and markets' interests, to a primarily social orientation.[19]

The proliferation of information intermediaries[20] that gather data, rate and rank companies according to their ESG performance is a result of the growing attention given to performance, ethical conduct and culture of the integrity of companies.

---

[16] Charlotte Villiers, "Global Supply Chain and Sustainability: The Role of Disclosure and Due Diligence Regulation", in *"The Cambridge Handbook of Corporate Law, Corporate Governance and Sustainability"* [Edited by] Beate Sjåfjell, Christopher M. Bruner, p. 551.

[17] Iris H.Y. Chiu, "Disclosure Regulation and Sustainability: Legislation and Governance Implications", in *"The Cambridge Handbook of Corporate Law, Corporate Governance and Sustainability"* [Edited by] Beate Sjåfjell, Christopher M. Bruner, p. 521.

[18] Cassandra Walsh and Adam J. Sulkowski, "A Greener Company Makes for Happier Employees More so Than Does a More Valuable One: A Regression Analysis of Employee Satisfaction, Perceived Environmental Performance and Firm Financial Value" (December 10, 2009). Available at https://ssrn.com/abstract=1521745.

[19] Ibid, p. 527 and Iris Chiu, "Unpacking the Reforms in Europe and UK Relating to Mandatory Disclosure in Corporate Social Responsibility: Instituting a Hybrid Governance Model to Change Corporate Behaviour?" *European Company Law*, Vol. 14, no. 5 (2017), p. 208.

[20] Examples of these intermediaries may be KLD, Thomson Reuters ASSET4, FTSE4Good Index, Dow Jones Sustainability Index, CDP, and MSCI.

The reporting duties require companies to identify the ESG risks involved in their business activities and the external consequences of materialization of such risks. Based on the outcome of such assessment, they disclose publicly what they have done to contribute to the mitigation of such risks and how that has decelerated any harmful consequences. The public commitment to follow certain sustainability standards, enables stakeholders to monitor compliance and to ensure accountability for the outcome of set standards.

Notwithstanding the above, reporting can only serve the transparency purpose if the information reported is accurate, comprehensible to the public (and hence not extremely technical or complex) and comparable. Otherwise, it may just become a formalistic[21] exercise and serve merely as a propaganda tool. The comparability depends, however, on harmonized guidance concerning parameters, metrics, KPI and terminology. Otherwise, each entity will make its interpretation of the rules and be tempted to manipulate the information disclosed according to their needs.

An important landmark in national law imposing disclosure duties is the UK Modern Slavery Act 2015.[22] Under this law, UK companies that supply goods or services and have a total turnover above a certain threshold[23] must issue an annual statement to the public identifying the steps taken (if any) to prevent situations of human trafficking, slavery or abusive employment practices on their supply chains and any parts of their own business in the UK or abroad. The particularity of this law was the ambition to extend the scope of the transparency requirement to cross-border entities and activities, and how it was designed to have an international reach.

Interestingly, the UK Modern Slavery Act does not require the reported information to have one specific direction: even a report stating that the company is not taking any action seems sufficient to comply with the law. Equally, the validity, accurateness and truthfulness of affirmative statements describing steps (allegedly) taken are not required to be independently verified by a third party, which does little to incentivize real transparency. This has led many authors to consider that the UK Modern Slavery Act has a modest impact.

At the European Union level, the Non-Financial Reporting Directive[24] ("NFRD") is an example of a departure from the traditional shareholder-centric approach to reporting, based primarily on financial information performance. It requires large public-interest companies to disclose to the public

---

[21] Jingchen Zhao, "Extraterritorial Attempts at Addressing Challenges to Corporate Sustainability", in *"The Cambridge Handbook of Corporate Law, Corporate Governance and Sustainability"* [Edited by] Beate Sjåfjell, Christopher M. Bruner, p. 29.

[22] https://www.legislation.gov.uk/ukpga/2015/30/contents/enacted.

[23] Currently, this threshold is £36 million (https://www.gov.uk/guidance/publish-an-annual-modern-slavery-statement#who-needs-to-publish-a-statement).

[24] Directive 2014/95/EU of the European Parliament and of the Council of 22 October 2014 amending Directive 2013/34/EU as regards disclosure of non-financial and diversity information by certain large undertakings and groups.

certain information[25] on their action and impact on society and the environment. In the same spirit of other reporting legislation, the main goal of NFRD is to ensure investor and consumer trust and to allow comparability of non-financial information released by companies in the European Union. The directive provides that the Member States should have procedures in place to enforce compliance with such reporting obligations.[26]

The NFRD is not exempt from critique and stakeholders have recently expressed their views in a public consultation.[27] In particular, the problems identified by the consultation respondents were that the reporting involves a significant administrative burden and costs due to the level of uncertainty of what needs to be reported and the lack of guidance/harmonization of the reporting criteria and standards. The responses also flagged the difficulties in obtaining information from business partners; the fluctuation of users' needs and problems related to the lack of harmonization of the rules across the Member States. This explains the general perception of the modest impact of NFRD.[28] In this regard, the amendments to the NFRD that are expected for the first semester of 2021 have significant challenges to address.

The Shareholder Rights Directive II[29] ("SRD II") introduced new ESG disclosure requirements to institutional investors and asset managers given the important role they play in the corporate governance and stewardship of listed companies, indirectly affecting companies. By imposing the creation and disclosure of engagement policies, SRD II aims to enhance long-term shareholder engagement taking ESG matters into account, issuer-investor dialogue and transparency in the voting process for listed companies.[30]

---

[25] Non-binding guidelines were subsequently released with the aim of harmonizing disclosure. See https://ec.europa.eu/transparency/regdoc/rep/3/2017/PT/C-2017-4234-F1-PT-MAIN-PART-1.PDF and https://eur-lex.europa.eu/legal-content/PT/TXT/HTML/?uri=CELEX:52019XC0620(01)&from=EN.

[26] The NFRD was transposed in Portugal in 2017 (by the Decree-Law 89/2017, of 28 July 2017 that amended the Portuguese Companies Code (articles 65, 66-B, 508-G, 451, 528 and 546) and the Portuguese Securities Code (article 245-A)) and applies to accounts published by large companies since 2018. In the same year of 2017, Law 62/2017, of 1 August 2017 introduced in Portugal the obligation of publicly listed companies and state-owned companies to prepare and publish equality plans and to meet targets for the representation of both genders in corporate bodies.

[27] The summary report of the public consultation is available here https://ec.europa.eu/info/law/better-regulation/have-your-say/initiatives/12129-Revision-of-Non-Financial-Reporting-Directive/public-consultation.

[28] As per the same public consultation, 70% of users of company reports believe that companies fail to disclose relevant information. Around 74% have concerns about how reliable the information is and 84% find that reported information is not comparable between companies.

[29] Directive (EU) 2017/828 of the European Parliament and of the Council of 17 May 2017 amending Directive 2007/36/EC as regards the encouragement of long-term shareholder engagement.

[30] The SRD II was transposed in Portugal in 2020 (by Law 50/2020, of 25 August amending the Portuguese Securities Code (articles 85, 93, 222-A, 359, 390, 392, 394,

In conclusion, despite the yet modest achievements of sustainability reporting initiatives, they are key to the European sustainability strategy. The European Commission reinforced that it is a priority for 2021 to level the playing field of non-financial reporting with financial reporting. The European Union has an ambitious legal framework on sustainable finance, including the Taxonomy Regulation[31] and the Sustainable Finance Disclosure Regulation[32] and intends to put forward a proposal for new legislation on sustainable corporate governance.

Due Diligence Initiatives

Due Diligence is an alternative tool to reporting, that has potential for progress in different industries and has particular relevance in relation to companies integrated in supply chains. Instead of driving action indirectly by demanding public commitment as it happens with disclosure; due diligence initiatives prescribe an inquisitive approach by target companies and a consequential response set in motion to address adverse circumstances or risks discovered.

The carrying out of due diligence depends on the active engagement from various actors within and beyond the company's sphere: the employees and members of corporate bodies from the headquarters, subsidiaries, suppliers and other companies of the supply chain; but also the competitors, consumers, NGOs and other activist groups and, ultimately, the communities in general.

Due diligence may also be seen as complementary to reporting initiatives, by ensuring that the publicly disclosed facts are underpinned by corporate action to guarantee the reliability of data used. The fieldwork involved in due diligence shall be based on ongoing two-way communication between companies and their business partners, in particular on meetings, hearings, consultation proceedings and questionnaires, but also inspections and local visits.

From the companies' perspective, due diligence is an opportunity to put in place measures to mitigate or remedy adverse externalities discovered (e.g. hazard removal or disengagement from a business partner involved in unsustainable practices), or to become better informed to monitor existing risks (e.g. risks associated with a product line or subcomponent). The conclusions obtained from investigations shall drive business decision-making for a sustainable purpose (e.g. decision on sources of supply or disinvestment in a certain country).

397, and 400, and introduced new articles 22-A, 26-A to 26-F, 29-B to 29-E, 245-C, 249-A to 249-D, and 251-A to 251-E), and the Legal Framework of Credit Institutions and Financial Institutions (article 211).

[31] COM (2018) 353 final, available at https://eur-lex.europa.eu/legal-content/EN/TXT/PDF/?uri=CONSIL:ST_14970_2019_ADD_1_COR_1&from=EN.

[32] Regulation (EU) 2019/2088 of the European Parliament and of the Council of 27 November 2019 on sustainability-related disclosures in the financial services sector.

The scenarios above are particularly relevant in the context of supply chains however, banks are also important actors in the crusade against unsustainable business conduct. Corporate lending and security underwriting are subject to "know your customer" (KYC) and, when relevant, to "know your customer's client" (KYCC), which increasingly include ESG criteria checking. It is in the Banks' best interest to remain vigilant of their clients' conduct to avoid providing services to clients with activities directly or indirectly causing, contributing to, or linked to significant adverse ESG impacts.

The success of due diligence initiatives depends on standard-setting guidance[33] and on the collaboration and responsiveness from the persons involved. If contributions are not honest, plagued by a lack of good faith or hidden interests, the exercise will be far from comprehensive. Consequently, it is crucial for the success of due diligence initiatives that companies focus on building strong relationships and developing empathy and trust in their supply chains.

Enforcing compliance with the due diligence requests by application of sanctions might improve responsiveness from business partners. Additionally, it might be more effective to target the actors at the top of the chain with the most control, power and influence.[34] This is due to the fact that liability loses effectiveness as it spreads alongside the supply chain and because targeting the companies at the top, behind complex supply chains, will concentrate efforts and reduce enforcement difficulties across multiple jurisdictions.

The European Union has progressed in regulating due diligence in the sectors acknowledged as more vulnerable to ESG violations.[35] There are also

---

[33] The OCDE has issued due diligence guidance such as OECD, "*OECD Guidelines for Multinational Enterprises*", OECD Publishing, 2011, http://dx.doi.org/10.1787/9789264115415-en; OECD, "*OECD Due Diligence Guidance for Responsible Supply Chains of Minerals from Conflict-Affected and High-Risk Areas: Third Edition*", Paris: OECD Publishing, 2016, http://dx.doi.org/10.1787/9789264252479-en; OECD, "*OECD Due Diligence Guidance for Responsible Supply Chains in the Garment and Footwear Sector*", Paris: OECD Publishing, 2018, https://doi.org/10.1787/9789264290587-en; OECD "*OECD Due Diligence Guidance for Responsible Business Conduct*", 2018; and OECD, "*Due Diligence for Responsible Corporate Lending and Securities Underwriting: Key considerations for banks implementing the OECD Guidelines for Multinational Enterprises*", 2019.

[34] Charlotte Villiers, "Global Supply Chain and Sustainability: The Role of Disclosure and Due Diligence Regulation", in "*The Cambridge handbook of corporate law, corporate governance and sustainability*" [Edited by] Beate Sjåfjell, Christopher M. Bruner, p. 565.

[35] Said EU regulation includes the conflict minerals regulation (Regulation (EU) 2017/821 OF THE EUROPEAN PARLIAMENT AND OF THE COUNCIL of 17 May 2017 laying down the supply chain due diligence obligations of Union importers of minerals or metals), the timber regulation (Regulation (EU) No 995/2010 of the European Parliament and of the Council of 20 October 2010 laying down the obligations of operators who place timber and timber products on the market), and voluntary approaches with third countries such as the staff working document (2017) regarding sustainable garment value chains and the Trade for Decent Work Project.

examples of national due diligence laws and ongoing projects[36] that point out that progress in this field is in the right direction.

At a national level, the French Due Diligence Act[37] requires that companies of a certain size[38] headquartered in France perform risk assessments of their international business activities and consequently implement due diligence plans throughout their supply chains, to prevent violations of human rights, fundamental liberties, health and safety of individuals and damage to the environment.[39]

Another example is the Dutch Child Labour Due Diligence Law [Wet Zorgplicht Kinderarbeid][40] that applies to all companies that sell or supply goods or services to Dutch consumers, irrespective of their home country. Such companies must investigate if any goods or services they use are produced with child labour and in the case of "reasonable suspicion", the company shall implement an action plan to solve the issue.

Despite the promising ambitions of the due diligence initiatives, it is broadly accepted[41] that impactful outcomes can only be achieved with the adoption of a stronger stance. In this regard, the European Commission committed to introduce legislation on mandatory corporate human rights and environmental due diligence in 2021.

3 HORIZONTAL COMPLIANCE

In the previous section, we have seen the ambitions and the bounds of the sustainability legal efforts. We will now consider how companies internalize sustainability aims and what they can do to go beyond what public bodies can regulate. It is in the interdependency between the ESG rules and corporate behaviour,[42] that compliance influence emerges.

---

[36] See the list of national human rights and environmental due diligence initiatives (as of 3 July 2020) at https://www.business-humanrights.org/en/latest-news/national-regional-movements-for-mandatory-human-rights-environmental-due-diligence-in-europe/.

[37] Loi n° 2017-399 du 27 mars 2017 relative au devoir de vigilance des sociétés mères et des enterprises donneuses d'ordre ("French Due Diligence Act"), available at https://www.legifrance.gouv.fr/eli/loi/2017/3/27/2017-399/jo/texte.

[38] Article L. 225-102-4 du code de commerce.

[39] For an analysis of the limitations of this law, see Véronique Magnier, "Old-Fashioned yet Innovative", in *"The Cambridge Handbook of Corporate Law, Corporate Governance and Sustainability"* [Edited by] Beate Sjåfjell, Christopher M. Bruner.

[40] Available (in Dutch) at https://zoek.officielebekendmakingen.nl/stb-2019-401.html, is expected to become effective in mid-2022.

[41] See the European Commission "Study on Due Diligence Requirements Throughout the Supply Chain" (January 2020).

[42] Benjamin van Rooij and D. Daniel Sokol, "Compliance as the Interaction Between Rules and Behavior" (March 28, 2020). (Introduction to Cambridge Handbook of Compliance). In B. Van Rooij and D. Daniel Sokol (Eds.), *"Cambridge Handbook of Compliance"*, Cambridge, UK: Cambridge University Press, 2021 (Forthcoming), UC

Under the traditional compliance perspective, rules shape behaviour and are designed to determine the direction of corporate response and action, leaving no space for voluntary enterprise. In this context, violation or non-compliance with the law is enforced by punishment. The underlying assumption behind this approach is that companies will only employ time, resources and capital to meet the rules only when "specifically required to do so by law" and when "they believe that non-compliance is likely to be detected and harshly penalized".[43] However, the behavioural mechanisms triggered by punishment may not always be the desired ones such as the legitimate duty to obey the law, dissuasion of violation or compliance reassurance. On the contrary, punishment may instigate organizations to elaborate on ways to circumvent the law or lead them to the realization that punishment is less onerous than compliance. This promotes a form of creative compliance,[44] while meeting the letter of law and deliberately frustrating its spirit.

A growing number of scholars is emphasizing the importance of the dynamics in the opposite direction: how corporate's response can shape the meaning and functioning of the rules. This is particularly relevant in the context of the ESG legal framework that is still evolving and hence more permeable to the results obtained.

Rules that are vague, ambiguous, complex and highly technical, leave considerable room for interpretation and are influenced by instrumental, political or normative processes at the corporate level.[45] This theory supports the idea that achieving ESG goals is beyond compliance, meaning that rules establish a foundation of principles and values, but it is by discretionary self-regulation that companies embody the rules in the form of codes and policies. And when internalizing such rules, companies will only commit to plans that are best suited for their size, nature, context and ambition; leaving for the internal control functions the responsibility to monitor compliance with planned actions.

Irvine School of Law Research Paper No. 2020-29, Available at SSRN: https://ssrn.com/abstract=3563295.

[43] Neil A. Gunningham, Dorothy Thornton and Robert A. Kagan, "Motivating Management: Corporate Compliance in Environmental Protection", *Law & Policy* (2005), p. 290.

[44] D. McBarnet and C. Whelan, "The Elusive Spirit of the Law: Formalism and the Struggle for Legal Control", *The Modern Law Review*, Vol. 54, no. 6 (1991), pp. 848–873. Retrieved February 11, 2021, from http://www.jstor.org/stable/1096920.

[45] Shauhin Talesh and Jérôme Pélisse, "How Legal Intermediaries Facilitate or Inhibit Social Change", in "*Studies in Law, Politics, and Society*" [Edited by] Austin Sarat, Vol. 79 (2019), pp. 111 et seq.

In the context of self-regulation, compliance uses alternative tools to a punitive threat. The dichotomy of comply/not comply is replaced by a "business construction of law and compliance model",[46] where legitimacy of law[47] overlaps the power of organizations.

The Sustainability Self-Regulation Stages

Various studies[48] have examined that the process of self-regulated sustainability develops through different stages and identified the typical characteristics of each of the stages. Despite variations determined by context, sector, industry, disposition to embrace change and compliance maturity of the company, we can typically describe this process in three stages:

Compliance Stage
The sustainability journey starts with the acknowledgement that there is a need for change.[49] This might be motivated by an exogenous reason (if the trigger comes from the outside, such as a legal requirement, an incident[50] or media pressure) or an endogenous process (driven by a progressive management team or employee activism).

In this preliminary stage, the focus is on compliance with the law through tactical actions. Due to the embryonic state of things, the actions taken on this phase are simple, easy and supported by existing structures that are merely adapted to accommodate a new purpose.

Throughout this stage, a company shall gradually shift from an unconscious and reactive approach to an increasingly informed approach. The pace of this transition depends on the ability of the main actor leading the change internally (which at this stage might not be a formally appointed sustainability officer) to gain buy-in from the key opinion formers to adhere and collaborate. It is therefore critical at this stage to obtain support from stakeholders representing the company's interests across the board, as a guarantee for the fast pollination of the sustainability awareness. Additionally, it may be effective

---

[46] Lauren B. Edelman and Shauhin A. Talesh, "Chapter 5: To Comply or Not to Comply—That Isn't the Question: How Organizations Construct the Meaning of Compliance", in *"Explaining Compliance: Business Responses to Regulation"* [Edited by] Christine Parker and Vibeke Lehmann Nielsen (2011), p. 116.

[47] Law stipulates the right thing to do.

[48] For an overview of such studies see Kathleen Miller and George Serafeim, in *"Chief Sustainability Officers: Who Are They and What Do they Do"*, p. 19, available at http://ssrn.com/abstract=2411976.

[49] Christoph Lueneburger and Daniel Goleman, "The Change Leadership Sustainability Demands", in MIT Sloan Management Review (Summer 2010), available at https://sloanreview.mit.edu/article/the-change-leadership-sustainability-demands/.

[50] The Nike sweatshop incident in the 90s is an example of a scandal that triggered internal change, for more information see Andrea Newel, *"How Nike Embraced CSR and Went From Villain to Hero"* (June 2015) in Tripple Pundit.

to display the benefits of the process and gain the trust of stakeholders, which can be achieved by prioritizing straightforward measures that deliver results relatively fast.

Efficiency Stage
This stage is about translating vision into action.[51] When a company acquires more knowledge on sustainability and puts in place a sustainability plan, it moves from a reactive approach to an increasingly strategic and proactive behaviour. Consequently, more staff tends to become involved in sustainability projects with bottom line impacts. Even though still small-scale, the actions around the topic are expected to gain more visibility.

This stage normally involves increasingly complex initiatives, which are regularly monitored and fine-tuned. More staff across the organization become directly responsible for sustainability performance and embed sustainability principles in their day-to-day business activities. The company's sustainability evolves to become progressively idiosyncratic.

The focus on this phase goes beyond law response and integrates stakeholders' demands and expectations. ESG values and principles become gradually part of the DNA of the company.

Innovation Stage
At this stage, companies already look to convert the outcome of investment made in sustainability, into benefit and other advantages for them (e.g. reputation and brand recognition).

The proactive attitude that characterizes the former stage evolves to a transformational approach and the focus becomes leveraging opportunities into growth. Sustainability is viewed not only from the cost and risk perspective but also as investment and opportunity. Taking the examples of Danone[52] and Proctor & Gamble,[53] we note that these companies have redesigned or adapted their products to meet ESG standards and evidence demonstrates that such transformation opened up their businesses to new opportunities. Similarly, Dow Chemical[54] reacted to public outcry by committing to embed innovation into their long-term sustainability strategy. The three companies have in common a business model based on circular economy principles, and

---

[51] Christoph Lueneburger and Daniel Goleman, "The Change Leadership Sustainability Demands", in MIT Sloan Management Review (Summer 2010), available at https://sloanreview.mit.edu/article/the-change-leadership-sustainability-demands/.

[52] Asad Ghalib, Farhad Hossain and Thankom Arun, "Social Responsibility, Business Strategy and Development: The Case of Grameen-Danone Foods Limited", *Australasian Accounting, Business and Finance Journal*, Vol. 3, no. 4 (2009).

[53] Peter White, "Building a Sustainability Strategy into the Business". *Corporate Governance*, Vol. 9, no. 4 2009, pp. 386–394.

[54] Robert G. Eccles, George Serafeim, and Shelley Xin Li. "Dow Chemical: Innovating for Sustainability", Harvard Business School Case 112-064, January 2012 (Revised June 2013).

that led them to move beyond the goal of "not harming" to pursuing a positive impact on the environment and society.

At this stage of maturity, companies recognize that incorporating sustainability in the company's strategy and practice, leads to short and long-term value delivered to stakeholders, turning "green into gold".[55] Research[56] shows that a strong ESG proposition improves corporate performance at various levels: (i) customer loyalty, as there is evidence that "green" brands are more attractive to customers and better placed to enter new markets[57]; (ii) operational and resource efficiency, as for example, companies investing in the reduction of water and energy consumption, or investing in waste recycling, are likely to cut related costs in the long-term; (iii) employee relations, considering that social credibility attracts and retains talent, and that employee satisfaction avoids strikes and is directly related with better performance[58]; (iv) supplier relations, as risk mitigation reduces product recalls or boycotts, and (v) media coverage.

Shaping Compliance from Within

Being a collective of individuals, companies need a strategy to act as one. Compliance programs play an important role by disseminating the "tone from the top" and promoting the awareness and dialogue about sustainability issues, by building the pillars of the corporate culture and preventing unethical or illegal individual behaviour.

Strategic- and sustainability-driven compliance programs shall resort to different mechanisms such as incentive schemes (e.g. remuneration), cognitive and organizational processes, behavioural ethics nudging, leveraging capacity for compliance (e.g. legal knowledge and self-control), and reducing the opportunity for violation. The effectiveness of these interventions reflects the compliance maturity of the company. We will look in more detail at some sustainability-related initiatives below.

---

[55] D.C. Esty and A.S. Winston, "*Green to Gold: How Smart Companies Use Environmental Strategy to Innovate, Create Value, and Build Competitive Advantage*", New Haven: Yale University Press, 2006.

[56] For an analysis of research on the connection between ESG and financial performance, see Pollman, Elizabeth, "Corporate Social Responsibility, ESG, and Compliance" (November 2, 2019). Forthcoming, Cambridge Handbook of Compliance (D. Daniel Sokol and Benjamin van Rooij, eds.), Loyola Law School, Los Angeles Legal Studies Research Paper No. 2019-35, available at SSRN: https://ssrn.com/abstract=3479723 or http://dx.doi.org/10.2139/ssrn.3479723.

[57] Kronthal-Sacco, Randi and Whelan, Tensie and Van Holt, Tracy and Atz, Ulrich, "Sustainable Purchasing Patterns and Consumer Responsiveness to Sustainability Marketing" (August 1, 2019). NYU Stern School of Business, available at SSRN: https://ssrn.com/abstract=3465669 or http://dx.doi.org/10.2139/ssrn.3465669.

[58] A. Edmans, "The Link Between Job Satisfaction and Firm Value, With Implications for Corporate Social Responsibility", *Academy of Management Perspectives*, Vol. 26 (2012), pp. 1–19.

Board Accountability and Delegation of Powers on a Sustainability Committee

At the governance level, a company demonstrates commitment to the prosecution of sustainability goals by ensuring that, collectively, the management body has competencies and insightful coverage of economic, environmental and social topics.

Additionally, if proportional to the size and nature of the company, the creation of a sustainability committee shall push the topic to the top of the management's agenda. Such committee has the advantage of promoting an integrated approach by gathering people across departments and areas of expertise, however, it is crucial to have executive directors as members in order to ensure collaborative work and a full integration of this committee with the business lines.

This consultative body shall have the expertise to define an internal sustainability strategy and set in motion its implementation at the organizational level. Moreover, as the full board has the last word and is accountable for the sustainable conduct of the company, the members of the committee have a key role at board meetings: they enrich the collective thinking by asking insightful questions, making suggestions and constructive critics, by offering perspectives, raising counterpoints and proposing alternatives.[59]

Appointment of a Chief Sustainability Officer ("CSO")

Research[60] shows that the pace and manner in which companies navigate through the sustainability development stages discussed above, improves with the formal creation of a function dedicated to ESG matters: the Chief Sustainability Officer ("CSO"). The identification of the need to have one resource exclusively allocated to defining a strategy for sustainability, assessing the company's needs, researching external sources, communicating with stakeholders and providing internal training, is the first step to accelerate the sustainability progression.

The role and power of the CSO depend on which stage of the sustainable development is the company, i.e. the level of penetration of the principles and the commitment demonstrated by actual investment made and other actions and initiatives.

In the sustainability infancy of the institution, the CSO tends to concentrate all the decisions and actions related to sustainability because in the early stages, the CSO is the driving force of the sustainable development process. It is

---

[59] Lynn S. Paine, *"Sustainability in the Boardroom"* in Harvard Business Review (July–August 2014), about Nike's experience that has led to the creation of a corporate responsibility committee.

[60] Dina Gerdeman, *"What Do Chief Sustainability Officers Do?"* in Harvard Business School Working Knowledge, available at https://www.forbes.com/sites/hbsworkingknowledge/2014/10/08/what-do-chief-sustainabilty-officers-do/?sh=446f1c9633ab.

likely, though, that at this stage the CSO will be seating in the hierarchy on a more junior level. However, as the organization matures into sustainability, more internal people become active actors for sustainability development and the CSO, being legitimately empowered,[61] will progressively step back for a coordination role.

Incentives

Incentives may take the form of human resources practices including recruitment, promotion, training, performance evaluation, remuneration and recognition. According to a recent study[62] one useful tool to promote real change in sustainability matters is the setting of sustainability-related incentives to directors and top management.

From what companies report, it is apparent that there is still room for progress in regards to tying compensation to sustainability. Individual compensation scorecards shall progressively include ESG metrics and performance shall not be exclusively measured by contribution given to own division or unit but also to enterprise-wide sustainability performance. This will foster cross-division collaboration that is key to achieve innovation on sustainability matters.

On the other hand, the structure of remuneration shall not encourage excessive sustainability risk-taking and shall be linked to a risk-adjusted performance (e.g. clawback in case sustainability risks are not timely identified or not properly addressed). Finally, short-term-based rewards shall progressively reflect a long-term outlook.

Internal Policies and Codes of Conduct

An important area of compliance is the norm generation. Corporate norms and codes internalize laws and regulations but also create the foundations for organizational procedures and ethical culture. As discussed before, internal norms and codes have the ability to fill the gaps left by ambiguous and complex laws.

In developing countries, where governments have more difficulties in enforcing laws and regulations - for instance on labour conditions—companies may play an essential role by putting in place codes of conduct and taking the lead to address poor or hazardous working conditions in the factories linked to their global supply chain or by demanding compliance with certain standards by their business partners. Undeniably, codes of conduct have the potential to steer conduct on a cross-border basis, in supply and value chains. In such

---

[61] French, John & Raven, Bertram (1959), "*The Bases of Social Power*".

[62] Mazars, "Responsible Banking Practices: Benchmark Study 2020" (2020), available at https://www.mazars.com/content/download/1012584/52903051/version// file/Mazars_Responsible_Banking_Practices.pdf.

scenarios, compliance is more about monitoring private, voluntary codes of conduct.[63]

However, some companies might perversely adopt codes of conduct for the wrong reasons. It is the case when the driver behind the formalization of a code of conduct is publicity only, or when the main purpose is to deviate the attention from other issues and to be used as a shield from scrutiny from authorities or the public.

Notwithstanding, we can find examples of internal policies of companies that can drive real change. It is the case of Goldman Sachs[64] that committed to taking public only companies with at least one diverse member on the board, at a cost of an estimated loss of $100M in fees. The more sceptical may argue that one board member is not enough to give diversity a voice on boards, but it is undeniable that the media attention and debate around this unprecedented policy in the financial sector was a step forward in corporate action fostering diversity.

Behavioural Ethics Nudging

"Nudge"[65] has been defined as a subtle intervention that changes human behaviour without creating economic incentives or banning other possibilities. With the intervention of nudging techniques, choices are influenced by how options are presented[66] and the conditions under which individuals make decisions.

Behavioural ethics nudging may be applied in corporate compliance to the extent that it prevents employees from acting self-interestedly[67] and creates the conditions for them to voluntarily make better choices. Depending on the degree of intrusiveness, ethical nudges used by employers include some mechanisms such as, amongst others, pop-up messages, reminders, requests to fill in forms certifying facts, visual cues and organization of office spaces.

---

[63] Scott Killingsworth, "The Privatization of Compliance" (May 29, 2014). RAND Center for Corporate Ethics and Governance Symposium White Paper Series, Symposium on "Transforming Compliance: Emerging Paradigms for Boards, Management, Compliance Officers, and Government" (2014), available at https://ssrn.com/abstract=2443887.

[64] The public announcement by the Goldman Sachs CEO David Solomon was made from the World Economic Forum in Davos in January 2020 and is available at https://www.cnbc.com/video/2020/01/23/goldman-sachs-ceo-ipo-diversity-squawk-box-interview.html.

[65] The term was first used by Richard H. Thaler and Cass R. Sunstein, in "Nudge: Improving Decisions About Health, Wealth, and Happiness", 2009 Print.

[66] For example nudging by the use of "opt-out" instead of "opt-in" choices, as further detailed in Haoyang Yan and J. Frank Yates, "Improving Acceptability of Nudges: Learning from Attitudes Towards Opt-in and Opt-out Policies", *Judgment and Decision Making*, Vol. 14, no. 1 (January 2019), pp. 26–39.

[67] Todd Haugh, "Nudging Corporate Compliance" (July 17, 2017). 54 *American Business Law Journal* 683 (2017), Kelley School of Business Research Paper No. 17-54, pp. 687–688, available at https://ssrn.com/abstract=3004074.

The challenge for the use of this tool in the corporate context is the ability of companies to predict employee organizational behaviour (which can be based on data gathered by compliance departments, as for example history of past wrongdoings, surveys, conclusions from monitoring actions and interviews) and to use that information to nudge behaviour towards compliance with sustainability goals. The key question for that purpose being "What leads employees to breach rules?". Employees may rationalize unethical or non-compliant behaviour based on several reasons that are, to a large extent, motivated by their cultural background. Such reasons vary from "everyone else is doing it", "no one will be harmed by this" or "this is necessary to meet my performance goals". Based on the knowledge of these intrinsic drivers, nudging will aim to tackle the natural inclination of human nature to follow the easiest route to solve problems or attend exclusively to self-interests.

There is a record that "green nudging" is already being used in some countries by policymakers as a complement to the traditional policy instruments and a substitute for coercive measures and economic tools (e.g. fiscal incentives, subsidies, taxes or fees).[68] The advantages of the use of this mechanism can also be achieved in the corporate context.

Companies' Approach to ESG Disclosure and Reporting

The reporting regulations force companies to disclose internal policies and procedures, indirectly increasing the public attention over such internal procedures and processes. Public scrutiny on how sustainable conduct is integrated into day-to-day processes, decision-making and workflows, has the ability to reveal any omissions and flaws that companies will aim to address or anticipate. This public focus on internal procedures will lead to an internal effort to enrich organization of processes at all levels of the hierarchy, what has the potential for a real impact on corporate behaviour.

Notwithstanding the above, research[69] shows that companies are mostly disclosing policies, not outcomes. According to data from 2019, NFRD reports broadly disclose existing policies (80% of reports surveyed) but out of 35% reports disclosing targets on climate change, only 28% disclose their outcome. Moreover, only 8% of reports describe outcomes of data privacy and cybersecurity policies. In relation to the reports that cover human rights issues (57%), only 4% specifically refer the results from actions undertaken to

---

[68] Oksana Mont, Matthias Lehner & Eva Heiskanen. (2014). "Nudging. A Tool for Sustainable Behaviour?", Swedish Environmental Protection Agency, p. 69, available at https://www.researchgate.net/publication/271211332_Nudging_A_tool_for_sustainable_behaviour.

[69] The Alliance for Corporate Transparency, "*Research Report 2019: An Analysis of the Sustainability Reports of 1000 Companies Pursuant to the EU Non-Financial Reporting Directive*", available at https://www.allianceforcorporatetransparency.org/assets/2019_R esearch_Report%20_Alliance_for_Corporate_Transparency-7d9802a0c18c9f13017d6864 81bd2d6c6886fea6d9e9c7a5c3cfafea8a48b1c7.pdf.

address those issues. Finally, just 14% of companies report Board engagement, and only 15% report a link between sustainability objectives and executive remuneration.

Companies' Approach to ESG Due Diligence

Companies shall use their international footprint to encircle violations of environmental laws and of human or labour rights. Through the insertion of clauses in contracts with business partners that grant rights to regular monitoring and the imposition of codes of conduct, companies exert influence over their business partners (subsidiaries, subcontractors, suppliers, distributors, intermediaries). However, complex supply chains involving multiple ties with different tiers of suppliers, forming an intricate supplier pyramid, makes control and enforcement of such supplier contracts and codes, less effective beyond the tier-1 supplier level.[70]

Expanding influence over the supply chain downwards to the lower levels of the hierarchy will require from companies the implementation of a strategic plan and greater spending in enhanced management and control systems. Outsourcing audit works locally or to NGOs[71] is a strategic approach to ensure a physical presence where the higher risks reside and an independent assessment, but might represent higher costs and requires regular monitoring of the outsourced services. Another alternative that may be considered is the implementation of worker hotlines and other whistleblowing mechanisms that enhance the communication channels and proximity with local workers. Moreover, gaining trust and collaboration from local workers might be facilitated by an impactful scheme of incentives and trade-offs.

Supply chain pressure shall also work more effectively over smaller trading partners if they fear losing business or bad publicity. Conversely, companies at the top of the chain that are "locked-in" supplier relationships, because either switching costs are high or they do not have the scale to keep suppliers captive, have less bargaining power and consequently will be less likely to persuade their suppliers to follow their rules.

Even though ticking the box with the minimum standard of intervention may ensure compliance with due diligence legal requirements, the challenge of compliance with ESG rules is precisely that it requires from companies going the extra mile to achieve real change, as demonstrated above. Additionally, companies face the dilemma of "going green" and lose competitiveness, in that many sustainable initiatives will most likely raise the production costs,

---

[70] Galit Sarfaty, "Shining Light on Global Supply Chains" (October 20, 2014). *Harvard International Law Journal*, Vol. 56 (2015), p. 431, available at SSRN: https://ssrn.com/abstract=2512417.

[71] See Richard Locke, Fei Qin, and Alberto Brause, "Does Monitoring Improve Labor Standards?: Lessons from Nike", MIT Sloan Working Paper No. 4612-06, July 2006 for a discussion on how monitoring efforts shall be combined with other interventions focused on tackling the root causes of poor working conditions.

slow down or even block internationalization and business activities. Undoubtedly, if companies impose increased procedural burden to their suppliers and requirements to source quality goods and services, they shall not be able to ensure large-scale swift production at low costs and that will have an impact on their margins and, ultimately, profit. Therefore, the effectiveness of due diligence initiatives is intertwined with the profit-making behaviour of companies. This tends to perpetuate the pressure over suppliers to keep the production costs low, what is normally achievable at the cost of low wages of workers, child labour and inhumane working conditions. This exploitative vicious cycle is also related with environmental harm caused by pollution and overutilization of resources.

Risk Management

Companies are increasingly facing the somewhat unpredictable,[72] complex and severe consequences of emerging risks such as climate change, resource depletion, migration and income inequality, just to name a few. Integrating ESG factors in enterprise risk assessment will account for non-financial risks and will reduce company's exposure to liabilities (resulting from litigation or sanctions) and loss of value (which may result from negative press or knock-on effect on shares in the case of listed companies) resulting from unprecedented events.

As there is no definition of ESG-related risks, companies must determine their own, based on their industry, sector, geography, mission, internal and external environment and business model.[73] The activities conducted by companies cause *impacts* (either negative or positive) on local communities, employees, resources or the environment and have *dependencies* on resources (human, social or natural) that may vary in the short, medium or long term. The consideration of the interconnection between impact/dependencies in the context of the business model will help companies map the spectrum of ESG risks and opportunities and manage them accordingly.

In what regards financial institutions, ESG risks ultimately entail prudential risks (e.g. credit risk, market risk, operational risk) for institutions as ESG factors distressing their clients and counterparties consequently impact

---

[72] ESG risks were once considered "Black Swans", rare and unpredictable outlier events that may cause extreme impacts as first defined by NASSIM NICHOLAS TALEB, in "*Fooled by Randomness: The Hidden Role of Chance in Life and in the Markets*" (2001). These risks are more common today but still advance rapidly. They are commonly inherent to the nature of the services or products and object of particular interest from media or society in general what amplifies the reputational consequences of their materialisation.

[73] The Committee of Sponsoring Organizations of the Treadway Commission (COSO) and World Business Council for Sustainable Development (WBCSD), "*Enterprise Risk Management: Applying Enterprise Risk Management to Environmental, Social and Governance-Related Risks*" (2018).

their financial performance or solvency.[74] The European Central Bank (ECB) recently published a guide on climate-related and environmental risks for banks,[75] explaining how it expects institutions to prudently manage and disclose such risks.

Despite the said advantages of following holistic risk management and covering ESG elements, numbers show that companies have not taken it fully on board yet. According to the World Business Council for Sustainable Development (WBCSD),[76] only 29% of companies integrate ESG issues on their risk management frameworks.

4 Conclusion

The analysis above leads us back to this chapter's motivating question: How can Compliance steer companies to deliver on ESG goals?

With less than ten years left for the implementation of the 2030 Agenda for sustainable development, only 21%[77] of world-class CEOs are convinced that business is playing a critical role in contributing to the SDG. Is the problem a lack of incentives (positive and negative) for companies to genuinely seek compliance with the rules and promote sustainability?

Experience has demonstrated that ESG goals cannot be met in light of marketing and public relations' strategies but need input from compliance, risk management and business perspectives. It is not enough motivation for a company just to seek a reputation as a sustainable business as much as it is not to avoid involvement in scandals related to human/social or environmental breaches. With the advent of social media and the power that movements/hashtags have, pressure from civil society may force companies to adopt certain behaviours, but the drive to change needs to come from within.

In this chapter, I aimed at unveiling the fragilities of the existing rules that promote sustainability and understand the origin of the humble outcomes achieved thus far, when analysed from a compliance perspective. I have pointed out internal mechanisms and initiatives that may assist companies in progressing in the sustainability path. The conclusion is that compliance with sustainability rules cannot per se eradicate violations of human rights or environmental harm, as there are challenges to address that go beyond compliance

---

[74] See European Banking Authority, "EBA Discussion Paper on Management and Supervision of ESG Risks for Credit Institutions and Investment Firms", of 30 October 2020 (EBA/DP/2020/03), p. 28.

[75] European Central Bank, *Guide on Climate-Related and Environmental Risks: Supervisory Expectations Relating to Risk Management and Disclosure*", November 2020, available at https://www.bankingsupervision.europa.eu/ecb/pub/pdf/ssm.202011finalguideonclimate-relatedandenvironmentalrisks~58213f6564.en.pdf.

[76] WBCSD is a global, CEO-led organization of over 200 leading businesses working together to accelerate the transition to a sustainable world.

[77] See latest 2019 United Nations Global Compact—Accenture Strategy CEO Study on Sustainability, "The Decade to Deliver: A Call to Business Action".

influence. Notwithstanding, a strong compliance culture has the merit to place the companies, as the addressees of such rules, at the forefront of the global fight towards a fair and sustainable world. For this purpose, stepping up the efforts that companies put in adhering to the ESG rules is crucial for the conversion of legal commands into a lawful conduct.

PART III

ESG in Particular Types of Companies

CHAPTER 17

ESG and Listed Companies

Abel Sequeira Ferreira

1 Sustainability: From "Why Not?" to "Must Have"

Climate-related disasters and extreme-weather events increased by approximately 80% over the last twenty years claiming 1.23 million lives, affecting 4.2 billion people and resulting in approximately US$2.97 trillion in global economic losses, seriously disrupting the World's economy.[1]

Over the past year, the Covid-19 pandemic led to devastating loss of human life and caused disastrous economic and social disruption; the full impact of this crisis is yet to be seen and assessed, but it is clear that we are looking at catastrophic figures, with likely hundreds of thousands of companies' bankruptcies and foreclosures and many millions of workers losing their former livelihoods.

---

[1] "The Human Cost of Disasters: An Overview of the Last 20 Years (2000–2019)", United Nations Office for Disaster Risk Reduction, 2020 (https://www.undrr.org/publication/human-cost-disasters-2000-2019).
See also the "The Global Risks Report 2021", World Economic Forum, January 2021 (https://www.weforum.org/reports/the-global-risks-report-2021): four of the top five risks by likelihood are directly related to climate concerns.

A. S. Ferreira (✉)
Lisbon, Portugal
e-mail: Abel.ferreira@aem-portugal.com

© The Author(s), under exclusive license to Springer Nature Switzerland AG 2022
P. Câmara and F. Morais (eds.), *The Palgrave Handbook of ESG and Corporate Governance*, https://doi.org/10.1007/978-3-030-99468-6_17

We face unprecedented challenges of transformation and recovery for our economies, our societies, and our way of life, making climate action and the rebuilt of a sustainable global economy even more urgent.

These challenges are interrelated, because when we address the Covid-19 crisis or when we discuss how to fight climate change, we are in both cases effectively looking for answers to our shared future on the planet.

Ours is a time of resilience.

Resilience is at the heart of our fight against the pandemic and resilience is equally at the core of sustainability, in doing today what is crucial to ensure the ability of future generations to also resiliently succeed tomorrow.[2]

We now have enough evidence regarding the need to simultaneously account for both nature and the economy, acknowledging how economic activity affects nature and how nature's negative impact on the economy affects society and different groups of people in different ways.

A revolution, and for sure a generational shift, will be needed to cope with the severity of climate change and Covid-19 financial implications, and to change economic activities in a way that allows us to achieve prosperity while not damaging or destroying the environment in the process.

In recent years, a substantial increase in sustainable investments and attention to environmental, social and governance ["ESG"] factors by institutional investors and asset managers has already taken place, and it seems obvious that ESG will be increasingly important for investors and creditors when making capital allocation decisions in the future.[3]

That is why we now refer to "sustainable finance", generally described as "the process of considering environmental, social and governance factors when making investment decisions, leading to increased longer-term investments into sustainable economic activities and projects".[4]

---

[2] The most commonly adopted concept of sustainable development, since the publication of the "Report of the World Commission on Environment and Development: Our Common Future", United Nations, 1987, Gro Harlem Bruntland (coord.), is defined as "development that meets the needs of the present without compromising the ability of future generations to meet their own needs" (https://digitallibrary.un.org/record/139811).

[3] See the Edelman Trust Barometer Special Report: Institutional Investors (https://www.edelman.com/research/investor-trust).

[4] R. Boffo & R. Palatano (2020), "ESG Investing: Practices, Progress and Challenges", OECD Paris (www.oecd.org/finance/ESG-Investing-Practices-Progress-and-Challenges.pdf).

Reaching climate neutrality by 2050 will imply unprecedented modernisation and transformation, and an unprecedent amount of investment as far as $50 trillion being estimated to reach worldwide "net zero" emissions by 2050; see BluePaper from Morgan Stanley Research, 2021 (https://www.morganstanley.com/ideas/investing-in-decarbonization).

The cost of inaction will be higher; for a good synthesis on the role of environmentally sustainable investment is as enabler of economic growth, see "G20 Note on Environmentally Sustainable Investment for the Recovery" (https://www.imf.org/en/Publications/Policy-Papers/Issues/2021/04/29/G20-Note-On-Environmentally-Sustainable-Investment-For-The-Recovery-460112).

That is also why the EU recognises that sustainability and the transition to a safe, climate-neutral, climate-resilient, more resource-efficient and circular economy are crucial to ensuring the long-term competitiveness of the continent's economy.[5]

And finally, that is why, in a context of expanding scrutiny regarding companies' purpose, culture and values, attention to ESG and sustainability issues is of unescapable importance for boards of directors setting up its long-term business strategy; and failing to address ESG opportunities would be detrimental to their companies' financial stability and reputation.

Listed companies and boards of directors in Europe, specially, must be aware of an overabundance of new laws, requirements and initiatives from different organisations, governments, regulators, investors and other stakeholders, and must familiarise themselves with an increasing collection of legislation and regulation, at global and European level, such as the UN Principles for Responsible Investment,[6] the UN Sustainable Development Goals,[7] the OECD's Guidelines for Multinational Enterprises on Responsible Business Conduct,[8] the Recommendations of the Task Force on Climate-related Financial Disclosures,[9] the EU Sustainable Finance Action Plan,[10] the European Green Deal,[11] the EU Taxonomy of Sustainable Activities,[12] the

---

[5] Position (EU) 8/2020 with a view to the adoption of a Regulation of the European Parliament and of the Council on the establishment of a framework to facilitate sustainable investment, adopted on 15 April (https://eur-lex.europa.eu/legal-content/EN/TXT/PDF/?uri=CELEX:52020AG0008(01)&rid=8).

[6] Available at https://www.unpri.org/pri/what-are-the-principles-for-responsible-investment.

[7] The 17 Sustainable Development Goals ["SDG"] address some of the most pressing ESG challenges such as ending poverty, fighting inequality and injustice, and tackling climate change, providing specific measurable targets to be achieved by 2030 (https://sdgs.un.org/goals).

[8] See https://www.oecd.org/corporate/mne/responsible-business-conduct-matters.htm.

[9] See https://www.fsb.org/work-of-the-fsb/financial-innovation-and-structural-change/climate-related-risks/.

[10] See https://eur-lex.europa.eu/legal-content/EN/TXT/?uri=CELEX:52018DC0097.

[11] See https://ec.europa.eu/info/strategy/priorities-2019-2024/european-green-deal_en.

[12] See https://ec.europa.eu/info/business-economy-euro/banking-and-finance/sustainable-finance/eu-taxonomy-sustainable-activities_en.

Non-Financial Reporting Directive and Guidelines on Climate-related Information,[13] or the Sustainable Finance Disclosure Regulation,[14] among many other ongoing developments and initiatives.[15]

This excess of initiatives, legislation and regulation will also likely drive a significant rise on ESG-related shareholder resolutions, shareholder activism, disputes and litigation.

Listed companies and boards have been preparing for such new context, namely by understanding how ESG risks can impact their business and operations and by ensuring that those risks are subject to adequate assessment and management, independently of their respective industry, as, for example, risks associated with climate change will impact all companies although in different manners.

Simultaneously, based on indications that *green* investments performed better during 2020 and were financially profitable, and that companies with stronger ESG adherence also seem to have done better during the pandemic, producing higher yields than similar corporations that are less ESG committed, large global investors and proxy advisors are increasingly promoting sustainability goals and campaigning for ESG objectives.[16]

It seems clear that Covid-19 will act as a catalyst for change and companies should anticipate the effects of an accelerating expansion of ESG, until now very concentrated on the scope of E (environmental), to S (social) matters, including increased focus on social risks as diversified as the ones related to human rights, supply chain management and transparency, consumer protection and labour and employment issues, including diversity, inclusion and working conditions.[17]

---

[13] See https://eur-lex.europa.eu/legal-content/EN/TXT/?uri=CELEX%3A32014L0095.

[14] See https://eur-lex.europa.eu/legal-content/EN/TXT/?uri=celex%3A32019R2088.

[15] Including a number of *soft law* measures to be considered by the financial system and companies, namely, the *Principles for Responsible Investment* (https://www.unpri.org), the *Green Bond Principles* (https://www.icmagroup.org/sustainable-finance/the-principles-guidelines-and-handbooks/green-bond-principles-gbp/), the *Social Bond Principles* (https://www.icmagroup.org/sustainable-finance/the-principles-guidelines-and-handbooks/social-bond-principles-sbp/), *The Sustainability Bond Guidelines* (https://www.icmagroup.org/sustainable-finance/the-principles-guidelines-and-handbooks/sustainability-bond-guidelines-sbg/), the *Green Loan Principles* (https://www.lsta.org/content/green-loan-principles/).

[16] "During 2020, 81% of a globally representative selection of sustainable indexes outperformed their parent benchmarks", "Blackrock: Larry Fink's 2021 letter to CEOs" (https://www.blackrock.com/corporate/investor-relations/larry-fink-ceo-letter).

Fidelity, for example, also reported that companies with the highest ESG ratings fared better than their less-sustainable peers during the pandemic-induced market sell-off between February 19 and March 26 of 2020 (https://www.fidelityinternational.com/editorial/article/proprietary-esg-ratings-prove-their-worth-3b1449-en5/).

[17] Regarding socially responsible investments, see https://sibdatabase.socialfinance.org.uk.

This is especially important for European listed companies, as the EU positioned itself as the first and faster mover in sustainability which, if companies fail to seize the business opportunities coming from sustainability, will only increase their compliance burden and probable competitive disadvantages to businesses in other continents.

Thus, ESG is no longer something that European listed companies and boards can view as a "why not?" or "nice to have".

This is why, as regulators and investors shift from soft questions to requirements on hard data concerning climate-related risks, social measurable goals, supply chains impacts and sensitive governance changes, and it becomes clear that all companies will be profoundly affected by the transition to a net-zero economy, listed companies already made ESG management and reporting on sustainability a part of their everyday business, and definitely a "must have".

Indeed, a renewed focus is expected on boards of directors themselves, and their duties of care, as the European Commission ["EC"] has put forward a "sustainable corporate governance initiative" under the scope of its European Green Deal, with the aim, in its own words, to ensure that "environmental and social interests are fully embedded into business strategies" and companies "focus on long-term sustainable value creation rather than short-term financial value" (which, in our view, they already are).[18]

The pillar G (*governance*) is therefore expected to be the subject of legislative intervention with large implications for companies, as this new initiative on sustainable corporate governance clearly seems to be an antechamber to new and likely deep changes to corporate law with probable unintended consequences.

This is also a confirmation that the discussion on sustainability and sustainable finance has another underlying theme: the ongoing larger debate regarding the nature and purpose of the company, and the renewed concerns with the apparent erosion of the democratic capitalism model of Western societies, against a background of technological, social, economic, political and historical transformations and a path forward of multiple uncertainties (definitely aggravated by the economic and social effects of the Covid-19 pandemic).

2 Sustainability and Capital Markets: A Bridge Too Far?

Acting proactively on ESG business and growth opportunities can help companies to prepare their long-term success and value creation.

Listed companies are fully aware of that, as shown by existing research and reporting, as they have to identify and manage a broad and diverse range of

---

[18] EC' Sustainable corporate governance consultation: (https://ec.europa.eu/info/law/better-regulation/have-your-say/initiatives/12548-Sustainable-corporate-governance/public-consultation).

stakeholders, contribute to the goals of their larger community, and generate positive returns, while being at the forefront of incorporating ESG factors into corporate strategies.[19]

Earlier attention to ESG has helped companies, for example, to grow investment returns, improve credit ratings, enter new markets with more sustainable products, decrease risk, reduce operational costs, built consumer preference, attract better talent, create stronger community relations or gain government support for their initiatives.[20]

One would expect, based on the work, capital and innovation, that companies have been channelling to sustainable investments and to address complex sustainability, ESG and SDG challenges, that those efforts would be recognised through a certain degree of regulatory pressure easing in order to allow companies a larger degree of strategic freedom and avoid stifling innovation.[21]

Unfortunately, this is not the case in Europe, and, in a context of increasing legislation and very complex regulation, listed companies are the first ones to sustain the high burden of implementing ESG scores (e.g. learning curve and implementation costs).

This is also the reason why the danger exists that sustainability may cause European listed companies to be unfairly penalised in comparison with privately owned companies across the same regions or industries.

Additionally, the excess of regulation is a problem not only for companies but also for capital markets.

According to the EC's baseline scenario, between 25 and 35% of companies were to expect a financing shortfall by the end of 2020 after exhausting

---

[19] Most recently, the Report "Running Hot - Accelerating Europe's Path To Paris", an analysis by Oliver Wyman and CDP Europe, confirms Europe's companies progress in their action on climate change; see https://www.oliverwyman.com/our-expertise/insights/2021/mar/running-hot.html?utm_source=twitter&utm_medium=social&utm_campaign=cdp-ow&utm_content=2021-mar&utm_id=cmp-11471-j9g9b9.

[20] Of course, this is not a novelty; the mostly misunderstood Milton Friedman said it long time ago: "It may well be in the long-run interest of a corporation that is a major employer in a small community to devote resources to providing amenities to that community or to improving its government. That may make it easier to attract desirable employees, it may reduce the wage bill or lessen losses from pilferage and sabotage or have other worthwhile effects." Milton Friedman, "A Friedman Doctrine—The Social Responsibility of Business is to Increase its Profits", *New York Times Magazine*, September 13, 1970 (http://umich.edu/~thecore/doc/Friedman.pdf).

[21] Currently, more than 1000 of the largest European companies set science-based emission reduction targets, with over 325 committed in line with the 1.5 °C trajectory and reaching net-zero by 2050: (https://www.wemeanbusinesscoalition.org/blog/paris-agreement-the-power-of-ingenuity-with-collaboration-is-insurmountable/).

Regarding the extraordinary efforts that will be required from all stakeholders to meet the SDGs in a Covid-19 pandemic context, and after the pandemic, see Dora Benedek et al., "A Post-Pandemic Assessment of the Sustainable Development Goals" (https://www.imf.org/en/Publications/Staff-Discussion-Notes/Issues/2021/04/27/A-Post-Pandemic-Assessment-of-the-Sustainable-Development-Goals-460076).

working capital and liquidity buffers, respectively, with these shares rising to 35% and 50%, respectively, in an adverse scenario.[22]

This means that around 180,000–260,000 European companies employing around 25–35 million people could experience a financing shortfall should the more adverse scenario materialise.

The corresponding liquidity shortfall could range between €350 billion and €500 billion in the baseline scenario, and between €650 billion and €900 billion in the adverse scenario.

The EC calculated an amount to at least €1.5 trillions between 2020 and 2021, for basic investment needs due to the pandemic impact, also additional investment needs to stabilise the public sector capital stock to GDP ratio, and finally investment needs for green transition, digital transformation and strategic investment.[23]

Banks, under increased pressure by probable corporate defaults and increase in non-performing loans, and constrained by banking-specific financial and prudential ratios, would not be able to help, and the same will be true for Member States faced with increasing government debt-to-GDP ratios.[24]

---

[22] European Commission, Staff Working Document accompanying the Communication on the Next Generation Plan, "Identifying Europe's Recovery Needs", 27.5.2020 SWD (2020) 98 final (https://ec.europa.eu/info/sites/info/files/economy-finance/assessment_of_economic_and_investment_needs.pdf).

[23] As recognised by Member States national governments, the vast majority of the required investment will occur on the private sector of the economy, sourced mainly by companies, industries and families, and also by social economy institutions.

Taking the example of Portugal, the overall aggregate investment needed to achieve carbon neutrality by 2050 is projected to be €1,017 billion, of which €930 billion were expected to be invested in any case as a result of the normal dynamics of modernisation of the economy (estimations pre-COVID-19). Given that Portugal is aiming to reduce emissions by more than 85% by 2050, the additional investment required to achieve carbon neutrality will be around €86 billion for the whole period, or between €2.1 and €2.5 billion per year (around 1.2% of GDP, pre-COVID-19).

Portugal's sustainability strategy is based on the Roadmap for Carbon Neutrality in 2050, Resolution 107/2019 of 1 July 2019 (https://www.portugal.gov.pt/download-ficheiros/ficheiro.aspx?v=%3D%3DBAAAAB%2BLCAAAAAAABACzMDexBAC4h9DR BAAAAA%3D%3D) which consubstantiates the long-term development strategy for low greenhouse gas emissions; in articulation with the RCN 2050, Portugal also approved the Integrated National Energy and Climate Plan for 2030 (https://ec.europa.eu/energy/sites/ener/files/documents/pt_final_necp_main_en.pdf), framed within the obligations arising from Regulation (EU) 2018/1999 of the European Parliament and of the Council of 11 December 2018 on the Governance of the Energy Union and Climate Action.

The need for private sector investment and capital market development was also emphasised by the Portuguese Think Tank for Sustainable Finance in its report "Guidelines to Accelerate Sustainable Financing in Portugal" and the "Letter of Commitment to Sustainable Financing in Portugal", which presents a set of recommendations for the Portuguese government and economy, and was signed by AEM—the association representing Portuguese listed companies (https://emitentes.pt/2019/07/26/aem-signs-letter-of-commitment-to-sustainable-financing-in-portugal/).

[24] The highest ratios of government debt to GDP at the end of the third quarter of 2020 were recorded in Greece (199.9%), Italy (154.2%), Portugal (130.8%), Cyprus

Thus, the recovery of the European economy depends mainly on the adaptability of the private sector, counting on more resilient and well capitalised companies and complex restructuring activities to restore competitiveness and to meet its ambitious sustainability goals.

In this context, it seems clear that European policies should ensure an environment in which companies are able to raise capital through public capital markets.

Public capital markets that serve the interests of both investors and issuers have a unique and irreplaceable role to play in financing a future sustainable economy and are of essence to provide access to companies wanting to raise capital, supporting them in overcoming current difficulties and affording deep transformation of their activity, and to all investors, retail and institutional, wishing to diversify their portfolios.

Coincidently, literature also shows that for given levels of economic and financial development and environmental regulation, CO_2 emissions *per capita* are lower in economies that are relatively more equity-funded.[25]

Promoting capital markets and sustainable finance are therefore mutually reinforcing projects which should be pursued jointly and with due regard to the need to meet the financing gap faced by European companies over the coming years and the resources available to companies listed on public markets.[26]

However, such an environment will not be achieved through more legislation, more complex regulation or further information requirements for companies.

As underlined by the most recent and complete study on European capital markets, the costs of becoming a public company have already risen considerably in recent decades, with regulatory costs associated with listing noted as particularly relevant for smaller issuers, and challenges associated with meeting regular financial reporting requirements, the time and cost associated with

---

(119.5%), France (116.5%), and Spain (114.1%) (https://ec.europa.eu/eurostat/documents/portlet_file_entry/2995521/2-21012021-AP-EN.pdf/a3748b22-e96e-7f62-ba05-11c7192e32f3).

[25] Ralph De Haas & Alexander Popov, "Finance and Carbon Emissions", ECB, Working Paper Series, 2019, (https://www.ecb.europa.eu/pub/pdf/scpwps/ecb.wp2318~44719344e8.en.pdf) and Ralph De Haas and Alexander Popov, 'Finance and Decarbonisation: Why Equity Markets Do It Better', ECB, Research Bulletin No. 64, 2019, https://www.ecb.europa.eu/pub/economic-research/resbull/2019/html/ecb.rb191127~79fa1d3b70.en.html).

[26] See the "European IPO Report 2020—Recommendations to Improve Conditions for European IPO Markets", Caroline Nagtegaal (Coord.), Abel Sequeira Ferreira et al., for an overview of the issues that companies, investors, exchanges and other market participants are facing in trying to promote companies' access to capital market financing and of the issues at stake to reverse the trend of declining numbers of initial public offerings (IPOs) on European markets, as well as policy recommendations to address these challenges (https://emitentes.pt/2020/03/02/european-ipo-report-2020/).

compliance and administration and excessive requirements to disclose sensitive information, among the main reasons for companies voluntarily choosing to delist.[27]

These are some of the reasons why capital markets in the EU have been continuously decreasing in size since the financial crisis therefore remaining increasingly smaller than in other major economies.

Indeed, the EU lags behind the US (and China) both in the use of equity to finance companies, which lack access to more and better equity, as well as in risk capital to fund innovation and growth at a time of massive recapitalisation needs.

Public equity markets in the EU shall match companies' demand for risk capital, to rebalance their financial structure and to sustain investments including the investments necessary to achieve the transition to a low-carbon economy, through a coherent high-level strategy, a stable regulatory framework and a strong set of incentives for capital markets access.

This idea of development of capital markets is, of course, very much engrained in the (new) Capital Markets Union ["CMU"] Action Plan, but without exhaustive assessment of the detrimental impacts of regulation, the streamlining of rules (in some cases rebuilding some regimes from the ground up), and finding the right balance between regulatory burden and investor protection, capital markets in Europe will remain a bridge too far between companies and investors.[28]

Worse than that, the urgency of well-developed capital and IPO markets in Europe is contradicted by some new announced regulatory requirements menacing to generate more and more complex regulation, further information requirements and new burdensome obligations that public companies will have to fulfil in order to be listed or remain listed.

In our view, if the EC insists on a philosophy of so-called *harmonisation* rather than *simplification* of the European legal framework it will continue to fail in its own goals of capital markets development and better market efficiency.

---

[27] European Commission, "Primary and Secondary Equity Markets in the EU—Final Report", Oxera, November 2020 (https://www.oxera.com/wp-content/uploads/2020/11/Oxera-study-Primary-and-Secondary-Markets-in-the-EU-Final-Report-EN-1.pdf).
Similar results were identified, specifically for the Portuguese market, in OECD (2020), *OECD Capital Market Review of Portugal 2020: Mobilising Portuguese Capital Markets for Investment and Growth*, OECD Capital Market Series, (http://www.oecd.org/corporate/OECD-Capital-Market-Review-Portugal.htm).

[28] In our view, the new action plan on the CMU lacks consistency, with regards both to the timeline and the focus of the proposals. "Capital markets union 2020 action plan: A capital markets union for people and businesses" (https://ec.europa.eu/info/business-economy-euro/growth-and-investment/capital-markets-union/capital-markets-union-2020-action-plan_en).
See Franklin Allen et al. (eds), *Capital Markets Union and Beyond*, MIT Press, 2019, and namely, Chapter 18, "The Politics of Capital Markets Union", by Wolf-Georg Ringe.

The recent past abundantly shows that European *harmonisation* always translates into the proliferation of new and more complex regulatory initiatives and the multiplication of new measures generating substantial additional costs for listed companies of all sizes.

3   ESG: Legal Framework

Overview of European Initiatives

European law and policies provide for the most important part of the regulatory context in which listed companies adopt their decisions related to environmental and social performance.

Below we shall briefly summarise the main aspects of the European strategy, agenda and initiatives regarding sustainability.

Inspired by the UN 2030 agenda and sustainable development goals and the Paris Climate Agreement, both adopted in 2015, the EC's Action Plan to Finance Sustainable Growth was first launched in 2018, setting the ground for the climate transition of the financial system.[29]

The Plan specifies three main ambitious objectives: (i) reorient capital flows towards sustainable investment, (ii) promote better management of financial risks arising from climate change, (iii) promote transparency and a long-term vision for the proper assessment of long-term value in financial and economic activity; the Plan also defines a set of ten main actions to be implemented, in order to substantiate these objectives, all with relevant impact for the future activity of listed companies.[30]

Following the launch of the Action Plan, its implementation resulted in an extensive set of legislative proposals from the EC, including numerous documents of utmost importance for listed companies.

The main pillars of the Action Plan are currently at varying stages of development, but all have taken significant steps forward within a very pressing timeline generating natural anxiety among market participants faced with

---

[29] The Paris Climate Agreement sets out a global action plan to put the world on track to avoid climate change by limiting global warming to well below 2 °C above pre-industrial levels and pursuing efforts to limit it to 1.5 °C (as this would significantly reduce the risks and impacts of climate change) (https://ec.europa.eu/clima/policies/international/negotiations/paris_en).

United Nations 2030 Agenda for Sustainable Development, (https://sustainabledevelopment.un.org/topics/sustainabledevelopmentgoals).

[30] The ten Actions are: (i) establishing an EU classification system for sustainability activities, (ii) creating standards and labels for green financial products, (iii) fostering investment in sustainable projects, (iv) incorporating sustainability when providing investment advice, (v) developing sustainability benchmarks, (vi) better integrating sustainability in ratings and research, (vii) clarifying institutional investors and asset managers' duties, (viii) incorporating sustainability in prudential requirements, (ix) strengthening sustainability disclosure and accounting rulemaking, (x) fostering sustainable corporate governance and attenuating short-termism in capital markets.

daunting amounts of sometimes unclear, confusing and overwhelming new regulatory burden.[31]

In 2019, the EC published the European Green Deal with the intention to provide an action plan to boost the efficient use of resources by moving to a clean, circular economy, restore biodiversity and cut pollution.

This new plan outlines the investments needed and available financing tools and explains how the EU intends to provide financial support and technical assistance to ensure a just and inclusive transition towards a green economy.

Further to that, in September 2020 the EC presented a "Communication Stepping up Europe's 2030 climate ambition - Investing in a climate-neutral future for the benefit of our people" (the 2030 Climate Target Plan), proposing to cut greenhouse gas emissions by at least 55% by 2030 setting Europe on a path to become climate-neutral by 2050—an economy with net-zero greenhouse gas emissions.[32]

Under the scope of the European Green Deal, a legislative proposal for a European Climate Law was adopted on March 2020 (later amended on September 2020 to introduce a target of 55% reduction of the EU's GHG emissions by 2030); a provisional political agreement was already reached between the Portuguese Presidency of the European Council's and the European Parliament's negotiators setting up the frame for the approval of the European Climate Law by the Council and Parliament.[33]

The European Climate Law intends to establish a legally binding EU-wide common for achieving climate neutrality making the goals stated for carbon neutrality in 2050 mandatory for member states.

It is also expected that a Renewed Sustainable Finance Strategy will be adopted in the first half of 2021; announced in the framework of the European Green Deal, the renewed strategy intends to contribute to its objectives, in particular to create and enable the framework for private investors and the

---

[31] Unclear, confusing, and overwhelming, even for regulators, as evidenced by a recent letter (https://www.esma.europa.eu/sites/default/files/library/jc_2021_02_letter_to_eu_commission_on_priority_issues_relating_to_sfdr_application.pdf).

[32] See the EC Communication "Stepping up Europe's 2030 Climate Ambition—Investing in a Climate-Neutral Future for the Benefit of Our People (COM/2020/562 final)" (https://eur-lex.europa.eu/legal-content/EN/TXT/?uri=CELEX:52020DC0562).
See also the 2030 Climate Target Plan (https://ec.europa.eu/clima/policies/eu-climate-action/2030_ctp_en) and the 2050 Long-Term Strategy (https://ec.europa.eu/clima/policies/strategies/2050_en).

[33] Proposal for a Regulation establishing the framework for achieving climate neutrality and amending Regulation (EU) 2018/1999 (European Climate Law) COM/2020/80 final (https://eur-lex.europa.eu/legal-content/EN/TXT/?qid=1588581905912&uri=CELEX:52020PC0080).
See also the Council' press release "European Climate Law: Council and Parliament Reach Provisional Agreement" (https://www.consilium.europa.eu/en/press/press-releases/2021/04/21/european-climate-law-council-and-parliament-reach-provisional-agreement/pdf).

public sector to facilitate sustainable investments: the strategy will aim for redirecting private capital flows to green investments and "for embedding a culture of sustainable corporate governance in private sector".[34]

Lastly, a Platform on Sustainable Finance was established in October 2020 as an advisory body that brings together sustainability experts from private stakeholders, financial, non-financial and business sectors, NGOs and civil society, academia and think tanks, as well as public and international institutions, with intention to foster cooperation for the creation of further sustainable finance policies and further development of the EU taxonomy.[35],[36]

---

[34] See https://ec.europa.eu/info/consultations/finance-2020-sustainable-finance-strategy_en.

[35] See https://ec.europa.eu/info/business-economy-euro/banking-and-finance/sustainable-finance/overview-sustainable-finance/platform-sustainable-finance_en.

In March 2021, the Platform published its "Report on Transition Finance", regarding how the Taxonomy can enable inclusive transition financing for companies and other economic actors working to improve their environmental impact and trying to identify a relevant concept and frameworks of transition in the context of climate change. See https://ec.europa.eu/info/files/210319-eu-platform-transition-finance-report_en.

[36] It is important to keep in mind that the above-mentioned initiatives are only some of the highlights of more than seventy major undergoing EU files and initiatives with an impact on sustainable finance and the life of companies.

Among others on which we will not expand in this chapter one must consider:

- The NFRD—Non-Financial Reporting Directive; the Directive 2014/95/EU of the European Parliament and of the Council of 22 October 2014 amending Directive 2013/34/EU as regards disclosure of non-financial and diversity information by certain large undertakings and groups (NFI Directive). As the EC carried out, during the first half of 2020, a public consultation on the revision of this Non-Financial Information Directive a new version is expected which should also incorporate the recommendations of the Task Force on Climate-related Disclosures regarding the assessment of the company's climate risks and the climate risks present in their supply chains.

- The SFDR—Sustainable Finance Disclosure Regulation; the Regulation (EU) 2019/2088, of 27 November 2019, on sustainability-related disclosures in the financial services sector, lays down harmonised rules for financial market participants and financial advisers on transparency with regard to the integration of sustainability risks and the consideration of adverse sustainability impacts in their processes and the provision of sustain ability-related information with respect to financial products. Among others, its main provisions include (i) the definition of sustainable investment relevant to the financial sector and (ii) information disclosure regarding policies for integrating sustainability risks in the investment decision-making process, and the negative impacts of investment decision on sustainability factors; see https://eur-lex.europa.eu/legal-content/EN/TXT/PDF/?uri=CELEX: 02019R2088-20200712&from=EN.

Regarding the SFDR see for further analysis Julien Froumouth and Joana Frade, Chapter 12 in this book.

Other relevant initiatives, subject to more detailed analysis in different chapters of this book, include the amendment to the Directive on shareholders' rights and rules for asset managers, creation of benchmarks and changes of the directives on the markets for financial instruments (MiFID II), alternative investments (AIFMD) and collective investment undertakings (UCITS).

Taxonomy Regulation[37]

Sometimes described as a new *green* language for companies, the EU Taxonomy Regulation for sustainable activities is the backbone of the EC's plan on sustainable finance and will require non-financial companies to outline the environmental sustainability of their activities, helping investors in distinguishing which investments contribute to the European environmental objectives.[38]

Large public-interest companies with over 500 employees and/or including listed companies will have to align key performance indicators to report which part of their turnover and expenditure is in line with the Taxonomy.

Although the Taxonomy is rightly seen as a progress and supposedly will help companies in their transition to sustainable businesses and activities, a number of difficulties for companies may already be identified.[39]

Recent reporting notes that in several aspects the Taxonomy is too strict and constraining, and its impact may be undermined by incomplete definitions,

---

Regarding the EU regime on asset managers, see also Tiago Santos Matias, Chapter 20 in this book.

[37] The Regulation (EU) 2020/852 on the establishment of a framework to facilitate sustainable investment and amending Regulation (EU) 2019/2088 (Taxonomy Regulation).

This Regulation establishes a classification system or taxonomy to the identification of economic activities that may be considered sustainable, based on environmental objectives, which effectively translates into a new unified language for sustainable finance, within the financial sector, essential to redirect investments towards sustainable projects and activities and achieve the climate goals of the EU.

Under the new Taxonomy sustainable economic activities are the ones that (i) contribute substantially to one or more environmental objectives (i.e. climate change mitigation; climate change adaptation; sustainable use and protection of water and marine resources; transition to a circular economy; pollution prevention and control and protection and restoration of biodiversity and ecosystems); (ii) does not significantly harm any of those objectives; (iii) is carried out in compliance with minimum safeguards; (iv) meets the technical screening criteria for environmental objectives.

A delegated act on the reporting of the EU Taxonomy alignment is expected to be adopted by June 2021, specifying the information companies will have to disclose on how, and to what extent, their activities align with the EU Taxonomy. The first disclosures by financial institutions are due as of the 1st of January 2022 for the first two environmental objectives. Non-financial companies are to comply with the associated criteria throughout 2022.

[38] Regarding the Taxonomy Regulation, see also Rui de Oliveira Neves, Chapter 13 in this book.

[39] In this point we follow closely the impressive Report "European Sustainable Finance Survey – 2020", Adelphi/ISS ESG, supported by the German Federal Ministry for Environment Nature Conservation and Nuclear Safety (https://sustainablefinancesurvey.de/sites/sustainablefinancesurvey.de/files/documents/european_sustainable_finance_survey_2020_final_2.pdf); for a perspective of concerns from the investment side, see the PRI—Principles for Responsible Investment study, 2020: "Testing the Taxonomy: Insights from the PRI Taxonomy Practitioners Group" (https://www.unpri.org/eu-taxonomy-alignment-case-studies/testing-the-taxonomy-insights-from-the-pri-taxonomy-practitioners-group/6409.article).

because the taxonomy insufficiently reflects life cycle and supply chain issues and does not reflect the current complexity of business practices.

Companies may find discrepancies between taxonomy-relevant and taxonomy-aligned revenue shares or their more significant activities not being classified as taxonomy-relevant activities, which makes achieving full compliance with the taxonomy very challenging.

This is because some taxonomy criteria are too ambitious, namely much more ambitious than common market standards, and/or non-verifiable.

Some activities do not fulfil substantial contribution criteria and, worse than that, there may be that a majority of revenues that substantially contribute to one of the taxonomy's environmental objectives may have a negative impact on another environmental objective, failing to meet the "do no significant harm" criteria, again because these seem too ambitious.

Another source of difficulties comes from the circumstance that the taxonomy may exclude, at least temporarily, many potentially relevant activities that contribute to climate change mitigation and adaptation, and a main focus on revenue also may exclude relevant activities that contribute to climate change mitigation or adaptation but do not generate revenue.

Among other difficulties, companies note a lack of clarity and resources available for applying the taxonomy, whose criteria in many cases remain unclear and based on vague definitions, and do not perceive a clear match between their economic activities and taxonomy-relevant activities.

Companies' difficulties will be aggravated by insufficient guidance on how to assess capital expenditures (CapEx) and operational expenditures (OpEx) regarding taxonomy alignment, and notorious misalignments between available data and mandatory disclosure, requiring substantial adjustments to internal processes for the more granular criteria of the taxonomy to be incorporated and disclosed.

Due to the above-mentioned difficulties and still unclear disclosure requirements, auditing and liability, companies may see the implementation timeframe as too short, and fear operational costs increase namely from adjusting data collection and sustainability disclosure processes to meet taxonomy criteria.

Companies' remarks that compliance with the taxonomy will be very challenging, should be subject to careful attention from the EC.

In this respect there is work to be done to add clarity and provide the necessary resources for clear understanding and accurately applying the taxonomy.

Particularly, the EC must help companies and its boards to navigate the overabundance of ESG reporting standards, helping avoid sustainability reporting fatigue, and must review existing regulation in order to ensure its coherence and decrease compliance burdens.

The EC must also work with companies in clarifying how the taxonomy will benefit the environment and the real economy, and how companies and their stakeholders will benefit from the taxonomy-based disclosure, promoting willingness among market participants to apply it.

Failing to do so and pushing for taxonomy compliance before companies have had a chance to adapt to the criteria might result in negative impacts, including diverting investment away from transitioning European activities, hurting the real economy.

ESG Ratings

The OECD has identified the need for more consistent, comparable and available data on ESG performance, as the main priority to better incorporate material ESG considerations into financial decision-making and align investments with sustainable, long-term value and we have been watching a proliferation of metrics and disclosure frameworks, oriented to assess material sustainability risk performance in companies, with extensive real-world application.[40,41]

However, it is also true that the use of different metrics and methodologies of weighing diverse indicators can easily return very different ESG scores for similar companies or even the same company, making it difficult to compare the performance of companies and assets against ESG risk factors.

Current market practices, from ratings to disclosures and individual metrics, present a fragmented and inconsistent view of ESG risks and performance, and new measures are needed to ensure that all market participants have access to consistent data, comparable metrics and transparent methodologies.[42]

This is one point where there is a real market failure, one that may prevent investors from accurately assessing ESG risks and this is where the EC must concentrate its efforts.[43]

---

[40] See OECD Business and Finance Outlook 2020—Sustainable and Resilient Finance (https://www.oecd-ilibrary.org/finance-and-investment/oecd-business-and-finance-outlook-2020_eb61fd29-en).

[41] R. Boffo & R. Palatano (2020), "ESG Investing: Practices, Progress and Challenges", report that almost 80% of total global market capitalisation is already composed of ESG-rated companies.

See also R. Boffo & R. Palatano (2020), "ESG Investing: Environmental Pillar Scoring and Reporting", OECD Paris (www.oecd.org/finance/esg-investing-environmental-pillar-scoring-and-reporting.pdf).

[42] OECD Business and Finance Outlook 2020.

[43] Notwithstanding the fact that a number of other institutions and bodies are already developing and establishing platforms or mechanisms to coordinate efforts and drive global standards (https://www.ifrs.org/projects/work-plan/sustainability-reporting/).

Also, at national level one can count several initiatives trying to help harmonising information disclosure; in the Portuguese case, for example, the regulator (CMVM) published a new "Non-financial information disclosure template for companies issuing securities admitted to trading on a regulated market", a non-binding report template, primarily addressed to regulated market issuers—those subject to non-financial reporting—and aiming to assist companies disclosing non-financial information, particularly in respect of information on environmental, social and governance factors, and help stakeholders to consult and use such information. See https://www.cmvm.pt/en/Legislacao/ConsultasPublicas/CMVM/Documents/Modelo%20de%20Informação%20não%20Financeira_EN.pdf.

4 Sustainable Corporate Governance: A Reflection

EC's Initiative on Sustainable Corporate Governance

The EC's objective and initiatives to focus on long-term value creation and the alignment of corporate and community objectives, and its proposals for integrating sustainability at companies' board level, improving sustainability performance and contributing to the 2050 net-zero objective, are mostly positive and deserve support.

Sustainability must be embedded in the purpose, strategy and culture of companies, establishing their strong commitment to ESG progress, and risk management and oversight mechanisms and procedures must be in place.

Transparency and disclosure are of the utmost importance for companies to manifest their commitment to ESG factors and ESG principles need to be properly embodied and driven through corporate bodies and all areas of governance.

However, this does not mean that new or different governance mechanisms are required, much less that we need such mechanisms to be consecrated by new hard law and regulation, as there is no real evidence of a failure in existing legislation and regulation to ensure that sustainability elements and ESG factors are given the importance they deserve by companies from strategy to operations.

And specially, at this stage, any regulatory options underlying undesirable political interference in business strategy, which may harm listed companies and European businesses willingness to adopt ESG factors and restrict competition and innovation, must be carefully avoided.

Although the generation of profits cannot be the unique or the ultimate purpose of a company, and clearly defined purpose and values, with a long-term strategy focus, must be in place, the definition of a company's strategy must stay within company's discretion as part of the economic and business freedom ensured by democratic legal systems of constitutional law.

Also, the fundamental role that companies play in society should be acknowledged and respected by politicians and lawmakers who have a duty to promote the best possible corporate environment for European firms to grow and create value for all.

In our view, it is in this context that the EC's "sustainable corporate governance" initiative has to be appreciated.

Ernst & Young "Study on Directors' Duties and Sustainable Corporate Governance"

As part of the European Green Deal, the EC's "sustainable corporate governance" initiative is presented as intending to ensure that environmental and social interests are fully embedded into business strategies, and that companies focus on long-term sustainable value creation rather than short-term financial value.

With this goal the EC launched the study on "directors' duties and sustainable corporate governance" prepared by an accounting firm, Ernst & Young ["EY"].[44,45]

The study raises a number of critical concerns, both regarding its methodology and its conclusions.

Indeed, everything about the EY' report is in itself a "case study" on how not to make a study and how not to consult about a study (as an example, the EC published the commissioned 160-pages study, inviting responses with a space limit of 4000... characters).

As noted by an extensive number of Authors, the study ignores basic academic norms of empirical research and suffers from severe methodological shortcomings, including the presentation of misleading figures.

The study also presents equally serious analytical deficiencies, failing to define essential concepts (namely, the concept of "short-termism" the main

---

[44] Study on directors' duties and sustainable corporate governance (https://op.europa.eu/en/publication-detail/-/publication/e47928a2-d20b-11ea-adf7-01aa75ed71a1/language-en).

[45] The EC also launched a "Study on due Diligence Requirements Through the Supply Chain", prepared by the British Institute of International and Comparative Law (https://op.europa.eu/en/publication-detail/-/publication/8ba0a8fd-4c83-11ea-b8b7-01aa75ed71a1/language-en).

The study is mainly focused on due diligence processes to address adverse sustainability impacts, such as climate change, environmental or human rights distress in along companies' operations and their value chain, by identifying and preventing relevant risks and mitigating negative impacts. This study is beyond the scope of our chapter, but we must at least refer that its methodology and conclusions seem to be balanced and accurate and sound enough to base the expected formal legislative process that will follow, although we have some doubts regarding the exact terms anticipated by the EC, and the wording used by the EP, in regard of its announced legislative initiative on mandatory company due-diligence obligations in their value-chains, especially if such initiative ignores that multiple aspects of that problem have already been strengthened by the new disclosure obligations on non-financial information.

On 10 March 2021, the European Parliament ["EP"] adopted a legislative initiative report setting out recommendations to the EC on corporate due diligence and accountability, including a draft directive. The report proposes the introduction of a mandatory corporate due diligence obligation to identify, prevent, mitigate and account for human rights violations and negative environmental impacts in business' supply chains. The EP also states its strong support for the EC's sustainable corporate governance initiative.

See Press Release: https://www.europarl.europa.eu/news/en/press-room/20210304IPR99216/meps-companies-must-no-longer-cause-harm-to-people-and-planet-with-impunity.

See Briefing: https://www.europarl.europa.eu/RegData/etudes/BRIE/2020/659299/EPRS_BRI(2020)659299_EN.pdf.

See Report: https://www.europarl.europa.eu/RegData/etudes/STUD/2020/654191/EPRS_STU(2020)654191_EN.pdf.

See Resolution: https://www.europarl.europa.eu/doceo/document/TA-9-2021-0073_EN.pdf.

Regarding this EP' Resolution, see for further analysis Guido Ferrarini, Chapter 2 in this book.

target at the core of the study) and to understand how corporate governance works, misrepresenting fundamental concepts of company law, ignoring the externalities that may negatively affect the environment and communities as a result of corporate actions, not differentiating different kinds of shareholders and stakeholders and showing a fundamental lack of understanding of the market economy.[46]

In addition, critics emphasise the low-quality, one-sided, evidence presented while ignoring high-quality evidence that suggests the opposite of EY preconceived and politically biased conclusions which seem to be oriented by a misguided distrust of European companies.[47]

---

[46] See, for example, "EC Corporate Governance Initiative Series: 'A Critique of the Study on Directors' Duties and Sustainable Corporate Governance prepared by Ernst & Young for the European Commission", by The European Company Law Experts Group (ECLE: e.g. Paul Davies (Oxford), Susan Emmenegger (Bern University), Guido Ferrarini (Genoa), Klaus Hopt (Max Planck, Hamburg), Adam Opalski (Warsaw), Alain Pietrancosta (Paris), Andrés Recalde (Autonomous University of Madrid), Markus Roth (Marburg), Michael Schouten (Amsterdam), Rolf Skog (Gothenburg), Martin Winner (Vienna University of Economics and Business), Eddy Wymeersch (Gent).) (https://www.law.ox.ac.uk/business-law-blog/blog/2020/10/ec-corporate-governance-initiative-series-critique-study-directors); Alex Edmans, "Diagnosis Before Treatment: The Use and Misuse of Evidence in Policymaking" (https://www.law.ox.ac.uk/business-law-blog/blog/2020/10/ec-corporate-governance-initiative-series-diagnosis-treatment-use-and); Jesse M. Fried and Charles C. Y. Wang, "Short-Termism, Shareholder Payouts, and Investment in the EU" (https://papers.ssrn.com/sol3/papers.cfm?abstract_id=3706499); Mark J. Roe, Holger Spamann, Jesse M. Fried and Charles C. Y. Wang, "The European Commission's Sustainable Corporate Governance Report: A Critique" (https://papers.ssrn.com/sol3/papers.cfm?abstract_id=3711652); John Coffee, "The European Commission Considers 'Short-Termism' (and 'What Do You Mean by That?)" (https://ecgi.global/news/european-commission-considers-"short-termism"-and-"what-do-you-mean-); The Nordic Company Law Scholars, "Response to the Study on Directors' Duties and Sustainable Corporate Governance by Nordic Company Law Scholars" (https://papers.ssrn.com/sol3/papers.cfm?abstract_id=3709762); Claus Richter, Steen Thomsen, and Lars Ohnemus, "A Response From the Copenhagen Business School" (https://www.law.ox.ac.uk/business-law-blog/blog/2020/10/ec-corporate-governance-initiative-series-response-copenhagen); Alexander Bassen, Kerstin Lopatta and Wolf-Georg Ringe, "The EU Sustainable Corporate Governance Initiative—Room for Improvement" (https://www.law.ox.ac.uk/business-law-blog/blog/2020/10/ec-corporate-governance-initiative-series-eu-sustainable-corporate); Søren Friis Hansen and Troels Michael Lilja, "Shareholder Primacy and Property Rights Connected to Shareholding" (https://papers.ssrn.com/sol3/papers.cfm?abstract_id=3744162).

More recently, and maybe with a more balanced perspective, Guido Ferrarini, Michele Siri, Shanshan Zhu, "The EU Sustainable Governance Consultation and the Missing Link to Soft Law", April 2021 (https://ecgi.global/working-paper/eu-sustainable-governance-consultation-and-missing-link-soft-law).

[47] A good example of a fair study that looks at similar problems but engages substantial opposite literature is ESMA's Report "Undue short-term pressure on corporations": https://www.esma.europa.eu/press-news/esma-news/esma-proposes-strengthened-rules-address-undue-short-termism-in-securities.

The study is clear in its objective of championing a certain further specific regulatory outcome with impact in Member States national law.[48]

EY states that there is a trend for listed companies within the EU "to focus on short-term benefits of shareholders rather than on the long-term interests of the company", and that "sustainability is too often overlooked by short-term financial motives".[49]

EY also states that to some extent corporate short-termism finds its root causes in current regulatory frameworks and market practices.

Essentially, the study fails to understand the present-day reality of companies and sustainability because it is oriented by unsupported and biased statements and preconceived assertions rather than following a dispassionate, impartial, rigorous and comprehensive account of the role played by capital markets and listed companies in the dynamic and innovative, yet under construction, new sustainable economy.

European listed companies are not dominated by a philosophy of short-termism, boards of directors do take into account ESG issues, and all shareholders' interests are not always short-termism centric, as all stakeholders' interests are not always long-term centric.

To try to state the opposite, without presenting unassailable convincing strong evidence, is only a way of trying to solve a false dichotomy based on misconceived and contradictory reasonings.

EC's Consultation on Sustainable Corporate Governance

Following the EY' report, the EC launched a public consultation (in fact, a questionnaire), on a proposed sustainable corporate governance framework.

Unfortunately, and although a majority of Authors and respondents to the Study, both from the academic and market communities, besides pointing its shortcomings warned the EC that such a document could not be the basis on which to support legislative action, the flaws of the study were basically ignored by the EC and therefore most of the study proposals are also present in the questionnaire as in the EC's Inception Impact Assessment.[50]

---

[48] Although the study also fails in assessing the current solutions provided by different Member States national law and their respective outcomes.

[49] The study proceeds by identifying seven key problem drivers: (i) directors' duties, and company's interest to favour the short-term maximisation of shareholder value; (ii) growing pressures from investors with a short-term horizon; (iii) companies lack a strategic perspective over sustainability; (iv) board remuneration structures that incentivise the focus on short-term shareholder value; (v) current board composition inadequacy to support a shift towards sustainability; (vi) insufficient stakeholder engagement and involvement in current corporate governance frameworks and practices and (vii) limited enforcement of the directors' duty to act in the long-term interest of the company.

[50] See https://ec.europa.eu/info/law/better-regulation/have-your-say/initiatives/12548-Sustainable-corporate-governance.

All the public Responses to the consultation are available here: https://ec.europa.eu/info/law/better-regulation/have-your-say/initiatives/12548-Sustainable-corporate-governance/feedback?p_id=8270916.

Also, the consultation was framed in a way that does not meet minimum standards of transparency and participation as respondents were limited by a structure and methodology, based on a "yes/no" choice "box ticking" approach, biased towards a legislative approach to which the EC seems already committed, and not allowed to freely express the full range of their views.

Overall, to say the least, the questionnaire is a disappointing document.[51]

But, more importantly, the set of 26 substantive questions presented by the EC allows an observer to draw a map of its intentions which may include legislative initiatives regarding: (i) the obligation of directors to balance the interests of all stakeholders, as part of their duty of care, instead of, in the EC's view, focusing on the short-term financial interests of shareholders, (ii) the obligation of identifying the interests that are relevant to the long-term success and resilience of companies, (iii) the obligation of companies and their directors to take into account stakeholder interests alongside the financial interests of shareholders, (iv) the mandatory integration of sustainability risks and opportunities into companies' strategies, (v) the consecration of a stakeholders role in the enforcement of the directors' duty of care (including through liability actions against directors), (vi) options to counter executive remuneration promoting a short-term focus, (vii) options to enhance sustainability expertise in the board.

The EC's Initiative on Sustainable Corporate Governance

Go wisely and go slowly. Those who rush stumble and fall.[52]

In our view, by acritically accepting the results of the study, the EC seems to completely ignore that all the main aspects it recommends covering are already sufficiently accounted-for by existing legal requirements, both of international and national origin, and *soft law* mechanisms like national corporate governance codes.[53]

And in fact, this is so because the current legal and regulatory framework reflects the measures adopted by EU legislators, particularly the ones approved since the last financial crisis towards integrating a closer vision to a

---

See also the EP' "Report on Sustainable Corporate Governance (2020/2137(INI))", Rapporteur: Pascal Durand, December 2020 (https://www.europarl.europa.eu/doceo/document/A-9-2020-0240_EN.pdf).

[51] "EC Corporate Governance Initiative Series: Comment by the European Company Law Experts Group on the European Commission's Consultation Document 'Proposal for an Initiative on Sustainable Corporate Governance', (https://www.law.ox.ac.uk/business-law-blog/blog/2020/12/ec-corporate-governance-initiative-series-comment-european-company).

[52] From William Shakespeare, "Romeo and Juliet".

[53] As shown by Ferrarini et al., "The EU Sustainable Governance Consultation and the Missing Link to Soft Law" (https://ecgi.global/working-paper/eu-sustainable-governance-consultation-and-missing-link-soft-law).

stakeholder model into EU company law and the very recent but numerous initiatives in sustainable finance, which, globally, provide a valuable framework to incentivize companies to progress in the direction of sustainable long term.

At the same time, the intentions of the EC regarding corporate law appear to directly contradict other EU initiatives, like the Taxonomy Regulation, the (upcoming revision of the) Non-Financial Reporting Directive, the Regulation on sustainability-related disclosures in the financial services sector and the upcoming Directive on corporate due diligence, all aimed at improving companies' commitment with disclosure, transparency and sustainability, or the CMU, with focus in further useful equity investment, and more specifically retail investments, into capital markets.

Thus, from the point of view of legal coherence there would be strong inconsistency in an initiative that would certainly reduce investors' incentives to provide equity and risk capital to companies, making companies capitalization even more difficult and problematic, furthermore when one knows that institutional investors and shareholders also have a leading role in exerting influence to get companies and directors behind ESG goals and efforts.

Consequently, it is of no surprise to see that, alongside some of the world's leading corporate governance scholars, the EC initiative also generated a rare consensus between market players *across the aisle*, as both business, exchanges and investors share a concern that the upcoming proposal on sustainable corporate governance, if going beyond the form of recommendations, would be counterproductive, paralysing the functioning of boards and hampering the ability of companies to act decisively to promote a sustainable transition.[54]

The reality is that the EC seems to ignore that the boards of directors of listed companies have been very proactive in approaching sustainability topics not as a compliance issue but as an integral part of the culture and strategy of companies, answering to investors and stakeholders' attention to sustainability and ESG.

One can accept that, as everything in life, companies could still do better and even more than they already do, and all signs are encouraging that they will do, but either the study or the consultation didn't provide any evidence or sensible suggestions in that regard.

In fact, an extensive set of international reports show that listed companies have been at the forefront of the sustainability progress and are fully aware that they will only ensure future growth if and when taking into account the complete ecosystem in which they operate, which is made up of many different stakeholders.[55]

---

[54] See the joint letter by EuropeanIssuers, Better Finance, ecoDa, European Family Businesses, Federation of European Securities Exchanges (FESE), Invest Europe and SMEUnited (http://www.europeanissuers.eu/publications-viewer#?id=1633).

[55] Moreover, arising from the observation of listed companies' reports a number of notes allow us to identify acknowledgement of the most common categories of risks, particularly, physical, financial, value-chain, technology, regulatory, litigation and reputational risks, as

Companies know that the success of a corporation is ultimately the result of work that embodies contributions from a range of different resource providers including investors, employees, creditors, customers and suppliers, and other stakeholders.[56]

For long, companies have been recognising that the contributions of stakeholders constitute a valuable resource for building competitive and profitable corporations and that it is in the long-term interest of corporations to foster wealth-creating cooperation among stakeholders.

That is the reason why, even in areas where stakeholder interests may not be legislated, many companies make additional commitments to stakeholders, knowing that concerns over corporate performance and reputation often require the recognition of broader interests.[57]

Companies do not need to be reminded to manage risks and opportunities, as well as the companies' interest and stakeholders' related interests, as these are well known essential elements already taken into account by boards when

---

also shown by plenty of evidence available in European/international reports as in national reports.

In this particular, we find the Portuguese experience a very impressive one; in Portugal as in most of other countries in Europe, listed companies, and its boards of directors', have been at the forefront of the process of incorporating ESG factors into corporate strategies as a way to best attract new investment towards sustainable, long-term value creation.

Specific literature regarding sustainability reporting by Portuguese listed companies is still limited but significant signs exist that companies are incorporating ESG considerations into their decision-making processes and operations, as well described and substantiate in their publicly available sustainability reports.

An empirical analysis of governance and sustainability reports published by Portuguese publicly traded companies during the year 2020 seems to offer useful evidence, testifying on how companies perceive and carry out the instructions received from the regulatory framework and opening a window to the reality of business and governance bodies practices, confirming that companies know, apply, and, in many cases, expressly associate sustainability elements with strategy, organisational culture, ethical behaviour, purpose, strategy, risk management, long-term investment, organisation continuity and the preservation of its global community.

This observation shows that most Portuguese listed companies are discussing sustainability commitments in view of their alignment with the UN SDG, and some are explaining their efforts to embody specific SDG; besides, annual reports already published in 2021 show great attention to social issues in the midst of the pandemic situation.

Duarte Calheiros, Rui Pereira Dias, Abel Sequeira Ferreira, and Mafalda de Sá, "Annual Monitoring Report of the Portuguese Corporate Governance Code" also confirms a high level, and continuous increase, of compliance with corporate governance recommendations (https://cam.cgov.pt/images/ficheiros/2020/relatorio_2020_en.pdf).

[56] As fully recognised by the G20/OECD Principles of Corporate Governance (https://www.oecd.org/corporate/principles-corporate-governance/).

That is also the true meaning and relevance of the Business Roundtable Statement on the Purpose of a Corporation, which, in our view, does not redefine but indeed reiterates Company purpose (https://system.businessroundtable.org/app/uploads/sites/5/2021/02/BRT-Statement-on-the-Purpose-of-a-Corporation-Feburary-2021-compressed.pdf).

See also Colin Mayer, Shareholderism versus Stake-holderism—A Misconceived Contradiction. A Comment on "The Illusory Promise of Stakeholder Governance" by Lucian Bebchuk and Roberto Tallarita, June 2020 (https://ssrn.com/abstract=3617847).

[57] G20/OECD Principles of Corporate Governance.

establishing the companies' business strategic plan, which is not and should not be a matter of law; therefore, lawmakers and regulators should not submit companies to new intrusive and unnecessary regulation.

The matter is all the more serious as some of the possible legislative initiatives to come may have far-reaching harmful impacts for companies and the European economy.

Paramount of that is the possible change in the concept of the company interest apparently in order to welcome an obligation of directors to (*equally?*) balance the interests of all stakeholders, as the first and foremost sign of an intended exacerbated multiplication of directors' duties.

Obviously, a new European legal obligation to "balance the interest of all stakeholders" would lead to corporate dilemma and decision-making stalemate, as any mandatory requirement to take into account the often contradictory and constantly changing interests of a large variety of possible stakeholders would probably result in an impossible or at least rather confusing legal obligation that would only contribute to fuel litigation.[58]

And the point is that there is no need, much less any urgency, to generate such a confusion through intervention into national company laws, as there is no evidence of a market failure or regulatory gap regarding board members fiduciaries duties in respect to the long-term sustainable orientation of companies' strategy: all Member States already provide mechanisms, although with different degrees of latitude, requiring directors to act in the interest of the company, further accepting that the interest of the company may involve a multitude of stakeholders such as shareholders, bondholders, employees, creditors, suppliers, customers and the community and public at large.[59]

---

[58] As evidenced by the classic Stephen Bainbridge hypothetical, in "A Duty to Shareholder Value", shareholders are not necessarily the ones with a short-term interest and stakeholders don't always have a long-term interest (https://www.nytimes.com/roomfordebate/2015/04/16/what-are-corporations-obligations-to-shareholders/a-duty-to-shareholder-value).

[59] Alongside European Regulations becoming automatically binding throughout the EU, Directives being incorporated into national law, and other international instruments being adopted by EU countries, there are as well national developments, for example, through corporate governance codes. This is the case in Portugal where the recently updated Corporate Governance Code presents a sharper new focus on sustainability, based on a comply or explain approach. The Code explicitly includes a new recommendation (R. IV.3.) on sustainability, unparalleled in the previous version, focusing on strategy and executive management, its relationship with the larger community of stakeholders and how this contributes to a company's long-term and sustainable success, and expressly calling companies for "promoting the long-term success of society" and "contributing to the community in general" (stakeholders are specifically referred to in R. I.1.1). See also, the first General principle, Principles IV.A., V.2.A. and V.2.B., and Recommendations V.2.7. and V.2.8 (https://cam.cgov.pt/images/ficheiros/2020/revisão_codigo_en_2018_ebook-05.11.2020.pdf).

Therefore, along with other main benchmark governance codes in Europe, the Portuguese Corporate Governance Code has the merit of addressing sustainability, and its importance, including references to the role of the company and its responsibilities before the community in general, and the need for consideration and articulation

Directors of well managed companies know that "stakeholders are their business" and that, from a business perspective, they are required to know their stakeholders, to develop mutually beneficial relationships with customers, suppliers and employees, and to carefully weigh and balance the interests of all stakeholders' interests in order to achieve success; therefore, there is no need for the introduction of new legal requirements to force directors to engage with the company's stakeholders.

To follow such path is unnecessary but also seems dangerous and potentially harming for companies, shareholders and stakeholders.

Translating these intentions into an array of new directors' duties will likely expose companies to: (i) a complexification of the board's decision-making process, (ii) directors' liability perceived as unlimited, (iii) a likely impracticability to set the limits of the business judgement rule, (iv) increased and continuous risk of litigation.

In such a scenario, characterised by high levels of legal uncertainty and management unpredictability, companies and shareholders will certainly be harmed by the detrimental effects of directors' increased aversion to risk, board slowness and administrative burden, resulting in performance decrease, affected competitiveness, disruption in access to capital and loss of attractiveness for European listed companies.

But the stakeholders themselves, besides being harmed by all these negative impacts will also see a very likely reduction of the potential multitude of interests that the board would otherwise be able to consider as the board would become limited by binding statutory requirements.

As a conclusion, a possible EC' initiative on sustainable corporate governance must be completely reconsidered before any further regulatory reform.[60]

And, in any case, such fundamental change to the heart of company law cannot be based on a Study as biased and incomplete as the one presented by EY; such a discussion, if necessary, would have to start from an informed and consensually accepted analytical work and evidence that does not exist at the moment.

---

between purpose, strategy and business culture, and the long-term sustained performance and sustainability of the company, common themes that companies must tackle in an articulated, consistent and transparent way. Sustainability integration into the Portuguese corporate governance and its reporting by companies will be subject to monitoring already in 2021. For a view on how corporate governance codes in Europe have been recently adapted to sustainability requirements, see Ferrarini et al., "The EU Sustainable Governance Consultation and the Missing Link to Soft Law" (https://ecgi.global/working-paper/eu-sustainable-governance-consultation-and-missing-link-soft-law).

[60] Alex Edmans, Luca Enriques, Jesse Fried, Mark Roe, Steen Thomsen, made a "Call for Reflection on Sustainable Corporate Governance", April 2021 (https://ecgi.global/news/call-reflection-sustainable-corporate-governance), emphasizing the need to separate the two issues conflated by EY and the EC: "the *horizon* (short-term vs. long-term) and the *objective* (shareholder value vs. stakeholder value)". Since publication, this Call has been co-signed by a large number of corporate governance scholars and researchers.

The same is true for any initiatives, ESG related or not, that may contribute to enlarge the number of obstacles and disincentives to invest in European companies, to increase the already huge compliance costs on companies, to create even more legal uncertainty for companies and their directors or to deepening the unlevel playing field vis-à-vis third-country competitors.

Facing a pandemic crisis that is going to last—with potentially long and stronger effects in the future of EU businesses' survival and growth—, this is the moment for listed companies to put all their energies and resources in their business activity, the recovery of the economy and the generation of employment, and not to be distracted by ideological or vain attempts to make an already incredibly difficult business environment even more complex for issuers, hindering innovation and growth in Europe.

A Portuguese Contribution

Nonetheless, and in spite no evidence at all is presented, the Study claims that, from among twelve Member States, Slovakia, Belgium, the Netherlands and... Portugal are the most short-term oriented ones.

One can easily disregard such assertion as one platitude more, as the analysis is presented ignoring all aspects of the Portuguese general economic situation before 2016, namely the impact of the financial crisis on earnings in listed companies of Portugal, and there is no information allowing readers to exactly understand what such a conclusion means as the study fails in both justification of methodology and the definition and implications of short-termism.

But, as for other jurisdictions, the fundamental question to address, and the most relevant in its articulation with the sustainability discussion would be the one regarding the purpose of the corporation in Portugal to investigate if there is a market failure in the legal framework (an aspect that was completely ignored by the EY' study).

In Portuguese law, this discussion concentrates on article 64 of the Companies Code, which focus on directors' legal duties, namely a duty of care and a duty of loyalty, the former interpreted within the scope of the business judgement rule.[61]

Referring to a duty of loyalty and diligence, article 64, 1, b), states that directors must act "in the interest of the company, taking into account the long-term interests of the shareholders, and considering the interests of other matters relevant to the sustainability of the company, as well as its employees, clients and creditors".

This rule therefore incorporates a broad definition of corporate purpose: directors must act and perform according to the company's interest but considering shareholders' long-term interests and other stakeholders' interests.

---

[61] See Article 72, 1, of the Portuguese Companies Code. Article 64 refers to "company interest" and not to "company purpose"; to know if the two concepts can technically be assimilated to the same reality would require further investigation which is beyond the scope of this chapter; nonetheless, in this chapter they are treated as similar concepts.

In the directors' decision-making processes, the (long-term and sustainable) corporate interest is the one that must prevail, followed by the long-term interests of shareholders (but not their short-term interests, which is not the same as to say that short-term shareholder interests do not deserve protection), and also considering (balancing) the interests of other subjects relevant to the sustainability of companies and communities, such as the interests of other stakeholders.

The wording of the provision also adds to the clarification of a certain sense of hierarchy: directors must "act" in the interest of the company, "take into account" the long-term interests of the shareholders and "consider" the interests of other matters relevant to the sustainability of the company, "as well" as other stakeholders.

This said, the way article 64 mandates directors and the board to act in the interest of the company as a whole and refers to the right way to articulate that interest with other potentially contradictory interests of different stakeholders, remains open to multiple interpretations and the same is true concerning the weight to be attributed to each conflicting interest or the appropriate combination of interests at any given moment.

It is our view that this provision establishes a mitigated shareholder model based on a specific hierarchy of interests.

However, regardless of the way we characterise the Portuguese model, what article 64 clearly does not do is to instruct or incentivise directors to pursue short-term benefits at the costs of long-term benefits or to somehow neglect the future of the company; indeed it provides for exactly the opposite, as just shown.

Irrespective of further discussion regarding the best interpretation of article 64, the important point to be made is that the provision already covers as primary interest to be respected the concerns with the sustainable development of the corporation.

At the heart of the provision is the reference to "matters relevant to the sustainability of the company" which acts as the decisive criteria to identify what is the best course of action for the corporation and, for example, to assess the relevance of the long-term interests of the shareholders.

And, at the same time, it is also clear that article 64 provides for the consideration of (sustainable) interests other than the shareholders ones, emphasising the interests of workers, customers and creditors, but, it seems, allowing to also consider the interests of suppliers or the community at large as the list of stakeholders is merely exemplary.

In conclusion, during the 2006 reform of the Companies Code, the Portuguese lawmakers had already found a balanced legal answer that rejects a view of short-term shareholder profit maximisation at any cost, to the challenges and questions that seem to concern the EC nowadays, clearly defining the duties that require directors to take stakeholders' interests into account when pursuing the interest of the company.

A similar investigation regarding the mechanisms to be found in jurisdictions of other Member States is beyond the scope of this chapter, but abundant literature shows that similar solutions prevail in other national jurisdictions and that no changes to national laws are necessary to make clear that short-termism, in the sense of deliberately neglecting the sustainability and future of the company or its community, is contrary to directors' duties.

What is true is that if the ongoing sustainable corporate governance initiative goes ahead as anticipated, it will have a profound disruptive impact in traditional legal systems of continental Europe, generating much more questions and problems than the current non-existing ones that the EC misconceivedly intends to solve.

As the Portuguese example shows, a more open legal definition of the corporate interest, in articulation with a similar approach developed in national corporate governance codes, can help to find a balance that practically solves the (artificial) debate regarding stakeholder *versus* shareholder interests, as far as the board of directors is emboldened to pursue long-term value creation while considering other interests relevant to the company business and its community.

And, at the same time, more innovative and dynamic mechanisms than traditional hard law are not only more flexible and effective in attending all interests concerned but also can be revised and updated with greater ease in the future, which is an important point to take into consideration when facing a legislative process as complex as the European one.

5 Board of Directors Involvement

A final brief word is due regarding board involvement, board structure and its relevance for sustainability.

European jurisdictions are more or less evenly divided between favouring one-tier board systems or two-tier boards, with a number of jurisdictions allowing both one and two-tier structures.[62]

The flexibility to choose between single or two-tier boards is consistent with EU regulation for European public limited-liability companies (*Societas Europaea*).[63]

Literature doesn't seem to show enough evidence of some board structure being best than the others in addressing climate risk and ESG, as this

---

[62] See OECD (2019), OECD Corporate Governance Factbook 2019 (www.oecd.org/corporate/corporate-governance-factbook.htm).

[63] Portugal, as it is the case also for Italy, has a hybrid system that allows shareholders to choose one of three mandatory governance models and provides for an additional statutory body mainly for audit purposes. For more detail: "Chapter 17 – PORTUGAL", Paulo Olavo Cunha and Cristina Melo Miranda, The Corporate Governance Review Tenth Edition (https://www.vda.pt/xms/files/05_Publicacoes/2020/Livros_e_Artigos/The_Corporate_Governance_Law_Review_POC_Cmiranda.pdf).

greatly depends on board composition and mainly on their industry or sector of activity and how affected it is by climate change.[64]

The important point to emphasise is that companies and boards should carefully examine their purpose, culture and strategy, and assess which is the better internal structure and procedures to address ESG issues in order to ensure that all aspects and changes are subject to detailed focus and research and are thoroughly aligned with business and operations.

In our view, the board of directors' most important responsibility is the company's strategy, including its future direction and competitive position.

Thus, in relation with sustainability, the board's strategic role entails (i) understanding the ESG ecosystem, (ii) recognising ESG risks and opportunities as structural changes, (iii) realising how ESG developments impact stakeholder expectations, (iv) guiding ESG strategy development and oversight that drives value to the company and its stakeholders, (v) preparing for continuous change in business and take advantage of competitive changes in the market, (vi) ensuring stakeholder engagement, (vii) adapting to increased and complex regulation and ESG compliance, (viii) devising and adopting risk management and internal control processes, (xix) anticipating detailed disclosure and reporting for ESG and sustainable products.

Additionally, no matter what their structure and composition, boards must also play a critical role in identifying priorities and establishing goals and objectives in regard to a number of topics of increasing reputational relevance such as diversity targets, community and societal impacts, supply chains and human rights.

6   Conclusions

Most large companies and listed companies in Europe are fully aware of ESG challenges and opportunities and have been investing resources to foster climate change mitigation and adaptation, social issues and best governance practices.

Companies are also very much aware that their ability to deliver long-term value is tied to its reputation within their communities and so, as well as

---

[64] Naturally, the two-tier board model, as used in Germany, being itself a reflection of stakeholder primacy and codetermination, allows companies to better embrace stakeholders integration and engagement; see Klaus Hopt & Patrick C. Leyens, "The Structure of the Board of Directors: Boards and Governance Strategies in the US, the UK and Germany", April 2021 (https://ecgi.global/working-paper/structure-board-directors-boards-and-governance-strategies-us-uk-and-germany). Pedro Jorge Magalhães, "Governo Societário e a Sustentabilidade da Empresa", Almedina, 2019, specifically looking at sustainability issues, also argues that for large corporations the German-inspired dualistic model could be advantageous and more tailored to fit the participation of the main stakeholders, as it ensures better participation of those which should be seen as long-term investors in society. But, as the Author also recognises, all models can be multifunctional and include the involvement of different stakeholders, and, although in theory the two-tier system may have some advantages, those same advantages can be obtained in one-tier structures through the application of rules on independent non-executive directors.

looking at ESG-related opportunities, companies are also paying particular attention to climate and other risks and the different ways they can impact their business and operations.

Even under extremely difficult conditions, during the COVID-19 pandemic, companies kept implementing climate solutions, leading the process of industry transformation, driving innovation and growth and fighting to keep long-term jobs.

Of course, companies can and will be required to do more and most of them have already committed their way forward on ESG, SDG and sustainability strategies.

However, for that purpose, the EC will have to make great progress towards simplification on its approach to sustainability regulation, in a way that promotes the development of capital markets as a true solution for companies' capitalization, increases the attractiveness of EU stock exchanges for European companies and encourages willingness among companies to apply the vast catalogue of requirements implied in its regulations.

The EC cannot go on launching too short and inadequate consultations and disregarding the opinions and evidence provided by market participants concerning the way its regulations are effectively hindering the real economy.

By acting on alternative facts and insisting in initiatives (like the "sustainable corporate governance initiative") the EC may create new obstacles for effective ESG impact-investing strategies and hopelessly slow down the real economy transition to a new climate economy.

This is the time for the EC to encourage the use of (its own) already existing reliable mechanisms that may help to scale up long-term oriented strategies and sustainable economic activities, and to act to correct true market failures (like the ones regarding its new Taxonomy Regulation or ESG ratings) or enhance shareholders engagement, not to wreak havoc and add to confusion and disruption, carelessly interfering with carefully crafted national company law developed over a long period of time.

It is also the time for the EC to offer further clarification on their initiatives and its implications, showing clear connections and benefits to new requirements and obligations, in order to help companies to better understand and willingly integrate them into their overall strategies.

CHAPTER 18

ESG in Growth Listed Companies: Closing the Gaps

Filipe Morais, Jenny Simnett, Andrew Kakabadse, Nada Kakabadse, Andrew Myers, and Tim Ward

1 INTRODUCTION

The 2020 Global Pandemic has provided impetus to the sustainability agenda around the globe. Governments, investors and companies have awakened to the urgency of accelerating the transition towards a sustainable economy. The EU has launched an ambitious Green Deal[1] promoting it as a paradigm shift for the European economy. The election of President Biden has re-instated the US in the Paris Agreement and is a key player in the global green transition. There is therefore a unique set of conditions that indicate a renewed momentum for the sustainability imperative.

---

[1] European Commission (December, 2019). The European Green Deal, Available from: https://eur-lex.europa.eu/legal-content/EN/TXT/?uri=CELEX:52019DC0640#document2.

F. Morais (✉) · J. Simnett · A. Kakabadse · N. Kakabadse · A. Myers
Henley Business School, University of Reading, Henley-on-Thames, UK
e-mail: f.morais@henley.ac.uk

J. Simnett
Tower Hamlets Community Housing, London, UK

T. Ward
The Quoted Companies Alliance (QCA), London, UK

© The Author(s), under exclusive license to Springer Nature Switzerland AG 2022
P. Câmara and F. Morais (eds.), *The Palgrave Handbook of ESG and Corporate Governance*, https://doi.org/10.1007/978-3-030-99468-6_18

In countries where this shift is more market-led such as the US and the UK, the responsible investment community and quoted companies have also shown renewed strength both in terms of the volume of capital allocated to sustainable business as well as on companies adopting meaningfully sustainability frameworks. Just at the time of writing, Vanguard, the worlds' second largest asset manager with 7 trillion dollars of assets under management and 42 other large investors, have pledged to slash emissions across their portfolios, committing to net zero by 2050.[2] However, from announcement to action there is a huge gap even among the world's largest investors and firms. For instance, the much-celebrated 2019 Business Roundtable announcement about redefining the purpose of business and committing to a more stakeholder-centric governance model has been heavily criticised as empty rhetoric (Bebchuk and Tallarita 2020). Moreover, significant financial flows are being increasingly invested in private markets which are largely escaping the sustainability agenda, whereas public companies cannot. According to McKinsey's Global Private Markets Review "private market Assets Under Management (AuM) grew ... $4 trillion in the past decade, an increase of 170 percent .... Over that same period, global public market AuM has grown by roughly 100 percent, while the number of US publicly traded companies has stayed roughly flat (but is down nearly 40 percent since 2000)" (McKinsey 2020).

In public markets, the large, listed companies generally in highly "dirty" businesses such as oil & gas, energy, mining, cement and certain types of manufacturing regularly make the headlines for good and not such good reasons and are front-of-mind for governments, regulators and other stakeholders when sustainability is discussed and policy and regulation crafted. This leaves a significant and important market segment out of the conversation and unsupported: small and mid-cap (SMID) companies, often referred to as growth companies.

These growth companies have specificities in terms of governance and stages of development that require closer attention and examination. They typically have less resources and critical mass and more short-term pressures to survive and grow, exposing boards more acutely to the dilemmas and trade-offs involved in sustainable growth.

Despite growth companies' importance, little is known about how they perceive the ESG agenda and how equipped and supported they feel to go on that journey.

In November 2020, the report "ESG in Small and Mid-Cap companies: Perceptions, Myths and Realities" (Morais et al. 2020) was published providing a picture of ESG adoption in UK small and mid-sized quoted companies.

---

[2] *Financial Times* (March, 2021). Vanguard pledges to slash emissions by 2030. Available from: https://www.ft.com/content/87becf56-a249-4133-a01b-1b4b3b604bd5.

In this chapter, the authors draw on their research carried out for this report to provide an examination of four critical gaps in ESG adoption in UK SMID companies recommending practical ways for company boards to close them.

The chapter is organised as follows. First, a brief overview of the state of affairs on ESG in small and mid-sized companies and key initiatives to support them across the world is undertaken. It follows an overview of two UK growth company markets: the London Alternative Investment Market (AIM) and the Aquis Exchange. The Henley Business School-QCA study (Morais et al. 2020) on ESG adoption in growth companies is introduced and the following sections identify and discuss four ESG gaps by sector found in the study. The chapter outlines and discusses key recommendations for growth companies in their efforts to start or strengthen their ESG adoption and therefore close the existing gaps. The chapter concludes with some recommendations for boards on how to close these gaps on ESG.

2   ESG in Small and Mid-Sized Quoted Companies: Key Figures and Initiatives

There remains a distinct skepticism across the investment community about investing in small and mid-cap companies. For many investors small means riskier, more volatile and a lack of dividend yield. Adding ESG to the equation just exacerbates in most cases the already existing skepticism. Small and mid-caps are slowly making progress on ESG adoption on both sides of the Atlantic, but without enjoying the same level of support and incentives of large and very large caps.

Hermes Investment Management, an early adopter and leading sustainability asset management company, recognises that the "small-and-mid cap (SMID) sector is less heavily explored on the sustainable investment map" (Hermes 2019). The reasons for this include the fact that "ESG assessments often rely on subjective judgments that don't immediately translate into top-line revenue or bottom-line profits in financial projection models", "the poorer quantity and quality of disclosure in SMIDs" and the difficulty to assess materiality in fast-growing companies (Hermes 2019, p. 2). In fact, only a small fraction of FTSE small and mid-cap companies report annually on key metrics such as "GHG intensity data", "GHG Scope 1", "water intensity", "percentage of women in the workforce" or "lost time incident rate", when compared to FTSE 100 and FTSE 250. Similar disparity in quantity and quality of reporting is observed in the US between the S&P500 and the Russell 2500. Furthermore, SMIDs tend to be more affected by key biases of ESG ratings such as the market-cap, disclosure, geographic, industry and reactivity bias (Doyle 2018).

In continental Europe, the picture is not much different. Since the beginning of the ESG data movement, SMIDs have been excluded from the development and structuring of the rating systems, even though these companies comprise 80% of listed companies in the continent. Furthermore, at least 40% of these companies operate in sectors with Europe's highest greenhouse gas emissions (Agriculture, Manufacturing, Transport, Construction, Electricity and Energy) and stringent legislation (Eurofi 2020). Despite this, there remains an "heterogeneity and lack of relevant information linked to the ecological transition demanded by the Green Deal and the Covid 19" (Eurofi 2020, p. 1). Much needs to be accomplished to support SMIDs on both sides of the Atlantic to progress the ESG agenda, not least enabling them to develop their capabilities to adopt ESG meaningfully.

Notwithstanding the delay in getting SMIDs on the journey, there are some initiatives that are emerging as potential platforms of self-regulation and support.

As part of the United Nations (UN) seventeen Sustainable Development Goals to be achieved by 2030, The Sustainable Stock Exchanges (SSE) initiative was launched in 2017 at the 57th United Nations Conference on Trade and Development (UNCTAD) in Thailand, with the publication of the report "The Role of Stock Exchanges in Fostering Economic Growth and Sustainable Development" (WFE and UNCTAD 2017). The aim is clear: to close the financing gap of SMEs via the creation and development of SME listings across the world. Underpinning a growth and sustainable development framework for stock exchanges was developed. The key policy priorities for stock exchanges are to (i) adopt and implement international standards of good governance; (ii) promote high-quality disclosure on material environmental, social and governance issues and; (iii) develop support programmes for SMEs aimed at strengthening management capacity and governance. Thus far, 66 out of the 106 stock exchanges tracked by this initiative have an SME listing. However, as we have seen in this section there remains significant challenges in achieving high-quality ESG disclosures and there is still too little support for SMEs to strengthen their management capacity and governance, even in some of the most sophisticated stock exchanges.

The next sections turn our attention to the London Alternative Investment Market and the Aquis Exchange as two examples of SME listings and introduce the Henley-QCA report.

3 THE LONDON ALTERNATIVE INVESTMENT MARKET (AIM) AND THE AQUIS EXCHANGE

The London AIM

AIM (previously called the Alternative Investment Market) is the junior market to the London Stock Exchange.[3] It is designated as an SME growth market in

---

[3] The London Stock Exchange AIM https://www.londonstockexchange.com/raise-finance/equity/aim.

EU parlance and also a multi-lateral trading facility (MTF). The exchange was launched in 1995 with ten companies' combined value of £82.2m to replace the Unlisted Security Market (USM) which had existed since 1980. As such, it responds to a gap in the market for founders and entrepreneurs to access company investment and fulfil their growth potential. This is especially important when growth companies cannot always afford a full listing on the London Stock Exchange main market which comes with regulatory requirements that AIM companies may struggle to meet. The 822 companies on AIM (by end of September 2020) are spread across 40 different sectors and originate from over 100 different countries. They have a combined market capitalisation of £130.6b, which is an all-time high for AIM.

AIM provides small and medium sized enterprises (SMEs) with a means of entry into a supportive advisory community, investors and private wealth or business angels who understand the needs of these types of businesses. Funds raised on AIM vary between £1m and £50m via an initial public offering (IPO) although some flotations have exceeded £100m in value. It is the most successful and established market for trading high-growth companies and as a result, is highly dynamic. The total funds raised during the year (by end of September 2020) was £4,085m. Since its inception, over 3,865 companies have raised over £115bn and it has become the world's most successful growth market, however today, very few companies use AIM as a springboard to the main market, fewer than 10 per annum and some companies move out of AIM back to private ownership. Mergers and acquisitions activity is one of the key reasons for companies leaving AIM with about 20 companies acquired during 2020. There are three indices for measuring the AIM, which are maintained by the FTSE Group and these are the FTSE AIM UK 50 index, the FTSE AIM 100 index and the FTSE AIM All-Share Index.

The advantage for SMEs is that AIM matches a diverse and knowledgeable investor base driven to provide capital and the young, but fast-growing companies who need capital to develop. AIM companies can therefore access institutional and retail investors and being based in London, it opens a door to a valuable pool of international capital. This support is critical for these fledgling businesses to succeed. By trading on AIM, these SMEs can use share capital as currency to make acquisitions as well as incentivise their leadership and employees. AIM also provides a regulatory framework to guide SMEs along their growth path.

The main market of the London Stock Exchange requires companies to have operated for a minimum of three years, have a market value of over £700,000 and able to float at least 25% of their share capital, leaving them sufficient funds to operate for a full year's trading. AIM does not impose these requirements and thus makes it an attractive marketplace for SMEs. As an exchange regulated market with much of the day-to-day regulation delegated to the Nominated Adviser retained by an AIM-listed company at all times, AIM does not apply heavy regulation on growth companies and relies on SMEs to largely self-regulate. However, they are required to prepare and file audited annual accounts. AIM companies are permitted to use the "plc" tag as a public limited company, a tag which is derived from company law and is a designation that shows a company has shares that can be purchased by the public and which has allotted share capital with a nominal value of at least £50,000. It should be noted that not all "plc" companies are listed. The AIM regulatory model is based on "comply or explain" and a range of principle-based rules.

The Aquis Exchange

The Aquis Exchange is a relative newcomer to the UK stock market and offers a virtual exchange through cloud-based technology.[4] It is an independent pan-European exchange operator and technology services provider. It was launched in 2012 and is now the 7th largest exchange in Europe. Aquis undertakes cash equities trading in 14 European markets and is authorised and regulated by the UK Financial Conduct Authority (FCA) and France's Autorité des Marchés Financiers to operate Multilateral Trading Facility businesses in the UK and in EU27, respectively. Since 2015, Aquis has also developed and licensed exchange software to third parties. In 2020, it became a Recognised Investment Exchange under Section 285 of the Financial Services and Markets Act (2000). Aquis has also received recent approval from the FCA to operate a bond trading market.

Aquis is a matching system which also provides trade surveillance technology. This virtual exchange with no physical trading floor is further differentiated in the market by its innovative subscription pricing model, which works by charging users according to the message traffic they generate, rather than a percentage of the value of each stock that they trade. This pricing model is claimed to significantly reduce the cost of trading. Aquis regulates the conduct of its members firms and issuers through monitoring trading and market activity. It has rules and guidance to promote a fair and transparent marketplace for growth companies to raise capital and trade securities. Aquis claims to have created a marketplace with lower toxicity and aggression when compared to other conventional trading venues in Europe.

The Henley-QCA Study on ESG Adoption in Growth Companies

To redress the distinctive lack of research on small and mid-sized quoted companies and their ESG adoption, the Quoted Companies Alliance[5] (QCA)

---

[4] The Aquis Exchange https://www.aquis.eu/.

[5] The Quoted Companies Alliance (QCA) is an independent membership organisation, championing the interests of approximately 1,250 SME quoted companies. The QCA has developed its own code of corporate governance, tailored to the needs of smaller growth companies, which has been adopted by about 90% of all companies on AIM in preference to the UK Corporate Governance Code, issued by the Financial Reporting Council. For more info: https://www.theqca.com/.

and Downing LLP, an asset manager from the City of London, asked a team of academics and researchers from Henley Business School to investigate the state of affairs in this market segment.

To this end 30 in-depth interviews were conducted with board directors and investors and two different versions of a survey were designed: one for companies and one for investors. The company survey was distributed to QCA corporate members, from which 100 completed and valid responses were obtained. Some 53% of companies had over 250 employees, and 47% had less than 250 employees.

Among these, 47% had market capitalisations of £100 million or more, with 6% having over a billion pounds in market capitalisation. Some 17% traded on the London Main Market and 83% on the Alternative Investment Market (AIM).

The surveyed companies cover a wide range of sectors and respondents held various roles: 30% were chairs/non-executives; 54% were CEOs, CFOs or other executives and 16% were company secretaries. The investor version of the survey was distributed to investors via YouGov—the international research data and analytics group—and 50 completed and valid responses were gathered.

A report entitled "ESG in Small and Mid-Cap companies: Perceptions, Myths and Realities" (Morais et al. 2020) was published in November 2020, and favourably received by business and investment communities.

The next sections focus on some of the findings from this early report obtained from the interviews and the company version of the survey data. Specifically, an analysis and discussion of four ESG adoption gaps identified in growth companies by sector is provided.

4 ESG Adoption Gaps in Growth Companies: A Sectorial Discussion

The Henley Business School—QCA study has concluded that UK small and mid-sized quoted company boards "need to become less reactive and take the sustainability agenda seriously - in all that it has to offer - competitive advantage, risk management and the attraction of long-term financial gain" (Morais et al. 2020, p. 38). In fact, there remain significant gaps in ESG knowledge, accountability, leadership and execution, and disclosure in this market segment that require bridging. Each gap identified is defined as follows:

- ESG Knowledge Gap: perceptions of ESG value.
- ESG Accountability Gap: clarity of accountability for ESG.
- ESG Leadership and Execution Gap: allocation of specific resource to ESG, and perceived barriers to execution.
- ESG Disclosure Gap: company self-rating of MSCI criteria.[6]

---

[6] Morgan Stanley Capital International or MSCI is a world-leading provider of ESG ratings. Data is standardised across 5 categories: (i) Environmental; (ii) Human Capital; (iii)

Sectorial data on each of these gaps reveals different stages of development (Table 1). Even accounting for materiality some sectors lag clearly behind while others are early adopters.

It is evident that "Real Estate and Construction" and to a less extent "Banking and Financial Services" are the sectors leading ESG adoption in UK small and mid-sized quoted companies. These sectors are followed by "Retail, Food, Travel and Leisure" and "Utilities, Oil & Gas, and Chemicals" who are found as average performers. The "Aerospace and Engineering" and "Technology and Comms" sectors are clearly lagging. Next, the findings on each ESG adoption gap are discussed in greater detail.

Table 1 ESG gaps in UK small and mid-caps by sector

ESG gaps[a]/sector	Banking and financial services	Utilities, oil and gas, chemicals	Real estate and construction	Retail, food, travel and leisure	Aerospace and engineering	Technology and comms
Knowledge	***	**	***	**	*	*
Accountability	***	*	***	**	**	*
Leadership and execution	***	**	***	**	*	*
Disclosure	**	**	***	***	*	*
Overall	***	**	***	**	*	*

[a]ESG gaps sectorial rankings represent companies' self-rating/evaluation across several questions presented in the survey. *Knowledge Gap* includes statements such as: (i) How knowledgeable of the potential impact of ESG on performance in promoting the company to investors? (% Very knowledgeable); (ii) Understand the positive impact that ESG can have on long-term financial performance (Average highest agreement); (iii) A limited awareness at Board level is preventing the company from effectively managing ESG risks and opportunities? (% Yes); *Accountability Gap* includes: (i) Clear responsibility for driving ESG (% Board/Next Level down); (ii) Does ESG form an integral part of company's strategy / vision? (% Yes). *Leadership and Execution Gap* includes: (i) When did the company determine that effective management of ESG risks and opportunities could impact on its long-term financial performance? (% No current impact); (ii) Number of factors preventing the company from effectively managing ESG risks and opportunities? (Average number); (iii) Understanding of impact of ESG helps inform the development of strategy and business model regarding performance (Average agreement). *Disclosure Gap* includes: (i) Does company disclose sufficient information to allow investors to understand material ESG risks and opportunities? (% Very much so); (ii) Volume of communication with investors / shareholders about ESG? (Average number of channels); (iii) Currently use standards to evaluate and report on ESG? (% Yes)
Key *** = Best; ** = Average; * = Worst
Source Adapted from Morais et al. (2020)

Social Capital; (iv) Business Model and Innovation and (v) Leadership and Governance. Together these five categories include 30 sub-criteria.

5 THE KNOWLEDGE GAP

ESG has fast become a very familiar concept with companies but there is a wide range of knowledge and awareness of what this concept means.[7] It tends to be narrowly defined based on individual projects and applied in a piecemeal fashion by silos. This lack of integration especially with strategy, risk and remuneration, appears to be a source of frustration for investors. Last year saw ESG catapulted up the corporate agenda, not least with the Covid-19 pandemic as a catalyst focussing directors on the environment, stakeholders and governance. The best-entrenched of the three pillars is governance, with knowledge of "E" mostly customised by sector but often limited to carbon emissions from business travel or office energy consumption and recycling. The "S" pillar is most often understood via diversity and the gender pay gap, driven by regulation with employee welfare and mental health highlighted by Covid-19. However, there is generally little consideration of other stakeholder groups in ESG, such as supply chain partners, customers or competitors. SMEs are less knowledgeable about ESG, with 19.6% of them self-rating as very knowledgeable, over two thirds as moderately knowledgeable and 10.9% of SMEs as no knowledge at all. This lack of holistic understanding is partly due to companies being in the early phases of their ESG journey, especially smaller companies typically focused on short-term growth and innovation. For many smaller firms, good understanding of ESG by their boards and executive directors is critical to decision making. It can also be a route to accessing long-term finance and gaining a competitive advantage in the marketplace.

In terms of sectorial differences, all sectors rated themselves over 90% moderately or very knowledgeable except banking and financial services. This figure could well be overstated if the question was interpreted as one pillar of ESG. Real estate and construction was the sector most aware of ESG during the past 12 months (31%), alongside utilities, oil & gas, chemicals (28%), followed by retail, food, travel & leisure (23%). Technology and communications were the sectors rated least knowledgeable over the past year (16%). However, during this period, awareness of ESG and its link to long-term financial performance is now appreciated by approximately 30% of SMEs. Many SMEs (19.5%) are not currently focussed on ESG but anticipate greater focus in the coming 1–2 years. By sector, the sectors forecasting more activity are banking and financial services (27%), utilities, oil & gas, chemicals (22%) and technology & communications (21%). However, three major economic sectors predict relatively little focus in the next 24 months: retail, food, travel & leisure (8%), aerospace & engineering (14%) and real estate & construction (15%). These findings are surprising in the light of both the aerospace & engineering (28.6%) and the utilities, oil & gas, chemicals (27.8%) sectors indicating that they are too focused on the short-term to understand the long-term impact

---

[7] Regarding the scope of ESG, see Paulo Câmara, The systemic interaction between corporate governance and ESG, Chapter 1 in this book.

of ESG on performance. It seems congruent that the banking and financial services sector was rated the least subject to short-termism (13.3%) and this sector predicts greater ESG activity in the coming two years.

6 THE ACCOUNTABILITY GAP

Accountability for ESG is optimum when owned by both the board for strategy and risk mitigation, but by the CEO and the executive team for implementation and delivery of the ESG strategy. However, the reality is that accountability is both diffused and varied across the board and executive team. Investors prefer the board and especially the chair to own ESG strategy and the CEO to drive execution. This ideal is not evident as the board is only seen to be the internal driver in 44.6% of cases with the executive team owning ESG in 38% of companies. The figures are even more disparate when comparing ESG accountability by the CEO (28.3%) and the Chair (12%). When compared to larger companies, there is greater ownership by the CEO and Chair in SMIDs but diminished board accountability in SMIDs. The CFO in SMEs is often the senior executive who is driving ESG (29%).

The importance of board ownership cannot be overstated, however there are some clear sector differences. The best sectors for having a clear board responsibility for driving ESG are the real estate & construction (100%) and banking & financial services (77.8%). The worst sectors for this board level responsibility were found to be technology & communications (66.7%) and utilities, oil & gas, chemicals (63.2%). In terms of integrating ESG into the company vision and strategy, the sectors which self-rated themselves as doing this best were again real estate & construction (84.6%) and banking & financial services (73.3%). Similarly, the worst sectors for board ownership were technology & communications (52.6%) and utilities, oil & gas, chemicals (55.6%). Focus on ESG in the past 12 months ranges between 16 to 31% across all sectors which does not suggest a high priority. Of those companies which will start to focus on ESG in the coming 1–2 years, the sector range is 8%–27%. These figures do not suggest any urgency in embracing the criticality of ESG for business sustainability.

7 THE LEADERSHIP AND EXECUTION GAP

ESG is not an operational matter: it really means a fundamental transformation of how the company does businesses in society. That is why it requires ownership from the board and execution from the CEO—exactly because it goes to the heart of the very purpose of the company and touches every level of the business. Despite the fact that the CEO is not the one ultimately responsible for integrating ESG across the business in many small and mid-sized quoted companies, it is an indisputable indicator as to how ESG is understood and that tends to be in a piecemeal manner, without the commitment of sufficient resources towards its implementation. Indeed, there is a notable lack of resources being allocated or hired by companies for ESG implementation, and an apparent lag between understanding, discussion and action. Of the companies surveyed, 62% claim that ESG is integral to their strategy and vision, but piecemeal approaches are in evidence, with key projects such as diversity or customer service built into strategy and therefore merely "ticking the box". Companies see ESG execution as fashionable—and therefore optional—whereas investors see it as an imperative for quality, reputation and credibility. Investors would like companies to define ESG more broadly and customise it to their businesses, but companies are still perceiving it as a compliance exercise and providing little quantification in their communications to stakeholders. Distinct sectoral differences are evident in terms of ESG maturity and the degree of integration: the real estate, construction, retail, travel and leisure sectors are clearly ahead in most ESG activities. These sectors use their ESG knowledge to inform the development of their strategy and business model regarding performance. Firms in these sectors tend to feel less the short-term performance pressures as detrimental to pursuing ESG, because they understand that ESG needs to be integrated in their business model in order to continue to be competitive and successful. Contrary to these sectors, Utilities, Oil and Gas, and Chemicals for example, are struggling to turn their business models—who are inherently unsustainable—into sustainable businesses. True ESG integration will require longer time horizons throughout which there will be a shift of resources from old business models to new ones. Most companies in such sectors are more concentrated in managing ESG risks to their existing business model, and simultaneously attempting to fundamentally transform their way creating value—by no measure an easy feat.

8   THE DISCLOSURE GAP

As part of the general trend of improved disclosure in corporate governance, SMEs need to be able to communicate an integrated ESG policy and strategy. However, this seems to be a challenge when it comes to telling an engaging and credible story around ESG as 37% of SMEs lack clarity on how to communicate ESG to their stakeholders. They find it easier to report high-quality disclosures on leadership and governance, more familiar areas, but less so on environmental factors. A word of warning from investors is that SMEs are not unduly burdened with reporting disclosure to the extent that they are distracted from being innovative and disruptive businesses. Just over two thirds of SMEs claim to have a formal ESG statement (67%) and 58% say that it is integrated into their company strategy or vision. Most smaller companies are utilising their websites and annual reports as vehicles for disclosure, but with only 31.7% using investor meetings and/or roadshows. Very few growth companies use any of the global standards to evaluate and report ESG. The use of standards, such as the United Nations Sustainable Development Goals (UN-SDG), Global Reporting Initiative (GRI) or Sustainability Accounting Standards Board (SASB) is generally low with only 6% of SMEs using them, claiming that they are unsuitable or irrelevant for their businesses. SMEs appear to primarily favour the development of sector-specific standards (36.6%), followed by 19.5% who say that a single standard across all SMEs would be most relevant and helpful. The sectors most in support of using standards are real estate & construction with retail, food, travel & leisure.

In distilling out disclosure by sector across E, S and G, we see that the most challenging environmental reporting is done least well by the aerospace & engineering on air quality and technology & communications sectors on carbon emissions. The best quality of disclosure on E is seen within the real estate & construction on carbon emissions and banking & financial services sectors on waste and wastewater management. On S or stakeholder disclosure, SMEs are rated worst in the technology & communications sector on human rights and community relations, and again in the aerospace & engineering sector on customer privacy as well as access and affordability. The best sector performer on S reporting is the banking & financial services sector, especially on employee health and safety and product quality and safety. Similarly, the banking & financial services sector was rated also best for G or governance disclosure quality on board oversight and the worst disclosure was in the aerospace & engineering sector on competitor behaviour. The investor perspective is for SMEs to go beyond reporting purely KPIs, but include progress against targets, a growth history, risk mitigation, succession planning, director remuneration, customer service and outcomes, reputation management, visibility of crisis management, especially with Covid-19, climate, independence and board diversity, all of which, they argue, contribute towards the sustainable competitive advantage which is critical for growing companies.

9 Closing the Gaps: Recommendations for Boards of Directors

Early in 2021 a study conducted by a researcher from NY Stern made its way to the *Financial Times* headlines echoing that "too many boardrooms are climate incompetent" and outlining a "striking lack of directors with expertise in climate change and ESG issues". In fact of the 1,188 board members from the 100 largest US corporations, only 0.2% had specific climate change expertise and only 6% had broader environmental experience (Whelan 2021). Of course we are not suggesting that SMIDs start hiring climatologists onto their boards, if the world's largest companies themselves are not. However, there are many activities that large and small companies can initiate to meaningfully adopt ESG. ESG is highly contextual, specifically when we are speaking of high-growth companies that constantly change in scope and scale. Each company will have to plot its own journey. However, there are a number of questions boards need to ask of themselves in order to begin to close the gaps identified in this chapter.

Table 2 Closing the ESG gaps: questions for boards

ESG gaps	Questions for boards
Knowledge	• How can boards and executive teams fill the gaps in their holistic knowledge about ESG?
	• How can boards and executive teams continue to grow their ESG knowledge?
Accountability	• How can the board and executive team drive accountability for ESG?
	• How can the board and executive team create some urgency to embrace ESG?
Leadership and execution	• How can the board meaningfully embed ESG in the company strategy and ultimately, in the business model?
	• How can ESG be central to the company's culture or "how the things are done around here"?
Disclosure	• How could the board and executive team improve the company "story" on ESG?
	• How can boards and executive teams ensure that ESG is integrated with the rest of their business?

Table 2, provides a number of key questions for boards to consider in their attempt to close knowledge, accountability, leadership and execution, and disclosure ESG gaps.

To fill the ESG knowledge gaps in the board and executive team, one approach would be to elect a board champion for each of E, S and G with each director focussed on ensuring that there is equal attention given to each pillar. Understanding of the sectorial benchmarks and what the most advanced companies on ESG in their sector are measuring. Similarly, speaking to ESG consultancies can help fill the knowledge gaps through board and executive training. This is a key topic for board and executive strategy days including discussion about growing or hiring ESG expertise for the small and mid-sized quoted companies if relevant. These actions should help ensure that ESG will provide competitive advantage. Additionally, it could consult with key stakeholder groups to find out what elements of ESG are critical to the growing SME within the sector. Look for best practice in terms of conferences, working groups, academic research and membership of trade associations which can educate on ESG in the sector.

To create accountability for ESG, the chair and CEO could start with agreeing on a business case to share with executive and non-executive board directors to explain the importance and role of ESG. Whoever is most knowledgeable about ESG or an ESG specialist could be invited to present to the board. How the organisation is going to manage it and what they wish to do needs to form part of an ESG strategy which is aligned to the corporate strategy. The Company Secretary needs to incorporate ESG into board planning and agendas. The board may also wish to hire ESG skills either as part of the executive or non-executive teams. Furthermore, The Chair and CEO could decide a plan to educate and obtain buy-in to an ESG strategy. It could include examining competitors' public disclosure on ESG. A business case could be easily constructed for the business and how engagement with ESG at board level could help grow the SME business, especially with key stakeholder groups such as customers, employees, suppliers and investors. Asking investors for guidance and how ESG can benefit growth is a key activity.

To focus leadership and execution on ESG, there needs to be whole board engagement with ensuring ESG is central to boardroom discourse, whether ESG is a separate conversation or whether it is a critical part of three conversations under strategy, risk and reputation. The challenge for leadership teams is to embed ESG in the board mindset which starts with the Board Chair as champion. ESG then ideally needs to be integrated into corporate strategy and the business model, so with reinforcement, it becomes central to the board and company culture over time.

To optimise disclosure on ESG, there is a requirement to improve the company "story" on ESG which incorporates the growth trajectory of SMEs and how ESG is contributing towards both their competitive advantage and their sustainability. Integrating ESG with the rest of the business and therefore illustrating this integration in annual reports and on company websites does not need to create a reporting burden on SMEs. The emphasis is on impact and customisation of the message, not merely adding volume to disclosure. It is also showing how ESG is driven by strategy and permeates throughout the company's operations.

Boards will depart from different positions in ensuring the transition to sustainable and responsible business models. While large corporates have a resource advantage to enable this transition, small and mid-sized quoted companies have the advantage of being more agile and adaptable. In fact, small and mid-sized companies typically grow very fast and are used to having to scale, transform and even re-invent themselves quickly to survive and prosper. Where many businesses see ESG as a constraint to growth, others see it as a competitive advantage and an enabler of growth. There is clearly an opportunity for smaller companies to integrate ESG in their frequent transformations and make it a source for attracting finance and creating growth opportunities.

Irrespective of the position from where the board is departing, it is fundamental that a roadmap to sustainability that incorporates ESG holistically is developed in conjunction with key stakeholders and shared publicly, with clear targets and accountabilities well-defined.

A final consideration is how best to communicate and report the ESG story for the company in a credible and integrated manner. Growing companies need to tell their "story" from an ESG perspective, especially regarding the environmental pillar. Collaborating with friendly investors and shareholders who understand the potential impact of ESG and know the company history is helpful input to this "story". Seeking out any relevant ESG standards for the sector and applying them is also critical. The setting of compelling targets and explaining progress and setbacks is vital to show development. Participating in brainstorming with the board and executive team can help focus on the integration of ESG into the SME business. Key stakeholder groups (customers, suppliers, employees, shareholders, investors) can contribute on how they wish to or have integrated ESG into the business and this is a consultative exercise which helps fill gaps in knowledge. Ensuring that ESG is integrated with risk management and with the activities of all board committees is a key activity. And finally, boards and executive teams should consider where ESG integration will contribute to competitive advantage in the strategic plan.

Acknowledgements The authors would like to thank The Quoted Companies Alliance (QCA) and Downing LLP for their financial support of the original study as well as for facilitating access to their membership and customer base, respectively.

References

Bebchuk, L.A., and Tallarita, R. (2020). The Illusory Promise of Stakeholder Governance. Discussion Paper No. 1052 12/2020, John M. Olin Center for Law, Economics, and Business, Harvard. Available from https://papers.ssrn.com/sol3/papers.cfm?abstract_id=3544978.

Doyle, T. (2018). Ratings That Don't Rate: The Subjective World of ESG Ratings Agencies, American Council for Capital Foundation. Available from https://accfcorpgov.org/wp-content/uploads/2018/07/ACCF_RatingsESGReport.pdf.

Eurofi. (2020). Sustainability Transition: SMIDs Challenges. ESG Report on Small & Mid-Caps, Eurofi Initiative. Available from https://esg-report-on-small-mid-caps_zagreb_april2020.pdf (eurofi.net).

Hermes Investment Management (2019). Go to the Source. ESG Integration in Small- and Mid-Cap Equity Investing. Available from https://www.hermes-investment.com/us/wp-content/uploads/2019/05/bd03387-small-and-mid-cap-esg-integration-paper-q2-2019-us-version-final.pdf.

McKinsey and Company (2020). McKinsey Global Private Markets Review 2020. A New Decade for Private Markets. Available from https://www.mckinsey.com/~/media/mckinsey/industries/private%20equity%20and%20principal%20investors/our%20insights/mckinseys%20private%20markets%20annual%20review/mckinsey-global-private-markets-review-2020-v4.ashx.

Morais, F., Simnett, J., Kakabadse, A., Kakabadse, N., and Myers, A. (2020, November). ESG in Small and Mid-sized Quoted Companies: Perceptions, Myths and Realities. Available from https://www.theqca.com/article_assets/articledir_442/221356/QCA_Research_Report_ESG_in_Small_and_Mid-Sized_Quoted_Companies.pdf.

WFE and UNCTAD (2017). The Role of Stock Exchanges in Fostering Economic Growth and Sustainable Development. Available from https://unctad.org/system/files/official-document/WFE_UNCTAD_2017_en.pdf.

Whelan, Tensie (2021, January 1). U.S. Corporate Boards Suffer from Inadequate Expertise in Financially Material ESG Matters. NYU Stern School of Business Forthcoming. Available from http://dx.doi.org/10.2139/ssrn.3758584.

World Business Council for Sustainable Development (2020). Board Directors' Duties and ESG Considerations in Decision-Making. Available from www.wbcsd.org.

CHAPTER 19

ESG and Banks: Towards Sustainable Banking in the European Union

Mafalda de Sá

1 Introduction

Environmental, Social and Governance (ESG) factors are today of unequivocal concern to the financial system, the banking sector included. Nevertheless, the environmental aspect has thus far been given greater attention.

Indeed, environmental sustainability is deemed an existential goal for humankind and the lack thereof as a threat to society as we know it. While this may seem like a scenario that will come about much further down the line, we are already facing some consequences.[1] At the same time, action to transition to a low-carbon economy will initially also have adverse effects.

All of this spills over to the economy, which in turn affects the financial system.[2] ESG has become a source of actual financial risk, and not merely

---

[1] The first climate change bankruptcy, as coined by the *Wall Street Journal*, has allegedly already occurred—Russel Gold, "PG&E: The First Climate-Change Bankruptcy, Probably Not the Last", 18 January 2019. Available at: www.wsj.com.

[2] The World Economic Forum (WEF) has for long identified climate change as one of the pressing and most impactful global risks to the economy and the financial system—see WEF, "The Global Risks Report", 16th ed., 2021. For a description of the impacts of environmental risks to the economy and the financial system, see NGFS, "A call for action, Climate change as a source of financial risk", April 2019, pp. 13–17; Patrick Bolton et al., "The green swan—Central banking and financial stability in the age of climate change", Bank for International Settlements, 2020, pp. 17–20.

M. de Sá (✉)
Faculty of Law, University of Coimbra, Coimbra, Portugal
e-mail: mafaldadesa@hotmail.com

reputational consequences for financial institutions. Simultaneously, the financial system itself affects the course of the economy and the state of the environment and the global community, either contributing or hindering sustainable development.

In this regard, strong sustainability accounts for what corporations need to do to foster it, and not solely its impact on corporations (so-called "business case for sustainability").[3] Likewise, sustainability must be understood in a broad sense, without neglecting the social and governance factors, but rather integrating them with environmental concerns.

The mutual interdependency intrinsic to strong sustainability is particularly true for the banking sector, given its crucial role in financing—and hence steering—the economy. In fact, the impact of ESG in this sector has two relevant dimensions: (i) there is an impact on banks themselves, affecting aspects such as their purpose, the information they provide, the risks to be managed and overall prudential implications; (ii) and there is a further impact arising from banks as lenders, through the pressure they exert over financed entities, thereby potentially triggering changes in other companies.

Therefore, banks are particularly exposed to ESG risks from their counterparties, but they may also influence them and impact the non-financial sector.

Although the relevance of banks in terms of exposure to risks and the role they may play in ESG is nowadays undisputed, a comprehensive understanding of these risks and corresponding opportunities are still open for debate.

The European Union (EU) is at the forefront of policy initiatives to address ESG. Within the ambitious sustainable finance plan, banks are no exception to expected profound changes and in fact, sustainability has become a matter of prudential concern.

In this chapter, we will analyse how banks are affected by ESG risks, as well as their role in sustainable development, and look into the EU's regulatory agenda for the banking sector, including the proposed Banking Package amending the Capital Requirements Regulation (CRR)[4] and the Capital Requirements Directive (CRD),[5] of October 2021.[6]

---

[3] Beate Sjåfjell/Christopher M. Bruner, "Corporations and Sustainability", in Beate Sjåfjell/Christopher M. Bruner (eds.), *Cambridge Handbook of Corporate Law, Corporate Governance and Sustainability* Cambridge University Press, Cambridge, 2019, pp. 3, 6. For some considerations on the evolution of corporate sustainability, see Rui Pereira Dias/Mafalda de Sá, "Deveres dos Administradores e Sustentabilidade", in Paulo Câmara (Ed.), *Administração e Governação das Sociedades*, Almedina, 2020, pp. 49 et seq.

[4] Regulation (EU) No 575/2013 of the European Parliament and of the Council of 26 June 2013 on prudential requirements for credit institutions and investment firms.

[5] Directive 2013/36/EU of the European Parliament and of the Council of 26 June 2013 on access to the activity of credit institutions and the prudential supervision of credit institutions and investment firms.

[6] The analysis is limited to credit institutions, without taking into account the specificities of the law applicable to investment firms.

2 THE INTERACTION BETWEEN ESG AND THE BANKING SECTOR

ESG as a Risk to Banks

The identification of financial risks deriving from ESG factors started by focusing not only on the environment, but exclusively on climate change, as a factor with the potential to negatively affect the value of financial assets and liabilities. However, climate change does not exhaust the sources of financial risk, which further include environmental degradation, such as air, water and land pollution, water stress, biodiversity loss and deforestation.

Only now are studies and policy action broadening their scope to all *environmental risks* posed by the exposure to activities that may potentially cause or be affected by environmental degradation and the loss of ecosystem services.[7] This is more in line with a full sustainability perspective based upon the respect to all planetary boundaries.[8]

There are two main categories of relevant financial risks: physical and transition.[9]

Physical risks relate to the occurrence of climate or other environmental events, such as floods or droughts, leading to economic costs and financial losses. They can either be extreme weather events (acute risks) or gradual shifts in climate patterns (chronic risks), as well as other types of environmental degradation. *Transition risks* derive from policy action to prevent the occurrence of such events, in the context of an adjustment towards a low-carbon economy. Transition matters include technological changes and the obsolescence that will come with it, as well as behavioural changes in consumers and investors preferences and demands (market or public sentiment). There is also a non-negligible *liability risk*, emerging from environmental as well as social factors, relating to the financial consequences of legal claims from parties who have suffered ESG-related loss or damage.[10]

---

[7] As defined in NGFS, "Guide for Supervisors Integrating climate-related and environmental risks into prudential supervision", May 2020, p. 9. A sign of this trend towards broadening the scope of "E" factors is the newly-formed Task Force on Nature-Related Financial Disclosures which aims at providing "a framework for corporates and financial institutions to assess, manage and report on their dependencies and impacts on nature" and building "awareness and capacity to reduce the negative impacts of the financial sector on nature and biodiversity"—see https://tnfd.info/.

[8] Johan Rockström et al., "Planetary Boundaries: Exploring the Safe Operating Space for Humanity", in *Ecology and Society*, 14(2), 2009; Will Steffen et al., "Planetary boundaries: Guiding human development on a changing planet", in *Science*, 347, 2015.

[9] FSB, "Stocktake of Financial Authorities' Experience in Including Physical and Transition Climate Risks as Part of Their Financial Stability Monitoring", July 2020, p. 2; NGFS, "Guide for Supervisors Integrating climate-related and environmental risks into prudential supervision", May 2020, pp. 10, 12–13.

[10] FSB, "Proposal for a disclosure task force on climate-related risks", November 2015, pp. 1–2; EBA, "Report on Management and Supervision of ESG Risks for Credit Institutions and Investment Firms", June 2021, pp. 39–40.

What is ultimately at stake is the transition to a low-carbon economy and the uncertainty surrounding the when and how, as well as the physical progression until then. The impact depends on whether it will be smooth or disorderly. At worst, we may face a too-late, too-sudden scenario, where policy action comes too late, physical events have materialised, and the economic adaptation struggles.[11]

The banking sector may be extensively affected by environmental risks, as well as social and governance risks, though a sector-specific analysis has only recently begun, mainly due to a lack of accurate data on banks' exposures, as well as to initial doubts on their prudential relevance. Today, there is growing consensus as to the fact that banks will indeed be affected by unsustainability.[12]

Risks arise from banks' exposures to their counterparties. There are several transmission channels which work as drivers of pre-existing financial risks—mostly identified in relation to environmental factors,—with consequences on balance sheets, portfolios, decreased profits and increased exposures.[13]

For instance, *credit risk* emerges when the counterparty is from a sector or a geography particularly vulnerable to physical risk, increasing the probability of default. *Market risk* may arise from an abrupt repricing of financial assets and asset stranding, following certain physical events or transition policy changes, with an impact on banks' balance sheets. There is also *operational risk*, mainly from legal claims against credit institutions, with additional reputational losses. Liability risks can also manifest as credit risks, when legal proceedings are filed against counterparties. Ultimately, other categories are at stake, such as business model risk and even liquidity risk.

Stranded assets are of particular concern in the transition to a carbon–neutral economy, given the tendency for revaluation of carbon-intensive assets following obsolescence. Once "stranded", these assets affect banks with direct exposures to sectors with high environmental risk, such as mining, oil and gas, and there is a risk of a "carbon bubble" effect.[14]

From a supervisory perspective, the European Commission, following the European Banking Authority (EBA), has adopted a broad, prudentially oriented definition of ESG risk as meaning "the risk of losses arising from

---

[11] ESRB, "Too late, too sudden: Transition to a low-carbon economy and systemic risk", February 2016.

[12] ECB, "Guide on climate-related and environmental risks, Supervisory expectations relating to risk management and disclosure", November 2020, pp. 11–12; ESRB, "Positively green: Measuring climate change risks to financial stability", June 2020, p. 45.

[13] For an analysis of the transmission of environmental risks into the banking sector, see ECB, "Guide on climate-related and environmental risks, Supervisory expectations relating to risk management and disclosure", November 2020, pp. 10–13; ESRB, "Positively green: Measuring climate change risks to financial stability", June 2020, pp. 23–24; EBA, "Report on Management and Supervision of ESG Risks for Credit Institutions and Investment Firms", June 2021, pp. 29–39.

[14] On the impact of stranded assets, Maria J. Nieto, "Banks and environmental sustainability: Some financial stability reflections", in *IRCCF Working Paper*. Available at www.ssrn.com, 2017, pp. 8–12.

any negative financial impact on the institution stemming from the current or prospective impacts of [ESG] factors on the institution's counterparties or invested assets".[15]

Furthermore, the EBA has taken significant steps towards a more comprehensive reading of sustainability, analysing the risks arising from social and governance factors—which are more difficult to foresee and do not fit into the physical and transition categories.[16] As regards *social* considerations, there are social consequences interlinked with environmental phenomena, in particular migration and labour conditions, as well as social risks on their own. Lack of implementation of social rules or practices, such as diversity, anti-discrimination, labour law and human rights, may lead to legal and reputational risks for counterparties, and subsequently affect the financing bank via credit risk and even reputational risk.

The *governance* factor is also relevant on its own, as there are governance risks deriving from inadequate governance practices by counterparties, such as poor codes of conduct or lack of action on anti-money laundering. These may have financial and non-financial consequences, which then lead to credit risk for banks. In addition, governance is fundamental in ensuring that counterparties actually integrate environmental and social concerns.

Apart from these ESG effects on each individual bank, they rapidly escalate to the whole banking sector, and the financial system at large. On the one hand, a rapid adjustment of asset prices to reflect either the materialisation of physical risks or the unexpected transition changes contends with financial stability.[17] On the other hand, the financial system itself may amplify some of the effects, through pro-cyclical behaviour by market participants and self-reinforcing reductions in bank lending and insurance provision.[18] Thus, ESG is viewed as a threat to financial stability and even identified as a potential systemic risk.[19]

---

[15] Article 1(1)(l), introducing a novel point (52d) to Article 4(1), of the Proposal of 27 October 2021 for a Regulation of the European Parliament and of the Council of 27 October 2021 amending Regulation (EU) No. 575/2013 as regards requirements for credit risk, credit valuation adjustment risk, operational risk, market risk and the output floor (Proposal to amend the CRR). See, also, EBA, "Report on Management and Supervision of ESG Risks for Credit Institutions and Investment Firms", June 2021, p. 33.

[16] Ibid., pp. 39–45.

[17] Pierpaolo Grippa/Jochen Schmittmann/Felix Suntheim, "Climate change and financial risk", in *Finance and Development*, December 2019, pp. 27–28.

[18] FSB, "The implications of climate change for financial stability", November 2020, pp. 17–25.

[19] NGFS, "The Macroeconomic and Financial Stability Impacts of Climate Change", June 2020; ECB, "Climate Change and Financial Stability", May 2019; ESRB, "Positively Green: Measuring Climate Change Risks to Financial Stability", June 2020. In the U.S., the American Commodities Futures Trading Commission (CFTC) has also stated that climate change could pose systemic risks to the U.S. financial system—CFTC, "Managing Climate Risk in the US Financial System", September 2020, pp. ii–iii.

The Role of Banks in ESG

While the banking sector is directly and indirectly affected by the above-mentioned ESG risks, it further plays an important role in supporting the economy's adaptation to greater sustainability. Banks have the ability to be part of the solution (as well as part of the problem) by channelling funds towards certain sectors, and by influencing the financed entities.

ESG factors can become opportunities, not only risks, giving rise to sustainable, green or responsible banking practices.[20] These practices are for some time now mainstream and no longer limited to social banks—the purpose of which is to further certain social goals, and not simply profit-maximisation, like general commercial banks. ESG represents commercial opportunities for banks to enter into different markets, attract new clients and have significant reputational gains. In doing so, credit institutions are also aligning with regulatory and societal expectations.

There is, in fact, an increasing ESG activism by banks in finding contractual arrangements which meet the demands for socially responsible, environmentally friendly financial services. Alongside sustainable capital market products, such as green bonds and social bonds,[21] there is a vast array of bank sustainable products. For instance, green and sustainability-linked loans are two types of financing via regular loan instruments, but with a specific ESG scope.

Green loans exclusively finance green projects, which need to provide clear environmental benefits.[22] *Sustainability-linked loans,* including other contingent facilities, such as guarantee lines, incentivise the borrower via predetermined sustainability performance objectives.[23] The difference lies in the use of proceeds specifically for a green investment in the former, which is not determinant of the latter. Sustainability-linked loans instead seek to improve the borrower's sustainability profile by measuring its performance against sustainability performance targets. Both of them require reporting duties, thereby enhancing transparency.

There are also *green deposits,* in which the funds are allocated to a given pool of eligible assets that fulfil certain environmental criteria. Deutsche Bank, for example, has recently announced the launch of green deposits for corporate

---

[20] On green banking practices, see Kern Alexander, *Principles of Banking Regulation,* Cambridge University Press, Cambridge, 2019, p. 352.

[21] Regarding ESG bonds, see Manuel Requicha Ferreira, Chapter 8 in this book.

[22] Loan Market Association/Asia Pacific Loan Market Association/Loan Syndications and Trading Association, "Green Loan Principles", December 2018. Available at www.icmagroup.org.

[23] Loan Market Association/Asia Pacific Loan Market Association/Loan Syndications and Trading Association, "Sustainability Linked Loan Principles", March 2019. Available at www.icmagroup.org.

clients, which must themselves have a certain level of ESG ratings.[24] Similarly, there are *green credit cards* that typically offer donations to non-profit organisations on the basis of a given percentage on purchases.[25]

Microfinance is another sustainability-related banking service. Traditionally seen as financing for those who do not have access to the traditional banking system, it is today a part of some banks portfolios and has been growing in the European Union. It involves the provision of financial services—such as loans, but also advisory services—to individuals and small businesses that would otherwise lack access to them, addressing many social concerns, some of them also pursuing environmental goals.[26]

In parallel with these—and many others not referred to[27]—contractual innovations, banks are acting on the basis of international guidelines which harmonise practices and provide a benchmark for the industry. While not binding, these initiatives rely on transparency, requiring reports on compliance, and reputational motivations.

The *Equator Principles*, mostly applicable to project finance, are a risk management framework for environmental and social risk in projects, providing a minimum standard for due diligence and monitoring to support responsible risk decision-making.[28] There are also *Principles for Responsible Banking*, a United Nations initiative providing the framework for a sustainable banking system, and helping the industry to demonstrate how it makes a positive contribution to society.[29] Green and sustainability-linked loans have been standardised by a group of loan market associations, establishing the core principles which these products should comply with.[30]

Altogether, the banking sector may have a two-folded positive impact on ESG. On one hand, the rise of ESG products directs funds to activities and sectors which have a positive impact on environmental and social matters, allowing funding that fosters sustainability innovation and development. On the other hand, in their capacity as lenders, banks are in a privileged position

---

[24] Deutsche Bank, "Deutsche Bank launches green deposits for its corporate clients", March 2021. Available at www.db.com.

[25] Rosella Carè, *Sustainable Banking—Issues and Challenges*, Palgrave Macmillan, London, 2018, p. 70.

[26] Davide Forcella/Marek Hudon, "Green Microfinance in Europe". *Journal of Business Ethics*, 135, 2016, pp. 445–449.

[27] For an overview of some sustainable banking products, see Rosella Carè, *Sustainable Banking—Issues and Challenges*, cit., pp. 65–74; UNEP Finance Initiative, "Green Financial Products and Services—Current Trends and Future Opportunities in North America, 2007, p. 15 et seq.

[28] See https://equator-principles.com.

[29] See www.unepfi.org.

[30] Loan Market Association/Asia Pacific Loan Market Association/Loan Syndications and Trading Association, "Green Loan Principles", December 2018, and "Sustainability Linked Loan Principles", March 2019, both Available at www.icmagroup.org.

to influence and lead to change in the financed entities. The actions of banks thus have a strong impact on the non-financial sector.

Notwithstanding the merits of this "doing well, by doing good" approach, the risk of greenwashing is evidently significant. Even if sustainable banking practices and green products are transparent and effective, no credit institution is fully stepping away from financing carbon-intensive industries. The announced pledges of commitment to net-zero greenhouse emission, as well as other social goals, are thus often met with scepticism.[31]

3   Sustainable Banking Regulation in the EU

The Need for Regulatory Intervention

From the above, it becomes clear why ESG in the banking sector has been in the spotlight in recent years. Faced with ESG risks, banks, individually and collectively, are at stake and will, in all likelihood, be affected. It is thereby contended that regulatory intervention is necessary and that central banks and supervisors have a role to play and a mandate to ensure the resilience of the financial system.[32]

Simultaneously, banks' increased presence in ESG financing and investing may require regulation to limit opportunities for greenwashing, as well as to provide conditions and incentives to foster the flow of funds to a more sustainable economy. Some barriers to these investments include information asymmetries, uncertainty about risk evaluations and lack of common standards.[33]

Despite the overwhelming acceptance of the need for sustainable regulation, the road is still filled with obstacles, including data availability, methodological challenges, difficulties in mapping of transmission channels, time horizon misalignments and lack of a clear and widely accepted taxonomy regarding "green" and "brown" assets.[34]

At the EU level, while sustainability has been the hot topic for some years, sustainable banking regulation is only now taking its first steps. The main focus has been on ESG as risks to be managed and internalised. Regardless, the underlying overall policy is also to ensure that banks play their part in sustainable finance.

---

[31] See, for example, Rainforest Action Network, "Banking on climate chaos—Fossil fuel finance report 2021", 2021. Available at www.ran.org.

[32] NGFS, "First Progress Report", October 2018, pp. 3, 12.

[33] Kern Alexander, *Principles of Banking Regulation*, Cambridge University Press, Cambridge, 2019, pp. 354–355, 357; OECD, "Promoting responsible lending in the banking sector: The next frontier for sustainable finance", in *Business and Finance Outlook 2020: Sustainable and Resilient Finance*. Available at www.oecd-ilibrary.org.

[34] BCBS, "Climate-related financial risks: A survey on current initiatives", April 2020, pp. 3–4.

The CRD V[35]/CRR II[36] package of 2019 specifically embedded ESG in the framework governing credit institutions, relying on further analysis and specification by the EBA. Sustainability has become a matter of prudential concern.

The EBA was thereby given a triple mandate, spanning across all three pillars of Basel III. Firstly, Article 449a of CRR II requires the disclosure of ESG risks and the banking supervisory authority is to specify them as part of the comprehensive technical standard on Pillar 3. As to Pillar 2, Article 98(8) of CRD V requires the assessment of the inclusion of ESG risks in the supervisory review and evaluation process (SREP), with an analysis of the mechanisms to be implemented by banks to manage those risks, i.e., governance and internal risk management. Finally, Article 501c of CRR II envisages an inquiry into whether there should be a dedicated treatment to Pillar 1 capital requirements. The EBA announced a staged approach to all of these mandates[37] and there have been significant developments, though still at an early stage.

More recently, in October 2021, the European Commission went a step further by proposing new amendments to all these provisions and others of both the CRD and the CRR.[38]

In the following subsections, we will briefly cover these three strands of regulation: ESG disclosure requirements; governance and risk management; and the possibility of amending capital requirements.

Disclosure Requirements and the Green Asset Ration

Disclosure has been the backbone of sustainability initiatives in the financial sector. Internationally, the TCFD recommendations[39] have become the standard-setter, followed by the Network of Central Banks and Supervisors for Greening the Financial System (NGFS) and the EU.

While transparency fulfils in itself the goal of fostering market discipline, in this particular area disclosure is also aimed at providing all market actors,

---

[35] Directive (EU) 2019/878 of the European Parliament and of the Council of 20 May 2019.

[36] Regulation (EU) 2019/876 of the European Parliament and of the Council of 20 May 2019.

[37] EBA, "Action Plan on sustainable finance", December 2019, pp. 6–7.

[38] Proposal to amend the CRR (see footnote 15 of Chapter 1) and Proposal of 27 October 2021 for a Directive of the European Parliament and of the Council amending Directive 2013/36/EU as regards supervisory powers, sanctions, third-country branches, and environmental, social and governance risks, and amending Directive 2014/59/EU (Proposal to amend the CRD). Besides sustainability, these proposals also intend to fully implement the Basel III agreement, and reinforce the supervisory powers of European authorities, topics which are outside of the scope of this Chapter.

[39] TCFD, "Recommendations of the Task Force on Climate-related Financial Disclosures", June 2017.

including supervisors, with a better understanding of the current environmental risk exposure. This information will then feed back into other policy proposals that are all still dependent upon better and more accurate data. Disclosure is hence pivotal.

Under the auspices of sustainable finance, a comprehensive and extensive transparency regime has been put in place. At the centre is the Taxonomy Regulation,[40-41] which addresses the first challenge: what is and is not sustainable. By establishing the criteria for determining whether an economic activity qualifies as environmentally sustainable, the goal is to have common language used by financial institutions, investors, and legislators. It is, for instance, applicable to the sustainability-related disclosures in the financial services sector regulation (SFDR),[42] laying out an extensive set of transparency duties. These apply to banks insofar as they act as financial market participants, providing portfolio management, or financial advisers, providing investment advice.[43]

Moreover, all banks are subject to the Non-Financial Reporting Directive (NFRD),[44] adopted already in 2014 in the context of the EU's action on corporate social responsibility. Since 2019 there is a subset of guidelines on reporting climate-related information,[45] which took into account the TCFD recommendations in the dedicated annex for banks and is still in force.

In addition, there are now two main legal bases for bank-specific ESG disclosures: Article 449a CRR and Article 8 of the Taxonomy Regulation.

Arising directly from the prudential framework, Article 449a CRR currently requires certain institutions to disclose to the public information on ESG risks, including physical and transition risks, from June 2022. The EBA is to ensure disclosure will cover comprehensive and comparable information on banks' risk profile, and there is an ongoing public consultation, the result of which will be integrated in a comprehensive Pillar 3 framework.[46] While work on this

---

[40] Regulation (EU) 2020/852 of the European Parliament and of the Council of 18 June 2020 (Taxonomy Regulation).

[41] Regarding the Taxonomy Regulation, see also Rui de Oliveira Neves, Chapter 13 in this book.

[42] Regulation (EU) 2019/2088 of the European Parliament and of the Council of 27 November 2019 (SFDR).

[43] Articles 2(1)(j) and 2(11)(c) of the SFDR.

[44] Directive 2014/95/EU of the European Parliament and of the Council of 22 October 2014 (NFRD). The European Commission committed to revising the NFRD as part of the action on the European Green Deal: A proposal for a Corporate Sustainability Reporting Directive (CSRD), amending the NFRD's reporting requirements, is currently under legislative discussion.

[45] European Commission, "Guidelines on non-financial reporting: Supplement on reporting climate-related information", 2019. These supplement the general guidelines on non-financial reporting—"Guidelines on non-financial reporting (methodology for reporting non-financial information)", 2017.

[46] EBA, "Draft Implementing Standards on prudential disclosures on ESG risks in accordance with Article 449a CRR", March 2021.

subject is still underway, the European Commission has already proposed an amendment to Article 449a, with two significant alterations[47]: firstly, while the CRR II imposed this disclosure obligation to large institutions with publicly listed issuances, the Proposal broadens its scope to all institutions; secondly, while originally the large institutions were expected to disclose annually for the first year, and biannually thereafter, now the distinction is between small and non-complex institutions and others, which have to disclose annually and semi-annually, respectively. Additionally, the Proposal requires all institutions to report to their competent authorities their exposures on ESG risks.[48]

These disclosure requirements will be particularly burdensome on small institutions, and respect for the principle of proportionality must be ensured. The rationale is that larger institutions are not alone in the exposure to ESG risks and that these risks are not necessarily proportional to an institution's size and complexity, being contingent upon other factors, such as geographical exposure.[49]

To the preceding question of what are the ESG risks that need to be reported, the Proposal provides uniform definitions of environmental, social, and governance risks,[50] essentially as any of those stemming from the current or prospective impacts of the relevant factors on the institutions' counterparties or invested assets.

As to Article 8 of the Taxonomy Regulation, it requires all entities subject to the NFRD to include information on how and to what extent their own business operations are associated with economic activities that qualify as environmentally sustainable. The specification of such information is detailed in a proposed Commission Delegated Regulation of July 2021,[51] which includes among the key performance indicators (KPIs) for credit institutions the green asset ratio, as previously proposed by the EBA.[52]

The *Green Asset Ratio* (GAR) is the proportion of a bank's assets invested in sustainable economic activities as compared to its total relevant on-balance-sheet assets. It shall be disclosed in aggregate terms (total GAR) and broken-down by environmental objective, type of counterparty and subset of transitional and enabling activities. This figure is intended to uncover the exposure of credit institutions, through their main lending and investment

---

[47] Article 1(189) of the Proposal to amend the CRR.

[48] Article 1(176) of the Proposal to amend the CRR, adding point (h) to Article 430(1).

[49] See Recital 40 of the Proposal to amend the CRR.

[50] Article 1(1)(l) of the Proposal to amend the CRR, adding points (52d) to 52(i) to Article 4(1).

[51] Proposal for a Commission Delegated Regulation, C(2021) 4987 final, 6 July 2021. The rules for credit institutions are laid out in Article 4 and Annexes V and XI of the Delegated Regulation.

[52] See EBA, "Opinion on the disclosure requirement on environmentally sustainable activities in accordance with Article 8 of the Taxonomy Regulation", February 2021.

activities,[53] to different economic activities as classified by the Taxonomy Regulation, displaying the extent to which their financing is (or is not) sustainably aligned.

Some consequences may be anticipated from the GAR. Firstly, it will allow stakeholders to assess a bank's portfolio from an ESG perspective, namely in terms of environmental risk exposure. Conversely, it will demonstrate how they are or not committed to sustainability and how they contribute to the development of green sectors through their lending options. Secondly, the calculation of the GAR is dependent upon an inquiry of the Taxonomy-alignment of the financed entities' activities. Due diligence will be required of both parties, since banks will need to evaluate their loan books and their counterparties will need to look into all of their activities so as to correctly qualify them.

It will be the first time that such a metric is mandatorily disclosed.[54] This framework is very ambitious, uncovering extremely relevant data from both financial and non-financial sectors, through the intermediary of credit institutions. Its efficacy will depend on the adequacy of the imposed methodology, as well as on the accuracy of the data provided.

Governance and Risk Management

If environmental risks are material to the soundness of banks, then disclosure by itself is insufficient and they need to be properly internalised and managed. This is foremost achieved through *governance* and *risk management*.

Already in 2019, the EBA had encouraged banks to act proactively in incorporating ESG considerations into their business strategy, risk management, internal control framework and decision-making processes.[55] Shortly after, the first policy document incorporating sustainability concerns was adopted, on the topic of loan origination and monitoring, calling on institutions to incorporate ESG factors in their credit risk appetite and risk management policies and procedures, as well as to develop environmentally sustainable lending policies and procedures.[56]

More recently, on the basis of the Article 98(8) CRD V mandate, the EBA has adopted recommendations on how institutions can embed ESG risks into their corporate governance, referring separately to business strategies and processes, internal governance, and risk management. Legislative amendments

---

[53] There will also be KPIs for off-balance-sheet exposures and for fees and commission for services other than lending.

[54] Huw Jones, "Banks in EU to publish world's first 'green' yardstick from next year", 1 March 2021, Reuters. Available at www.reuters.com.

[55] EBA, "Action plan on sustainable finance", December 2019, p. 16.

[56] EBA, "Guidelines on loan origination and monitoring", May 2020, pp. 26–27.

to the CRD/CRR framework are also proposed, so as to incorporate these recommendations.[57]

Highlight is given to business strategies, suggesting that banks introduce ESG concerns in the establishment of long-term objectives, limits and engagement policies, as well as adjusting their products to the Taxonomy standards. As to internal governance, guidance is provided in regard to the role of the management body and the committees on ESG matters, calling for "tone from the top" and proper allocation of responsibilities; internal control framework, including ESG-related tasks for risk management, compliance and internal audit functions; and remuneration policies which should consider ESG indicators and ESG risk-related objectives and limits. Lastly, for risk management, an area in which there are still data and methodologic shortfalls, banks should nevertheless be proactive and fully integrate ESG risks in their respective frameworks (including in the Internal Capital Adequacy Assessment Process—ICAAP).

The banking package proposed in October 2021 transforms some of these recommendations into binding rules for credit institutions and supervisors, once the proposal for a directive is approved and transposed by the Member States.

First of all, the Commission proposes that the SREP include an assessment of exposures to ESG risks as well as of the institutions' processes for dealing with them.[58] Secondly, in what concerns the risk management framework, there is a formal recognition of the forward-looking nature of ESG to be addressed across the short, medium and long-term, applicable to the assessment of internal capital and to governance arrangements related to risk.[59] A new provision is proposed specifically on strategies and processes for the identification and management of ESG risks, in a time frame of at least ten years and for which subsequent EBA guidelines will ensure consistency in criteria and methodology.[60]

Thirdly, the governance of the management body is directly called upon via two proposals. On the one hand, it shall approve and review risk strategies and policies resulting from the impact of ESG factors and develop specific plans and quantifiable targets to monitor and address those risks.[61] On the other hand, in the context of other changes to the rules on suitability of members of the management body, it shall collectively have knowledge and awareness of

---

[57] EBA, "Report on Management and Supervision of ESG Risks for Credit Institutions and Investment Firms", June 2021, pp. 80 et seq.

[58] Article 1(24) of the Proposal to amend the CRD, adding paragraph 9 to Article 98.

[59] Respectively, Article 1(12) and Article 1(13) of the Proposal to amend the CRD, altering Articles 73(1) and 74(1). See also recital 32 of the same Proposal.

[60] Article 1(17) of the Proposal to amend the CRD, adding Article 87a.

[61] Article 1(14)(a) and (b) of the Proposal to amend the CRD, altering Article 76(1) and (2).

ESG factors and their potential risks.[62] Management will hereby be expected to approve plans on ESG and, a priori, to actually have knowledge on ESG. The type of qualifications relevant to understand these novel risks might differ from what is traditionally expected of bankers.

Lastly, a concrete supervisory power is to be introduced so as to require institutions to reduce the risks arising from the "misalignment with relevant policy objectives of the Union and broader transition trends relating to [ESG] factors",[63] coupled with a power to assess and monitor banks' practices and plans on ESG risk management, including "the progress made and the risks to adapt their business models to the relevant policy objectives of the Union or broader transition trends towards a sustainable economy."[64]

The wording of the latter two provisions raises some questions. It is one thing to require banks to internalise, manage and mitigate ESG risks, in conformity with the institution's overall strategy. It is quite another to expect banks to align with Union's policies. Presented as a prudential risk to both individual banks and financial stability,[65] what is inevitably also at stake is the role the banking sector is expected by the EU—and society at large—to play in the transition to a greener economy *in addition to* managing risks arising from ESG factors.

Another tool that assists in internalising risks is *stress testing*. To better understand the extent of banks' exposures to environmental risks, many have been advocating for climate or carbon-stress tests.[66] The EBA also has a mandate to develop stress testing processes and scenario analyses, including to assess the impact of ESG risks,[67] and has launched, in May 2020, a pilot sensitivity exercise on climate risk to test current methodologies and data availability.[68] The ECB is presently conducting an economy-wide climate stress test and in 2022 there will be a supervisory climate stress test of individual

---

[62] Article 1(19) of the Proposal to amend the CRD, adding Article 91(4).

[63] Article 1(26) of the Proposal to amend the CRD, adding point (m) to Article 104. Similar wording is also used in Article 1(14)(b) of the Proposal to amend the CRD, altering Article 76(2): "risks arising in the short, medium and long-term from the misalignment of the business model and strategy of the institutions, with the relevant Union policy objectives or broader transition trends towards sustainable economy in relation to [ESG] factors.".

[64] Article 1(17) of the Proposal to amend the CRD, adding Article 87a(4).

[65] See recital 33 of the Proposal to amend the CRD.

[66] ESRB, "Too late, too sudden: Transition to a low-carbon economy and systemic risk", February 2016, pp. 15–17. See also Deloitte, "The Predictive Power of Stress Tests to Tackle Climate Change", 2020. Available at www2.deloitte.com.

[67] Under Article 98(8) of CRD V and Articles 23 and 32(2)(e) of Regulation (EU) No 1093/2010 of the European Parliament and of the Council of 24 November 2010 (EBA Regulation).

[68] For a comprehensive report of the results of this exercise, see EBA, "Mapping climate risk: Main findings from the EU-wide pilot exercise", May 2021.

banks.[69] Besides, banks are already expected to evaluate the appropriateness of their own stress tests and incorporate material environmental risks.[70] There are, nevertheless, methodological limitations to risk assessment in general and stress testing in particular: in ESG risks, the time horizon is not compatible with regular timelines; historical data does not capture the type of risks at stake; and tests will necessarily be less accurate and granular than usual.[71] Acknowledging the relevance of such tool to be applied in a consistent manner across Europe, the Commission proposed that the three financial supervisory authorities jointly develop guidelines on ESG stress testing methodologies and long-term considerations, which should begin with climate-related factors.[72]

The internalisation and management of these new risks—through governance in general and risk management in particular—must bear its specificities in mind, given the timeframe for their materialisation is different, as are the methodologies required to adequately study and respond to them.

From a corporate governance perspective, it is clear that the aim is to ensure that current practices take sustainability concerns seriously, including ESG in the core functioning and organisation of credit institutions through already existing governance mechanisms and, in the expected near future, mandatorily, through banking legislation. Additionally, the recommendations and proposals are expressly linked to a long-termism perspective, as sustainability is by nature incompatible with short-term decision-making and policies.[73]

ESG may therefore help consolidate governance practices with a focus on the long term. This, however, will need to be aligned with the peculiarities of bank corporate governance,[74] a body of mostly mandatory rules that draw upon classic governance mechanisms, but more intrinsically risk-oriented and based on considerations exogenous to the company, given banks' risk exposure

---

[69] Luis de Guindos, "Shining a light on climate risks: the ECB's economy-wide climate stress test", 18 March 2021. Available at: www.ecb.europa.eu.

[70] ECB, "Guide on climate-related and environmental risks: Supervisory expectations relating to risk management and disclosure", November 2020, pp. 42–43.

[71] On the methodological difficulties of climate stress tests, see Maria J. Nieto, "Banks and environmental sustainability: Some financial stability reflections", cit., pp. 19–22; Alexander Lehmann, "Climate risks to European banks: a new era of stress tests", in *Bruegel Blog*. Available at www.bruegel.org, 2020.

[72] Article 1(25) of the Proposal to amend the CRD, adding paragraph 4 to Article 100; a third paragraph is also proposed, establishing a prohibition on institutions or any third parties acting in a consulting capacity from activities that can impair stress testing, such as benchmarking, exchange of information, agreements on common behaviour, or optimisation of their submissions in stress tests.

[73] On calls for long-termism in the financial system, see European Commission, "Action Plan: Financing Sustainable Growth", 2018, Action 10, p. 11; EBA, "Report on undue short-term pressure from the financial sector on corporations", December 2019.

[74] On bank corporate governance, see in general Klaus J. Hopt, "Corporate governance of banks after the financial crisis", in Eddy Wymeersch/Klaus J. Hopt/Guido Ferrarini (Ed.), *Financial Regulation and Supervision—A Post-Crisis Analysis*, Oxford University Press, 2012; Paulo Câmara, "O Governo dos Bancos: Uma Introdução", in Paulo Câmara (Ed.), *A Governação dos Bancos nos Sistemas Jurídicos Lusófonos*, Almedina, 2016.

and the extensive negative externalities arising from a poorly managed credit institution. Both ESG and financial stability draw from the need to account for a wider set of stakeholders and benefit from a long-term perspective on banks' management. Nevertheless, this compatibility does not necessarily mean that their goals will always align. Looking at ESG factors as a risk to mitigate through internal governance, it is easily in favour of financial stability; however, the wider sustainability perspective in which banks play a crucial role in sustainable development might at times be at odds with the stability of the financial system, at least in a transition phase.

Capital Requirements

The last level of banking regulatory tools that is on the table for sustainability purposes is the most prudentially sensitive of all: capital requirements.

Adjusting these requirements could involve one of two possibilities, or a combination of both: higher flexibility in relation to "green" assets or more stringent rules on "brown" assets.

The former is known as the *green supporting factor* and it entails decreasing the capital calculated against green assets, which has been criticised for many reasons.[75] The first obstacle is the absence of precise definitions of sustainable assets. Even with the Taxonomy Regulation already in force, there is still a long way to go, though it is hoped that further disclosure will provide helpful data for categorisation. Additionally, greener is not always safer. Capital requirements are aimed at safeguarding financial stability and should not be subverted for the sake of other public policies, if there is no financial risk supporting such decision. There is a further risk of regulatory arbitrage, giving banks the room to design complex financing structures with some green assets, simply to decrease capital weights. Overall, even from the standpoint of steering capital flows towards sustainable sectors, it is contended that other tools may be better placed in doing so, such as tax law.

The second possibility would be more stringent rules increasing capital requirements in relation to non-sustainable assets, through a *brown penalising factor*. It would incentivise banks to decrease their exposures to environmental risks through increased asset risk weights on given assets, rendering banks safer and the unsustainable activities more expensive to fund.[76]

---

[75] On this proposal and the main criticism against it, see Kern Alexander/Paul Fisher, "Banking Regulation and Sustainability", in Frits-Joost B. van den Boezem/Corjo Jansen/Ben Schuijling, *Sustainability and Financial Markets*, Wolters Kluwer, 2019, pp. 20–22, 26; Jacob Dankert et al., "A Gren Supporting Factor—The Right Policy?", in *SUERF Policy Note*, Issue no. 43, 2018, pp. 2–5; Gábor Gyura, "ESG and bank regulation: moving with the times", in *Economy and Finance*, 7(4), 2020, pp. 380–381; High-Level Expert Group on sustainable finance (HLEG), "Financing a Sustainable European Economy", July 2017, p. 32.

[76] See Maria J. Nieto, "Banks and environmental sustainability: Some financial stability reflections", cit., pp. 22–23; Jay Cullen/Jukka Mähönen, "Taming Unsustainable Finance, The Perils of Modern Risk Management", Beate Sjåfjell/Christopher M. Bruner

At the EU level, the introduction of either a green or a brown factor has been proposed[77] and even discussed during the legislative procedure of CRR II, although it was not part of the original proposal. An agreement was finally reached to approve the novel Article 501c of the CRR on the prudential treatment of environmental and social exposures.

Instead of a mandatory rule, the end result was to merely mandate EBA, after consulting the ESRB, to assess *"whether a dedicated prudential treatment of exposures related to assets or activities associated substantially with environmental and/or social objectives would be justified"* until June 2025. Given the sensitivity of the topic as well as the need to first have a solid classification of assets, this task was left for last in EBA's staged approach.[78] Nevertheless, the European Commission is now proposing to bring forward by two years, until June 2023, EBA's deadline for a report, further omitting the carefully drafted paragraph that the Commission "shall, *if appropriate*" submit a legislative proposal.[79]

This debate brings about the issue of using banks and prudential regulation to further other public policies, which is not unheard of, nor new in the field of capital requirements. CRR II amended the already existent Small and Medium-Sized Enterprises (SME) supporting factor in article 501 and introduced the new infrastructure factor in article 501a, while the CRR "quick fix", in light of the Covid-19 pandemic, even anticipated the entry into force of these new rules, so as to "incentivise institutions to increase much-needed lending".[80]

Contending with mandatory rules on institutions' own funds, specially through a green asset ratio, would theoretically be one of the measures that would do the most to increase banks' positive impact on sustainability, leading banks to fund such activities. However, this ought not be done at the expense of the risk-oriented nature of prudential law, nor of financial stability—which in itself is a condition for sustainability.[81]

Although ultimately sustainability and financial stability go hand-in-hand, they do not fully overlap and there is still overall uncertainty and lack of

---

(eds.), *Cambridge Handbook of Corporate Law, Corporate Governance and Sustainability* Cambridge University Press, Cambridge, 2019, pp. 111–112.

[77] The idea was put forward by the HLEG—see "Financing a Sustainable European Economy", July 2017, p. 32—and integrated in the Commission's action plan—see "Action Plan: Financing Sustainable Growth", 2018, p. 9.

[78] EBA, "Action Plan on Sustainable Finance", December 2019, pp. 10–11, 13.

[79] Article 1(202) of the Proposal to amend the CRR. There is also a proposed change in wording, referring to the prudential treatment of exposures to assets or activities "subject to impacts from" environmental and/or social factors, instead of "associated substantially with". It appears to be a welcome clarification that what is at stake is the (potentially negative) *impact* that ESG may have on assets or activities, and not merely their association.

[80] Recital 19 of Regulation (EU) 2020/873 of the European Parliament and of the Council of 24 June 2020 (CRR "quick fix").

[81] Kern Alexander/Paul Fisher, "Banking Regulation and Sustainability", cit., pp. 23–24.

data.[82] From a risk perspective, the brown penalty seems more in line with both financial stability and sustainability concerns, encouraging banks to diminish exposure to assets that are more consensually seen as both risky and unsustainable.

Regardless of the future amendments to Pillar 1, there may be an alternative route to adjusting bank capital—through Pillar 2 bank-specific additional requirements. These are imposed by supervisors when, following supervisory review, they find that mandatory capital is not enough in light of a bank's particular risk profile, which could include sustainability risks.[83]

The sensitivity and uncertainty around these measures will keep capital amendments last in the regulatory agenda, although the Commission's proposal signals a greater urgency on this matter, as pressure for timely action on sustainable finance continues to increase.

4 THE WAY FORWARD

Despite the changing regulatory landscape, growing expectations from supervisors and the market, and the "greening" of banking services and products, it seems that banks' understanding and integration of ESG factors and risks still has a long way to go.

According to a recent market study, there has been some progress in terms of governance structures and disclosures, but less so in actually integrating ESG into the risk management framework and implementing strategies for sustainability.[84] The same findings were reached at the EU level, where the incorporation of ESG risks into strategies, governance and risk management was found to be incipient and with divergent approaches.[85] Even in what concerns disclosure, there is more ESG information, but it is still not comprehensive and does not comply with applicable guidelines,[86] sometimes

---

[82] For an example of a quantitative analysis of the potential impact of both alternatives on European banks, estimating that the green supporting factor would not significantly lead to capital savings and that the brown penalty would have a greater impact, but without destabilising banks' balance sheets, see Jakob Thomä/Kyra Gibhardt, "Quantifying the potential impact of a green supporting factor or brown penalty on European banks and lending", in *Journal of Financial Regulation and Compliance*, 27(3), 2019.

[83] Asserting that national authorities, even in the context of the SSM, have significant discretionary room in their SREP to require banks to take certain sustainability risks into account and to require them to hold additional Pillar 2 capital on the basis thereof, Bart Bierman, "Sustainable Capital: Prudential Supervision on Climate Risks for Banks", in Frits-Joost B. van den Boezem/Corjo Jansen/Ben Schuijling, *Sustainability and Financial Markets*, Wolters Kluwer, 2019, p. 145.

[84] Mazars, "Responsible banking practices—benchmark study 2021", 2021, p. 27.

[85] Adrienne Coleton et al., "Sustainable Finance: market practices", in *EBA Staff Paper Series* no. 6, 2020.

[86] ECB, "Report on institutions' climate-related and environmental risk disclosures", November 2020, p. 2; ECB, "Risk assessment for 2021", January 2021. Available at www.bankingsupervision.europa.eu.

simply being part of non-financial reports and seen from a mere corporate responsibility perspective.[87]

Since so far a proper understanding of sustainability is still a work in progress, bank disclosures will be of extreme value, especially when the Green Asset Ratio comes into force. This will provide insight into the exposures of the institutions that most finance the European economy, and simultaneously information on non-financial sectors. The data hence collected, combined with the Taxonomy Regulation, will be paramount to any further regulation, to ensure a more adequate internalisation of ESG risks, and to foster market transparency and a levelled playing field for sustainable banking products.

With the emerging sustainable banking law in the EU penetrating the core of banking regulation, banks should prepare for extensive compliance requirements in the near future—as signalled by the European Commission's proposed Banking Package –, while also keeping up with the expectations of investors, depositors and the public in general against greenwashing.

Central banks and supervisors are also expected to take on a role of their own in positively contributing to sustainability. According to Christine Lagarde, the ECB "should assess its potential role in the transition" to a carbon–neutral economy,[88] and monetary instruments, such as quantitative easing, could incentivise banks' investment in green assets.[89] These entities are, much like the supervised institutions, not immune to greenwashing criticism for still favouring carbon-intensive industries.[90]

Overall, the "business case" approach is dominant in the response to ESG as both a risk and an opportunity: regulation is imposed due to the perception of ESG as risks for banks, and banks' own initiatives are a response to market demands. From a governance perspective, while the incorporation of ESG might lead banks and other companies to be managed with a long-term perspective, there seems to be a need for *reconceptualising time and risk*.

---

[87] Adrienne Coleton et al., "Sustainable Finance: market practices", cit., p. 40.

[88] Christine Lagarde, "Climate change and central banking", Keynote speech at the ILF conference on Green Banking and Green Central Banking, 25 January 2021. Available at: www.ecb.europa.eu. Expecting the ECB to engage pro-actively in the inclusion of climate change in its prudential supervisory mandate, René Smits, "SSM and the SRB accountability at European level: room for improvements?". Available at europarl.europa.eu, 2020, pp. 33–35.

[89] Jay Cullen/Jukka Mähönen, "Taming Unsustainable Finance, The Perils of Modern Risk Management", cit., pp. 112–113.

[90] Balazs Koranyi/Philippa Fletcher, "Greenpeace paragliders land on ECB building in protest", Reuters, 10 March 2021. Available at www.reuters.com. Similarly, the European Commission's decision to award Blackrock a contract to oversee the development of the integration of ESG factors in the banking prudential framework was highly criticized and subject to an inquiry by the European Ombudsman—see European Ombudsman, "Report on the inspection and meeting concerning the European Commission's decision to award a contract to BlackRock on the development of tools for the integration of ESG-factors into the EU banking prudential framework", September 2020.

True sustainability demands an intergenerational *long-term* perspective. As stated already back in 1987, sustainable development requires humankind "to ensure that it meets the needs of the present without compromising the ability of future generations to meet their own needs".[91] That this is extremely difficult to implement is well captured by Mark Carney's "Tragedy of the Horizon".[92] Although there are calls for long-termism, the timescale of banks' governance mechanisms needs recalibrating.[93]

Likewise, the very notion of relevant and material *risk* cannot be the same. Current struggles to find appropriate methodologies and metrics in risk management and stress testing have been designated as an "epistemological obstacle", requiring a conceptual break for new approaches to be found.[94]

Furthermore, ESG needs to be equated with discussions around corporate purpose and sustainable value creation.[95] Banks may be particularly targeted by renewed societal expectations of their role in sustainable development. Credit institutions are often called upon in moments of crisis or of particular public interest, as demonstrated during the pandemic.[96] In spite of banks' indispensable role in positively impacting ESG, a rush to sustainability should not contend with financial stability, both being indispensable. In the coming years, the tension between competing goals during a transitory phase will be one of the great challenges of sustainable finance.

---

[91] United Nations, "Report of the World Commission on Environment and Development: Our Common Future", 1987.

[92] "We don't need an army of actuaries to tell us that the catastrophic impacts of climate change will be felt beyond the traditional horizons of most actors—imposing a cost on future generations that the current generation has no direct incentive to fix"—Mark Carney, "Breaking the Tragedy of the Horizon—climate change and financial stability", speech at Lloyd's of London, 29 September 2015.

[93] In this regard, it is interesting to note the proposals to include references to a "short, medium and long term time horizon" for risk management, which, for ESG, should be of at least 10 years—see Article 1(17) of the Proposal to amend the CRD, adding Article 87a(2).

[94] Patrick Bolton et al., "The green swan—Central banking and financial stability in the age of climate change", Bank for International Settlements, 2020, pp. 20–22.

[95] See Beate Sjåfjell, "Realising the Potential of the Board for Corporate Sustainability", in Beate Sjåfjell/Christopher M. Bruner (eds.), *Cambridge Handbook of Corporate Law, Corporate Governance and Sustainability* Cambridge University Press, Cambridge, 2019, pp. 705–707; Japp Winter, "Addressing the Crisis of the Modern Corporation: The Duty of Societal Responsibility of the Board", 2020. Available at: www.ssrn.com.

[96] On the amendment to the European banking regime in light of the Covid-19 pandemic, see European Commission, "Interpretative Communication on the application of the accounting and prudential frameworks to facilitate EU bank lending, Supporting businesses and households amid COVID-19", April 2020.

REFERENCES

Alexander, Kern, *Principles of Banking Regulation*, Cambridge University Press, Cambridge, 2019.
Alexander, Kern/Fisher, Paul, "Banking Regulation and Sustainability", in Frits-Joost B. van den Boezem/Corjo Jansen/Ben Schuijling, *Sustainability and Financial Markets*, Wolters Kluwer, 2019, pp. 7–33.
BCBS, "Climate-Related Financial Risks: A Survey on current initiatives", April 2020.
Bierman, Bart, "Sustainable Capital: Prudential Supervision on Climate Risks for Banks", in Frits-Joost B. van den Boezem/Corjo Jansen/Ben Schuijling, *Sustainability and Financial Markets*, Wolters Kluwer, 2019, pp. 129–162.
Bolton, Patrick et al., "The Green Swan—Central Banking and Financial Stability in the Age of Climate Change", Bank for International Settlements, 2020.
Câmara, Paulo, "O Governo dos Bancos: Uma Introdução", in Paulo Câmara (Ed.), *A Governação dos Bancos nos Sistemas Jurídicos Lusófonos*, Almedina, 2016, pp. 13–61.
Carney, Mark, "Breaking the Tragedy of the Horizon—Climate Change and Financial Stability", speech at Lloyd's of London, 29 September 2015.
CFTC, "Managing Climate Risk in the US Financial System", September 2020.
Coleton, Adrienne et al., "Sustainable Finance: market practices", *EBA Staff Paper Series* no. 6, 2020, pp. 1–47.
Cullen, Jay/Mähönen, Jukka, "Taming Unsustainable Finance, The Perils of Modern Risk Management", Beate Sjåfjell/Christopher M. Bruner (eds.), *Cambridge Handbook oCorporate Law, Corporate Governance and Sustainability*, Cambridge University Press, Cambridge, 2019, pp. 100–113.
Dankert, Jacob et al., "A Gren Supporting Factor—The Right Policy?", *SUERF Policy Note*, 43, 2018, pp. 1–8.
Deloitte, "The Predictive Power of Stress Tests to Tackle Climate Change", 2020. Available at www.deloitte.com.
Deutsche Bank, "Deutsche Bank Launches Green Deposits for Its Corporate Clients", March 2021. Available at www.db.com.
Dias, Rui Pereira/Sá, Mafalda de, "Deveres dos Administradores e Sustentabilidade", in Paulo Câmara (Ed.), *Administração e Governação das Sociedades*, Almedina, Coimbra, 2020, pp. 33–85
EBA, "Action Plan on Sustainable Finance", December 2019.
EBA, "Report on Undue Short-Term Pressure from the Financial Sector on Corporations", December 2019.
EBA, "Guidelines on Loan Origination and Monitoring", May 2020.
EBA, "Mapping climate risk: Main findings from the EU-wide pilot exercise", May 2021.
EBA, "Draft Implementing Standards on prudential Disclosures on ESG Risks in Accordance with Article 449a CRR", March 2021.
EBA, "Opinion on the Disclosure Requirement on Environmentally Sustainable Activities in Accordance with Article 8 of the Taxonomy Regulation", February 2021.
EBA, "Report on Management and Supervision of ESG Risks for Credit Institutions and Investment Firms", June 2021.
ECB, "Climate Change and Financial Stability", May 2019.
ECB, "Guide on Climate-Related and Environmental Risks, Supervisory Expectations Relating to Risk Management and Disclosure", November 2020.

ECB, "Risk Assessment for 2021", January 2021. Available at www.bankingsupervision.europa.eu.

ESRB, "Too Late, Too Sudden: Transition to a Low-Carbon Economy and Systemic Risk", February 2016.

European Commission, "Guidelines on Non-Financial Reporting (Methodology for Reporting Non-Financial Information)", 2017.

European Commission, "Action Plan: Financing Sustainable Growth", 2018.

European Commission, "Guidelines on Non-Financial Reporting: Supplement on Reporting Climate-Related Information", 2019.

European Commission, "Interpretative Communication on the Application of the Accounting and Prudential Frameworks to Facilitate EU Bank Lending, Supporting Businesses and Households Amid COVID-19", April 2020.

European Commission, Proposal of 27 October 2021 for a Directive of the European Parliament and of the Council amending Directive 2013/36/EU as regards supervisory powers, sanctions, third-country branches, and environmental, social and governance risks, and amending Directive 2014/59/EU.

European Commission, Proposal of 27 October 2021 for a Regulation of the European Parliament and of the Council of 27 October 2021 amending Regulation (EU) No 575/2013 as regards requirements for credit risk, credit valuation adjustment risk, operational risk, market risk and the output floor.

European Ombudsman, "Report on the inspection and meeting concerning the European Commission's decision to award a contract to BlackRock on the development of tools for the integration of ESG-factors into the EU banking prudential framework", September 2020.

Forcella, Davide/Hudon, Marek, "Green Microfinance in Europe", *Journal of Business Ethics*, 135, 2016, pp. 445–459.

FSB, "Proposal for a Disclosure Task Force on Climate-Related Risks", November 2015.

FSB, "Stocktake of Financial Authorities' Experience in Including Physical and Transition Climate Risks as Part of Their Financial Stability Monitoring", July 2020.

FSB, "The Implications of Climate Change for Financial Stability", November 2020.

Gold, Russel, "PG&E: The First Climate-Change Bankruptcy, Probably Not the Last", 18 January 2019. Available at: www.wsj.com.

Grippa, Pierpaolo/Schmittmann, Jochen/Suntheim, Felix, "Climate Change and Financial Risk", *Finance and Development*, December 2019, pp. 26–29.

Guindos, Luis de, "Shining a Light on Climate Risks: The ECB's Economy-Wide Climate Stress Test", 18 March 2021. Available at: www.ecb.europa.eu.

Gyura, Gábor, "ESG and Bank Regulation: Moving with the Times", *Economy and Finance*, 7(4), 2020, pp. 366–385.

HLEG, "Financing a Sustainable European Economy", July 2017.

Hopt, Klaus J., "Corporate Governance of Banks After the Financial Crisis", in Eddy Wymeersch/Klaus J. Hopt/Guido Ferrarini (Ed.), *Financial Regulation and Supervision—A Post-Crisis Analysis*, Oxford University Press, 2012, pp. 337–367.

Jones, Huw, "Banks in EU to Publish World's First 'Green' Yardstick from Next Year", 1 March 2021, Reuters. Available at www.reuters.com.

JPMorgan, "JPMorgan Chase Adopts Paris-Aligned Financing Commitment", 6 October 2020. Available at www.jpmorganchase.com.

Koranyi, Balazs/Philippa Fletcher, "Greenpeace paragliders land on ECB building in protest", Reuters, 10 March 2021, available at www.reuters.com.

Lagarde, Christine, "Climate Change and Central Banking", Keynote Speech at the ILF Conference on Green Banking and Green Central Banking, 25 January 2021, available at: www.ecb.europa.eu.

Lehmann, Alexander, "Climate Risks to European Banks: A New Era of Stress Tests", *Bruegel Blog*. Available at www.bruegel.org, 2020.

Loan Market Association/Asia Pacific Loan Market Association/Loan Syndications and Trading Association, "Green Loan Principles", December 2018. Available at www.icmagroup.org.

Loan Market Association/Asia Pacific Loan Market Association/Loan Syndications and Trading Association, "Sustainability Linked Loan Principles", March 2019. Available at www.icmagroup.org.

Mazars, "Responsible Banking Practices—Benchmark Study 2021", 2021.

NGFS, "First Progress Report", October 2018.

NGFS, "A Call for Action, Climate Change as a Source of Financial Risk", April 2019.

NGFS, "Guide for Supervisors Integrating Climate-Related and Environmental Risks into Prudential Supervision", May 2020.

NGFS, "The Macroeconomic and Financial Stability Impacts of Climate Change", June 2020.

Nieto, Maria J., "Banks and Environmental Sustainability: Some Financial Stability Reflections", *IRCCF Working Paper*. Available at www.ssrn.com, 2017, pp. 1–34.

OECD, "Promoting Responsible Lending in the Banking Sector: The Next Frontier for Sustainable Finance", in *Business and Finance Outlook 2020: Sustainable and Resilient Finance*. Available at www.oecd-ilibrary.org.

Rainforest Action Network, "Banking on Climate Chaos—Fossil Fuel Finance Report 2021", 2021. Available at www.ran.org.

Rockström, Johan et al., "Planetary Boundaries: Exploring the Safe Operating Space for Humanity", *Ecology and Society*, 14(2), 2009.

Rosella, Carè, *Sustainable Banking—Issues and Challenges*, Palgrave Macmillan, London, 2018.

Sjåfjell, Beate, "Realising the Potential of the Board for Corporate Sustainability", in Beate Sjåfjell/Christopher M. Bruner (eds.), *Cambridge Handbook of Corporate Law, Corporate Governance and Sustainability* Cambridge University Press, Cambridge, 2019, pp. 696–709.

Sjåfjell, Beate/Bruner, Christopher M., "Corporations and Sustainability", in Beate Sjåfjell/Christopher M. Bruner (eds.), *Cambridge Handbook of Corporate Law, Corporate Governance and Sustainability* Cambridge University Press, Cambridge, 2019, pp. 2–11.

Smits, René, "SSM and the SRB Accountability at European Level: Room for Improvements?". Available at europarl.europa.eu, 2020.

TCFD, "Recommendations of the Task Force on Climate-Related Financial Disclosures", June 2017.

Thomä, Jakob/Gibhardt, Kyra, "Quantifying the Potential Impact of a Green Supporting Factor or Brown Penalty on European Banks and Lending", *Journal of Financial Regulation and Compliance*, 27(3), 2019, pp. 380–394.

UNEP Finance Initiative, "Green Financial Products and Services—Current Trends and Future Opportunities in North America, 2007.

United Nations, "Report of the World Commission on Environment and Development: Our Common Future", 1987.

WEF, "The Global Risks Report", 16th ed., 2021.
Will Steffen et al., "Planetary Boundaries: Guiding Human Development on a Changing Planet", *Science*, vol. 347, 2015.
Winter, Japp, "Addressing the Crisis of the Modern Corporation: The Duty of Societal Responsibility of the Board". 2020, Available at www.ssrn.com.

CHAPTER 20

The EU Asset Managers' Run for Green

Tiago dos Santos Matias

1 THE RISE OF ESG

The strive for concrete and significant progresses in protecting environmental and human rights, sided by the attempts to foster environmental, social and governance (ESG) standards have been around for several years, but while several instruments have been adopted at the international and at the EU level, the latter is pushing forward, by introducing several legal measures that may have significant impact.

In fact, the increasingly and unpredictable consequences of climate change and resource depletion on our planet have urged the adoption of a more sustainable economic and social model and, considering this, ESG issues in investing have become increasingly noted.

The present text will focus, mainly, on collective investment undertakings, and not on portfolio management, foreseen in paragraph (4) of the Section A of the Annex I of the Directive 2014/65/EU, on markets in financial instruments (MiFID II).

T. dos Santos Matias (✉)
International Affairs and Regulatory Policy Department, Securities Market Commission (CMVM), Lisbon, Portugal
e-mail: tsmatias@gmail.com

© The Author(s), under exclusive license to Springer Nature Switzerland AG 2022
P. Câmara and F. Morais (eds.), *The Palgrave Handbook of ESG and Corporate Governance*, https://doi.org/10.1007/978-3-030-99468-6_20

Therefore, despite the different geopolitics'[1] approaches, numerous countries from around the world endorsed a more sustainable way forward, by adopting the Paris agreement on climate change (Paris Agreement),[2] in 12 December 2015.[3]

The Paris Agreement was a significant milestone in the multilateral climate change process, representing the first binding agreement that brings all nations into a common cause to undertake ambitious efforts to combat climate change and adapt to its effects.

At the same time, in 2015, the United Nations (UN) defined 17 Sustainable Development Goals,[4] inscribed in the UN 2030 Agenda for Sustainable Development,[5] representing a worldwide commitment to achieve a sustainable development in three dimensions—economic, social and environmental.

Recognized in the EU Treaties,[6] sustainability has been a priority for the European Union, and, therefore, these UN initiatives were aligned with the European Commission's initiative on sustainable development that laid the foundations for a European framework, which has placed ESG considerations at the centre of the financial system in order to support the greening of Europe's economy.

To achieve this, the European action plan aimed to *(i)* reorient capital flows towards sustainable investment in order to reach sustainable and inclusive growth; *(ii)* assess and manage financial risks stemming from climate change, resource depletion, environmental degradation and social issues; and *(iii)* foster transparency and long-termism in financial and economic activity.[7]

Since then ESG represented a shift in the investors and investment paradigm, representing a voluntary link between sustainability and financial

---

[1] On the "geopolitics of green" see, among others, Mark Leonard and Jeremy Shapiro "*The geopolitics of the European Green Deal*", European Council on Foreign Relations (available at https://ecfr.eu/publication/the-geopolitics-of-the-european-green-deal/).

[2] https://unfccc.int/process-and-meetings/the-paris-agreement/the-paris-agreement.

[3] While other, as the USA, have decided to rejoin (https://edition.cnn.com/2021/02/19/politics/us-rejoins-paris-agreement-biden-administration/index.html).

[4] The 17 sustainable development goals provided qualitative and quantitative objectives for the future. The goals and their underlying information are available at the United Nations Department of Economic and Social Affairs Sustainable Development website (available at https://sdgs.un.org/goals).

[5] See https://sdgs.un.org/2030agenda.

[6] See, among others, art. 3.3 of the Treaty on the European Union (TEU) and the role of environmental and social issues in international cooperation (article 21 TEU).

[7] For additional information on the "*Action Plan: Financing Sustainable Growth*" see the European Commission "*Communication from the Commission to the European Parliament, the European Council, the Council, the European Central Bank, the European Economic and Social Committee and the Committee of the Regions*", COM(2018) 97final. Concomitantly, the European Green Deal emphasizes the EU commitment with sustainability (https://ec.europa.eu/info/strategy/priorities-2019-2024/european-green-deal/actions-being-taken-eu_en).

services, duly supported by the UN Paris Agreement in 2016[8] and, since 2018, by the European Commission's Action Plan of Financing Sustainable Growth.[9]

In this context, the EU has adopted a sustainable finance standard setting agenda, which emphasizes how the asset management industry can foster the run for green.

In fact, the EU strategy on sustainable finance has set out a roadmap covering all relevant actors in the financial system, including *(i)* establishing a common language for sustainable finance, *i.e.* a unified EU classification system—or taxonomy—to define what is sustainable and identify areas where sustainable investment can make the biggest impact; *(ii)* creating EU labels for green financial products to be based on this EU classification system, allowing investors to easily identify investments that comply with green or low-carbon criteria; *(iii)* clarifying the duty of asset managers to take sustainability risks and factors into their organization, operations, product governance processes and in its risk and portfolio management; *(iv)* requiring insurance and investment firms[10] to advise clients on the basis of their preferences on sustainability; *(v)* incorporating sustainability in prudential requirements: banks and insurance companies are an important source of external finance for the European economy; *(vi)* enhancing transparency in corporate reporting, by revising the guidelines on non-financial information to further align them with the recommendations of the Financial Stability Board's Task Force on Climate-related Financial Disclosures (TCFD).

2 THE ROLE OF ASSET MANAGERS AS INSTITUTIONAL INVESTORS

Collective investment undertakings[11] have long represented one of the most democratic and recognized means to raise capital from the public, who typically invest their savings and benefit from their undertakings being managed by specialized entities—the fund managers—under the principle of risk-spreading and, moreover, the duty of managers to act on behalf of the investor's interests.

The key duties of investment fund managers, under the AIFM Directive and UCITS Directive, are to perform discretionary portfolio and risk management[12] on behalf of their investors.

---

[8] https://unfccc.int/process-and-meetings/the-paris-agreement/the-paris-agreement.

[9] https://ec.europa.eu/info/publications/sustainable-finance-renewed-strategy_en.

[10] Including asset managers when duly authorized to provide MiFID II investment advice, as ancillary services.

[11] In the present study the expression "collective investment undertakings" and "investment funds" will be used indistinctively.

[12] For which they charge the fund, under Article 4(1)(w) and Annex I AIFM Directive and Annex II UCITS Directive.

To the investors, who are unable to, significantly, influence the investment fund management, whose fiduciary nature is obliged to comply with the investment funds constituting documents, the fundamental right is to exit the investment fund[13] that acts as a substitute of control.

In fact, investment funds units may, at the request of holders, be repurchased or redeemed, directly or indirectly, out of those undertakings' assets.

Emerged in the second half of the eighteenth century,[14] several studies demonstrate that the investment funds are the institutional investors who hold and manage increasingly more assets.[15,16] This makes investment funds a major force in capital markets.[17]

Alongside with the growth of large institutional investors came the expectation that a new highly specialized and well-resourced professional shareholders would, through their informed and effective use of their shareholder rights, foster good corporate governance in companies in which they invest.[18] While some fund managers are actively engaged in its investment strategies, others rely on passive management, mostly made against a benchmark, resorting to

---

[13] In this line of thought see John Morley and Quinn Curtis, *"Taking Exit Rights Seriously: Why Governance and Fee Litigation Don't Work in Mutual Funds"*, Yale Law Journal (2010), p. 84, which also compares investors existing right in investment funds with the rights of corporations shareholders. In this point, i.e. how the exit right compares with the rights of shareholders see John Armour, Henry Hansmann and Reinier Kraakman, *"The Essential Elements of Corporate Law"*, ECGI Law Working Paper no. 134/2009, p. 8, that find that the *"rule of 'liquidation protection'—provides that the individual owners of the corporation (the shareholders) cannot withdraw their share of firm assets at will, thus forcing partial or complete liquidation of the firm, nor can the personal creditors of an individual owner foreclose on the owner's share of firm assets"*.

[14] More precisely in the Netherlands, in accordance K. Geert Rouwenhorst, *"The Origins of Mutual Funds"*, Yale ICF Working Paper No. 04/48 (available at http://ssrn.com/abstract=636146).

[15] See, among other, Serdar Çelik and Mats Isaksson (2014), *"Institutional investors and ownership engagement"*, in OECD Journal: Financial Market Trends, 2013/2, p. 97, according to whom *"[i]n OECD countries, pension funds, investment funds and insurance companies have in the last decade more than doubled their total assets under management from USD 36 trillion in 2000 to USD 73.4 trillion in 2011. The largest increase among the three categories of traditional institutions has been for investment funds that have increased their assets under management by 121%."*; and, OECD (2017), Institutional Investors Statistics 2009-2016, (available at https://read.oecd-ilibrary.org/finance-and-investment/oecd-institutional-investors-statistics-2017_instinv-2017-en#page1).

[16] On the geographical location of investment funds, at the global scale, see Hugo Moredo Santos (2011), *Um governo para os fundos de investimento*, in Governo das Organizações, Almedina, p. 373.

[17] See Tiago dos Santos Matias, *"O olho do dono engorda o cavalo: algumas questões atuais dos Fundos de Investimento enquanto Investidores Institucionais"*, in Acionistas e Governação de Sociedades (2019), Almedina.

[18] It was precisely these prospects that were reflected in Principles II.F and II.G, added in 2004 to the *OECD Principles of Corporate Governance,* to promote the disclosure of voting policies, managing conflicts of interest and co-operation between investors.

indexing and proxy advisers, without properly monitoring the corporate governance of the investee companies and failing to assure the expected institutional shareholder engagement.[19]

This also relates with the issue of whether institutional investors have the right incentives to be, or to become, long-term investors, which may be crucial for ESG, especially in the "G" field.[20]

To this end, the EU lawmakers recently revised the European rules[21] to foster the long-term shareholder engagement, applicable to asset managers and institutional investors, including investment firms, fund managers and UCITS self-managed funds,[22] allowing the Member States to exempt from its scope of application the investment funds, as issuers of securities. However, such exemption does not cover the transparency obligations applicable to institutional investors and asset managers.

These rules emphasize the relevance of the role that Institutional Investors may have in fostering long-term engagement,[23] considering that greater involvement of shareholders in corporate governance is one of the levers that can help improve the financial and non-financial performance of companies.

---

[19] Passive managers tend to focus in offering low costs to investors and are, mostly, remunerated on the basis of the volume of assets under management and not of the performance of the investee companies, promoting the increase of size of the assets under management which are them investment in indices. These circumstances do not provide for institutional shareholder engagement, who become free riders, with engagement costs being borne by other shareholders.

[20] See "*The Role of Institutional Investors in Promoting Good Corporate Governance*", (2011), OECD.

[21] The Directive (EU) 2017/828 (The Shareholder Rights Directive II—SRD II), amending Directive 2007/36/EC as regards the encouragement of long-term shareholder engagement.

[22] As defined in point (1) of article 4(1) of MiFID II that provides portfolio management services to investors, an AIFM (alternative investment fund manager) as defined in point (b) of article 4(1) of AIFM Directive that does not fulfil the conditions for an exemption in accordance with article 3 of that Directive (i.e. excluding sub-threshold AIFMs) or a management company as defined in point (b) of Article 2(1) of UCITS Directive, or an investment company that is authorised in accordance with UCITS Directive, provided that it has not designated a management company authorised under that Directive for its management (*i.e.* self-managed).

[23] See, among others, "*OCDE Corporate Governance—The Role of Institutional Investors Promoting Good Corporate Governance*" (2011), (available at http://www.oecd.org/corporate/ca/corporategovernanceprinciples/theroleofinstitutionalinvestorsinpromotinggooddcorporategovernance.htm); "*Collective Investment Schemes as Shareholders: Responsibilities and Disclosure—A report of the technical Committee of the International Organization of the Securities Commissions*" (2003), (available at https://www.iosco.org/library/pubdocs/pdf/IOSCOPD129.pdf); and, Panel 6, "*Sound Management in Collective Investment Schemes*" at the XXIVth Annual Conference of IOSCO held in Lisbon, Portugal, on May 28, 1999. Greg Tanzer of the Australian Securities and Investments Commission and chair of the Technical Committee's Standing Committee on Investment Management presented the paper "*The Role of Collective Investment Schemes as an Institutional Investor in the Management of Listed & Other Public Companies.*" (available at https://www.iosco.org/library/pubdocs/pdf/IOSCOPD158.pdf).

This includes ESG factors, in particular as referred to in the Principles for Responsible Investment (PRI),[24] supported by the UN.[25]

Therefore, even if these rules may, in a way, constitute an overlap to the ones set in the UCITS and AIFM Directives,[26] considering that the fund managers were already required under the UCITS and AIFM Directives to report on investment activities, portfolio composition, turnover costs and conflicts of interests and, moreover, to develop an adequate and effective strategy for determining when and how voting rights attached to instruments held in the managed portfolios were exercised to the exclusive benefit of the investment funds concerned and its investors,[27] the SRD II adds on the existing rules, clearly supporting the responsible investment by fund managers.

In fact, SRD II imposes on asset managers the duty to report information, on an annual basis, to investors for whom they act either with the annual report[28] or in periodic communications[29]; in order to disclose the use of proxy and how the investment decision process evaluates of medium to long-term performance of the investee companies, including non-financial performance. In this sense, this information should indicate whether the asset manager adopts a long-term oriented and active approach to asset management and if, and to what extent, it takes ESG factors into account.

The mentioning of the PRI in the SRD II recitals when referring to institutional investors is a relevant sign, especially considering that each PRI signatory is required to state that it recognizes and embraces its duty to act in the best long-term interests of its beneficiaries and, moreover, that in his fiduciary role, it believes that ESG issues can affect the performance of investment portfolios (to varying degrees across companies, sectors, regions, asset classes and through time), recognizing that applying the PRI may better align investors with broader objectives of society.

---

[24] See https://www.unpri.org/.

[25] See recital 14 of the SRD II.

[26] Supporting this line of thought, considering that, in the space of asset management, SRD II rule may represent an *"unnecessary duplication of duties for asset managers"* is *EFAMA's Views on the European Commission's legislative proposal for a Directive amending Directive 2007/36/EC as regards the encouragement of long-term shareholder engagement and Directive 2013/34/EU as regards certain elements of the corporate governance statement—'Revision Shareholders' Rights Directive'* (October 2014), p. 2, by EFAMA, available at https://www.efama.org/Publications/Public/Corporate_Governance/14-4068_FinalPositionPaperSRDII_290914.pdf).

[27] See article 37 of the Delegated Regulation 231/2013 and article 21 of the Directive 2010/43/EU.

[28] See article 68 of the UCITS Directive and article 22 of the AIFM Directive.

[29] See article 25(6) of MiFID II.

3 EUROPEAN INVESTMENT FUNDS GOVERNANCE AND THE FIDUCIARY DUTY

The European origin of investment funds was, probably, a good premonition of what has been one of the most trusted investment funds regime worldwide.[30]

Considering the investment funds characteristics, the European legislator has been taking steps to enable fund managers to do well by doing good,[31] through the growing regulatory requirements towards the integration of ESG factors in their investment strategies and risk management[32] and by improving institutional investors disclosure requirements on how they integrate such factors. Additionally, the European Commission adopted a package of measures aimed at establishing a unified EU classification system of sustainable economic activities—the "taxonomy".

Fiduciaries have long considered that fiduciary duty required them to only consider financial interests of beneficiaries.

However, fiduciary principles impose on fiduciaries a duty of care and a duty of loyalty towards beneficiaries, where the duty of care requires fiduciaries to exercise skill and prudence when looking after the beneficiaries' assets and the duty of loyalty requires fiduciaries to manage funds in the beneficiaries' interests, and not their own, in order to provide beneficiaries with the benefits of such management.

On the other hand, the fiduciary relationship and its underlying duties have evolved through time, as the fiduciary duty is a dynamic concept,[33] that must be interpreted it accordance with the existing changes. Therefore, the

---

[30] Especially after the implementation of Council Directive 85/611/EEC, of 20 December 1985, on the coordination of laws, regulations and administrative provisions relating to undertakings for collective investment in transferable securities (UCITS), which led to Directive 2009/65/EC of the European Parliament and of the Council of 13 July 2009, on the coordination of laws, regulations and administrative provisions relating to UCITS (UCITS Directive).

[31] See, among other, Oliver Falck and Stephan Heblich "*Corporate social responsibility: Doing well by doing good*", Business Horizons, 50(3), May–June 2007, p. 247–254.

[32] On the risk management of investment funds see Dirk A. Zetzsche & David Eckner (2012), "*Risk Management*", in The Alternative Investment Fund Managers Directive, Wolters Kluwer.

[33] See Edward J. Waitzer and Douglas Sarro in "*Pension Fiduciaries and Public Responsibilities: Emerging Themes in the Law*" (2013), Rotman International Journal of Pension Management, 6(II), p. 28 who also consider that "*fiduciary duty is a dynamic concept – on that has responded to changing contexts and world views, but is firmly rooted in clear and enduring legal principles*". Advocating for the fulfilment of the fiduciary duty, John C. Boogle, in "*The Fiduciary Principle: No Man Can Serve Two Masters*", The Journal of Portfolio Management, 2009, p. 16, states that "*self-interest, unchecked, is a powerful force, but a force that, if it is to protect the interests of the community of all of our citizens, must ultimately be checked by society*" who cites (p. 24) Justice Harlam Fiske Stone to express its views that "*[there is] nothing more vital to our own day than that those who act as fiduciaries in the strategic positions of our business civilization, should be held to those standards of scrupulous fidelity which [our] society has the right to demand*".

raising relevance of new risks and the impacts they pose, alongside with the beneficiaries' investment purposes, must be duly considered.[34]

The European investment funds are characterized by a fiduciary relationship between investors, fund managers and depositaries, which is at the heart of fund governance and represents a fundamental difference to corporations, ruled by the corporate and commercial law.

The investment funds governance is clearly defined by the separation of "investment" and "management", and by the fiduciary relationship established between the so called "investment triangle"[35] or, under UCITS and AIFM Directives, the "investment quadrangle"[36]; making European funds governance a key advantage of investment funds in what could represent the most democratic and effective mean to promote sustainable investing, managed by specialized and resourced managers that will be required to consider ESG risks in their investment decisions, and subject to the depositaries safekeeping of the fund's assets and its oversight and control duties.

In this tripartite relationship, all relations between the investors and the fund manager, on one hand, and the investors and depositaries, on the other, are of a fiduciary nature; and these three parties relationships justify that investors rely on the oversight of the depository towards the management carried out by the fund manager, as per its constituting documents and, therefore, its right to exit the fund is the crucial mean towards the control of both the fund manager and the depositary.

The investment fund governance, and its difference from corporate governance, is rightly explained by the theory of "the four stages of capitalism".[37]

---

[34] On issues that arise from ESG factors consideration in the investment decision process and the fiduciary duty, of loyalty, under which "the sole interest rule" imposed by the fiduciary duty requires a manager to only consider the financial interests of the investors, i.e. the investment funds unit holders, see Max M. Schanzenbach and Robert H. Sitkoff "*Reconciling Fiduciary Duty and Social Conscience: The Law and Economics of ESG Investing by the Trustee*", Stanford Law Review, 72, February 2020, p. 381.

A different and setting stone view is presented by the report "A legal framework for the integration of environmental, social and governance issues into institutional investment", produced in 2005 for the Asset Management Working Group of the UNEP Finance Initiative by Freshfields Bruckhaus Deringer, which sets out three circumstances where integrating ESG factors into investment decision-making was permissible under ERISA: *(i)* as tie-breakers of the financial characteristics of alternative investment choices were equivalent; *(ii)* could, and should, be considered as an integral part of the investment decision when they were relevant to the financial performance of an investment; and, *(iii)* could be taken into account when there was a consensus amongst the beneficiaries about doing so.

[35] The so called "investment triangle" represents the translation of the German expression "das Anlagedreieck", as it is called by Nils Seegebarth (2004), in "*Haftung der Depotbank im Investment-Dreieck*", Peter Lang.

[36] See Sebastiaan Hoogiemstra (2009), "*Towards Modernization of the Luxembourg Legal Form "Toolboox" for Funds*", JurisNews, Larcier, Vol. 8 – No. 3-4, p.137 (available at https://ssrn.com/abstract=3509731).

[37] See Robert Charles Clark, "*The Four Stages of Capitalism: Reflections on Investment Management Treatises*", Harvard Law Review, 94(3) (1981), p. 561.

The first stage is the age of entrepreneurs, who promoted, invested, and managed, with no split between ownership and control.

In the second stage with the split of entrepreneurship into ownership and control came the management professionalization of modern corporations,[38] and with it the agency costs addressed by corporate law that provided legal personality, limited liability, transferable shares and investor ownership.[39] Limited liability and legal personality led to asset partitioning that enhanced the corporations' creditworthiness, enabling shareholders to limit their liability and to benefit from voting rights to discipline managers.

The third stage resulted from the split of ownership into capital supplying and active investment, and the latter led to the professionalization of the investment function. It was precisely this separation between investments and investments management that resulted in a fiduciary relationship between investment managers and investors.

In the fourth stage, the current one, the split of capital supply into the possession of beneficial claims and the decision to save, resulted in the professionalization of the savings-decision function. The stage of savings planner.[40]

The four stages of capitalism provide a clear and objective evidence that, one, both commercial and corporate law is not designed to resolve the fiduciary relationship between investors and fund managers and, two, clearly supports the advantages that fund governance provides in fostering ESG and the role that asset management may have.

The European legislator has, with the UCITS and AIFM Directives, implemented a specific regulatory framework for investment funds where it is clearly stated that, in the context of their roles, the management company and the depositary must act independently and solely in the interest of the investors, *i.e.* unitholders.

It is this fiduciary duty[41] that has been commonly interpreted as imposing fund managers solely to pursue profit maximization, without incorporating

---

[38] As Robert Charles Clark correctly considered, the separation of the investment decision from the decision to provide capital for investment is one of the most striking institutional developments in our century (*"The Four Stages of Capitalism: Reflections on Investment Management Treatises"*, Harvard Law Review, 94(3) (1981), p. 564).

[39] See John Armour, Henry Hansmann and Reinier Kraakman, "The Essential Elements of Corporate Law", ECGI Law Working Paper no. 134/2009, p. 8.

[40] Whose agency issues between pension boards and beneficiaries are addressed in the Directive (EU) 2016/2341 of the European Parliament and of the Council of 14 December 2016 on the activities and supervision of institutions for occupational retirement provision (IORPs), as amended (IORPD II).

[41] In accordance with Keith L. Johnson (*"Introduction to Institutional Investor Fiduciary Duties"*, International Institute for Sustainable Development, February 2014, available at https://www.iisd.org/publications/introduction-institutional-investor-fiduciary-duties) the application of fiduciary standards vary significantly across different jurisdictions, cultures or contexts, concluding that there is no single definition of the fiduciary duty, who also states that *"fiduciary duties focus on process and behaviour, rather than investment outcomes."*

the interest of the beneficiaries in other objectives. Nevertheless, as mentioned before, the fiduciary duty is a dynamic concept.

While, in one hand, it seems unrealistic to consider that the externalities will not affect the investee value, the sustainability risk and its corresponding impact should be assessed and duly considered, especially considering its materiality.

Regarding the concept of materiality, the discussion around it seems to be, finally, coming to an end, with the growing consensus that materiality is double and dynamic.

In fact, double materiality recognizes that companies should report on impacts that are financially material, influencing enterprise value, and on impacts that materially affect sustainability on the ESG level.

Second, it should be considered dynamic as it is subject to, expected and unexpected, changes. This was acknowledged by the European Commission, in June 2019, in the *"Guidelines on non-financial reporting: supplement on reporting relating information"*[42,43] and by the Global Reporting Initiative (GRI),[44] the Sustainability Accounting Standards Board (SASB).[45,46]

This entails the conflict between "value driven" and "values driven" approaches to the fulfilment of the fiduciary duty that led to the report "A

---

[42] Guidelines on non-financial reporting: Supplement on reporting climate-related information (2019/C 209/01), available at https://eur-lex.europa.eu/legal-content/EN/TXT/PDF/?uri=CELEX:52019XC0620(01)&from=EN.

[43] According to article 19a of the Directive 2014/95/EU (the Non-Financial Reporting Directive - NFRD), a company is required to disclose information on ESG to the extent that such information is necessary for an understanding of the company's development, performance, position and impact of its activities relating to, as a minimum, environmental, social and employee matters. The European Commission in the section 2.2. of the mentioned Guidelines stated that the reference to the "company's development performance and position" indicates financial materiality, in the broad sense of affecting the value of the company. This perspective is typically of most interest to investors. And, therefore, the reference to "impact of the company's activities" indicates environmental and social materiality. This perspective is typically of most interest to citizens, consumers, employees, business partners, communities and civil society organizations.

[44] In its universal exposure draft, available at https://www.globalreporting.org/standards/media/2605/universal-exposure-draft.pdf.

[45] In its proposed changes to the SASB conceptual framework https://www.sasb.org/wp-content/uploads/2020/08/Invitation-to-Comment-SASB-CF-RoP.pdf.

[46] And, later, in September 2020 by the five reporting standards organizations: CDP, CDSB, GRI, IIRC, and SASB in their two papers *"Statement of Intent to Work Together Towards Comprehensive Corporate Reporting"* (available at https://29kjwb3armds2g3gi4lq2sx1-wpengine.netdna-ssl.com/wp-content/uploads/Statement-of-Intent-to-Work-Together-Towards-Comprehensive-Corporate-Reporting.pdf) and *"Reporting on enterprise value"* (available at https://impactmanagementproject.com/structured-network/global-sustainability-and-integrated-reporting-organisations-launch-prototype-climate-related-financial-disclosure-standard/).

legal framework for the integration of the environmental, social and governance issues into institutional investment"[47] which, while it recognizes that conventional investment analysis focuses on value, in the sense of financial performance, concludes that the growing recognition of the links between ESG factors and financial performance enables the integration of "*ESG considerations into an investment analysis so as to more reliably predict financial performance is clearly permissible and is arguably required in all jurisdictions*".[48]

Therefore, the dynamic nature of the fiduciary duty is precisely the reason why the European legislator has explicitly foreseen the integration of sustainability risks in the investment decision process and disclosure rules on how such integration is undertaken.

4 THE INTEGRATION OF ESG FACTORS IN EUROPEAN INVESTMENT FUNDS RISK MANAGEMENT

Therefore, in order to clarify the need to incorporate sustainability risks in their due-diligence processes and assess and manage the sustainability risks stemming from their investments along with all other relevant risks such as market, interest and credit risk[49] the European legislator has introduced several regulatory changes to the EU existing framework.

Nevertheless, proportionality was duly considered, and, in this context, sustainability risks are required to be captured by the due-diligence process and risk management systems in a way and to the extent that is appropriate to the size, nature, scope and complexity of their activities and the relevant investment strategies pursued.

In this context, the upcoming changes to UCITS and AIFM Directives will cover:

(i) Senior Management accountability:
   Clarifies that senior management is accountable for the integration of sustainability risks.

---

[47] Produced by the law firm Freshfields Bruckhaus Deringer, in October 2005, for the Asset Management Working Group of the UNEP Finance Initiative. See footnote 35.

[48] With a, seemingly evolving but still, somehow (considering the "Law and Economics of Environmental, Social, and Governance Investing by a Fiduciary", 2018, Harvard Discussion Paper n.º 971), opposing view see Max M. Schanzenbach and Robert H. Sitkoff who, while rejecting the claims that the duty of prudence either does or should require trustees to use ESG factors, conclude that American Fiduciary Law permits ESG investing if "*1) The trustee reasonably concludes that ESG investing will benefit the beneficiary directly by improving risk-adjusted returns, and 2) the trustee's exclusive motive for ESG investing is to obtain this direct benefit.*" (*Reconciling Fiduciary Duty and Social Conscience: The Law and Economics of ESG Investing by the Trustee*", Stanford Law Review, 72, February 2020, p. 382.

[49] See ESMA 34-45-569 Consultation paper "*Integrating sustainability risks and factors in the UCITS Directive and AIFMD*".

(ii) Knowledge and competence:
Fund managers are required to possess resources and expertise in order to integrate the sustainability risks.
(iii) Organizational requirements:
Organizational procedures, systems and controls must ensure the integration of sustainability risks in the investment and risk management processes.
(iv) Conflicts of interest:
The assessment of potential conflicts that may damage the interests of the investment funds undermanagement must consider conflicts that may arise as a result of the integration of sustainability risks in your processes, systems and internal controls.
(v) Due Diligence:
Requires due diligence to be foreseen in written policies and procedures in order to assure the consideration of sustainability risks when electing and monitoring investments.
(vi) Risk Management:
Explicit inclusion of the sustainability risks in the risk management policies and procedures.

As mentioned, fund managers have two core duties, the portfolio[50] and the risk management, on behalf of their investors.

In this context, the depositary oversight duties towards the fund manager required him to ensure that the latter is investing the fund's assets in compliance with the investment policy as laid down in the fund's constitutive documents.[51]

Additionally, conduct of business rules specify the general duty of loyalty and care that fund managers and depositaries have towards investors, resulting in the implementation of the separation between "investments" and "management", where organizational requirements, such as the risk management[52] are expressly set out.

Nevertheless, questions remain on materiality and measurability of ESG factors.[53]

---

[50] Contrary to MiFID II portfolio management, the investment funds portfolio management cannot be influenced by the investor's investment directions, as it only and strictly complies with the investment funds constituting documents.

[51] These characteristics of the collective portfolio management contrast with the individual investment relationships, such as MiFID II discretionary mandates where a custodian is appointed for merely safekeeping the investor assets.

[52] See article 15 AIFM Directive and article 51 UCITS Directive.

[53] According to Claire Woods (2011), "*The Environment, Intergenerational Equity & Long-Term Investment*", Worchester College, University of Oxford (available at https://ora.ox.ac.uk/objects/uuid:30dd270b-3f0f-4b8b-979e-904af5cb597b), the narrow interpretation of the fiduciary duty remains influential, partly, as a result of behavioural biases

5 THE EUROPEAN TRANSPARENCY AND METRICS REGIME

Created to regulate sustainability-related disclosures to investors, the Regulation (EU) 2019/2088 (SFDR) is applicable to fund managers in two dimensions, *(i)* at the fund manager level, referring to its obligations, *(ii)* at a product level, relation to obligations applicable to all the managers' financial products, with or without express ESG focus, and to the fund manager financial products with an express ESG focus.[54]

On the fund managers' level, managers are required to implement a policy to integrate sustainability risks in the investment decision process, for which purpose it must implement a due-diligence policy or explain why it does not consider such adverse impacts.[55] Lastly, for this purpose, managers are also required to reflect the sustainability risks on its remuneration policy.

In this regard, despite several interpretations that small AIFM would not fall under the SFDR remit,[56] the SFDR[57] is applicable to "financial market participant", including AIFM's, that are defined in the AIFM Directive, covering all entities who manage AIFs, without any distinction between registered or authorized under the mentioned Directive.

On the product level, with regards to the obligations applicable to financial products, managers are required to assess products impact of sustainability risks on expected returns or explain why it does not consider such adverse impacts. Additionally, all marketing materials are required to be reviewed in order to comply with the sustainability-related disclosures requirements.

On the other hand, when considering financial products with an express ESG focus, additional disclosures are required where environment or social characteristics are promoted, sustainable investment is a purpose, or the reduction of carbon emissions is targeted.

Additional duties will apply if the investment funds are labelled with different shades of green.

In fact, should the investment fund be labelled as a "light green product", as a result for promoting environmental or social characteristics, or as a

---

among fiduciaries, who tend towards inertia upon the lingering uncertainty about fiduciary duty and doubts about the materiality and measurability of ESG factors.

[54] Regarding this framework, see also Julien Froumouth and Joana Frade, ESG reporting, Chapter 12 in this book.

[55] For managers with less than 500 employees on a group consolidated basis, which leaves us to consider that proportionality could have been greatly considered by the European legislator, in this regard.

[56] Contributing to these views are the frequent lack of clearness around the SFDR. In fact, in this specific regard, the SFDR expressly includes in the list of entities within its scope managers of EuVECA and EuSEF funds, which can be managed by both sub-thresholds AIFM and licensed managers. In light of this, some views have been expressed that such reference would not be necessary if the European legislator intended to apply SFDR to sub-thresholds AIFM.

[57] See article 2(1) SFDR.

"dark green product"—the different shades of green –, for promoting sustainable investment,[58] additional disclosure requirements will apply to *(i)* the investment fund constituting documents (including its prospectus or private placement memorandum), *(ii)* publication on the investment fund or fund manager website and *(iii)* in its annual report.

The flowchart below exemplifies how the different shades of green are determined.

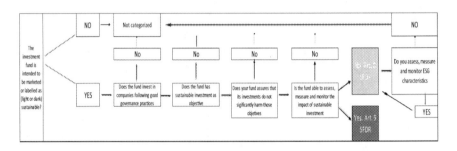

In this context, the Taxonomy Regulation[59] was established in order the create the world's first classification system for sustainable economic activities, to achieve a common language, enabling investors to invest in projects and economic activities with a substantial positive impact on climate and the environment.

Therefore, the Taxonomy Regulation foresees new disclosure obligations on the part of fund managers offering financial products as environmentally sustainable. These disclosure obligations supplement the rules on sustainability-related disclosures laid down in the SFDR.

After collecting the necessary data, carrying out its analysis, and, finally, classifying the environmentally sustainable nature of an investment,[60] the fund manager analysis must be converted into SFDR disclosures[61] providing for transparency and credibility, reassuring investors that their investment is genuinely green and not simply greenwashed.

Nevertheless, the combination of the SFDR and the Taxonomy Regulation haven't been perfect and the difficulties arising from their interpretation, even prior to its application, present a significant challenge.[62]

---

[58] See articles 8 and 9 SFDR.

[59] See Regulation (EU) 2020/852.

[60] In accordance with the criteria described above and article 3 of the Taxonomy Regulation.

[61] In accordance with articles 5 and 6 of the Taxonomy Regulation.

[62] Regarding the building blocks and the six building blocks, especially the Taxonomy Regulation see Dirk A. Zetzsche & Linn Anker-Sørensen, "Regulating Sustainable Finance in the Dark", p.5 (available at: https://ssrn.com/abstract=3871677).

In fact, not only the interpretation of key rules, such as the already famous articles 8 and 9 of the SFDR[63] and article 8 of the Taxonomy Regulation[64] has proven to be quite a challenge, but also the magnitude of the task and the time constraints prevented the EU lawmakers of having the regulatory technical standards ready in due time, while the level one legislation entered into force.

In addition to this unintended complexity, there are still some missing pieces concerning ESG metrics, relating to the lack of agreed definitions, labels and metrics at the EU—and the international—level which hampers the implementation of a harmonized approach to sustainable finance.[65]

6 Conclusion

The introduction of a renewed legal framework, based on several building blocks, is intended to, by the ruling role of law, support (mandatory) sustainable investing by the beneficiaries through their fiduciaries.

Nevertheless, should beneficiaries pursue prosocial goals one could assume that they will elect fiduciaries that uphold such investment strategies.[66]

Considering this, institutional investors and, among them, investment funds represent an opportunity towards promoting sustainability.

However, to elect the "right" fiduciaries, beneficiaries need standardized definition and information on sustainability. On the other hand, beneficiaries need to be conscient of the business models under which their fiduciaries operate, meaning that the fiduciaries who undertake an active management have strong incentives to manage through the exit, but far less to engage with investees on sustainability, while fiduciaries pursuing passive management strategies target low cost or cost saving strategies that disables them to actively engage with investees. Additionally, short-term financial returns may challenge the sustainability returns long runs.

---

[63] See, among other, the letter of the Chair of the Joint Committee of the ESAs "Priority issues relating to SFDR application" (JC 2021 02), the "*Joint ESA Supervisory Statement on the application of the Sustainable Finance Disclosure Regulation*" (JC 2021 06) and the Autorité des Marchés Financiers (AMF) guidance on the "*Implementation of the SFDR regulation for asset management companies as of March 10, 2021*".

[64] See ESMA's "*Advice on Article 8 of the Taxonomy Regulation – Final Report*" (ESMA30-379-471).

[65] See "*Aggregate Confusion: The Divergence of ESG Ratings*" by Florian Berg, Julian F Kölbel, Roberto Rigobon, who investigated the divergence of ESG ratings, based on data from six prominent rating agencies - namely, KLD (MSCI Stats), Sustainalytics, Vigeo Eiris (Moody's), RobecoSAM (SP Global), Asset4 (Refinitiv), and MSCI IVA, have decomposed the divergence into three sources: (*i*) different scope of categories, (*ii*) different measurement of categories, and (*iii*) different weights of categories, and found that scope and measurement divergence are the main drivers, while weights divergence is less important.

[66] See Oliver Hart and Luigi Zingales, "*Companies Should Maximize Shareholder Welfare Not Market Value*" (2017), Journal of Law, Finance and Accounting, 2, p. 247, who argue that where shareholders are prosocial the company and asset managers should pursue policies consistent with the preferences of their investors.

In this context, the European legal framework is designed to support a more informed investment decision making by beneficiaries, with regards to sustainability, by *(i)* foreseeing, on a comply-or-explain basis, greater transparency on investment funds voting policies and their application,[67] *(ii)* greater disclosure on what sustainable investment means and it represents to financial products being distributed,[68] requiring institutional investors to justify the sustainability label, and *(iii)* providing for a framework that defines environmentally sustainable activities and investments therein,[69] by setting two screening criteria whose technical standards will be developed and applied.

And last, but not least, by expressly imposing the integration of ESG factors in the investment decision process of fund managers, such framework implicitly recognizes that fiduciary duties are, in fact, dynamic.

Once again, the EU has chosen to lead the way, providing for a legal framework that will allow its fiduciaries, especially fund managers, to be at the forefront of the integration of ESG factors and, at the same time, allowing its beneficiaries to understand how and to what extent that integration is made by fiduciaries and investees themselves,[70] becoming the standard-setter in the long run for the different shades of green.[71]

---

[67] See the SRD II.

[68] See the SFDR.

[69] See the Taxonomy Regulation.

[70] See the NFRD.

[71] On the possible outcomes of this strategy, see Danny Busch, "Sustainable Finance Disclosure in the EU Financial Sector", EBI Working Paper no. 70/2021, *in* Sustainable Finance in Europe: Corporate Governance, Financial Stability and Financial Markets, 2021, Palgrave Macmillan, p. 442, who argues that *"the EU is not island. There are roughly two opposite scenario's. In a pessimistic scenario the more lenient or even non-existent sustainability agenda of other geopolitical powers gives them a competitive edge that is detrimental to the EU. In a positive scenario, the EU becomes a global standard-setter in the area of sustainability."*

CHAPTER 21

ESG, State-Owned Enterprises and Smart Cities

José Miguel Lucas

'Sustainability is the principle of ensuring that our actions today do not limit the range of economic, social, and environmental options open to future generations'.
—*John Elkington*[1]

1   Introduction

Currently, we have to consider two significant realities as determinants for the survival of humankind: sustainability and technology. In this chapter, we will analyse how Environmental, Social and Governance (ESG) factors influence and are influenced by State-Owned Enterprises (SOEs) and municipalities. In this latter case, which, due to the growing use of digital technology and the need to find more innovative ways to manage problems, are progressively

---

[1] *'Cannibals with Forks—The Triple Bottom Line of 21st Century Business'*, (1999), p. 20, similar to the concept presented by the United Nations' (UN) Brundtland Commission Report from 1987, mentioned bellow in this chapter.

This document (and the positions reflected in it) results from the author's understanding and not from the position of his current employer, Banco de Portugal, neither representing nor binding, therefore, this institution.

J. M. Lucas (✉)
University of Lisbon, Lisbon, Portugal
URL: https://www.linkedin.com/in/josemiguellucas

© The Author(s), under exclusive license to Springer Nature Switzerland AG 2022
P. Câmara and F. Morais (eds.), *The Palgrave Handbook of ESG and Corporate Governance*, https://doi.org/10.1007/978-3-030-99468-6_21

retrofitting cities sustainability or creating smart cities.[2] In that analysis, it is undeniable that technology (especially in the case of smart cities), is helping towards sustainability but also leading to the need of discussing the risks and preventing possible adverse effects.

With this background, the current work addresses the following research questions: (i) How do ESG factors influence SOEs and smart cities? (ii) How is it possible to measure/benchmark ESG factors, and does that ranking affect SOEs and smart cities? (iii) What are the risks related to technology for SOEs and smart cities, and how can innovation zones help?

In general, companies long recognised that it was necessary to adopt measures towards sustainability. That need has been more evident since the beginning of the century, and the recent contexts of investment marked by instability favoured sustainable, more responsible corporate practices, especially the ones that consider the integration of ESG factors. These factors involve analysing how key sustainability-related issues can be incorporated into the risk-return profiles of companies, increasing their financial returns[3] and allowing identifying new opportunities.[4]

The ability to measure ESG actions — policies, behaviours, training and investments — using specific metrics or ratings has allowed investors to benchmark ESG performance and made the process of evaluating companies' potential in terms of proactive ESG behaviour more precise. Thus, although disclosing information about most ESG factors is voluntary, looking at benchmarks has become a standard requirement for key stakeholders, especially investors.

---

[2] There is no agreed definition of the concept of smart city (HAFEDH CHOURABI, et al., *'Understanding smart cities: an integrative framework'*, 2012), but it can be considered as an urban area in which priority is given to the use of new technologies to increase the quality of life and the efficiency of the processes in six aspects: economy, mobility, people, environment, life/social, and governance (SARA FERNANDES, *'Smart cities'*, 2017; ANTHONY TOWNSEND, *'Smart Cities: Big Data, Civic Hackers, and the Quest for a New Utopia'*, 2013 and the International Telecommunications Union (ITU) in *'Smart sustainable cities: An analysis of definitions'* (2014). In the current work, we have considered that 'smart cities' are currently based on existing organisations, which in administrative law would correspond to local governments, normally municipalities (STEVE BERNARDIN et al., *'La ville intelligente sans les villes? Interopérabilité, ouvertures et maîtrise des données publiques au sein des administrations municipales'* (2019)). It remains unclear whether the current administrative model is sufficient, what adaptations it will have to undergo or whether it needs to be reinvented in the light of what one can already predict as challenges in the future.

[3] GORDON CLARK et al., *'From the stockholder to the stakeholder: how sustainability can drive financial outperformance'* (2014); ARTHUR HUGHES et al., *'Alternative ESG ratings: how technological innovation is reshaping sustainable investment'* (2021).

[4] For example, the 'Ayr' project by Ceiia, a Portuguese centre for engineering and product development, allows citizens to exchange credits obtained with carbon dioxide (CO_2) savings for the products of partners, who in turn use those credits to offset their emissions (https://www.ceiia.com/ayr).

Currently, ESG factors are primarily used as indicators of the non-financial performance of privately-owned enterprises (POEs)[5] — the idea is to help investors integrate non-financial, sustainability considerations into their investment process, for instance assessing funds based on their ESG risks and opportunities.

However, we defend that SOEs[6] and smart cities could also benefit from implementing good practices regarding ESG factors and their respective benchmarks.[7] We consider these two types of entities interesting because, on the one hand, SOEs generally operate in sectors with high social presence and have a significant (national or even international) dimension and impact. On the other hand, smart cities tend to be closer to population and benefit from ESG factors, namely to attract people and investment. According to the United Nations (UN), cities are significant contributors to climate change.[8] According to UN Habitat, cities consume two thirds of the world's energy and a significant portion of global CO2 emissions.

Both SOEs and smart cities may ensure quicker results regarding sustainability than central administration[9] and are likely to reveal a more significant commitment to sustainability.[10] Consequently, that could lead to more benefits from using benchmarking for ESG factors. Furthermore, guidance on sustainability could mean more efficiency,[11] serving the public interest and increasing public goodwill. In that vein, it must be recalled that SOEs are

---

[5] Regarding the scope of ESG dimensions, see PAULO CÂMARA, Chapter 1 in this book.

[6] On 29 September 2020, the OECD released its annual report, '*OECD Business and Finance Outlook 2020—Sustainable and Resilient Finance*' (2020c), Sect. 6 of which (i.e. 'State-Owned Enterprises, Sustainable Finance and Resilience') contains important recommendations regarding SOEs.

[7] ISMAIL ÇAĞRI ÖZCAN, '*Determinants of environmental, social, and governance reporting of rail companies: does state ownership matter?*', (2020).

[8] United Nations (UN) '*The New Urban Agenda Illustrated*' (2020).

[9] Municipalities/smart cities tend to be smaller organisations than central administrations, closer to citizens and more agile in implementing new strategies. So, they tend to have a more responsive performance than central administrations — for an example, see RUI RIO, '*A conciliação da gestão com a política: um exemplo concreto*' (2014).

[10] PO HSUAN HSU et al., '*Leviathan inc. and corporate environmental engagement*', (2020).

[11] The term *efficiency* is used in a broader sense—i.e., to describe what allows obtaining the greatest results with the least resources (BRIGITTA JAKOB, '*Performance in Strategic Sectors: A Comparison of Profitability and Efficiency of State-Owned Enterprises and Private Corporations*', (2017)) —, therefore, the cost of production per unit and subsequent profit can be appropriate criteria of assessment. However, the profit earned by an SOE is not an especially appropriate criterion of its efficiency, because although a POE would focus on profit and SOEs should as well, SOEs usually have other interests, namely regarding public interest (FORFAS, '*The Role of State Owned Enterprises: Providing Infrastructure an Supportin Economic Recovery*' (2010) and JOSÉ MIGUEL LUCAS, '*Empresas públicas e corporate governance: da definição da prossecução do interesse público ao controlo externo efectuado pela supervisão*' (2016).

required to achieve public interest objectives. Smart cities should also pursue the public interest, especially by assuring better decisions and providing a higher quality of life for their citizens, city dwellers and 'stakeholders'. Beyond that, there is a growing range of investments that aim to capitalise on the opportunities presented by the new generation of smart cities offering sustainable ways of living, and those investments rely on the performance of companies that invest in areas like smart infrastructures.

In contrast, SOEs and even smart cities that fail to adapt might lose interest and become obsolete, as may happen with POEs.[12] Beyond that, many major coastal cities with more than 10 million people are already under threat as over 90% of urban areas are coastal.[13]

In this context, we will begin in the next section with an overview of the context of ESG factors. Subsequently, the research topics will be addressed in Sect. 3. It will analyse the risks of technology and how innovation zones can help, namely by providing an overview of Portugal's framework regarding innovation zones (Zonas Livres Tecnológicas—ZLT)).[14]

---

[12] BLACKROCK, Inc., '*Our approach to sustainability - Blackrock investment stewardship*', (2020).

[13] UNITED NATIONS (UN), '*Habitat, Climate action for cities—innovate cities*', (2021).

[14] Unlike several studies addressing the economic and technological aspects of the digital transformation, the current work has a legal perspective. Albeit without referring to philosophy of law (e.g. regarding the role and impact of technology) or theory of law (e.g. regarding the advantages and disadvantages of regulation), we defend that defining criteria, objectives and indicators that facilitate the implementation of solutions requires legal support, as does identifying potential risks that may hinder their implementation. It may also be necessary to enforce obligations or impose consequences if implementation is not accomplished.

2 ESG Factors, SOEs and Smart Cities

Origins and Development

In the 1990s, the concept of the triple bottom line (or three Es 'concept' of sustainable development[15]) was introduced to represent the equity-related, economic, and environmental/ecological aspects of sustainability,[16,17] pursuing the idea of full cost accounting. By 2004, investors had begun adopting the term *ESG integration* to highlight the systematic inclusion of those factors into traditional financial analysis. In particular, the term *ESG factors* is attributed to a 2005 study, 'Who Cares Wins', conducted by a group of financial institutions at the request of the United Nations (UN) in the context of the UN Global Compact.[18]

---

[15] The concept of 'sustainable development' tends to be associated with countries and policies, as well as with criteria such as social justice, responsible economic growth and environmental preservation, thus being related to other terms such as 'sustainable development goals' (SDGs), 'corporate sustainability' ' and 'sustainable finance'.

The ideas about sustainable development have roots in the 1960s and 1970s with similar concepts such as green development and eco-development (WILLIAM ADAMS, '*Green development: environment and sustainability in the Third World*', 2001). According to the UN Brundtland Commission Report from 1987, sustainable development is 'development that meets the needs of the present without compromising the ability of future generations to meet their own needs'.

Sustainable development does not require eliminating the economic–financial factor (i.e. profit); instead, it demands the consideration of environmental and social impacts in how economic and financial returns are generated. The triple bottom line expands the traditional reporting framework to take into account also (i) social and (ii) environmental performance in addition to (iii) financial performance.

[16] Sustainability is a core concept on which most objectives converge (JOHN ZINKIN, '*The challenge of sustainability: corporate governance in a complicated world*', 2020) and that can be applied to every system and process, from personal habits to the survival of the human race. Regarding the concept of sustainability, according to the International Union for Conservation of Nature (IUCN), United Nations Development Programme (UNDP) and World Wide Fund for Nature (WWF) in '*Caring for the Earth: A strategy for sustainable living*' (1991), *sustainability* means 'improving the quality of life whilst living within the carrying capacity of supporting ecosystems'. We believe that sustainability results from planning and assessing potential adverse effects, namely regarding ESG factors. Ideally, that proactive planning and assessment should avoid those adverse effects, or prevent that they become irreversible. Thereby this would maintain a system or process working sustainably: i.e. efficiently and indefinitely without running out of resources (JOHN ZINKIN, ' *The challenge of sustainability: corporate governance in a complicated world*', 2020; WILLIAM ADAMS, '*Green development: environment and sustainability in the Third World*', 2001; MARGARET ROBERTSON, '*Sustainability: principles and practice. Earthscan*', 2014).

[17] JONH ELKINGTON, '*Cannibals with forks—the triple bottom line of 21st-century business*' (1999); ROBERT GOODLAND et al., '*Environmental sustainability: Universal and Non-Negotiable*' (1996) and RUTH JEBE, '*The convergence of financial and ESG materiality: taking sustainability mainstream*', (2019).

[18] At the same time, the UN Environment Programme (UNEP) Finance Initiative published '*The Materiality of Social, Environmental, and Corporate Governance Issues to Equity Pricing*' to outline how ESG-related issues could be integrated into mainstream

Since then, the economic pillar of the triple bottom line was considered to represent not only profit but also compliance, risk management and proper governance, meaning that it became somewhat converted into a governance pillar, adding to environmental and social concerns. From companies' perspective, we can consider ESG factors as indicators of contributions to sustainable development and of the integration of corporate social responsibility (CSR),[19] which was often related to disengaged departments with limited resources.[20]

Following the crisis caused by the COVID-19 pandemic, sustainability has become, again, a top priority for governments and public opinion, mainly because major pre-pandemic ESG problems remain to be solved. In that scenario, it is natural that citizens' pressure on companies, both private and public, and on existing municipalities or smart cities may intensify to ensure that sustainability is considered. From that perspective, SOEs and smart cities have ESG-related issues to address: the world is currently at the dawn of a new era regarding the use of certain types of technology, namely 5G and the Internet of Things (IoT). These tools are able to help organisations to become more sustainable. To seize that opportunity and recognise and mitigate the inherent risks involved, public managers of SOEs and smart cities need to be proactive and qualified.

Influence of ESG Factors

Common Aspects
When considering how ESG factors can affect SOEs and smart cities, it should be remembered that the values of efficiency, transparency and sustainability underlying ESG factors are of utmost relevance to those entities.

Because governments, as shareholders of SOEs, and municipalities/smart cities are accountable to citizens, a publicly shared strategy for sustainability and reporting on ESG performance can be highly valuable.

---

investment analysis. Those developments laid the groundwork for creating the 'Principles for Responsible Investment (PRI)', published in 2006.

[19] INÊS VIEIRA et al., '*Corporate governance, ethics and social responsibility: comparing continental European and Anglo-Saxon firms*' (2010).

[20] A commitment to CSR may imply, for instance, an obligation to public reporting about the business' substantial impact for the improvement of the environment. And we recall that the contribution should be positive, because companies often already contribute in negative ways — to name a few, corruption, pollution and violations of human, workers' and consumers' rights (RUI PEREIRA DIAS et al., '*Deveres dos administradores e sustentabilidade*' 2020). Ideally, apart from solely addressing the internalisation of ESG impacts by companies (i.e. 'weak sustainability'), there should also be positive externalities ('strong sustainability'). In this work, terms such as *CSR* and *sustainability* are used as 'umbrella constructs', as presented by PAUL HIRSCH et al. '*Umbrella Advocates Versus Validity Police: A Life-Cycle Model. Organization Science*' (1999), each as 'a broad concept or idea used loosely to encompass and account for a broad set of diverse phenomena' (JEAN-PASCAL GOND et al., '*Corporate social performance disoriented: saving the lost paradigm*' 2010) covering the three ESG dimensions.

Such measures enable citizens to scrutinise information about ESG performance. Moreover, they can deliver hard-to-quantify benefits beyond ones that concern the ESG factors, including public goodwill, a better reputation and the visibility of their sound, prudent management. ESG factors can also be an opportunity to generate value, with cost savings and better outcomes from products or public services offered to citizens and, in that way, contribute positively to the public interest and sustainability of the planet.

The adopted measures regarding ESG factors can also ensure access of SOEs and SC to financing on better terms. For SOEs in particular, the actions may concern equity or investment. In contrast, for smart cities, generally as public bodies with public funding, a better ESG rating may, for instance, lead to investment in funds of the shares of companies that offer goods and services that can help improve the quality of life in that city.

From a public procurement perspective, it can also be relevant to assess the ESG performance of companies that are expected to implement (and earn significant revenues with) smart and sustainable solutions or, in contrast, to exclude companies that are not investing in ESG performance.[21]

Impact of ESG Factors in SOEs

In general, SOEs are bound to pursue the objectives defined by the government, which themselves are subject to the public interest,[22] which tends to reflect in each one of the ESG factors.[23]

Regarding the environmental factor, there are externalities that many governments seek to overcome with taxes. In contrast, in SOEs, it is up to the state, the shareholder itself, to take charge of this concern, adopting the best practices to avoid environmental risks and benefit from environmental opportunities to generate shareholder value.

Second, as for social responsibility, SOEs, as a rule, are the most involved in the quality of employment, health, safety, training and professional development, diversity, human rights and social responsibility. Additionally, one of

---

[21] As (potential) smart cities, municipalities are interested in attracting private investors and signing contracts with sustainable companies.

[22] See PHILIPPE BANCE et al., *'Serving The General Interest With Public Enterprises— New Forms Of Governance And Trends In Ownership'* (2015); JOÃO SALIS GOMES, *'Interesse público, controle democrático do estado e cidadania'* (2010); SALOMÃO FILHO, *'Interesse público, interesses sociais e parâmetros de prossecução no estado social e democrático de direito'* (2014). Public interest is a complex concept that does not necessarily correspond to the will of the majority of people or is aligned with the concept of social solidarity. In any case, it is grounded in a positive, sovereign constitutional system with a complex balance of subjective rights (STEFANO MORONI, *'Towards a reconstruction of the public interest criterion'* 2004).

[23] CHAN LEONG, *'The role of state-owned enterprises in environmental, social, and governance issues'* (2017).

the more popular ESG strategies includes ways investors can implement the principles (of socially) responsible investment (PRI).[24]

Social responsibility must not be mistaken with compliance with current legislation, nor stating to have good practices without incorporating the principles of social responsibility in the organisation's management. Neither carrying out actions aiming only at (immediate) financial return, without an authentic culture and behaviour, which should be familiar to the whole organisation.

Finally, regarding governance, it must be recalled that the public sector tends to be characterised by bureaucracy and inefficiency, vested interests, corruption and/or authoritarian rulers.[25] In this context, the State should act as an informed, active shareholder to ensure that the governance of SOEs is conducted in a more transparent, more accountable way. This may be accomplished namely by introducing greater transparency into the rationales,[26] professionalising government ownership, strengthening commercial orientation and developing more robust, more independent boards[27], as well as reducing corruption and political interference.[28] Thus, ÖZCAN concluded, regarding publicly traded rail companies, that there had been a positive association between the disclosure of ESG information and financial performance.[29]

Regarding technology usage, some authors argue that soon technology can replace human board members in POEs,[30] with the participation of algorithms

---

[24] ERICK MEIRA DE OLIVEIRA, 'Corporate social responsibility and firm performance: a case study from the Brazilian electric sector', (2015).

[25] ALEXANDER PANEZ PINTO et al., 'The future is public: towards democratic ownership of public services', (2020).

[26] OECD, 'Implementing the OECD guidelines on corporate governance of state-owned enterprises: review of recent developments', (2020b).

[27] SIMON WONG, 'The state of governance at state-owned enterprises', (2018).

[28] FILIPPO BELLOC, 'Innovation in state-owned enterprises: reconsidering the conventional wisdom', (2014).

[29] See ISMAIL ÇAĞRI ÖZCAN, 'Determinants of environmental, social, and governance reporting of rail companies: does state ownership matter?', (2020). Moreover, according to the same author, company size, financial leverage, board size and percentage of independent directors are positively linked with ESG disclosure, whereas higher profitability and tangibility ratios, as well as being established in a country that observes common law, tend to decrease the disclosure of ESG performance. In the regime applicable to SOEs in Portugal, there is vast regulation regarding ESG factors (CARLA TEIXEIRA, 'Relatórios de sustentabilidade: que futuro? o papel dos auditores e da auditoria nesse futuro', 2011), and some companies have to publish sustainability reports according to portuguese Decree Law No. 89/2017, of 28 July.

[30] See MARK FENWICK et al., 'The "unmediated" and "tech-driven" corporate governance of today's winning companies', (2017). In 2014, Hong Kong-based venture capital firm Deep Knowledge announced the intelligent system 'VITAL ' (Validating Investment Tool for Advancing Life Sciences) as a new member of its board of directors, adding value in capturing and processing information with positive impacts on the quality of investment decisions (FLORIAN MÖSLEIN, 'Robots in the boardroom: artificial intelligence and corporate law', 2018). Subsequently, the Finnish company Tieto also

in the boards of directors.[31] The same could apply to SOEs. Currently, legal systems tend to limit functions to natural or legal persons with legal personality,[32] making it impossible for that solution to exist. However, a teleological interpretation about the end of such rules (enabling more efficient management, especially when this kind of technology can imply benefits on ESG factors) could conclude that it is possible or that it carries advantages.[33] For the time being, it seems that subjective knowledge, such as intuition, is still considered essential; however, what we currently believe to be analytical, 'cold' and limited decision-making process may soon change.[34]

Impact of ESG Factors in Smart Cities

In smart cities, in order to (i) increase the quality of life; (ii) to be more efficient in the use of resources; (iii) to create more opportunities and be more competitive; (iv) to increase wealth and (v) to create jobs, the first step is to know citizens and be aware of what happens in the municipality's area. That knowledge allows to reduce the asymmetry of information. This task is much more straightforward when using technology since it can collect and treat much information at the same time, especially now that we are facing a new era with 5G technology, which will allow faster IoT (i.e., machine-to-machine (M2M) communication).

Regarding the environmental factor, according to an analysis by the Stockholm Environment Institute for the Coalition,[35] by implementing a bundle of currently available technologies and practices, six countries could cut annual emissions from key urban sectors by 87–96% by 2050. This could imply e.g., making infrastructure more efficient, incorporating renewable energy into buildings, using different materials to build infrastructure and improving transportation; however, that would require a significant investment.

---

appointed an artificial intelligent system called 'Alicia T' as a manager with voting powers (UGO PAGALLO, '*Vital, Sophia, and co., The quest for the legal personhood of robots*', 2018).

[31] LUKAS RUTHES GONÇALVES et al./PEDRO DE PERDIGÃO LANA, '*Novas tecnologias, problemas de informação e governança corporativa na União Europeia e Brasil*', (2020).

[32] Furthermore, (FLORIAN MÖSLEIN, '*Robots in the boardroom: artificial intelligence and corporate law*', 2018) and UGO PAGALLO, '*Vital, Sophia, and co., The quest for the legal personhood of robots*', (2018) sustain that directors or shareholders who choose to use the programs can be held responsible if they do not take the necessary precautions, even when they only ratify an algorithmic decision.

[33] LUKAS RUTHES GONÇALVES et al., '*Novas tecnologias, problemas de informação e governança corporativa na União Europeia e Brasil*', (2020) and SHERLY ABRAHAM, '*Information technology, an enabler in corporate governance*', (2012).

[34] FLORIAN MÖSLEIN, '*Robots in the boardroom: artificial intelligence and corporate law*', 2018, pp. 649–650; ANTÓNIO RICCIULLI et al., '*Análise de investimentos, racionalidade económica e processo de decisão empresarial*', (2011).

[35] COALITION FOR URBAN TRANSITIONS, '*Seizing the urban opportunity*', (2021).

For instance, to save resources regarding real estate, municipalities could do a lot more to profit from the buildings they own by investing in smart and efficient reconstruction, allowing municipalities to save costs in the long term and be more sustainable. The same idea applies to (policies regarding) forests, because a large part of cities' outskirts can be profitable and, at the same time, contribute to decarbonisation. However, it is necessary to have an updated land register and establish the correct incentives for landowners. Technology can also be helpful in asset tracking, inventory control, security, individual tracking, and energy and water conservation.[36]

Regarding the social dimension, cities must be inclusive and have a good quality of life indicators and assessments based on safety, health, environment, mobility, access to services, housing, and employment opportunities. We agree with ANGELIKA TOLI et al. that this dimension includes community autonomy and citizen well-being. Moreover, 'an urban environment can [only] be sustainable when social equity, conservation of the natural environment and its resources, economic vitality and quality of life are achieved'.[37] As we will see in Sect. 3.2., smart cities can serve as laboratories for innovation and transformative change to motivate their citizens to conduct improvements and engage in sustainable habits.

Regarding the governance factor,[38] it must be taken into account that there are ESG issues associated with multiple stakeholders, high levels of interdependence, conflict of interests and competing objectives, and social and political complexity. To solve some of these issues, technology may also increase citizen participation, notably allowing platforms that allow to reach more people and form a common will, namely using electronic voting of the proposals being discussed.

ESG Factors' Ratings

As mentioned, using metrics or ratings allows investors to benchmark/compare ESG performance and to evaluate the potential sustainability strength of companies. However, the existence of an ESG rating is not indicative of how or whether ESG factors will be integrated into a fund or a company. So, inherent challenges to those benchmarks became obvious.

---

[36] According to the Portuguese Court of Auditors the state could save more than one billion euros from buildings that it occupies (TRIBUNAL DE CONTAS, '*Auditoria à inventariação do património imobiliário do estado, relatório n.º 16/2020, 2.ª secção*' 2020). An assessment of the profitability of assets as well as a relocation of structures from the State to properties owned by the state could imply greater efficiency and reduce occupancy costs.

[37] ANGELIKI MARIA TOLI et al., '*The Concept of Sustainability in Smart City Definitions*' (2020).

[38] ISABEL CELESTE FONSECA et al., '*Smart cities vs. smart(er) governance: cidades inteligentes, melhor governação (ou não)*', (2019).

Traditionally, ESG ratings have also been developed by human research analysts following proprietary methodologies to analyse industry research reports and companies' disclosures, amongst other sources, to identify companies' ESG credentials.[39] Such analysis demanded investment reports to collect information and conduct research to confirm compliance with governance procedures and good practices.[40] However, technological innovation in data scraping and artificial intelligence (AI) has improved that process. Nevertheless, according to ARTHUR HUGHES,[41] the alternative rating model, with both types of providers co-existing and possibly collaborating, may complement instead of substituting the conventional one. Even so, it may not be apparent to rating companies which ESG criteria to request[42] or how to assign scores, many of which can be subjective.[43] Moreover, although ESG ratings can eliminate negative behaviour, they do not necessarily generate innovation or invert climate change.[44]

Last, regarding the risks of ESG ratings, HESTER PEIRCE[45] has claimed, amongst other arguments, that because the three factors are evaluated together, it is difficult to establish reliable metrics, meaning that some companies may be unfairly stigmatised. On top of that, the same company may be treated differently across time.[46] To prevent such situations, standardisation has been proposed in some initiatives, including the Sustainable Finance Disclosures Regulation (SFRD) and the European Union Taxonomy Regulation, both aimed explicitly at tackling greenwashing and improving the

---

[39] See ARTHUR HUGHES et al., '*Alternative ESG ratings: how technological innovation is reshaping sustainable investment*', (2021). The authors have compared a set of traditional ratings with an alternative AI-based set of ESG ratings that summarized publicly available data sources based on public perceptions. The authors concluded that whereas alternative ESG ratings make major promises for how technology can reform sustainable investing, there are risks that remain due to concerns over how transparent AI can be.

[40] The process of generating ESG ratings involves identifying the sustainability-related issues that are financially material to the entity, usually referred to as 'key issues', thereby maintaining that good ESG performance prompts better financial performance, although only if companies focus on their particular key issues (RUTH JEBE, '*The convergence of financial and ESG materiality: taking sustainability mainstream*', 2019).

[41] ARTHUR HUGHES et al., '*Alternative ESG ratings: how technological innovation is reshaping sustainable investment*' (2021).

[42] SOYOUNG HO, '*ESG reporting is all the rage, but companies are unsure about which reporting standards to use*', (2020).

[43] AARON CHATTERJI et al., '*Do ratings of firms converge? Implications for managers, investors, and strategy researchers*' (2014).

[44] TORSTEN EHLERS et al., 'Green bonds and carbon emissions: exploring the case for a rating system at the firm level*Green bonds and carbon emissions: exploring the case for a rating system at the firm level*', BIS quarterly review, (2020).

[45] HESTER PEIRCE, '*Scarlet letters: remarks before the American enterprise institute*' (2019).

[46] JAMES MACKINTOSH, '*Is Tesla or Exxon more sustainable? It depends whom you ask*', (2018); KATE ALLEN, '*Lies, damned lies and ESG rating methodologies*', Financial Times, 6 December (2018).

accountability of ESG disclosure in financial markets.[47] Tensions nevertheless exist concerning whether standardisation is desirable, given investors' constant search for competitive advantages to maximise returns.[48]

Another risk is that ESG investment is used as a smokescreen for investors to follow business-as-usual strategies under the guise of sustainability.[49] In a different light, however, it may be an effective compromise made to solve the world's most pressing problems, one that vitally bridges sustainability and financial markets.

Aside from the mentioned challenges, ESG ratings allow positive developments in fulfilling ESG criteria imposed to improve visibility, especially if the organisations are proactive before their peers and have a comprehensive message planned for their stakeholders. Above and beyond that, ESG activism pressures states and public managers for change.[50]

The consequences hardly seem real for organisations that continue to lack a vision for improvement in those three pillars.[51] However, as younger generations enter the labour force and take a stand, the implications may become quite real for organisations that do not act.

There are also rankings to assess how close smart cities are to their objectives.[52] The OECD, for instance, has created a 'Better Life Index', with 11 topics. Those topics reflect what has been 'identified as essential to well-being in terms of material living conditions (housing, income, jobs) and quality of

---

[47] See Regulation (EU) 2019/2088 of the European Parliament and of the Council of 27 November 2019 on sustainability-related disclosures in the financial services sector.

[48] ARTHUR HUGHES et al., *'Alternative ESG ratings: how technological innovation is reshaping sustainable investment'*, (2021).

[49] MAGALI DELMAS et al., *'The drivers of greenwashing'*, (2011), 54, 64–87. See also PAULO CÂMARA, chapter 1 in this book.

[50] TORSTEN EHLERS et al., *'Green bonds and carbon emissions: exploring the case for a rating system at the firm level'* (2020), JOSÉ MIGUEL LUCAS, *'Deveres dos gestores públicos: orientação para a qualidade, especialmente em tempos de crise'* (2020) and HELENE SAGSTAD et al., 'The Impact of State Ownership on Companies' Sustainability An Empirical Analysis of the ESG Scores of Companies in the EU/EEA' *(2019)*, according to whom 'SOEs perform significantly better than non-SOEs when it comes to ESG scores and ESG scores increase with the size of the share owned by the state', defending that 'the results can be explained by shareholders' effect on companies' sustainability and governments' promotion of sustainability through policies and expectations for companies in their ownership. Moreover, as investors, the state often has a more long-term perspective than private actors, and thus prioritises sustainable development of the company over time'.

[51] Even so, action has been taken. According to BLACKROCK, INC., *'Our approach to sustainability - Blackrock investment stewardship'*, (2020), the world's largest asset manager, 'There are two primary categories of our voting actions: holding directors accountable and supporting shareholder proposals. In 2020, we took voting action against those companies where we found corporate leadership unresponsive to investors' concerns about climate risk or assessed their disclosures to be insufficient given the importance to investors of detailed information on climate risk and the transition to a low-carbon economy'.

[52] ADEOLUWA AKANDE et al., *'The Lisbon ranking for smart, sustainable cities in Europe'*, (2019).

life (community, education, environment, governance, health, life satisfaction, safety and work-life balance)'[53]. We agree with SARA FERNANDES[54] who mentions that it is necessary to plan, to manage and to govern sustainably, maximising economic opportunities and minimising environmental damage since these are the significant challenges countries will face in the coming decades. Public resources need to be better used, and natural assets exploited consciously and responsibly.

To ensure the proper implementation of ESG actions, the sequence of 'plan', 'do', 'check' and 'act' — that is, the Deming cycle or PDCA — could help to obtain the needed results. More specifically, ESG factors and their benchmarks and rankings could be implemented into the 'check' part of the sequence and allow assessing what is wrong or can be improved and, in turn, making the necessary changes.

Indeed, there are four stages established in ISO9001: 'planning' (i.e. identifying the desired results of the process and what will be addressed in the measures to be issued); 'doing' (i.e. implementing what was planned); 'checking' (i.e. monitoring and measuring processes, products and services) and 'acting' (i.e. taking actions to improve performance as needed).

In that way, the PDCA sequence stimulates the continuous improvement of people and processes and may also allow testing possible solutions on a small scale and in a controlled environment—in an innovation zone, for instance—and preventing recurring mistakes.

3   THE ROLE OF TECHNOLOGY REGARDING ESG FACTORS

Risks

Technology may not answer all the issues regarding sustainability that ESG factors address, but it helps, namely, offering the possibility of a sustainable generation of wealth and increased competitiveness. Simultaneously, it is necessary to be aware and mitigate some of those that we perceive as the most significant risks involved (analysed next, without a hierarchical concern).[55]

First, the concern regarding cybersecurity and personal data treatment. Especially with IoT, open data and big data analysis technologies (e.g., artificial intelligence) it is possible to invoke safety and security reasons. One can see the advantages of installing all sorts of sensors, like gunshots, accidents

---

[53] OECD, *'Executive summary', in how's life? 2020: Measuring well-being'*, (2020a).

[54] SARA FERNANDES, *'Smart cities'*, (2017).

[55] Ideally, one could even think of a virtuous circle model in which universities were even more connected to industry, seeking to solve current issues or challenges, which, in their turn would generate a better quality of life for citizens and new challenges for universities.

and noise sensors, which raises issues about security and privacy.[56] Specifically when a large number of data, users and financial movements are involved, there must be a growing concern. It is expected that companies and municipalities/smart cities will come under significant scrutiny crisis and that inadequate management of citizens' personal data can result in illicitness, loss of business and litigation. The universe of SOEs and smart cities, must have disclosed and transparent data (the information on how it works should be available and be easy to access), and the way it is presented should be appealing and understandable, so that citizens may feel encouraged to participate.[57]

Second, increasingly, the public perception of risk and fear of diseases has been influencing public policy development regarding technology usage, including in smart cities. i.e., beyond the 'invisible hazards', like radiation, according to CALESTOUS JUMA,[58] society tends to reject new technologies when they substitute, rather than augment, our humanity. Likewise, there is higher resistance to new technologies when the public perceives that, although there are risks likely to be widespread, but the benefits of new technologies will only accrue to a small section of society. EMILIO MORDINI sustains that more than fear, wonder and curiosity, can enable the integration of new concepts into mental schemes.[59] We tend to agree with this understanding, adding safety as a starting point.

Third, it can be considered that there is an 'excess of information'. We prefer to sustain that we have to be more selective regarding the attention paid to specific topics (in some cases, it will be diverted consciously, in others unconsciously).[60] It seems that content creation is a new trend and that the way SOEs and smart cities communicate with citizens must be adapted to that.

Forth, according to NANCY ODENDAAL,[61] technology can improve the quality of life for citizens, but it can also increase inequalities and promote a digital divide. We cannot ignore people who are not familiar with these new

---

[56] See GASPARE D'AMICO et al., '*Understanding Sensor Cities: Insights from Technology Giant Company Driven Smart Urbanism Practices*', (2020). For example, if the Portuguese transportation SOEs 'Carris' installed sensors on its buses, it could know, without needing to identify them, where they are going to and strengthen the network, without prejudice to maintaining a regular public service.

[57] SCIENCES PO (Paris). '*Smart cities: l'innovation au cœur de l'action publique?*', (2016).

[58] CALESTOUS JUMA, '*Innovation and its enemies: why people resist new technologies*', (2016).

[59] EMILIO MORDINI, '*Technology and fear: is wonder the key?*', (2008).

[60] ANTHONY TOWNSEND, '*Smart Cities: Big Data, Civic Hackers, and the Quest for a New Utopia*', (2013).

[61] NANCY ODENDAAL, '*Information and communication technology and local governance: understanding the difference between cities in developed and emerging economies*', (2003).

technologies.[62] Adaptations to the technology must be made with caution in order not to fall into the generational exclusions.

Quoting the film director GODFREY REGGIO: 'technology is not neutral'. Beyond that, even the choice of one scenario or option over the other, made by technology, cannot serve all the purposes of everyone involved. We must remember that the concept of 'the best solution' may vary and will always be associated with the objective defined and the criteria for achieving it.[63]

Fifth, technology, by itself, can be pointed in ESG reports as an issue to improve, to achieve a higher ranking, taking into account, for instance, energy consumption or potential future debris caused by decommissioning and the impact that it may have on employment or governance. Besides the possibility of current workers being replaced by artificial intelligence and automation (situation peaking in periods of economic recession). As a solution for this situation, progress has been made with continuous training and improved workers' health and safety conditions to have a more qualified workforce to help ensure long-term sustainability. SOEs and smart cities that are prepared will be rewarded. In contrast, SOEs and smart cities that fail to adapt, may become obsolete.

Finally, there is a rising difficulty, with a lack of appropriate standards and guidance, that clearly define roles and responsibilities, as well as an understanding of critical requirements.[64] Thus, an adequate and flexible framework is essential, and, in that vein, the following section regarding innovation zones may provide some light.[65]

Regulation: The Example of Innovation Zones

Implementing technology and offering new and better products and services may be an optimal solution for SOEs and smart cities, to have a higher score in ESG rankings. However, as referred, one of the risks is a public misperception of what is being developed, namely regarding cybersecurity, personal data treatment or increased inequalities. As such, the policy context is critical to understanding the use of information systems in correct ways. Therefore,

---

[62] That is why, for example, electronic summons to meetings did not dispense other notifications, namely in newspapers (JORGE COUTINHO DE ABREU, '*Responsabilidade civil dos administradores de sociedades*', 2010).

[63] LUKAS GONÇALVES et al., '*Novas tecnologias, problemas de informação e governança corporativa na União Europeia e Brasil*', (2020); OLIVEIRA ASCENSÃO, '*Propriedade intelectual e internet*' (2006).

[64] MORTA VITUNSKAITE et al. , '*Smart Cities and Cyber Security: Are We There Yet? A Comparative Study on the Role of Standards, Third-Party Risk Management and Security Ownership*' (2019).

[65] However, to ensure that disasters are prevented, and conflicts avoided, a legal framework is needed. Furthermore, one of the most challenging parts of defining a legal framework is precisely preventing and solving potential conflicts of interest and reducing asymmetry of information.

a public manager should not bet on innovation without a normative drive addressed in policy.[66]

In some cases, a simple trial run will be sufficient; in others, it will be necessary to have a larger scale to test new regulations. In this case, it is needed to create an 'innovation zone', to prevent or avoid some of the risks mentioned in the previous section. We sustain that creating an adequate framework can reconcile freedom to innovate with safety, at local level or at nation-wide level.

A framework regarding technologies seeks to regulate the introduction and ongoing management of potential risk hazards and self-verification of safety. Above all, it should be flexible and be able to evolve. As analysed by DIERK BAUKNECHT et al.,[67] there are already innovation zones to test regulation models.

In Portugal, there is a national strategy regarding innovation zones called 'Zonas Livres Tecnológicas' (ZLT).[68] The main ideas are to (i) create or delimit geographic zones; (ii) promote and facilitate the realisation of research, demonstration and testing activities in a natural environment of technologies; (iii) create knowledge and intellectual property with a Portuguese base and (iv) attract investment that will be converted into employment and revenue. Furthermore, the model does not rely on tax incentives.

First, according to this scenario, in the categories created by the abovementioned authors,[69] ZLTs will be included in pilot projects and innovation labs with regulatory support that establishes exemptions. Considering Decree-Law nr. 67/2021, of 30 July, there can be a need for specific legal or regulatory frameworks or reviews depending on the needs and sectors concerned, ZLTs seem to consider innovation in regulation, therefore, rather than a cause.[70]

Second, regarding the type of projects involved, ZLTs will include all technologies, even non-physical ones, connected with the market.

---

[66] HAFEDH CHOURABI et al., *'Understanding smart cities: an integrative framework'*, (2012); JOHN EGER et al., *'Technology as a tool of transformation: e-cities and the rule of law?'*, (2010).

[67] DIERK BAUKNECHT et al., *'Experimenting with policies: regulatory innovation zones as a tool for sustainability transitions'*, (2020).

[68] There are several examples throughout the world such as 'Keystone Innovation Zones' (KIZs) in Pennsylvania (USA); 'Innovation Zones' in Nevada (USA), 'Zona de Inovação Sustentável', in Brazil; and SINTEG ordinance in Germany developing model solutions for the energy supply of the future, allowing companies to test new technologies and procedures under real-world conditions.

[69] DIERK BAUKNECHT et al., *'Experimenting with policies: regulatory innovation zones as a tool for sustainability transitions'*, (2020).

[70] According to which there are four types of regulatory experiments: (i) pilot project/innovation lab; (ii) pilot project/innovation lab with regulatory support (establishing regulatory exemptions); (iii) regulatory innovation zone (to test regulatory innovations); and (iv) system innovation zone (regarding interaction bet projects and regulation). As a small country, Portugal hs a significant challenge regarding the number of people available to test new technologies. However, considering that it will be possible to test some online technologies, that hurdle can perhaps be overcome by focusing on more than a single region.

Third, considering governance, it currently seems that there will be a national 'supervision authority' and the projects will be subject to public tender. To better promote cooperation, we would prefer that ZLTs were autonomous and inserted in a national network to add value, thus promoting territorial cohesion based on the constitution of hubs and avoiding potential competing against one another.[71]

Considering that, similar to the abovementioned ranking for ESG, there are rankings for innovation published by the World Economic Forum and the European Union, we expect that these ZLTs will be able to measure the achievement of goals previously set, specifically regarding innovation.[72]

4 Conclusions

In conclusion, we sustain that, after the lockdowns and the Covid-19 pandemic, as well as other major global unpredicted events, we will not return to the world that we once knew (i.e., before 2020). The world has changed faster than ever and it has evolved beyond our perception. What we considered to be the future, namely with the revolution triggered by 5G and IoT, is already becoming a reality. Sustainability has become, again, a top priority for governments and public opinion, mainly because major pre-pandemic ESG problems remain to be solved.

Although the origin of ESG factors relates to financial analysis and private investment, we conclude that today, such concerns could be amplified beyond non-financial performance indicators for investments and apply to SOEs (which generally operate in sectors with high social presence and have a significant (national or even international) dimension and impact) and (smart) cities (which are significant contributors to climate change, namely regarding energy consumption and CO2 emissions.

Regarding how ESG factors influence SOEs, it is up to the state, the shareholder itself, to generate value and act as an informed, active shareholder, to ensure that the governance of SOEs is conducted in a more transparent and efficient way. Sustainability underlying ESG factors is of utmost relevance to those entities and can be highly valuable. Regarding smart cities, a ESG strategy may also increase the quality of life, a more efficient in the use of resources, as well as the the creation of more opportunities and wealth. An ESG strategy may also ensure access of SOEs and smart cities to financing on better terms: there is a growing range of investments that aim to capitalise on the opportunities presented by the implementation of technology in SOEs and smart cities and from a public procurement perspective, it can also be

---

[71] Regarding financing, the promotion and coordination of the ZLTs and the respective budgetary framework are ensured by the Portugal Digital Mission structure, i.e., public funding of ZLTs, assuming that promoters will finance projects.

[72] DIERK BAUKNECHT et al., '*Experimenting with policies: regulatory innovation zones as a tool for sustainability transitions*', (2020).

relevant to assess the ESG performance of companies that are expected to implement smart and sustainable solutions or, in contrast, to exclude companies that are not investing in ESG performance. There has to be a clear vision regarding sustainability and the intention to move proactively towards that goal. As such, we need additional commitment from the public managers (either in SOEs or municipalities), who need to be qualified, proactive, adapt and take immediate action in defining and implementing a model that allows to benefit from the new opportunities, following best practices, namely regarding ESG factors, benefiting from a regulated and adequate application of technology. Such measures enable citizens to scrutinise information about ESG performance and can deliver hard-to-quantify benefits beyond ones that concern the environment and social aspects, including public goodwill, a better reputation and the visibility of their sound, prudent management.

Aside from the mentioned challenges, ESG ratings allow to benchmark/compare and, for instance, to conclude, according to some studies, that SOEs perform significantly better than non-SOEs when it comes to ESG scores and ESG scores increase with the size of the share owned by the state. ESG ratings also allow to assess positive developments in meeting ESG criteria imposed to improve visibility, especially if the organisations are proactive before their peers and have a comprehensive message planned for their stakeholders. Beyond that, the sequence of 'Plan, Do, Check, and Act' (Deming cycle or PDCA), could be very useful to establish goals, assess what is wrong and correct it in due time.

Regarding the risks related to technology for SOEs and smart cities, we defend that although technology can help reduce information asymmetry and increase efficiency, it should be considered that we keep evolving technologically (namely in nanotechnology or artificial intelligence) without publicly questioning and discussing the continuous development vortex, making technological evolution somewhat unpredictable. As such, the debate regarding those questions and the risks highlighted could be relevant to regulate and adopt precautions regarding possible risks and adverse effects. Having no answer does not mean we cannot ask questions and an adequate and flexible framework is essential.

Finally, implementing technology and offering new and better products and services may be an optimal solution for SOEs and smart cities, to have a higher score in ESG rankings. In order to prevent risks, regulation will need to be transparent and flexible to accommodate new technologies or approaches to maintain trust amongst stakeholders. As such, we defend that innovation zones could indeed be a solution in order to foster innovation and fresh thinking regarding solutions that contribute to ESG factors and sustainability, whilst simultaneously analysing and preventing potential adverse effects that may hinder the goal of sustainability.

References

Abraham, S. E. (2012). Information technology, an enabler in corporate governance. *Corporate Governance: The International Journal of Business in Society, 12*(3), 281–291. https://doi.org/10.1108/14720701211234555

Adams, W. M. (2009). *Green development: Environment and sustainability in the third world*, Routledge. https://ysrinfo.files.wordpress.com/2012/06/green-development-environment-and-sustainability-in-developing-world.pdf

Akande, A., Cabral, P., Gomes, P., & Casteleyn, S. (2019). The Lisbon ranking for smart sustainable cities in Europe. *Sustainable Cities and Society, 44*, 475–487.

Allen, K. (2018). 'Lies, damned lies and ESG rating methodologies', *Financial Times*, 1–3. https://www.ft.com/content/2e49171b-a018-3c3b-b66b-81fd7a170ab5

Ascensão, J. de O. (2013). Propriedade intelectual e internet. *Direito Da Sociedade Da Informação - Associação Portuguesa Do Direito Intelectual (APDI), VI*, 1–62.

Backhaus, J. G. (1994). Assessing the performance of public enterprises: A public-choice approach. *European Journal of Law and Economics, 1*(4), 275–287. https://doi.org/10.1007/BF01540701

Bance, P., & Obermann, G. (2015). Serving The general interest with public enterprises—new forms of governance and trends in ownership. *Annals of Public and Cooperative Economics, 86*(4), 529–534. https://doi.org/10.1111/apce.12101

Bauknecht, D., Heyen, D. A., Gailhofer, P., & Gailhofer, P. (2020). *Experimenting with policies: Regulatory innovation zones as a tool for sustainability transitions*. https://www.oeko.de/fileadmin/oekodoc/WP-Regulatory-Innovation-Zones.pdf

Belloc, F. (2014). Innovation in state-owned enterprises: Reconsidering the conventional wisdom. *Journal of Economic Issues, 48*(3), 821–848. https://doi.org/10.2753/JEI0021-3624480311

Bernardin, S., & Jeannot, G. (2019). La ville intelligente sans les villes? Interopérabilité, ouvertures et maîtrise des données publiques au sein des administrations municipales. *Reseaux, 218*(6), 9–37. https://doi.org/10.3917/res.218.0009

Bernier, L., Florio, M., & Bance, P. (2020). *The Routledge Handbook of State-Owned Enterprises*. https://doi.org/10.4324/9781351042543

Blackrock Inc., & Boss, S. (2020). Our Approach to Sustainability. Blackrock Inc. https://corpgov.law.harvard.edu/2020/07/20/our-approach-to-sustainability/#more-131469https://www.blackrock.com/corporate/literature/publication/our-commitment-to-sustainability-full-report.pdf

Brundtland, G. H. (1987). Our common future: Report of the world commission on environment and development. *International Affairs, 64*(1), 126–126. https://doi.org/10.2307/2621529

Budiman, A., Lin, D., & Singham, S. (2009). Improving performance at state-owned enterprises. *McKinsey Quarterly*, May, 1–5.

Chander, A. (2017). The racist algorithm, *Michigan Law Review, 115*(6), 1023.

Chatterji, A. K., Durand, R., Levine, D. I., & Touboul, S. (2016). Do ratings of firms converge? Implications for managers, investors and strategy researchers. *Strategic Management Journal, 37*(8), 1597–1614. https://doi.org/10.1002/smj.2407

Chen, C. (2016). Solving the puzzle of corporate governance of state-owned enterprises: The path of the Temasek model in Singapore and lessons for China. *Northwestern Journal of International Law and Business, 36*(2), 303–370.

Chourabi, H., Nam, T., Walker, S., Gil-Garcia, J. R., Mellouli, S., Nahon, K., Pardo, T. A., & Scholl, H. J. (2012). Understanding smart cities: An integrative framework.

2012 45th Hawaii International Conference on System Sciences, 2289–2297. https://doi.org/10.1109/HICSS.2012.615

Clark, G. L., Feiner, A., & Viehs, M. (2014). From the stockholder to the stakeholder: How sustainability can drive financial outperformance. *SSRN Electronic Journal*. https://doi.org/10.2139/ssrn.2508281

Coalition for Urban Transitions. (2021). *Seizing the Urban Opportunity: How national governments can recover from COVID-19, tackle the climate crisis and secure shared prosperity through cities, insights from six emerging economies* (pp. 4–16). Coalition For Urban Transitions. https://urbantransitions.global/wp-content/uploads/2021/03/Seizing_the_Urban_Opportunity_WEB-1.pdf

Coutinho de Abreu, J. M. (2010). *Responsabilidade civil dos administradores de sociedades* (2nd ed.). IDET, Cadernos no. 5. Almedina.

D'Amico, G., L'Abbate, P., Liao, W., Yigitcanlar, T., & Ioppolo, G. (2020). Understanding sensor cities: Insights from technology giant company driven smart urbanism practices. *Sensors (Switzerland)*, 20(16), 1–24. https://doi.org/10.3390/s20164391

Delmas, M. A., & Burbano, V. C. (2011). The drivers of greenwashing. *California Management Review*, 54(1), 64–87. https://doi.org/10.1525/cmr.2011.54.1.64

Dias, R. P., & Sá, M. (2020). Deveres dos administradores e sustentabilidade. In *Administração e governação das sociedades* (1.a Edição, pp. 33–85). Almedina. https://www.seer.ufrgs.br/index.php/ppgdir/article/view/118025/64254

Eger, J. M., & Maggipinto, A. (2010). Technology as a tool of transformation: E-Cities and the rule of law. *Information Systems: People, Organizations, Institutions, and Technologies—ItAIS: The Italian Association for Information Systems*, 23–30. https://doi.org/10.1007/978-3-7908-2148-2_4

Ehlers, T., Mojon, B., & Packer, F. (2020, September). Green bonds and carbon emissions: Exploring the case for a rating system at the firm-level. *BIS Quarterly Review*, 31–47.

Elkington, J. (1999). Cannibals with forks: The triple bottom line of 21st century business. *Choice Reviews Online*, 36(07), 36-3997–36-3997. https://doi.org/10.5860/choice.36-3997

Fenwick, M., Kaal, W. A., & Vermeulen, E. P. M. (2017). *The unmediated and tech-driven corporate governance of today's winning companies*, 48

Fernandes, S. (2017). *Smart cities—inclusão, sustentabilidade, resiliência* (1st ed.). Glaciar.

Filho, S. A. A. I. (2014). Interesse público, interesses sociais e parâmetros de prossecução no estado social e democrático de direito. *Revista Do Instituto Do Direito Brasileiro*, 3(2), 1143–1166.

Fonseca, I. C., & Prata, A. R. A. (2019). Smart cities vs. Smart(er) governance: Cidades inteligentes, melhor governação (ou não). *Questões Atuais de Direito Local*, 24 (out.-dez. 2019), 19–38.

Forfas, A. (2010). *The role of state owned enterprises: Providing infrastructure and supporting economic recovery*. Forfas. https://books.google.pt/books?id=CBEaMwEACAAJ

Gomes, J. (2013). Interesse público, controle democrático do estado e cidadania. *Handbook de Administração Pública*, 353–380.

Gonçalves, L. R., & Lana, P. de P. (2020). Novas tecnologias, problemas de informação e governança corporativa na União Europeia e Brasil. *Revista de Estudos Jurídicos Do Superior Tribunal de Justiça*, 1(1) (jul./dez., 2020), 412–435.

Gond, J. P., & Crane, A. (2010). Corporate social performance disoriented: Saving the lost paradigm? *Business and Society, 49*(4), 677–703. https://doi.org/10.1177/0007650308315510

Goodland, R., & Daly, H. (1996). Environmental sustainability: Universal and non-negotiable. *Ecological Applications, 6*(4), 1002–1017. https://doi.org/10.2307/2269583

Hirsch, P. M., & Levin, D. Z. (1999). Umbrella advocates versus validity police: A life-cycle model. *Organization Science, 10*(2), 199–212. https://doi.org/10.1287/orsc.10.2.199

Ho, S. (2020). Esg reporting is all the rage, but companies are unsure about which reporting standards to use. *Thomson Reuters.* https://tax.thomsonreuters.com/news/esg-reporting-is-all-the-rage-but-companies-are-unsure-about-which-reporting-standards-to-use/

Hsu, P.-H., & Liang, H. (2017). Leviathan inc. and corporate environmental engagement. *Darden business school, Working Paper No. 2960832; ECGI - Finance Working Paper No. 526/2017.* Elsevier BV. https://doi.org/10.2139/ssrn.2960832

Hughes, A., Urban, M. A., & Wójcik, D. (2021). Alternative ESG ratings: How technological innovation is reshaping sustainable investment. *Sustainability, 13*(6). https://doi.org/10.3390/su13063551

ITU, I. T. U., Kondepudi, S. N., Singh, G. N., Agarwal, N. K., Kumar, R., Singh, R. ... Jain, A. (2014). *Smart sustainable cities: An analysis of definitions.* International Telecommunication Union's (ITU) focus group on smart sustainable sites—Working Group 1. https://www.itu.int/en/ITU-T/focusgroups/ssc/Documents/Approved_Deliverables/TR-Definitions.docx

IUCN (World Conservation Union), UNEP (United Nations Environment Programme), & WWF (World Wide Fund for Nature). (1991). *Caring for the Earth: A strategy for sustainable living.* https://portals.iucn.org/library/efiles/documents/cfe-003.pdf

Jakob, B. (2017). Performance in strategic sectors: A comparison of profitability and efficiency of state-owned enterprises and private corporations. *The Park Place Economist, 25*(1), Article 8). https://digitalcommons.iwu.edu/parkplace/vol25/iss1/8

Jebe, R. (2019). The convergence of financial and ESG materiality: Taking sustainability mainstream. *American Business Law, 56*(3), 645–702. https://doi.org/10.1111/ablj.12148

Juma, C. (2006). Innovation and its enemies: Why people resist new technologies. *Economist* (1st ed., Vol. 378). Oxford University Press. https://doi.org/10.1093/acprof:oso/9780190467036.001.0001

Leong, C. F. (2017). The role of state-owned enterprises in environmental, social, and governance issues. *CFA Institute Market Integrity Insights.* https://blogs.cfainstitute.org/marketintegrity/2017/07/10/the-role-of-state-owned-enterprises-in-environmental-social-and-governance-issues/

Lucas, J. M. (2016). Empresas públicas e corporate governance: Da definição da prossecução do interesse público ao controlo externo efectuado pela supervisão. *Revista de Direito das Sociedades, a.8*(1), 7–45.

Lucas, J. M. (2020). Deveres dos gestores públicos: Orientação para a qualidade, especialmente em tempos de crise. *Administração e governação das sociedades* (1a Edição, pp. 229–277). Almedina.

Mackintosh, J. (2018). Is Tesla or Exxon more sustainable? It depends whom you ask. *Wall Street Journal*, 8–10. https://www.wsj.com/articles/is-tesla-or-exxon-more-sustainable-it-depends-whom-you-ask-1537199931.

Mordini, E. (2007). Technology and fear: Is wonder the key? *Trends in Biotechnology*, 25(12), 544–546. https://doi.org/10.1016/j.tibtech.2007.08.012

Moroni, S. (2004). Towards a reconstruction of the public interest criterion. *Planning Theory*, 3(2), 151–171. https://doi.org/10.1177/1473095204044779

Möslein, F. (2018). Robots in the boardroom: Artificial intelligence and corporate law. *Research Handbook on the Law of Artificial Intelligence*, 649–669. https://doi.org/10.4337/9781786439055.00039

Odendaal, N. (2003). Information and communication technology and local governance: Understanding the difference between cities in developed and emerging economies. *Computers, Environment and Urban Systems*, 27(6), 585–607.

OECD (Organisation for Economic Co-operation and Development). (2020a). *Executive summary, in how's life? 2020: Measuring well-being*. OECD Publishing.

OECD (Organisation for Economic Co-operation and Development). (2020b). *Implementing the OECD guidelines on corporate governance of state-owned enterprises: Review of recent developments*. OECD Publishing.

OECD (Organisation for Economic Co-operation and Development). (2020c) *OECD business and finance outlook 2020—sustainable and resilient finance*. OECD Publishing.

Oliveira, E. M. de. (2015). *Corporate social responsibility and firm performance: A case study from the Brazilian electric sector* (Issue April) [PhD Thesis, Pontifícia Universidade Católica do Rio de Janeiro].

Özcan, I. Ç. (2020). Determinants of environmental, social, and governance reporting of rail companies: Does state ownership matter?. In *New Trends in Public Sector Reporting* (pp. 153–173). https://doi.org/10.1007/978-3-030-40056-9_8

Pagallo, U. (2018). Vital, Sophia, and Co.—The quest for the legal personhood of robots. *Information (Switzerland)*, 9(9). https://doi.org/10.3390/info9090230

Peirce, H. M. (2019). *Scarlet letters: Remarks before the American enterprise institute* (pp. 1–11). https://www.sec.gov/news/speech/speech-peirce-061819

Pinto, A. P., Pettersen, B., Dalby, K. D., Bayas, B., Planas, M., Piqué, A., Fontaíña, M., Putri, D., Sundby, E., Kane, M. T., Goeke, M., Grant, M., Manahan, M. A., Petitjean, O., Brøgger, P., Georg, M., Boulos, R. C., Ramsay, R., Jordan, K., & Philip, K. (2020). The future is public: Towards democratic ownership of public services'. In S. Kishimoto, L. Steinfort, & O. Petitjean (Eds.), (p. 258). https://www.tni.org/files/publication-downloads/futureispublic_online_def.pdf

Ricciulli, A. & Martins, A. (2011). Análise de investimentos, racionalidade económica e processo de decisão empresarial. *Boletim de ciências económicas da faculdade de direito da universidade de coimbra (FDUC), LIV*, 265–301.

Rio, R. (2014, July–September). A conciliação da gestão com a política: um exemplo concreto. In *Questões atuais de direito local, 3*, 7–18.

Robertson, M. (2014). *Sustainability principles and practice*'. https://doi.org/10.4324/9780203768747

Rodrigues, S. (2019). *Relatórios de sustentabilidade em Portugal e Espanha: sua publicação e auditoria* [Master degree, Instituto Politécnico de Leiria]. https://iconline.ipleiria.pt/bitstream/10400.8/4052/1/25_03_2019_Disserta{\c{c}}{\~{a}}o_Final.pdf

Sagstad, H., Schiefloe, M. S., & Valasek, J. (2019). *The impact of state ownership on companies' sustainability—an empirical analysis of the ESG scores of companies in the EU/EEA* [PhD Thesis, Norwegian School of Economics]. https://openaccess.nhh.no/nhh-xmlui/bitstream/handle/11250/2609955/masterthesis.PDF?sequence=1&isAllowed=y

Sciences Po (Paris)-Chaire Mutations de l'action publique et du droit public and Collectif Berger-Levrault. (2016). *Smart cities: l'innovation au cœur de l'action publique? Actes du 2e rendez-vous annuel de la Cité des smarts cities organisé le 29 septembre 2015'*. In G. B. L. Presses de Sciences Po (Ed.), Berger-Levrault.

Sjöström, E. (2008). Shareholder activism for corporate social responsibility: What do we know? *Sustainable Development*, 16(3), 141–154. https://doi.org/10.1002/sd.361

Sterman, J. (2002). All models are wrong: Reflections on becoming a systems scientist. *System Dynamics Review*, 18(4), 501–531. https://doi.org/10.1002/sdr.261

Stiglitz, J. E., Sen, A., & Fitoussi, J.-P. (2009). *Report by the commission on the measurement of economic performance and social progress*. The Commission.

Teixeira, C. (2011). Relatórios de sustentabilidade: que futuro? o papel dos auditores e da auditoria nesse futuro [PhD Thesis]. In *Dissertação de Mestrado*. https://repositorio.ipl.pt/bitstream/10400.21/2521/1/Relatorios.deSustentabilidade.pdf?fbclid=IwAR23bGbGbQPNjX65LTh_wJRy4cHtsRgMyvTEsxjvaKh8aqqadajEAzInynU

Toli, A. M., & Murtagh, N. (2020). The concept of sustainability in smart city definitions. In *Frontiers in Built Environment*, 6. https://doi.org/10.3389/fbuil.2020.00077

Tõnurist, P., & Karo, E. (2016). State-owned enterprises as instruments of innovation policy. *Annals of Public and Cooperative Economics*, 87(4), 623–648. https://doi.org/10.1111/apce.12126

Townsend, A. (2013). *Smart cities: Big data, civic hackers, and the Quest for a New Utopia'* (1st Ed.). W. W. Norton & Company.

Tribunal de Contas (Portugal). (2020). *Auditoria à inventariação do património imobiliário do estado—relatório no. 16/2020, 2.ª secção* (p. 95). Tribunal de Contas (Portugal). https://www.tcontas.pt/pt-pt/ProdutosTC/Relatorios/RelatoriosAuditoria/Documents/2020/rel16-2020-2s.pdf

UNEP Financial Initiative. (2004). The materiality of social, environmental and corporate governance issues to equity pricing. *Geneva, June*.

United Nations (UN). (2020). *The New Urban Agenda Illustrated—United Nations Human Settlements Program*. United Nations Human Settlements Programme (UN-Habitat). https://unhabitat.org/sites/default/files/2020/12/nua_handbook_14dec2020_2.pdf

United Nations (UN). (2021). *Habitat, climate action for cities—innovate cities*. https://unhabitat.org/fr/node/142315

Urueña Gutiérrez, B. (2004). *Cómo medir la eficiencia de las empresas públicas autonómicas, un estudio de casos con aplicación a Castilla y León* (Valladolid: Secretariado de Publicaciones e Intercambio Científico de la Universidad de Valladolid, Ed.; 1st Editio). Universidad de Valladolid.

Vieira, I. S., Jorge, M. J., Rafael, N. M., & Canadas, P. (2010). *Corporate governance, ethics and social responsibility: Comparing continental European and Anglo-Saxon firms*. 1–26. https://www.fep.up.pt/conferencias/10seminariogrudis/

Vieira,In{\'{e}}s(Leiria)-PAPER_CorporateGovernance,EthicsandSocialResponsibility.pdf

Vitunskaite, M., He, Y., Brandstetter, T., & Janicke, H. (2019). Smart cities and cyber security: Are we there yet? A comparative study on the role of standards, third-party risk management and security ownership. *Computers and Security, 83*, 313–331.

Wong, S. C. Y. (2018). The state of governance at state-owned enterprises. In *The state of governance at state-owned enterprises*. International Finance Corporation. https://doi.org/10.1596/29533

Zinkin, J. (2020). *The challenge of sustainability: corporate governance in a complicated world'*. In *The Challenge of Sustainability*. https://subzero.lib.uoguelph.ca/login?URL=?url=https://www.proquest.com/docview/2434964247?accountid=11233%0Ahttps://oculgue.primo.exlibrisgroup.com/openurl/01OCUL_GUE/01OCUL_GUE:GUELPH?genre=book&issn=&title=The+Challenge+of+Sustainability&volume=&iss

PART IV

Final Conclusions

CHAPTER 22

Conclusion: ESG and the Challenges Ahead

Paulo Câmara and Filipe Morais

1 ESG Changing Landscape

This Handbook has examined the ESG movement, its content and manifestations and its potential impact in different types of companies. ESG impact presents mainly two sides as it involves both financial firms and investee companies. The potential for impact however goes beyond these levels as it potentially affects the whole corporate landscape, in systemic terms—in terms described as a "cascade effect."[1]

ESG assembles different investment criteria thereby reflecting the need for an integrated approach in sustainability and corporate governance issues. ESG also proves that in an interconnected world, decisions taken by investors can have a powerful and far-reaching influential effect.

---

[1] See Paulo Câmara, Chapter 1 in this book.

P. Câmara (✉)
Faculty of Law/Governance Lab, Portuguese Catholic University (UCP), Lisbon, Portugal
e-mail: pc@servulo.com

F. Morais
Henley Business School, University of Reading, Henley-on-Thames, UK
e-mail: f.morais@henley.ac.uk

The growth of ESG in recent years has been exponential in such terms that represent "a game changer," as Guido Ferrarini points out.[2] The flow of ESG channelled investment is continuing to increase, as the last years have shown a sharp rise in ESG investment and ESG products.

There are reasons to believe that other signs of evolution may appear in the future. In other words, at present ESG is not a point of arrival, a fixed and definitive instant in time. It is rather a point of departure, a cultural and a regulatory process in motion. This is particularly clear in environmental objectives, where progress is required at global level, but also applies to social and corporate governance goals.

The factors driving change are multiple and promise to produce structural impacts as the very foundations of how we organise socio-economic activity are being transformed. In addition to government regulation, the most powerful drivers nclude investor activism, consumer pressure, civil society and pressure groups demands and the ensuing reputational and competitive concerns that these bring to company boards.

In the U.S., the number of proxy statement disclosures with references to selected keywords such as "sustainability,", "ESG," "climate change," and "human capital" more than doubled between 2018 and 2020.[3]

Investor activism and engagement on ESG matters is also increasing. A recent paper that collected the views of 70 top executives of 43 global institutional investing firms, including the world's three biggest asset managers (BlackRock, Vanguard, and State Street) and giant asset owners such as the California Public Employees' Retirement System (CalPERS), the California State Teachers' Retirement System (CalSTRS), and the government pension funds of Japan, Sweden, and the Netherlands, has noted that without exception ESG investing is at the very top of the agenda, concluding that:

> A sea change in the way investors evaluate companies is under way. Its exact timing can't be predicted, but it is inevitable. Large corporations whose shares are owned by the big passive asset managers and pension funds will feel the change the soonest. But it won't be long before mid-cap companies come under this new scrutiny as well.[4]

Despite this, there are dissenting voices that argue that large investors have to take it more seriously. For example, BlackRock's former Head of Sustainability Investing, Tariq Fancy, responsible for incorporating ESG across the investment giant's $8.7 trillion investment activities has criticised his former

---

[2] See Guido Ferrarini, *Sustainable Governance and Corporate Due Diligence*, Chapter 2 in this book.

[3] Papadopoulos, K., and Araujo, R., *ESG Management and Board Accountability*. Harvard Law School Forum on Corporate Governance, (2020).

[4] R. G. Eccles/S. Klimenko, *The Investor Revolution: Shareholders are getting serious about sustainability*. Harvard Business Review, May–June Issue (2019), p. 116.

employer accusing it of over-inflating the impact of current ESG practice. Mr Fancy commented in an April 2021 interview:

> I looked at CEO tenure: It is the shortest it has been in decades. CEO pay is the highest it has been in decades. That was worrying. The system works according to incentives and self-interest. And if their incentives and self-interest are on the next five years, and there's a problem whose impacts may become more acute 20 and 30 years away, then it's unlikely they'll act quickly enough or aggressively enough to address what is clearly in the long-term public interest.[5]

Executive compensation arrangements[6] are a critical area for ensuring ESG becomes embedded in executive decision-making. Despite this, a recent study by PricewaterhouseCoopers (PWC) and London Business School, has shown that nearly many FTSE 100 firms still have a piecemeal approach to integrating ESG factors in executive compensation, with 45% of measures used not linked to material ESG factors.[7] A further study in Belgium conducted by the PWC and The Diligent Institute, revealed that "Performance is still measured largely against financial criteria" with "financial KPIs representing 74% of the weighting of [short term incentives] STIs, while it reaches 88% for [long term incentives] LTIs" and "non-financial KPIs [which] typically reflect long-term performance and objectives (especially when linked to sustainability), [were found to] be more often integrated in STIs."[8]

But while both investors and firms have still a long way to go to create the right incentives for ESG to truly be integrated at all levels of the firm and the investment chain, other powerful drivers are also pushing for change. For example, customers have been found to increasingly becoming a key driver by small and mid-sized quoted firms in the UK[9] as well as large listed firms in the UK, Germany and the Netherlands.[10]

In the coming decade public pressure for companies to engage in sustainable activities generally and to adopt ESG factors into their strategy and business model will grow considerably and companies will have to step-up their efforts.

Academic research has been a key contributor to raising awareness and sparking debate about sustainability challenges among the governmental and

---

[5] Peter McKillop, *BlackRock's former head of sustainable investing says ESG and sustainability investing are distractions*, GreenBiz (28 April 2021).

[6] See Inês Serrano Lopes, Chapter 15 in this book.

[7] Gosling, T., Guymer, C.H., O'Connor, P., Harris, L., and Savage, A. *Paying well by paying for good*, PricewaterhouseCoopers and London Business School, (2021).

[8] PWC and Diligent Institute, *2021 Corporate Governance and Executive Pay report*, (2021), available from: https://www.pwc.be/en/news-publications/2021/pwc-s-corporate-governance-and-executive-pay-report-2021.html.

[9] See Filipe Morais et al. Chapter 18 in this book.

[10] Mooij, S., *Company's ESG efforts: Catalysts and Inhibitors*. Working Paper, (May 2018), Smith School of Enterprise and the Environment, Oxford University.

civil society as well as on the investment and business communities. Indeed, ESG has also been a very prolific field for academic research. This trend undoubtedly will persist. Scientific studies play a major role in understanding and measuring ESG problems, in devising solutions and in contributing for adequate metrics and the progress of regulatory guidelines regarding these matters.

Furthermore, the integrated nature of ESG determines the importance of an interdisciplinary flow of academic studies, combining namely economic, financial, accounting, legal, climate and social research. This is a unique historical moment where scientific contributions may make a difference in supporting decisions to be taken both by investors and policy makers.

2 Main Ecosystem Gaps

As ESG flourished these last years a vast amount of ESG data became available. However, ESG data is not comparable between different markets and, because of the risk of greenwashing, ESG data is not always reliable which is why a uniform taxonomy of sustainable activities is of critical importance. However, the devil hides in the details, and as level 2 EU rules on Taxonomy of Sustainable Activities is finalised it remains the subject of opposing views driven by ideological and economic interests. Just recently a draft of EU sustainable finance taxonomy has been under fire for being too strict and not strict enough, with gas and nuclear at the heart of the controversy. There are some encouraging steps from leading ESG disclosure standard-setters to respond to investors call for harmonised and reliable ESG information to be provided in a global scale. In late 2020, the Sustainability Accounting Standards Board (SASB) and the International Integrated Reporting Council (IIRC), have merged to create the Value Reporting Foundation (VRF) which has announced in November 2021 it will further merge with Climate Disclosure Standards Board (CDSB) to create the International Sustainability Standards Boards (ISSB). Furthermore, service providers are also increasingly important as ESG data sources - namely ESG rating services and ESG assurance services—and, as IOSCO recently concluded,[11] some regulation may be required with respect to these new ESG actors.

The growing influence of ESG lies not only in the perception of the financial and societal value attached to sustainable investments but also in the recognition of the risks that ESG factors may display in relation to investee companies. This context will inevitably force a further refinement of risk management techniques (in terms of risk prevention, measurement, and reporting) regarding ESG factors.

As it has been discussed in Chapter 18 in small and mid-sized, but in many ways also in the larger caps, there remain important gaps at the board level.

---

[11] IOSCO, *Environmental, Social and Governance (ESG) Ratings and Data Products Providers. Final Report*, (Nov 2021), 32–33.

These have been identified as the knowledge gap, the accountability gap, the leadership and execution gap, and the disclosure gap.[12] While the knowledge gap is significantly narrower in the larger caps, boards themselves still lack sufficient critical mass to assure investors and regulators that appropriate oversight of ESG efforts is taking place. There is also a significant trend to consider ESG from the risk perspective and more education is for boards to also assume a more entrepreneurial role and support real transformation of business models rather than mitigation of ESG impacts. For example, a recent study of European banks listed in STOXX Europe 600, between 2008 and 2019, confirmed the approach of banking authorities, of focusing on bank ESG risks, more than ESG opportunities, in order to "force" banks into adopting a new ESG business model, at this early stage of transition.[13] Indeed, the risks and associated costs of climate change, for example, are being well-demonstrated, as Miguel A. Ferreira and Beate Sjåfjell illustrate,[14] but the upside and the opportunities around climate and social innovation remain a challenge.

The accountability gap will remain an issue until there is clarity on the role and responsibilities of directors and where the fiduciary responsibility lies and how is it to be exercised. The regulatory clarity as to what are "the firms' precise obligations"[15] is still to be achieved as the fierce debate between shareholder primacy and stakeholder-ism continues.[16] The exact nature of directors' duties[17] in relation to sustainability and ESG is still contested and uncertain and so is how they are to be exercised. The failure of directors to recognise the changing nature of the director role and the emerging duty of societal responsibility has also been attributed to learning anxiety (as Jaap Winter explains[18]). Moreover, it is still not clear how the businesses judgment rule will impact the scrutiny of sustainability-driven board decisions[19] and the response may not be the same for every jurisdiction. There is therefore still a significant board education and literacy to achieve as to their precise role and duties, how to exercise them and how to overcome barriers to sustainably transform

---

[12] See Filipe Morais et al., Chapter 18 in this book.

[13] La Torre, M., Leo, S., Panetta, I.C., Banks and environmental, social and governance drivers: Follow the market or the authorities?. *Corporate Responsibility and Environmental Management*, (2021), 28(6): 1620–1634.

[14] See Beate Sjåfjell, Chapter 3 and Miguel A. Ferreira, Chapter 7 in this book.

[15] See Guido Ferrarini, Chapter 2 in this book.

[16] See namely Beate Sjåfjell, Chapter 3 (considering shareholder primacy a 'deeply entrenched norm') and Luca Enriques, Chapter 6 (clarifying the link between shareholder primacy and ESG), both in this book.

[17] See Chapter 3 by Beate Sjåfjell and Chapter 17 by Abel Sequeira Ferreira, in this book.

[18] See Jaap Winter, Chapter 5 in this book.

[19] See Bruno Ferreira and Manuel Sequeira, Chapter 16 in this book.

and prosper. Clarity of the regulatory framework[20] will help directors overcome anxieties, but both regulation and mind-set are likely to co-evolve until sustainability concerns become more and more mainstream and unavoidable as a license to operate and prosper.

The execution gap will also likely remain for the foreseeable future. Capital markets—which require significant reform if they are to be part of the solution—need also to develop a new set of rules that will release companies from shorter-term concerns (such as quarterly reporting, which is no longer mandatory in the EU) that hinder sustainable long-term value creation and find ways to capture value other than the share price. Unilever has famously stopped to its quarterly reporting when the UK Financial Conduct Authority made this optional in 2014. However subsequent studies found that stopping mandatory quarterly earnings has not had an effect on curbing a shorter-term orientation by companies.[21] This does not mean ending quarterly reporting is not beneficial, but just that a more systemic change to re-align incentives and checks and balances in the capital markets is required.

Most companies are "waiting to see" and playing second-mover advantage type of strategies, before fully committing its resources to more fundamental transformations. This is delaying the speedy execution and transformation at the company level that is required.

While the disclosure/information gap will also improve with time—as discussed above—it is clear that there will be still a long way to produce the systemic and cultural change that is required.

3 Main Regulatory Gaps

Climate change topics have a cross border nature because emissions of GHG in some jurisdictions can spread their negative and longstanding effects across other geographies.[22]

While ESG concerns are global, "hard law" regulation regarding ESG has been mainly the product of national initiatives—such is notably the case for countries like the UK and the US. The main exception comes from the European Union that has taken the lead following the European Green Deal (2019) and its multiple subcomponents with an ambitious regulatory programme covering all Member States.

---

[20] See Chapter 11 in this book by Gabriela Figueiredo Dias on the regulation and the regulator's role.

[21] Rajgopal, S., *What Would Happen if the U.S. Stopped Requiring Quarterly Earnings Reports?*, Harvard Business Review (2018).

[22] For the qualification of climate change as a public good, see William Nordhaus, *Climate change: the ultimate challenge for economics*, American Economic Review vol. 109 (2019), 1991-ff. See also Henry Shue, *The pivotal generation: Why we have a moral responsibility to slow climate change right now* (2022), New Jersey.

The EU has maintained a close dialogue with other counterparts in sustainable finance matters, such as the International Platform on Sustainable Finance, the G20 and the G7.[23] However, it is uncertain if the EU ESG approach will be followed by other jurisdictions. That is particularly important in climate change matters, where the EU only accounts for 17% of total emissions. The European Commission stated that "through the EU Taxonomy Climate Delegated Act, the economic activities of roughly 40% of listed companies, in sectors which are responsible for almost 80% of direct greenhouse gas emissions in Europe, are already covered." This means that overall, under EU law, we are solely targeting at 5.44% of the greenhouse gas emissions globally.

In sum, regulation is being set in several jurisdictions, but harmonisation is now required at a global level. As a WEF representative stated: "*A new phase of sustainability, underpinned by reinforced global law, is required.*"[24]

Further ahead, the main regulatory challenges posed concern international uniformity of ESG reporting, enforcement of disclosure accuracy, refinement of ESG-linked remuneration KPI, collective ESG engagement and application to SME.

On the one hand, due to the proliferation of ESG standard-setters, harmonisation is clearly required in the field of ESG disclosure. In spite of the relevance of any ESG item, given the need to migrate to a Net Zero carbon economy, it is particularly relevant to reach an international agreement regarding corporate disclosure on direct (scope 1) and indirect (scope 2 and 3) GHG emissions.[25] ESG reporting should be based on comparable metrics worldwide and should meet high standards of enforcement in order to prevent flawed disclosure. ESG enforcement is particularly relevant in human rights disclosures, where asymmetries around the globe are still striking.[26]

On the other hand, progress is required on the front of ESG-linked remuneration because there is a need to have adaptive KPI as a means to enforce and incentivise the accomplishment of ESG objectives.

Other of the areas where further progress is required is collective ESG engagement, in order to foster effective change in relation to investee companies. The proliferation of investors' fora, ESG shareholders committees and shareholders associations will clearly facilitate concerted action in relation to ESG matters.

---

[23] European Commission, *Communication regarding EU Taxonomy, Corporate Sustainability Reporting, Sustainability Preferences and Fiduciary Duties: Directing finance towards the European Green Deal*, COM(2021) 188 final, (21 April 2021).

[24] Maksim Burianov, *If the world is serious about sustainability, it must embark on a new era of global law*, WEF (November 2021).

[25] Patrick Bolton/ Stefan Reichelstein/Marcin Kacperczyk/Christian Leuz/Gaizka Ormazabal/Dirk Schoenmaker, *Mandatory corporate carbon disclosures and the path to net zero* (2021).

[26] See Ana Rita Campos, *The role of companies in promoting human rights*, Chapter 9 in this book.

Finally, most of the regulatory reforms are centred around financial firms and listed companies. The wide spectre of SME is left behind, but it is important that the ESG cultural and transformative process also embraces smaller firms and family firms. The need for proportionate and simple solutions for these companies is pressing.

Ideally, this trend for further global regulatory convergence should be also followed by a consolidation of international standard-setters. The international institutional architecture of ESB-related norms should be based on simplicity and not on duplication or redundancy. A relevant example in this sense comes from the International Sustainability Standards Board (ISSB), that has absorbed existing organisations including the Climate Disclosure Standards Board (CDSB) and the Value Reporting Foundation.

4   Preparing the Forthcoming ESG Agenda

While taking advantage of decades of CSR thinking, ESG is still under a development stage, as there are many challenges ahead.

Activist Greta Thunberg made a poignant appeal to what she refers as "cathedral thinking" to address climate change.[27] Using the same metaphor, ESG is surely a cathedral in the making, but not a final and completed work yet.

The COP 26 meeting in Glasgow (2021) revealed the difficulty in turning countries' pledges into effective action. ESG poses the same sort of challenges at companies' level. Effectiveness stands as the core indicator, both at financial firm level and at the investee company level. ESG intentions must translate into action and ESG objectives must be embedded in each firm's purpose, its strategy, business model and culture.

Company boards need to be significantly more proactive at infusing the companies they oversee with the new ESG operating philosophy. Many boards will be incapable of doing so on their own volition because of short term focus and lack of ESG literacy; furthermore, in some cases a revamp of ESG board strategy will mean recognising that directors are unfit to serve on the board and thereby calling out their own replacement. This is why starting to educate the board about the case for sustainability and what it means for their company is so important, alongside building the right board capabilities to provide steer and oversight to the executive implementation of ESG factors. Having a strong team of specialised, trusted advisors and creating policies and routines to proactively engage with enlightened investors and other stakeholders to begin to understand ESG risks and opportunities are other practices boards need to examine or re-examine as it might be the case. Creating strong internal accountability mechanisms perhaps with the

---

[27] Greta Thunberg, *Speech to the European Parliament* (16 April 2019), available at www.europarl.europa.eu.

creation of an ESG board committee or equivalent governance solutions[28] and reporting structures is another important step, as is the development of materiality assessments and integrating these factors into well-thought-out executive compensation arrangements that are reflexive of a long-term sustainability strategy. Trackable reporting of performance-related measures related to ESG factors is much need as is a plausible and convincing strategic narrative to puts that performance into the wider context and its relevance for the value creating process for the company.

But companies will require also that society provides them with a continues stream of trained future leaders that have sustainability as a core value. The role of education in general and business schools in particular cannot be overestimated. The new cadre of leaders needs to be fundamentally different in how they conceive the purpose of the company in society, how they think of strategy and how they relate to the wider set of stakeholders. This can only be developed through business education that is responsible and equips future leaders with the right mind-set and capabilities. Initiatives such as the UN Global Compact Principles of Responsible Management Education[29] (PRME) which have currently 850 signatories—primarily business schools— are important in embedding the UN Sustainable Development Goals in the management education programmes curricula. However, the number of signatories is still relatively small and many of these signatories have still a long way to go to truly embedding sustainability principles across their programme portfolio.

As a conclusion, much still needs to be accomplished at political, legal, regulatory and company levels.

These challenges demand considerable effort and require inclusive dialogue between all relevant parties (regulators, stakeholders, financial sector, academics, and companies)—but they surely represent an avenue worth pursuing given the higher purpose they aim to achieve. This clearly stands out as a priority task for the next decades. For the sake of our planet, our communities, and our future generations it is imperative that, building on credible and science-based foundations, this path progresses and that global efforts converge in order to push the ESG agenda forward.

---

[28] See Paulo Câmara, Chapter 1 in this book.

[29] More information on the PRME, its objectives, work and signatories can be found by visiting Principles for Responsible Management Education—UNPRME.

Index

A
Agency theory, 117, 119, 121
AIFMD, 6, 32, 37, 104, 227, 340
Asset managers, xii, 3, 8, 10–12, 15, 18, 29, 30, 93, 94, 98, 100, 107, 131, 134, 135, 178, 221, 223–225, 227, 245, 246, 311, 330, 338, 340, 341, 360, 365, 401, 403, 404, 413, 426, 442

B
Banks, xi, 3, 8, 11, 19, 32, 34, 35, 116, 139, 154, 174, 194, 215, 279, 287–289, 295, 296, 298, 300, 313, 325, 335, 376, 378–384, 386–394, 401, 445
B corps, 121
Benefit companies, 169, 182, 269
Blue bonds, 152, 163–165
Board duties, x, xi, 20–22, 40, 178, 179, 181, 270
Board of directors, 105, 116, 175, 179–181, 289, 295, 297, 298, 300, 303, 355, 356, 422
Business judgement rule, 272, 276, 352
Business Roundtable, 8, 9, 76, 278, 294, 360

C
Carbon dioxide (CO_2), 126, 127, 336, 416, 417, 431

Climate change, ix, 3, 7, 8, 13, 15, 21, 31, 33, 34, 39, 53, 59, 65–67, 70, 72, 75, 81, 93, 95, 102, 128, 138–140, 144–146, 150, 151, 158, 162, 163, 184, 205, 206, 213, 217, 234, 235, 243, 245–247, 249–252, 254–257, 261, 263, 265, 279, 282, 293, 322, 324, 330–332, 334, 338, 340–342, 345, 356, 370, 375, 377, 378, 394, 399, 400, 417, 425, 442, 445–448
Climate finance, xi, 138
Climate risk, 21, 66, 95, 106, 235, 340, 355, 388, 426
Company law, xi, 61, 64, 65, 71–75, 77, 82, 105, 192, 194, 212, 214, 346, 349, 351, 352, 357, 364
Compliance, xi, 10, 12, 21, 28, 42, 44–46, 48, 49, 52, 55, 56, 73, 76, 83, 84, 118, 154, 155, 158, 160, 169–171, 175–177, 179–183, 185, 186, 196, 202, 205, 213, 227, 228, 231, 233, 257, 261, 273, 277, 281, 282, 285, 307–311, 313–316, 318, 320–323, 325, 333, 337, 341–343, 349, 353, 356, 369, 381, 387, 393, 410, 420, 422, 425
COP 26 Glasgow, 7
Corporate governance, x–xii, 5, 7, 10, 11, 14, 18, 20, 23, 24, 39–41, 72, 74, 75, 77–79, 84, 117, 120, 121, 124, 125, 170, 173, 191, 192, 194,

198, 199, 207, 208, 211, 214, 215, 217, 253, 255, 257, 263, 268, 278, 281, 284, 288, 291, 294–296, 302, 306, 311, 312, 333, 338, 340, 345–352, 355, 357, 364, 367, 369, 386, 389, 402, 403, 406, 422, 441, 442
Corporate interest, 269, 354, 355
Corporate purpose. *See* Purpose
Corporate social responsibility (CSR), 8, 15, 35, 36, 43, 44, 71, 133, 135, 151, 156, 160, 170, 171, 178, 191, 192, 214, 232, 278, 283, 292, 306, 384, 420, 448
COVID-19, 9, 32, 85, 140, 184–186, 291, 292, 299, 329, 330, 332–334, 357, 367, 370, 391, 394, 420, 431

D

Disclosure, x, xi, 5–7, 10, 14, 16, 19–29, 36, 38–40, 42, 46, 47, 49, 71, 88, 89, 91–93, 96–99, 101, 103, 105, 109, 129, 156, 158, 160–162, 172, 173, 175, 193–204, 206–210, 213–215, 224, 225, 227, 232, 235, 238–242, 245–248, 250–252, 254, 255, 258–262, 264, 278, 279, 281, 282, 287, 299, 309–312, 338, 340–345, 349, 356, 361, 362, 365, 369–372, 383–386, 390, 392, 393, 402, 405, 409, 411, 412, 414, 422, 425, 426, 442, 444–447
Due diligence, x, xi, 6, 10, 16, 42, 45, 47–57, 72, 83, 84, 93, 171, 173–177, 179, 181–183, 185, 186, 196, 201–204, 210, 211, 213, 214, 242, 309, 312–314, 323, 324, 345, 349, 381, 386, 409, 410
Duty of societal responsibility, 20, 115, 116, 122–125, 127–129, 445

E

ESG data, 19, 26, 97, 221, 236, 246, 361, 444
ESG financial products, 4, 10, 160
ESG in general, xi, 17, 34, 239, 280

EU Benchmarking Regulation, 6
EU Corporate Sustainability Reporting Directive (CSRD), 6, 10, 25, 26, 30, 129, 199, 244, 245, 254, 384
EU Green Deal, 6, 24, 60, 65, 67, 72, 84, 92, 124, 193, 199, 244, 250, 259, 306, 331, 333, 339, 344, 359, 384, 400, 446
EU IORP II, 6
EU Law, xi, 26, 56, 62, 63, 81, 192, 215, 447
European Commission (EC), x, 10, 24, 25, 27, 29, 32, 38, 50, 60, 62, 71, 72, 81, 84, 85, 87–90, 93, 96, 116, 124, 150, 151, 164, 173, 174, 192–194, 200, 202, 211, 218, 220, 221, 238–245, 249, 253–255, 258, 261, 262, 265, 270, 278, 279, 281, 295, 296, 306, 312, 314, 333–335, 337–345, 347–349, 352, 354, 355, 357, 359, 378, 383–385, 389, 391, 393, 394, 400, 405, 408, 447
European Commission Action Plan on Financing Sustainable Growth, 5, 35, 71, 279
European Green Deal Investment Plan, 6
EU Sustainable Finance Disclosure Regulation (SFDR), 6, 10, 11, 16, 27, 28, 36, 39, 88, 91–93, 98, 105, 107, 161, 162, 193, 200–207, 209, 214, 215, 221, 238–243, 246, 254, 287, 292, 296, 312, 332, 340, 384, 411–414
EU Taxonomy Regulation, x, 6, 14, 19, 27, 30, 33, 50, 71, 88–90, 93, 96, 98, 99, 101, 102, 109, 158, 159, 162, 193, 199, 204–206, 209, 214, 221, 238, 243, 246, 255, 263–265, 290, 292, 312, 341, 349, 357, 384–386, 390, 393, 412, 413, 425

F

Fiduciary duties, 11, 28, 93, 161, 179, 218, 224, 226, 227, 250, 267, 269, 272, 278, 280, 281, 405–411, 414
Financial system, 4, 101, 139, 151, 217–219, 223, 228, 235, 279, 332,

INDEX   453

338, 375, 376, 379, 382, 389, 390, 400, 401
FinTech, 108
Fit & proper, 226

G
Global warming, 55, 60, 138, 150, 338
Green assets ratio (GAR), 385, 386, 391, 393
Green bonds, 150, 152–158, 160–165, 217, 227, 251, 290, 380
Greenhouse gas (GHG) reduction, 90. *See also* Carbon dioxide (CO_2), Paris Agreement
Growth listed companies, 359

H
Human rights, x, xi, 6, 14, 25, 45, 47–54, 56, 59, 61–63, 67, 69, 70, 72, 79–82, 84, 159, 167–186, 196, 197, 212, 213, 218, 234, 242, 247, 251, 253, 281, 314, 322, 325, 332, 345, 356, 370, 379, 399, 421, 447

I
Institutional investors, 7, 8, 10, 13, 34–36, 45, 95, 98, 100, 106, 116–118, 134, 135, 207, 221, 263, 292, 311, 330, 338, 349, 402–405, 413, 414
Investment funds, 4, 160, 223, 227, 241, 401–407, 409–411, 413, 414

L
Litigation, 21, 28, 120, 124, 251, 307, 324, 332, 349, 351, 352, 428

N
Non-Financial Reporting Directive (NFRD), 25, 29, 46, 47, 66, 71, 72, 75, 83, 89, 99, 162, 173, 192–194, 198, 199, 204–207, 214, 215, 221, 235, 237–239, 242–244, 258, 310, 311, 322, 332, 340, 349, 384, 385, 408, 414

P
Paris Agreement, 3, 7, 15, 65, 137, 150, 263, 306, 359, 400, 401
Principles for Responsible Investment (PRI), 38, 45, 247, 331, 332, 341, 404, 420, 422
Profit, 8, 41–43, 46, 70, 73, 78, 95, 98, 106, 115, 131–134, 136, 151, 168, 170, 178, 184, 226, 268, 278, 293, 324, 334, 344, 354, 361, 378, 407, 417, 419, 420, 424
Prudential requirements, 107, 108, 194, 338, 376, 401
Purpose, 8, 9, 11, 15, 18, 19, 22, 38, 42–44, 56, 64, 75–78, 82–84, 88, 91, 92, 100, 121, 122, 132, 134, 136, 158, 163, 168, 210, 212, 213, 215, 218, 219, 223–226, 234, 237, 242, 252, 253, 257–259, 261, 262, 273, 278, 282, 285, 287, 290–296, 298, 299, 302, 307–310, 312, 316, 321, 322, 326, 331, 333, 344, 350, 352, 353, 355–357, 360, 368, 376, 380, 390, 394, 406, 411, 429, 448, 449

R
Regulation, x–xii, 6, 10, 11, 14–16, 25, 28, 33, 36, 39, 41, 44–46, 50, 56, 64, 71, 74, 88, 89, 92, 94, 100–106, 108, 109, 117, 118, 122, 154–157, 159, 161, 171, 176, 183, 193, 194, 197, 200, 206, 208, 209, 213, 217, 219, 221, 224, 226, 238, 239, 241, 245–247, 252–264, 270, 278, 280, 282, 288, 290, 307, 309, 312, 313, 320, 322, 331, 332, 334–337, 339, 341, 342, 344, 351, 355–357, 360, 363, 367, 376, 379, 382–384, 388, 391, 393, 405, 412, 418, 422, 426, 430, 432, 442, 444, 446, 447
Regulators, 22, 27, 87, 91, 92, 96, 101–103, 106–109, 160, 219, 222, 223, 225–228, 232, 239, 241, 309, 331, 333, 339, 351, 360, 445, 449
Remuneration, xi, 10, 14, 16, 20, 28, 33–40, 91, 92, 105, 108, 109, 117,

119, 120, 149, 151, 184, 202, 207, 211–213, 218, 225, 263, 287–303, 318, 320, 323, 347, 348, 367, 370, 387, 411, 447
Reporting, xi, 10, 29, 30, 46, 49, 66, 72, 75, 76, 83, 89, 99, 100, 108, 145, 155, 156, 168, 171, 173, 174, 176, 177, 180, 195, 197, 199, 206, 221, 222, 232–238, 241, 243, 244, 246–250, 254, 262, 263, 273, 282, 292, 303, 309–312, 322, 333, 336, 341–343, 350, 352, 356, 361, 369, 370, 372, 380, 384, 401, 408, 411, 419, 420, 422, 425, 432, 444, 446, 447, 449
Risk, xi, 14, 17, 23, 26, 27, 31–34, 37, 40, 44, 52, 60, 66–70, 76, 85, 91, 92, 94–97, 100, 101, 104, 106–109, 124, 125, 127, 129, 138, 139, 151, 154, 156, 162, 168, 171, 175, 176, 179, 201, 222, 228, 232, 235, 241, 248, 251, 254, 263, 273, 275, 281, 282, 284, 285, 291, 295, 296, 299–301, 314, 317, 318, 320, 324, 334, 337, 343, 349, 352, 367, 368, 370, 372, 375, 377–379, 381, 382, 384, 386–394, 401, 408, 409, 416, 426, 428, 430, 444, 445
Risk management, x, 10, 12, 14, 18, 20, 23, 32, 34–36, 47, 50, 52, 56, 67, 83, 84, 89, 93, 94, 100, 101, 105–107, 135, 179, 224, 228, 254, 269, 289, 291, 325, 344, 350, 356, 365, 373, 378, 381, 383, 386–389, 392, 394, 401, 405, 409, 410, 420, 444

S
Shareholder primacy, xi, 73–76, 80, 82, 115, 117, 120, 121, 279, 280, 294, 445
Shareholder Rights Directive II (SRD II), 6, 311, 403, 404, 414
Smart cities, xii, 4, 416–418, 420, 421, 423, 424, 426, 428, 429, 431, 432
Smart regulation, xi, 108
Socially responsible investment (SRI), 45, 95, 151, 332

Stakeholder capitalism, 218, 226, 278, 284
Stakeholders, x, 8–10, 18–20, 22, 32, 36, 40, 43–46, 73, 76, 77, 80, 117, 121, 123, 124, 170, 171, 173, 178, 179, 181, 182, 185, 197, 198, 207, 212–215, 220, 225–227, 232, 233, 236–238, 240, 242, 243, 247, 248, 268, 271, 279–281, 283–285, 289–291, 294, 295, 297, 302, 310, 311, 316–319, 331, 334, 340, 342, 343, 346–356, 360, 367, 369–373, 386, 390, 416, 418, 424, 426, 432, 448, 449
State-owned enterprises (SOE), 4, 19, 415–418, 420–423, 428, 429, 431, 432
Stewardship, 8, 12, 36, 45, 311
Stewardship code, x, 8, 15, 16, 45, 233
Sustainability, ix–xii, 5, 6, 10–12, 14–16, 22, 23, 25, 29, 30, 32, 34, 36, 38–42, 45, 46, 56, 60–73, 75–80, 83–85, 88–109, 116, 124, 151, 152, 154, 156, 157, 159, 161, 162, 164, 165, 193, 199–203, 206–214, 219, 221–228, 232–240, 243, 244, 246–255, 257–260, 263–265, 273, 278, 280–282, 287–289, 292–303, 306–308, 310, 312, 314, 316–320, 322, 323, 325, 330–334, 336, 338, 340–345, 347–357, 359–361, 365, 368, 372, 375–377, 379–384, 386, 389–394, 400, 401, 408–417, 419–422, 424–427, 429–432, 441–443, 445–449
Sustainability Accounting Standard Board (SASB), 29, 235, 369, 408, 444
Sustainable Finance Action Plan (SFAP), 88–93, 95, 96, 98, 102, 103, 238, 245, 254, 331
Sustainable governance, xi, 50

T
Task Force on Climate-related Financial Disclosures (TCFD), 66, 235, 248, 250, 265, 279, 331, 383, 384, 401

U

Undertakings for collective investment in transferable securities (UCITS), 6, 32, 93, 97, 201, 227, 241, 245, 258, 340, 401, 403–407, 409, 410

United Nations (UN), ix, 5, 21, 45, 47–51, 60, 72, 80, 149, 150, 163, 167–169, 171, 172, 174, 177, 179–181, 183–185, 205, 234, 247, 257, 306, 330, 338, 362, 381, 394, 400, 404, 417–419, 449

United Nations Principles of Responsible Investment (UNPRI), 3, 12, 17, 24

United Nations Sustainable Development Goals (UN-SDGs), 3, 60, 61, 84, 350, 369, 449

USA, 26, 400, 430

W

World Economic Forum (WEF), 9, 21, 29, 35, 65, 69, 70, 268, 321, 375, 431, 447

Printed in the United States
by Baker & Taylor Publisher Services